TECHNOLOGY OF CEREALS

AN INTRODUCTION FOR STUDENTS OF FOOD SCIENCE AND AGRICULTURE

O.W.L.

Other Pergamon titles of related interest

TECHNOLOGY OF CEREALS

AN INTRODUCTION FOR STUDENTS OF FOOD SCIENCE AND AGRICULTURE

FOURTH EDITION

N. L. KENT

*Sometime Scholar of Emmanuel College, Cambridge
Formerly at the Flour Milling and Baking Research Association,
St. Albans and Chorleywood*

and

A. D. EVERS

*Flour Milling and Baking Research Association,
Chorleywood*

PERGAMON

U.K.   Elsevier Science Ltd, The Boulevard,
          Langford Lane, Kidlington, Oxford OX5 1GB, U.K.

U.S.A.   Elsevier Science Inc., 660 White Plains Road,
          Tarrytown, New York 10591-5153, U.S.A.

JAPAN   Elsevier Science Japan, Tsunashima Building Annex,
          3-20-12 Yushima, Bunkyo-ku, Tokyo 113, Japan

First edition 1966

Reprinted 1970

Second edition 1975

Reprinted (with corrections) 1978, 1980

Third edition 1983

Reprinted (with corrections) 1984

Fourth edition 1994

Library of Congress Cataloging in Publication Data

Kent, N. L. (Norman Leslie)
Technology of cereals: an introduction for students of food science
and agriculture/N. L. Kent and A. D. Evers. – 4th ed.
p. cm.
Includes bibliographical references and index.
1. Cereal products. 2. Grain. I. Evers, A. D. II. Title.
TS2145.K36   1993

664'.7 – dc20

British Library Cataloguing in Publication Data

A catalogue record for this book is available
from the British Library.

ISBN 0 08 040833 8 hardcover

ISBN 0 08 040834 6 flexicover

Printed in Great Britain by BPC Wheatons Ltd, Exeter

Contents

Preface to the Fourth Edition

THE PRINCIPAL purpose of the fourth edition is to update the material — including the statistics — of the third edition, while maintaining an emphasis on nutrition and, in particular, the effects of processing on the nutritive value of the products as compared with that of the raw cereals.

However, some new material has been introduced, notably sections dealing with extrusion cooking and the use of cereals for animal feed, and the section on industrial uses for cereals has been considerably enlarged.

A change in the fourth edition, which readers of earlier editions will notice, is the order in which the material is presented. Instead of devoting a separate chapter to each of the cereals, other than wheat, chapters in the fourth edition are devoted to distinct subjects, e.g. dry milling, wet milling, malting, brewing and distilling, pasta, domestic and small scale processing, feed and industrial uses, in each of which all the cereals, as may be appropriate, are considered. Besides avoiding a certain degree of repetition, we feel that this method of presentation may give a better understanding of the subject, particularly as to how the various cereals compare with each other.

Acknowledgements

WE WOULD like gratefully to acknowledge the help we have received from numerous colleagues or former colleagues at the Flour Milling and Baking Research Association, thinking particularly of Brian Eves and Barbara Stapleton in the Library, Miss Brenda Bell, Dr Norman Chamberlain, Bill Collins, Dr Philip Greenwell, Brian Stewart and Dr Robin Guy. Staff of Member Companies of the Association have also been generous with advice.

We thank the firms who most generously provided data for Tables 11.1 and 11.2, and those who provided pictures, and to the editors of journals, other individuals and publishers who have kindly allowed us to reproduce pictures or data from their publications.

The picture of ergotized rye (Fig. 1.4) is Crown Copyright and is reproduced with the permission of the Controller of Her Majesty's Stationery Office. The Controller of H.M.S.O. has kindly given permission for data from Crown Copyright publications to be quoted.

St Albans, Herts

November, 1992

N. L. KENT

A. D. EVERS

Preface to the First Edition

THIS INTRODUCTION to the technology of the principal cereals is intended, in the first place, for the use of students of Food Science. A nutritional approach has been chosen, and the effects of processing treatments on the nutritive value of the products have been emphasized. Throughout, both the merits and the limitations of individual cereals as sources of food products have been considered in a comparative way.

I am greatly indebted to Dr T. Moran, C.B.E., Director of Research, for his encouragement and advice, and to all my senior colleagues in the Research Association of British Flour-Millers for their considerable help in the writing of this book. My thanks are also due to Miss R. Bennett of the British Baking Industries Research Association and Mr M. Butler of the Ryvita Co. Ltd who have read individual chapters and offered valuable criticism, and particularly to Professor J. A. Johnson of Kansas State University and Professor J. Hawthorn of The University of Strathclyde, Glasgow, who have read and criticized the whole of the text.

I wish to thank the firms which have supplied pictures or data, viz. Henry Simon Ltd, Kellogg Co. of Great Britain Ltd, and also the authors, editors and publishers who have allowed reproduction of illustrations, including the Controller of H.M.S.O. for permission to reproduce Crown copyright material (Fig. 38, and data in Tables 1, 22, 26, 55 and 72).

Research Association of British Flour-Millers,
Cereals Research Station,
St Albans, Herts.
July 1964

N. L. KENT

Abbreviations, Units, Equivalents

AACC	American Association of Cereal Chemists
ACP	acid calcium phosphate
ADA	azodicarbonamide
ADD	activated dough development
ADF	acid detergent fibre
A.D.Y.	active dried yeast
an.	annum
ARFA	Air Radio Frequency Assisted
b	billion (10^9)
B.C.	before Christ
BFP	bulk fermentation process
BHA	butylated hydroxy anisole
BHT	butylated hydroxy toluene
BOD	biological oxygen demand
b.p.	boiling point
BP	British Patent
B.P.	British Pharmacopoeia
B.S.	British Standard
B.S.I.	British Standards Institute
Bz	benzene
cap.	head (capitum)
CBN	Commission of Biological Nomenclature
CBP	Chorleywood Bread Process
Cent.	central
cf.	compare
Ch.	Chapter
CMC	carboxy methyl cellulose
COMA	Committee on Medical Aspects of Food Policy
concn	concentration
COT	Committee on Toxicology
C_s.	coarse

C.S.I.R.	Council for Scientific and Industrial Research
CSL	calcium stearoyl-2–lactylate
CTAB	cetyl trimethylammonium bromide
CWAD	Canadian Western Amber Durum wheat
CWRS	Canadian Western Red Spring wheat
CWRW	Canadian Western Red Winter wheat
CWSWS	Canadian Western Soft White Spring wheat
CWU	Canadian Western Utility wheat
D	dextrorotatory
DATEM	di-acetyl tartaric esters of mono- and di-glycerides of fatty acids
d.b.	dry basis
DDG	dried distillers' grains
D.H.	Department of Health
DHA	dehydro ascorbic acid
d.m.	dry matter
DNA	deoxyribonucleic acid
DR	Democratic Republic (Germany)
DRV	dietary reference value
D.S.S.	Department of Social Security
E	east
EAR	estimated average requirement
EC	European Community
edn.	edition
EP	European Patent
F.A.O.	Food & Agriculture Organisation of the United Nations
F.D.A.	Food & Drug Administration (of the U.S.A.)
FFA	free fatty acid

FMBRA	Flour Milling & Baking Research Association	N	normal
FR	Federal Republic (Germany)	N	nitrogen
G	giga (10^9)	NABIM	National Association of British & Irish Millers
GATT	General Agreement on Tariffs and Trade	N.B.	nota bene
GC	Grade Colour	NDF	neutral detergent fibre
gg	grit gauze	NIR(S)	near infrared reflectance (spectroscopy)
GMS	glycerol mono-stearate	No.	number
hd	head	p., pp.	page, pages
HDL	high density lipoprotein	P.A.D.Y.	protected active dried yeast
HFCS	high-fructose corn syrup	PAGE	polyacrylamide gel electrophoresis
HFSS	high-fructose starch syrup	Pat.	Patent
H-GCA	Home-Grown Cereals Authority	propn	proportion
HPLC	high performance liquid chromatography	pt	part(s)
HPMC	hydroxy propyl methyl cellulose	RDA	recommended daily amount (of nutrients)
HRS	Hard Red Spring wheat	rDNA	recombinant DNA
HRW	Hard Red Winter wheat	RF	radiofrequency
HTST	high temperature short time	r.h.	relative humidity
I.A.D.Y.	Instant Active Dried Yeast	RNI	Reference Nutrient Intake
ICC	International Association of Cereal Science and Technology	r.p.m.	revolutions per minute
ISO	International Organisation for Standardisation	RVA	rapid viscoanalyser
		S	south
IUPAC	International Union of Pure and Applied Chemistry	SAP	sodium aluminium phosphate
		SAPP	sodium acid pyrophosphate
k	kilo (10^3)	SDS	sodium dodecyl sulphate
L	laevorotatory	SEM	solvent extraction milling
LDL	low density lipoprotein	SGP	starch granule protein
LRNI	Lower Reference Nutrient Intake	SI	Statutory Instrument
LSD	lysergic acid	sp., spp.	species
m	milli (10^{-3})	sp.gr.	specific gravity
μ	micro (10^{-6})	Sr	strontium
M	molar	SRW	Soft Red Winter wheat
MAFF	Ministry of Agriculture, Fisheries & Food	SSL	sodium stearoyl-2–lactylate
		ssp.	subspecies
max.	maximum	TD	tempering-degerming
m.c.	moisture content	temp.	temperature
Med.	medium	t.v.p.	textured vegetable products
Midds	middlings	U.K.	United Kingdom
min.	minimum	U.S.A.	United States of America
mol.	molecular	USDA	United States Department of Agriculture
M_r	relative molecular mass	USP	United States Patent
M.R.C.	Medical Research Council	UV	ultraviolet
MRL	maximum residue level	vac.	vacuum
MW	microwave	v/v	volume for volume
n	nano (10^{-9})	W	west

w.	wire bolting cloth	°	degree
w/w	weight for weight	<	less (fewer) than
WHO	World Health Organisation	>	greater (more) than
wt	weight	$\not>$	not more than
yr	year	%	percentage

UNITS

ac	acre (43,560 ft^2)	lb	pound
atm	atmosphere	m	metre
A$_w$	vapour pressure	mM	millimolar
bar	steam pressure	mg/kg	milligrammes per kilogramme (=ppm)
Bé	Baumé (hydrometer scale)		
Btu	British thermal unit	µg/kg	microgrammes per kilogramme (=ppb)
bu	bushel (8 imperial gal)		
°C	degree Celsius (centigrade scale)	Mha	mega hectare (10^6 ha)
cal	calorie	MHz	mega hertz (10^6 Hz)
Cal	Calorie (kcal)	MJ	mega joule (10^6 J)
Ci	curie	MN	mega-newton (10^6 N)
cm	centimetre (10^{-2}m)	Mt	mega-tonne (10^6 t)
cwt	hundredweight (112 lb)*	min.	minute (time)
cwt (U.S.)	hundredweight (100 lb)	ml	millilitre (10^{-3} l.)
°F	degree Fahrenheit*	mm	millimetre (10^{-3} m)
FN	Falling Number	µg	microgramme (10^{-6} g)
ft	foot, feet	µm	micrometre (10^{-6} m)
FU	Farrand unit	N	newton (unit of force)
g	gramme	nm	nanometre (10^{-9} m)
gal	gallon (imperial)	oz	ounce
GJ	gigajoule (10^9 J)	pCi	picocurie (10^{-12} Ci)
h	hour	ppb	parts per billion (µg/kg)
ha	hectare (10^4 m^2)	ppm	parts per million (mg/kg)
hl	hectolitre (10^2 l)	psi	pounds per square inch
h.p.	horse power	q	quintal (10^2 kg)
Hz	hertz	rad	unit of radiation
in.	inch	rev	revolution
i.u.	international unit	sec	second (time)
J	joule	sk	sack (280 lb of flour)
kcal	kilocalorie (10^3 cal)	t	metric tonne (10^3 kg; 2204lb)
kg	kilogramme (10^3 g)	ton	long ton (2240 lb)*
kJ	kilojoule (10^3 J)	ton (U.S.)	short ton (2000 lb)
kN	kilonewton (10^3 N)	W	watt
kW	kilowatt (10^3 W)	Wh	watt-hour
l.	litre	yd	yard

* Abolished in the U.K., 31 December 1980.

EQUIVALENTS

Metric units, viz. units of the SI (Système Internationale d'Unités), have been used throughout this book, but for the convenience of readers, particularly in those countries in which metric units are not adopted, some conversion factors are presented below.

Further information may be obtained from a National Physical Laboratory booklet *Changing to the Metric System*, by Pamela Anderton and P.H. Bigg, London: H.M.S.O., 1967. See also *Flour Milling* by J.F. Lockwood, Stockport: Henry Simon Ltd, 4th edition, 1960, Appendix 14.

Cereal crop yields are given in quintals per hectare (q/ha) in preference to the SI unit of kilogrammes per 100 square metres (kg/100 m^2) which is numerically equivalent to quintals per hectare.

Length
1m = 39.37 in.
1m = 3.281 ft
1m = 1.0936 yd

Volume
1 l. = 1.76 pint
1 l. = 0.22 gal
1 l. = 0.0275 bu
1 m^3 = 1000 l.
1 m^3 = 220 gal

Density
1 g/cm^3 = 62.5 lb/ft^3
1 kg/m^3 = 0.0625 lb/ft^3
1 kg/hl = 0.802 lb/bu

Area
1 m^2 = 10.76 ft^2
1 ha = 2.471 ac

Mass
1 g = 0.0353 oz
1 kg = 2.205 lb

1 t = 0.984 ton
1 t = 10 q

Mass per unit area
1 t/ha = 0.398 ton/ac
1 q/ha = 0.79 cwt/ac

Concentration
1 mg/100g = 4.54 mg/lb
1 mg/100g = 0.0447 oz/sk
1 mg/100g = 0.3576 oz/ton
1 mg/100g = 0.3518 oz/t
1 mg/lb = 0.00985 oz/sk
1 mg/lb = 0.0788 oz/ton
1 μg/g = 1mg/kg = 1 ppm
1 μg/kg = 1 ppb

Energy
1 J = 0.239 cal
1 kJ = 0.945 Btu
1 MJ = 0.278 kWh
1 kcal = 1.163 Wh

Power
1 kW = 1.341 h.p.

Pressure
1 kg/cm^2 = 14.232 lb/in.^{2}
1 kg/cm^2 = 0.968 atm
1 kN/m^2 = 0.145 lb/in.^{2}
1 kN/m^2 = 0.0099 atm

Dressing surface
1 m^2/24 h/100 kg (wheat)
   = 469.5 ft^2/sk/h (flour)

Roller surface
1 cm/24 h/100 kg wheat
   = 16.7 in./sk/h flour
   (milled to 73% extraction rate)
1 in./196-lb barrel of flour
   (U.S.A.) = 1.43 in./sk
1 in./100-lb cwt flour (U.S.A.)
   = 2.8 in./sk

Temperature
°F = (°C × 1.8) + 32
°C = (°F − 32) × ⅚

1

Cereal Crops: Economics, Statistics and Uses

Cereals

Cereals are the fruits of cultivated grasses, members of the monocotyledonous family Gramineae. The principal cereal crops are wheat, barley, oats, rye, rice, maize, sorghum and the millets.

Cereals have been important crops for thousands of years; indeed, the successful production, storage and use of cereals has contributed in no small measure to the development of modern civilization.

World crops

Area

The area occupied by the eight cereals averaged 665 million ha over the 3-year period of 1969–1971, and increased to 719 M ha on average in the period 1979–1981. There was a slight fall, to an average of 691 M ha, in the period 1987–1989. Between 1965 and 1989 the area under wheat, barley, maize, rice and sorghum has shown a small increase, whereas the area under oats, rye and millet has decreased slightly. The total area occupied by cereals in 1987–1989 was 5.3% of the entire land surface of the world. The annual world area and production and the average world yield of the individual cereals over the period 1960–1989 are shown diagrammatically in Fig. 1.1.

Production

The world production of wheat, barley, oats, rye, rice (paddy), maize, sorghum and millet was estimated at 1233 million tonnes (Mt) in 1969–1971, rising to 1652 Mt in 1981, and to 1784 Mt in 1987–1989.*

The total world production of the eight major cereals in 1987–1989 would have been sufficient to provide approximately 350 kg of cereal grain per head per annum, or about 960 g per head per day, if shared equally among the entire world population. This is a slight reduction on the figure estimated for 1981, which was 370 kg of cereal grain per head per annum, indicating that the increase in total world production is not quite keeping pace with the increase in world population.

The average human consumption of cereals is only about one half of these figures, as a variable proportion is used for other purposes, mainly animal feed, industrial processing, and seed, and there is considerable wastage. Thus, in 1984–1986, of a world total domestic supply of 1677 Mt of all cereals, 49% was used for human food, giving an average consumption of 171 kg per head per annum, 37% was used for animal feed, 10% for processing and other uses, and 4% for seed.

* Crop data for earlier years have been derived from *Grain Crops* and *Grain Bulletin* by permission of the Commonwealth Secretariat; those for later years form *F.A.O. Production Year Books* and *F.A.O. Trade Year Books* or *F.A.O. Food Balance Sheets*, and from *H-GCA Cereal Statistics*.

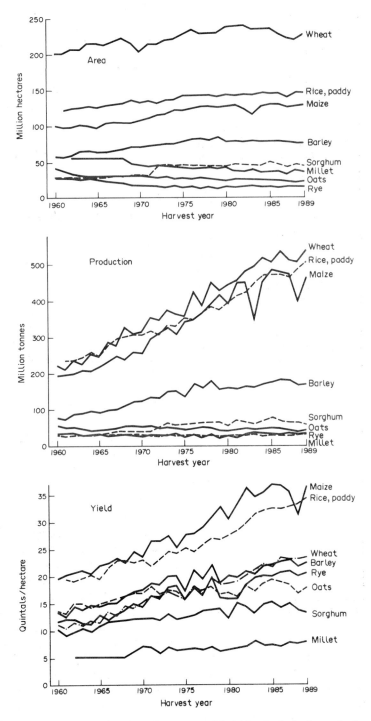

FIG 1.1 Annual total world area and productivity, and average world yield, of principal cereals since 1960. (Sources: *Grain Crops* Commonwealth Secretariat; U.S.D.A.; F.A.O.; EUROSTAT.)

Yield

The proportion of the total harvested area contributed by each of the eight cereals is similar to, or greater than, the proportion of the total production, except for maize and rice (see Table 1.1). This is because the yields of wheat, barley, oats, rye and sorghum do not vary greatly among themselves (and that for millet is very low), whereas the yields of maize and paddy rice are about 1.7 times the average yield of all the other cereals (apart from millet). Taking all the cereals together, the average yield for the whole world increased progressively from 18.1 q/ha in 1969–1971 to 22.5 q/ha in 1981, and to 25.8 q/ha in 1987–1989. Between 1969 and 1989, the yield of maize increased by 11.8 q/ha (from 24.7 to 36.5 q/ha), largely through the use of hybrid maize (cf. p. 99), and that of rice (paddy) increased by 12.1 q/ha (from 22.6 q/ha in 1969 to 34.7 q/ha in 1989), whereas that of all other cereals together increased by only 3.9 q/ha on average (from 13.8 to 17.7 q/ha). The percentage distribution of the world cereals area and production and the world average yields for each cereal over the period 1987–1989 are shown in Table 1.1

TABLE 1.1
World Cereal Area, Production and Yield for the Period 1987–1989*

Cereal	Percentage of total area	Percentage of total production	Average yield (q/ha†)
Wheat	32	29	23.3
Barley	11	10	22.6
Oats	3	2	17.8
Rye	2	2	20.6
Rice (paddy)	21	27	33.8
Maize	19	25	34.5
Sorghum	7	3	14.0
Millet	5	2	7.8

* Data derived from *F.A.O. Production Year Book* (1990).
† N.B. 1 q/ha = 0.79 cwt/ac.

As sources of carbohydrate related to land use, rice, producing 38.1 × 10⁶ kJ/ha, ranks first among cereal grains, followed by maize with 33.9 × 10⁶ kJ/ha and wheat with 27.7 × 10⁶ kJ/ha. These figures are obtained by multiplying the 1989 average world yield of milled rice (2.43 t/ha), corn meal (2.04 t/ha) and white wheat flour (1.72 t/ha) by the respective figures for energy content in kJ/g (15.7 for milled rice, 16.6 for corn meal, 16.1 for white wheat flour). As regards food protein yield, rice, producing 0.18 t/ha of protein (in milled rice), is second only to wheat, producing 0.19 t/ha (in white flour), followed by oats, producing 0.15 t/ha (in oatmeal).

Wheat

Cultivation

Wheat is grown throughout the world, from the borders of the arctic to near the equator, although the crop is most successful between the latitudes of 30° and 60° North and 27° and 40° South. In altitude, it ranges from sea level to 3050 m in Kenya and 4572 m in Tibet. It is adaptable to a range of environmental conditions from xerophytic to littoral.

Wheat grows best on heavy loam and clay, although it makes a satisfactory crop on lighter land. The crop repays heavy nitrogenous manuring.

Wheat flourishes in subtropical, warm temperate and cool temperate climates. An annual rainfall of 229–762 mm, falling more in spring than in summer, suits it best. The mean summer temperature should be 13°C (56°F) or more.

The seed is sown in late autumn (winter wheat) or in spring (spring wheat). Winter wheat can be grown in places, e.g. northwestern Europe, where excessive freezing of the soil does not occur. The grain germinates in the autumn and grows slowly until the spring. Frost would affect the young plants adversely, but a covering of snow protects them and promotes tillering. In countries such as the Canadian prairies and the steppes of Russia that experience winters too severe for winter sowing, wheat is sown as early as possible in the spring, so that the crop may be harvested before the first frosts of autumn. The area of production of spring wheat is being extended progressively northwards in the northern hemisphere by the use of new varieties bred for their quick-ripening characteristics.

Times of sowing and harvesting of the wheat crop in the various growing countries are naturally dependent upon local climatic conditions; wheat is being harvested in some country in every month of the year. However, the storage facilities in most wheat-growing countries are adequate to permit the best part of a year's harvest being stored; thus, the British miller can buy wheat from any exporting country at almost any time of the year. The times of harvest for the principal wheat-growing countries are shown in Table 1.2.

TABLE 1.2
Times of Wheat Harvest

Country	Harvest time
India	February
China	May
Italy	June–July
France	June–July
U.S.A.	May–September
Former Soviet Union	July–September
Canada	July–September
England	August–September
Australia	October–January
Argentina	November–January

Area, production, yield

Between 1965 and 1989 the world wheat area showed a small increase (215–227 M ha) while wheat production doubled, from 261 to 537 Mt per annum, reflecting the increase in world average yield over the period, from 12 to 23.6 q/ha. This increase has been due to the use of more highly yielding varieties, the greater use of fertilizers, and improved husbandry.

The area under wheat, the production, and the average yield in the principal wheat-producing countries and regions of the world, for four selected periods (1956/61, 1969/71, 1979/81 and 1986/88) are shown diagrammatically in Fig. 1.2. In the period 1986/88 the former Soviet Union and China each produced 18% of the world crop, Western Europe 15.3%, India/Pakistan 12%, the U.S.A. 11.2%, Eastern Europe 9.1%, Canada 5% and Turkey 4%.

The yield varies considerably among producing countries and regions, and is related to the water supply and the intensity of cultivation. In 1990

in Ireland an average yield of 79.7 q/ha was obtained, and 76.5 q/ha in the Netherlands; yields of 42–75 q/ha were general in other northwestern European countries in the same year. However, in more primitive agricultural communities, and in countries with less favourable climatic conditions, yields are still around 12 q/ha. The improvements in wheat yields since 1956/61 are particularly striking in Europe (both Western and Eastern), China, and India/Pakistan (see Fig. 1.2).

Wheat yield also depends upon the type of wheat sown: winter wheat (autumn-sown), with a longer growing period than spring wheat, normally produces a higher yield than spring-sown wheat (cf. p. 79). The yield of durum wheat (cf. p. 79), which is grown in drier areas, is lower than that of bread wheat.

The present yield of wheat in the U.K. (71 q/ha in 1991) is over three times the pre-war figure of 23 q/ha. In the U.S.A., yields have increased from 9 q/ha pre-war to 25 q/ha in 1987.

The capacity for cereal production continues to increase due to the use of higher-yielding varieties, and by changes in husbandry. The ultimate aim of the grower is to obtain the maximum yield of 'millable' wheat, just as it is of the plant breeder, even when he directs his attention towards the breeding of varieties which are resistant to drought, frost and diseases (Percival, 1921).

Both the yield and the quality of the wheat crop are affected by conditions of soil, climate and farm management. The yield of flour obtainable from the wheat during milling is dependent upon the degree of maturation — the extent to which individual grains are filled out with endosperm. Premature ripening, sometimes brought on by high temperatures prevailing in the later part of the season, produces shrivelled grain, which is of high protein content because relatively more protein than starch is laid down in the endosperm during the early stages of ripening, whereas the reverse holds during the later stages.

The effect of treatment with nitrogenous fertilizers depends on the time of application and the availability of nitrogen in the soil. Nitrogen taken up by the wheat plant early in growth results in increased tillering (see Ch. 2) which can result in

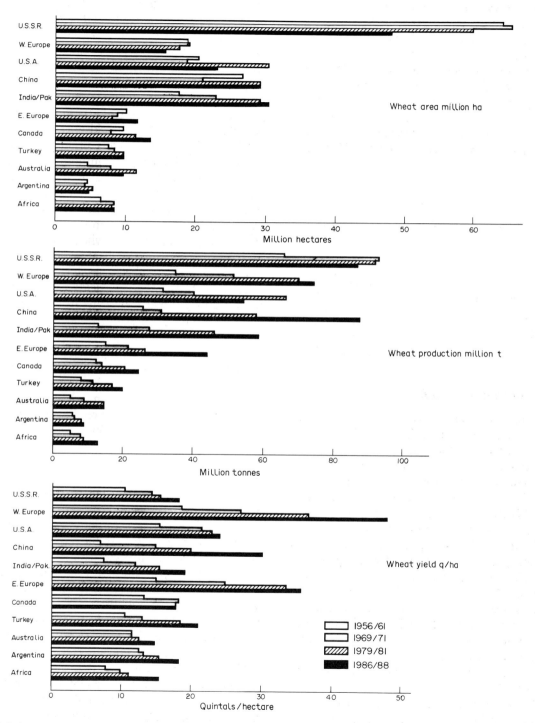

FIG 1.2 Area, production and yield of wheat in the main producing countries and regions of the world in 1956–1961, 1969–1971, 1979–1981 and 1986–1988. (Sources: *Grain Crops* (Commonwealth Secretariat); F.A.O.).

increased grain yield. If prolonged tillering occurs, the last ears formed may not ripen but produce small immature grains. Nitrogen taken up after heading is laid down as additional protein in the seed, with a consequent improvement in nutritive value and often baking quality also. Possible ways of making nitrogen available at a late stage of growth are the early application of slow-acting fertilizers or the late application of foliar sprays, e.g. urea, possibly by means of aircraft.

Green Revolution

This is an expression used to describe the rapid spread of high-yielding varieties of wheat and rice in many developing countries, particularly in Asia. The threat of a massive famine in these countries has been averted, at least in the short run, through the success of the Green Revolution.

Characteristics of the new varieties of wheat and rice, and of their use, are: higher, often doubled, yield of grain per unit of area, combined with a similar protein content (thus giving the possibility of a doubled yield of protein); a larger return of grain per unit of fertilizer applied and per man-hour of labour expended; higher yield of protein per unit of irrigation water; early maturation; less sensitivity to day length, giving greater flexibility in planting time and the possibility of two or even three crops per year.

The effect of the Green Revolution in India, for example, has been to increase the amount of wheat available to the total population from 23.4 kg per person in 1967 to 52.5 kg per person in 1984–1986.

The most rapid exploitation of the Green Revolution occurs in areas where the land is most productive, and where a high proportion of the land is already irrigated. Other regions are less well suited to benefit from the Green Revolution.

Crop movements

Over the period 1977/78 to 1990/91, 16–22% (average 19%) of the entire world wheat crop was exported from the producing country to other countries. Of the total exports of wheat, as grain, in 1990, 33% was provided by European countries, 28% by the U.S.A., 18% by Canada, 12% by Australia, 6% by Argentina, 1% by the former Soviet Union, and 1% by Saudi Arabia.

About nine-tenths of the wheat exports are in the form of unmilled grain, the remainder as flour. The major exporters of wheat flour in 1990 were France (23.5% of the total), Italy (17.8%), U.S.A. (12.9%), Belgium/Luxembourg (8.6%) and Germany FR (7.3%). American flour goes to a large number of countries, but mostly to Egypt (55% of total U.S. flour exports in 1989/90) and the Yemen. Much of the flour exported from Canada goes to Morocco and Cuba. France has big markets for flour in her former African territories.

European imports accounted for 16% of the world movement of wheat in 1990. Other large importers of wheat in 1990 were the former Soviet Union (15%), China (13%), Japan (6%), Egypt (6%), India/Bangladesh/Pakistan (4%), Iran (4%), Algeria (3%) and Korea (3%).

Wheat flour is imported principally by Egypt, Libya, Syria, Cuba, Hong Kong, Cameroon and Yemen. These countries absorbed about 53% of the total trade in wheat flour in 1989/90. Imports of flour decrease sharply when a domestic flour-milling industry is established: this happened about 1960 in the Philippines, 1977 in Egypt, 1980 in Sri Lanka.

Utilization

Data for the domestic utilization of wheat in certain countries are shown in Table 1.3.

Wheat quality

'Quality' in the general sense means 'suitability for some particular purpose'; as applied to wheat, the criteria of quality are:

- yield of end product (wheat, for the grower; flour, for the miller; bread or baked goods, for the baker, etc);
- ease of processing;
- nature of the end product: uniformity, palatability, appearance, chemical composition.

These criteria of quality are dependent upon the variety of wheat grown and upon environment

TABLE 1.3
Domestic Consumption of Common Wheat in Certain Countries

Country	Year	Total domestic usage (thousand t)	Percentage of total consumption				Source of data
			Human food	Industrial usage*	Feed	Seed	
World	1984/86	507,691	66.5	6.7	20.2	6.6	2
EC	1989/90	58,080	53.8	4.6	36.6	5.0	1
Bel./Lux.	"	1710	61.6	19.3	17.1	2.0	1
Denmark	"	1946	17.0	—	78.1	4.9	1
France	"	11,671	43.8	5.0	45.0	6.2	1
Germany, FR	"	10,118	41.8	5.9	49.3	3.0	1
Greece	"	1790	82.1	—	6.7	11.2	1
Ireland	"	710	42.3	12.7	43.0	2.0	1
Italy	"	10,310	80.0	0.6	12.6	6.8	1
Netherlands	"	1995	53.0	20.5	25.2	1.3	1
Portugal	"	1050	81.9	—	14.3	3.8	1
Spain	"	5413	67.1	1.1	24.9	6.9	1
U.K.	"	11,367	43.6	5.1	48.2	3.1	1
Australia	1984/86	3587	43.4	10.3	22.2	24.1	2
Canada	"	5618	37.0	—	41.0	22.0	2
Japan	"	6140	80.7	9.5	9.6	0.2	2
New Zealand	"	360	78.3	0.6	17.5	3.6	2
Turkey	"	19,020	50.6	24.4	16.3	8.7	2
U.S.A.	"	29,828	59.7	—	32.0	8.3	2
Former Soviet Union	"	96,204	40.1	10.7	39.2	10.0	2

* Including waste.
Sources: (1) EC Commission Documents, via HGCA Cereal Statistics (1991a); (2) F.A.O. Food Balance Sheets, 1990.

— climate, soil and manurial or fertilizer treatment. Within the limits of environment, quality is influenced by characteristics that can be varied by breeding, and is further modified during harvesting, farm drying, transportation and storage.

Quality requirements

Wheat passes through many hands between the field and the table: all those who handle it are interested in the quality of the cereal, but in different ways.

The *grower* requires good cropping and high yields. He is not concerned with quality (provided the wheat is 'fit for milling' or 'fit for feeding') unless he sells the grain under a grading system associated with price differentials (cf. p. 88).

The *miller* requires wheat of good milling quality — fit for storage, and capable of yielding the maximum amount of flour suitable for a particular purpose.

The *baker* requires flour suitable for making, for example, bread, biscuits or cakes. He wants his flour to yield the maximum quantity of goods which meet rigid specifications, and therefore requires raw materials of suitable and constant quality.

The *consumer* requires palatability and good appearance in the goods he purchases; they should have high nutritive value and be reasonably priced.

Field damage to wheat

The yield of wheat may be reduced, and its quality impaired, by the attack of various fungal and animal pests in the field.

Rusts

These are fungal diseases caused by species of the genus *Puccinia*. Yellow or Stripe Rust (*P. striiformis*) and Brown or Leaf Rust (*P. recondita tritici*) sometimes occurring in the west of Britain, are particularly troublesome in the U.S.A., Canada and Argentina, and generally in countries with a hot climate.

Rusts exist in many physiological races or forms, and from time to time new races arise to which hitherto resistant strains of cereals may be susceptible. Thatcher wheat (cf. p. 83) was resistant to Stem Rust when released, but proved to be susceptible to race 15B in 1950. Selkirk is a variety, bred for Canadian and U.S. HRS areas, which is resistant to Stem Rust 15B.

Yellow Rust is spread by air currents, and attacks cereal plants in favourable weather in May and June in central and western Europe. Bright orange-yellow patches of spots appear on the leaves; the patches increase in size and in number and eventually prevent photosynthesis occurring in the leaves, and the plant starves. In a bad attack, 80–90% of potential yield may be lost. Immunity to rust, a varietal character, was bred into wheat by Biffen, using Rivet (*Triticum turgidum*), Club (*T. compactum*) or Hungarian Red (*T. aestivum*) as the immune parent. The character for immunity was recessive, appearing in one-quarter of the plants which, however, bred true for immunity in the F$_1$ and subsequent generations. Yellow Rust can be controlled by treatment with benodonil (Calirus) and by a mixture of polyram and tridemorph (Calixin) or by growing resistant varieties.

Some improvement in resistance to rust in wheat has been achieved by incorporating part of a chromosome derived from rye. However, doughs made from flour of such substitution lines may display an undesirable degree of stickiness (cf. p. 211) (Martin and Stewart, 1991).

Common bunt, stinking smut

This is a disease caused by the fungus *Tilletia caries*. The fungus enters the plant below ground, becomes systemic, and invades the ovaries. As the grain grows, it becomes swollen and full of black spores. Bunted grains are lighter in density than normal grains and can be separated from the latter at the cleaning stage by aspiration or flotation (see Ch. 5). Bunt imparts an unpleasant taint of rotten fish (due to trimethylamine) to the flour and gives it an off-white colour. The disease is satisfactorily controlled by seed dressing with organo-mercury compounds (not currently permitted in the U.K.), copper carbonate, or formaldehyde.

Loose smut

The fungus *Ustilago nuda* infects wheat plants at flowering time. The disease is of little importance to the miller, but is of concern to the grower because infected plants fail to produce seed. It can be controlled by seed treatments with, for example, hot water, hot formaldehyde, or benomyl with thiram (Benlate T) which induce the formation of quinones which are fungi-toxic.

Mildew

The fungus *Erysiphe graminis* infects the leaves of cereal plants during warm humid weather in April–June, later producing greyish white patches of spores, or 'mildew'. The leaf surface becomes obliterated by the fungus, reducing or preventing photosynthesis, and the plants become unable to develop normal grains. Even a mild attack reduces the yield. Systemic fungicides are useful against mildew, which is best controlled, however, by the growing of resistant varieties.

New varieties of cereals resistant to mildew in Britain include Atem and Triumph (barleys), and Fenman and Torfrida (wheats). However, most of the varieties of spring and winter wheat recommended by the National Institute of Agricultural Botany in Britain are moderately resistant to mildew. Mildew-resistant oats have been bred at the Welsh Plant Breeding Institute, Aberystwyth.

Take-all; Eye-spot

Gaeumannomyces graminis and *Pseudocercosporella herpotrichoides*, the fungi causing take-all and eye-spot diseases, live in the soil, and may survive on straw or stubble for a year or more. Plants affected by these diseases have empty or half-filled ears, and prematurely ripened or shrivelled grains. Take-all may be controlled by suitable crop succession. Eye-spot on winter wheat is controllable by treatment with benomyl, carbendazim, or a formulation of thiophanate-methyl (Cercobin).

Other fungal diseases of wheat are Leaf Spot, caused by *Mycosphaerella graminicola* (= *Septoria tritici*), Glume Blotch (*Leptosphaeria nodorum* = *Septoria nodorum*) and Flag Smut (*Urocystis tritici*). A systemic fungicide recommended against *Septoria* on wheat is Tilt, a triazole. Flag Smut can be controlled by seed treatment and by the growing of resistant varieties.

Eelworm

Wheat may be attacked by the eelworm, *Anguina tritici*, the grains becoming filled with the worms, which are of microscopic dimensions. Infected grains are known as 'ear cockle' (not to be confused with 'corn cockle', the seeds of the weed *Agrostemma githago*).

Wheat bug

Bugs of the species *Aelia rostrata* and *Eurygaster integriceps* attack the wheat plants and puncture the immature grains, introducing with their saliva a proteolytic enzyme which modifies the protein, preventing the formation of a strong gluten (cf. p. 200). Flour milled from buggy wheat gives dough that collapses and becomes runny if more than 5% of attacked grain is present. Steam treatment of the attacked wheat for a few seconds (BP No. 523,116) is beneficial in inactivating the enzymes, which are localized near the exterior of the grain. The baking properties of flour milled from buggy wheat are improved by increasing the acidity of the dough, since the proteolytic enzyme in the bug saliva has an optimum pH of 8.5. Wheat bug damage is generally restricted to crops grown in the former Soviet Union, the Mediterranean littoral, eastern Europe and the Near East.

Wheat blossom midge

The damage caused by the midge *Sitodiplosis mosellana* varies greatly with year and locality. The female midge lays eggs in the wheat floret. The feeding larvae use part of the plant juices for their development; in consequence, infested grains become shrivelled. Secondary effects are reduced germination capacity and seed weight, increased alpha-amylase activity, and poorer baking quality of the flour.

Thrips

The larvae of tiny insects of the genus *Haplothrips* and of other genera in the order Thysanoptera ('fringe wing'), known as thrips, frequently attack the developing inflorescence of wheat in western Europe, but generally complete their development and leave the plants before the grain matures. In cold, wet harvest years, however, the adult insects occasionally fail to escape from the plant, and become occluded in the crease (cf. p. 40: see Fig. 1.3). Attack by thrips apparently does not affect milling or baking quality of the wheat, but insects remaining in the crease are fragmented during milling and contribute towards the insect fragment count of the milled flour (Kent, 1969).

Rustic Shoulder Knot moth

The larvae of the Rustic Shoulder Knot moth (*Apamea sordens*) feed on the developing wheat grains in the field. The young larvae penetrate the grains at the brush end and hollow them out. The fully developed larva may attain a length of 28 mm. Secondary effects of heavy attack by *A. sordens* are loss of flour yield, discoloration and an increased micro-organism count in the flour (due to infection of the exposed endosperm surfaces by fungi). Attack, in Britain, is more prevalent in Scotland and the north of England than in the south.

Harvesting

Dormancy

After the wheat appears to be ripe, it needs a further period of maturation before it is capable of germination (cf. p. 36); during this period the wheat is said to be 'dormant'. Dormancy is a valuable characteristic conferring a degree of resistance to sprouting at harvest time. The factor appears to be related to enzymic activity; however, not all varieties show a period of dormancy, and the factor appears to be linked genetically to redness of bran colour.

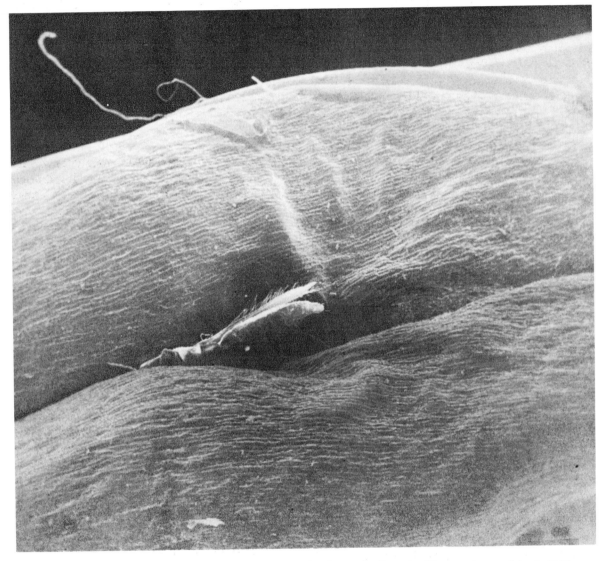

FIG 1.3 Female thrips insect partly buried in the ventral crease of a wheatgrain. (Scanning Electon micrograph by A. D. Evers, reproduced with permission of the copyright holder Leica Cambridge Ltd.)

Harvesting by binder

Wheat can be safely harvested by binder at moisture contents up to 19%, stooked in the field, and stored in ricks, where it will dry with the minimum of deterioration. However, harvesting by binder is no longer practised in the U.K.

Combine harvesting

When harvested by combine harvester, a machine which both cuts the stems and threshes the grain, the moisture content of the wheat should not exceed 15% for immediate storage, or 19% if the wheat can be dried promptly. Correct setting of the harvester to give efficient threshing coupled with minimum mechanical damage to the grain is important. Since 1943, when combine harvesters were first used in Britain, an increasing proportion of the wheat crop has been harvested by this means every year (cf. p. 87). The number of combines in use in the U.K. in 1979 was 55,000, but had fallen to 50,980 by 1989, notwithstanding a small increase in the area laid down to cereals: 3,862,000 ha in 1979, 3,903,280 in 1989.

Harvesting hazards

Cold weather at harvest time may result in imperfect ripening, or in delayed ripening. If rain follows ripening, fungal infection of the chaff, spreading to the seed, may occur.

Sprouting in the ear

The tendency to sprout, or germinate, in the ear depends on varietal characteristics and on atmospheric conditions. Hot dry weather hastens maturation; if followed by rain while the crop is still in the field, the conditions favour sprouting. Wheat is less likely to sprout in a wet harvest if the season is cool.

Badly sprouted wheat is not of 'millable quality', but mildly sprouted wheat may be described as 'millable' and yet have an undesirably high activity of certain enzymes, particularly *alpha*-amylase (cf. pp. 67 and 199), because the damage may not be visible. The (Hagberg) Falling Number test for *alpha*-amylase activity is described on p. 184.

Mycelium of fungi such as *Aspergillus*, *Penicillium*, *Alternaria* and *Cladosporium* is frequently present in and within the pericarp of sound wheat. In wet harvesting conditions, growth of mycelium within the pericarp may be sufficiently prolific to cause discoloration and spoilage of the milled flour. Fungal-infected wheat is still millable, although the flour quality will be inferior. Some improvement of flour colour may result from repeated dry scouring of the wheat and, in the milling process, by increased draught on the purifiers (cf. p. 156).

Wheat harvested wet needs drying to prevent mould development, but overheating damage may ensue if the grain is dried too rapidly at too high a temperature (cf. p. 114). The damage may not be obvious until the milled flour is baked. Heat damage to the protein of wheat may be detected by the Turbidity test, which estimates the content of soluble proteins (Harrison, *et al.*, 1969; cf. p. 185).

Barley

Cultivation

Barley is grown in temperate climates mainly as a spring crop and has geographic distribution generally similar to that of wheat. Barley grows well on well-drained soils, which need not be so fertile as those required for wheat. Both winter barley (autumn sown) and spring barley are grown, a higher yield being obtained from the winter barley. In the U.K., spring barley predominated until recently — 93% of the total barley crop was spring sown in 1970 — but the proportion of winter barley is increasing, and reached 62% of the total barley crop in 1991. In that year, the average yield of winter barley in the U.K. was estimated at 59.6 q/ha, that of spring barley 47.9 q/ha; thus, winter barley provided 65.8% of the total crop (H-GCA, 1991b).

Barley may be attacked in the field by various insect pests, such as wireworm, Hessian fly, frit fly and aphids, and by fungi such as *Helminthosporium*, smut (*Ustilago*), mildew (*Erysiphe graminis hordei*), leaf rust (*Puccinia graminis*), leaf scald (*Rhyncosporium secalis*). Leaf scald is particularly prevalent in wet years.

Area, production, yield

The world area laid down to barley increased from 40 M ha in 1937–1940 to 76 M ha in 1989, with production increasing from 83 Mt (annual average 1961–1966) to 171 Mt in 1991/92. In 1991/92, contributions to world production were: the EC (12 countries) 29.8% (of which France 6.3%, Spain 5.4%, Germany FR 8.3%, the U.K. 4.5%), former Soviet Union 25.2%, Canada 7.6%, the U.S.A. 5.9%, and Eastern Europe 8.5%.

The highest yields are obtained in intensively cultivated areas. In 1991, an average yield of 53 q/ha was obtained in the U.K., 60 q/ha in Belgium/Luxembourg, 61 q/ha in France, and 52 q/ha in Denmark and Germany FR. Yields were lower in New Zealand (44 q/ha in 1989), Canada (25 q/ha in 1990) and the U.S.A. (26 q/ha in 1990), while in India, Pakistan and the Middle East countries yields ranged between 6 and 23 q/ha in 1989.

Crop movement

The proportion of the world crop of barley that moved in world commerce rose from 5% in 1937–1940 to 12.6% in 1988. In 1988, the largest exporters were France (with 23% of the total exports), Canada (13%), U.K. (13%) and the U.S.A. (10%).

European countries were the principal importers of barley until 1961, when Chinese imports of over 1 Mt matched those of the U.K. (0.97 Mt) and Germany FR (0.96 Mt). In 1989/90 the principal importers of barley were the former Soviet Union (4.3 Mt), Saudi Arabia (2.8 Mt), Japan (1.1 Mt), Libya (0.6 Mt) and China (0.5 Mt).

Utilization

The principal uses for barley are as feed for animals, particularly pigs, in the form of barley meal (see Ch. 15), for malting and brewing in the manufacture of beer, and for distilling in whisky manufacture (see Ch. 9). There is little use for barley as human food in Europe and North America, but it is widely used for this purpose in Asian countries. Even there, however, its use as human food is declining as preferred grains become more plentiful. Domestic usage of barley in recent years in certain countries is shown in Table 1.4.

TABLE 1.4
Domestic Utilization of Barley

Country	Year	Total domestic usage (thousand t)	Percentage of total consumption				Source of data
			Human food	Industrial usage*	Feed	Seed	
World	1984/86	173,420	5.3	15.0	73.1	6.6	2
EC	1989/90	37,904	0.3	17.1	77.8	4.8	1
Bel./Lux.	"	899	0.4	42.9	54.5	2.2	1
Denmark	"	3629	—	5.5	90.1	4.4	1
France	"	4387	0.4	6.5	85.8	7.3	1
Germany, FR	"	9285	0.4	25.6	71.2	2.8	1
Greece	"	774	—	6.5	87.8	5.7	1
Ireland	"	947	—	12.7	82.8	4.5	1
Italy	"	2497	0.4	10.0	86.0	3.6	1
Netherlands	"	818	0.6	37.3	61.4	0.7	1
Portugal	"	170	—	23.5	66.5	10.0	1
Spain	"	8388	—	7.4	84.8	7.8	1
U.K.	"	6110	0.2	30.3	65.9	3.6	1
Australia	1984/86	1027	0.6	31.6	53.9	13.8	2
Canada	"	7777	0.2	5.8	88.0	5.9	2
Japan	"	2442	5.9	38.2	55.4	0.4	2
New Zealand	"	436	0.5	48.2	41.7	9.6	2
U.S.A.	"	10,482	1.5	29.6	64.6	4.2	2
Morocco	"	2453	61.0	9.4	21.7	7.8	2
China	"	3104	61.6	28.9	6.0	3.5	2
India	"	1781	72.8	8.7	12.0	6.5	2
Ethiopia	"	1044	79.4	14.9	—	5.7	2

* Including waste.
Sources: (1) EC Commission Documents, via H-GCA Cereal Statistics (1991a); (2) F.A.O. Food Balance Sheets, (1990).

Human consumption

The use of barley for human food (other than for beer) is relatively small in the developed countries. Thus, in the U.K., from a total domestic utilization of 6.1 Mt in 1989/90, only about 12,000 t of barley products were used for human food. Data for human consumption of barley products (pearl barley, malt) for certain countries in 1984/86 are given in Table 1.5.

Relatively high consumers are grouped into regions: the Far East (Korea DPR and Republic), the Middle East (Ethiopia, Afghanistan, Iran, Iraq, Yemen Arab Republic). In these countries much of the barley is consumed as pearled grain for soups, as flour for flat-type bread, and as ground grain to be cooked and eaten as porridge.

However, the consumption of wheat products exceeds that of barley products in all the countries mentioned. There is no country in the world in which the diet is based exclusively, or even mainly, on milled barley products.

TABLE 1.5
Human Consumption of Barley Products 1984/86 Average
(kg/hd/yr)

Morocco	68.3	Peru	5.1
Ethiopia	19.0	Finland	3.1
Algeria	17.6	Japan	1.2
Iran	14.4	Bel/Lux	1.2
Afghanistan	14.4	U.S.A.	0.7
Iraq	11.5	Germany, FR	0.6
Libya	9.9	Netherlands	0.5
Tunisia	9.9	New Zealand	0.5
Bulgaria	8.2	Canada	0.5
Yemen, Arab Republic	7.3	Australia	0.4
Poland	6.6	Denmark	0.4
Norway	6.3	France	0.4
Korea, DPR	6.1	U.K.	0.3
Korea, Rep.	5.5	Italy	0.2

Source: Food and Agriculture Organisation, Food Balance Sheets 1984–86, Rome 1990.

Oats

Cultivation

The oat crop is widely cultivated in temperate regions; it is more successful than wheat or barley in wet climates, although it does not stand cold so well. In the U.K., oats are grown extensively in Scotland and the north of England where better-quality crops are obtained than in the south. In 1990 about 50% of the U.K. crop was 'winter oats' (sown in the autumn), the remainder 'spring oats'.

Area, production, yield

The world area under oats was about 45 M ha in 1956/61 and since then steadily declined to a low of 22 M ha in 1988, i.e. by 51%, increasing slightly to 23.7 M ha in 1989. The decline affected most of the major producing countries, and particularly the U.S.A., which contributed only 12% of the total world area in 1989 as compared with 32% in 1946–1951. Other major contributors to the world area in 1989 were the former Soviet Union (48.5%), Canada (6.6%), Australia (4.8%) and Poland (3.6%). In the U.K., oats occupied a larger area than wheat until 1960; subsequently, the area sown to oats steadily decreased, and by 1991 was only 5.3% of that sown to wheat.

The decline in production since 1956/61 has been only 27%, viz. from 59 to 42 M t annually, because of the steady increase in average yield. Production has declined more steeply in the U.S.A. than in the other major producing countries on account of area restriction. Total world production in 1989 was 42.6 M t. The principal producing countries, with their contribution to the total, were the former Soviet Union 40%, U.S.A. 13%, Canada 8%, Poland 5.6%, Australia 4% and Germany FR 3.7%.

Pre-war yields of 24–26 q/ha in the Netherlands had increased to 50 q/ha by 1991. The increase in yield over the same period in the U.K. has been from 20 to 48 q/ha, and in Belgium/ Luxembourg from 20 to 42 q/ha. Yields in Denmark and Germany FR in 1991 were 48 q/ha. In some countries, however, yields are much lower, e.g. 14 q/ha in Australia and Argentina in 1989. The average yield in the former Soviet Union remained almost constant at about 8.8 q/ha from 1937–1940 until 1965, but increased to 15.3 q/ha in 1978, and reached 16.6 q/ha in 1986 (14.8 q/ha in 1989).

Crop movement

The bulk of the oat crop is consumed on the farm where it is produced. Only 4% of the total crop entered world commerce in 1988. Exports of oats have, however, doubled from 0.75 Mt in 1937–1940 to 1.67 Mt in 1988. Over this period, Argentina, U.S.A., Canada and Australia have been the biggest exporters, contributing at least 75% of the total world exports between 1955 and 1966, but less since then (54% in 1988), with an increasing contribution coming from Sweden (13% in 1988) and France (8% in 1988).

European countries were the biggest importers of oats in 1961–1962, and accounted for 85–95% of total imports. Germany FR, Netherlands and Switzerland were the individual countries taking the largest quantities. In 1988, the U.S.A. was the largest importer (52% of total imports), with Switzerland taking 7.5% of the total. The U.S.A. has changed from being an exporter in 1981 to a big importer in 1988. Exports were nearly eight times the size of imports in 1981 in the U.S.A., but by 1988 imports (0.82 Mt) were over 100 times as large as exports.

This change may reflect increased interest in the nutritional value of oats for the human diet, particularly as regards its effect on blood cholesterol (see Ch. 14).

Utilization

A small proportion of the oat crop is milled to provide products for the human diet: oatmeal for porridge and oatcake baking, rolled oats for porridge, oat flour for baby foods and for the manufacture of ready-to-eat breakfast cereals, and 'white groats' for making 'black puddings' —a popular dish in the Midlands of England. Most of the crop, however, is used for animal feeding,

TABLE 1.6
Domestic Utilization of Oats

Country	Year	Total domestic usage (thousand t)	Percentage of total consumption				Source of data
			Human food	Industrial usage*	Feed	Seed	
World	1984/86	48,919	5.3	7.4	78.2	9.1	2
EC	1989/90	4878	10.0	2.3	82.5	5.2	1
Bel./Lux.	"	147	4.1	—	92.5	3.4	1
Denmark	"	150	20.0	—	76.7	3.3	1
France	"	798	1.3	1.2	91.5	6.0	1
Germany, FR	"	2044	7.8	3.9	85.4	2.9	1
Greece	"	62	3.2	—	83.9	12.9	1
Ireland	"	106	14.2	—	81.1	4.7	1
Italy	"	371	—	—	90.6	9.4	1
Netherlands	"	80	31.2	2.5	65.0	1.3	1
Portugal	"	100	—	—	80.0	20.0	1
Spain	"	489	0.8	0.6	89.0	9.6	1
U.K.	"	531	44.1	3.2	48.4	4.3	1
Australia	1984/86	1381	3.1	7.0	77.3	12.6	2
Brazil	"	151	88.0	2.7	—	9.3	2
Canada	"	2788	2.8	—	92.0	5.2	2
China	"	497	63.5	4.6	25.0	6.9	2
Finland	"	1124	2.2	1.2	87.7	8.9	2
Germany DR	"	646	20.9	3.7	71.5	3.9	2
Korea DPR	"	182	72.0	4.4	19.8	3.8	2
Poland	"	2640	2.3	5.3	82.5	9.9	2
Sweden	"	1173	2.5	6.0	83.2	8.3	2
U.S.A.	"	7343	8.6	—	84.8	6.6	2
Former Soviet Union	"	20,808	2.1	14.1	72.3	11.5	2

* Includes waste.
  Sources: (1) EC Commission Documents, via H-GCA Cereal Statistics (1991a); (2) F.A.O. Food Balance Sheets (1990).

although increased mechanization on farms has reduced the quantity of oats required for feeding horses.

The domestic utilization of oats in various countries is shown in Table 1.6.

In 1978, out of a total of 148,000 t used for human consumption and industrial purposes in the U.K., milled oat products (flour, meal, groats, rolled oats and flakes) produced in the U.K. mills amounted to only 46,000 t. However, domestic consumption of oatmeal and oat products is increasing in the U.K.: from 0.42 oz/head/week in 1984 it has risen to 0.63 oz/head/week in 1988 (MAFF Household food consumption and expenditure, 1989).

Human consumption

Data for the human consumption of milled oat products in certain countries, 1984–1986 average, are shown in Table 1.7.

In general, consumption of milled oat products is very small in comparison with consumption of milled wheat products. Thus, in the thirteen countries listed in Table 1.7, average consumption figures for oat products and wheat products were 3.3 and 80.2 kg/head/year, respectively, in 1984–1986. This low level of consumption can be regarded as an indication of the relatively minor importance of milled oat products in the human diet.

TABLE 1.7
Consumption of Milled Oat Products, 1984–86
Average (kg/head/year)

Australia	2.8	Ireland	2.7
Canada	3.0	Korea, DPR	0.4
Denmark	5.1	Norway	2.5
Finland	5.2	Sweden	3.5
Germany, DR	8.1	Former Soviet Union	1.6
Iceland	3.0	U.K.	2.7
		U.S.A.	2.6

Source: Food and Agriculture Organisation, Food Balance Sheets 1984–86, Rome, 1990.

Rye

Cultivation

Rye (*Secale cereale* L.) is a bread grain, second only to wheat in importance, and the main bread grain of Scandinavian and eastern European countries (see Ch. 8).

On good soil, rye is a less profitable crop than wheat, but on light acid soil it gives a more satisfactory yield. The rye crop nevertheless benefits from manurial treatment. It is more resistant than wheat to most pests and diseases (although it is more liable to attack by ergot: see below), and can better withstand cold. Thus, rye tends to be grown on land just outside the belt which gives the most satisfactory return to the wheat crop, such as areas of northern and eastern Europe that have a temperate climate.

Rye is occasionally infested by Mildew (*Erysiphe graminis* f.*secale*), Stalk Rust (*Urocystis occulta*), Stem Rust (*Puccinia graminis*) — which can be dangerous and Brown Rust (*Puccinia dispersa*). Types of rye resistant to Stem Rust and Brown Rust have been bred (Starzycki, 1976).

Ergot and ergotism

Ergot is the name given to the sclerotia of the fungus *Claviceps purpurea* (Fr.) Tul., which infects many species of grasses and is particularly liable to infect rye in humid summers (see Fig. 1.4). Wheat, barley and oats are also attacked, but comparatively rarely.

Ergot has been associated with rye because the latter was generally grown on soils which were too poor to give a useful crop of other cereals but which provide suitable conditions for *Claviceps*. Rye grown on good land, from fresh seed, is probably no more liable than wheat or barley to become ergotized.

Ergot is a toxic contaminant: when consumed in quantity it causes gangrenous ergotism, a disease which was known as 'Holy fire' or 'St Anthony's Fire' in the 11th to 16th centuries — although its connection with ergot was not then known.

The fungus *Claviceps* infects the flower of the rye plant and invades the seed as it develops. Eventually the whole of the tissue of the seed is replaced by a dense mass of fungal mycelium, and the seed grows to a large size, protruding from the ear. The sclerotia of *Claviceps*, average length 14.6 mm and thickness 6.5 mm, are brittle when dry, dull greyish or purple-black in colour

FIG 1.4 Spikes of rye showing ergot sclerotia. (Photo by W. C. Moore. Reproduced from Ministry of Agriculture, Fisheries and Food Bulletin. No. 129, *Cereal diseases*, with permission of the Controller of H.M.S.O.)

on the outside, dull pinkish white within. They consist of a pseudo-parenchyma of closely-matted fungal hyphae. Alkaloids present in ergot sclerotia include ergotoxine and ergotamine, both of which have an active principal known as lysergic acid (LSD), produced when ergot ferments. This compound causes hallucinations, agitation and the other symptoms associated with 'St Anthony's Fire'.

Ergot tolerances in grain have been established in many countries. Wheat and rye are graded 'ergoty' if they contain 0.3% of ergot in the U.S.A. (cf. p. 86), 0.33% ergot in Canada.

The EC Intervention quality standard 1989 sets a maximum of 0.05% of ergot in wheat and rye. In non-EC countries with a high rye bread consumption the maximum limit for ergot in rye is generally 0.2%.

The former Soviet Union in 1926 fixed 0.15% as the maximum harmless quantity of ergot in flour. Flour produced in Germany and in the U.S.A. sometimes contains 0.1% of ergot, and an objection has not been made to this concentration.

There was a mild epidemic of ergotism in Manchester, England, in 1927 among Jewish immigrants from central Europe who lived on rye bread. This bread was made from flour reported to contain 0.1–0.3% of ergot.

The safe limits of ergot in flour would appear to be about 0.05% (Amos, 1973). With a daily consumption of 400g of bread made from flour containing this concentration of ergot, the intake of ergot would be 0.14 g per day, well below that usually prescribed medicinally (to assist childbirth, by its effect on unstriped muscle of the pregnant uterus), but, of course, continued over a long period of time.

Area, production, yield

The world area under rye has fallen steadily from 42.5 million ha per annum during the period 1937–1940 to 16 million ha in 1989. The decline in area has been greatest in the former Soviet Union, where the place of rye and oats has been taken by wheat and barley. Over this period, the former Soviet Union has accounted for 43–67% of the total world rye area, the only other major producing country being Poland (10–22% of the world area).

Decline of 62% of the world rye area between 1937–1940 and 1989 is matched by world production decline of 17.5% (from 40 to 33 million t). During the period 1981 to 1986 the former Soviet Union's share of the world production averaged 44.5%, that of Poland 24.5%, and that of Germany (FR and DR) 12.4%, but in the period 1987–1989 the Soviet Union's share increased to an average of 54.6% of the total, while that of Poland decreased to 18.4% and that of Germany (FR and DR) to 11.1%.

In the U.K., rye is grown mostly in East Anglia and Yorkshire as a winter crop. The total area in the U.K. was about 7280 ha in 1989, and the national production about 0.2% of that of wheat.

In the U.S.A., rye is grown chiefly in the upper North Central States, principally in North Dakota, where emigrants from rye-growing countries of eastern Europe have settled. In 1989 the area of the U.S. rye crop (0.19 million ha) was 0.75% of the wheat area. The U.S. rye crop was 1.0% of the world rye crop in 1989.

Yields reached 54.8 q/ha in Switzerland in 1984, and 50–52 q/ha in 1988/89, and yields ranged 36–50 q/ha in other western European countries practising intensive cultivation: Austria, Belgium/Luxembourg, Denmark, Germany FR, Netherlands, Sweden and the U.K. In the two principal producing countries, however, yields were only 26.1 q/ha in Poland and 17/6 q/ha in the former Soviet Union in 1989.

Crop movement

Some 2–4% of the world's total rye crop moves in international commerce. The former Soviet Union was formerly the biggest exporter, her share of the total exports steadily increasing to 62% in 1962. By 1970, however, her share of exports had fallen below 30%, and by 1977 she had become an importer of rye. In 1988, 27% of the total exports were supplied by Denmark, 22% by Germany FR, and 17% by Canada. In 1988, the major importer was Japan, with 33% of the total, followed by Germany FR with 15%. Korea Republic took 7%, and European countries — Czechoslovakia, Finland, Hungary, Netherlands, Norway, the U.K. — each took 3–6% of the total.

Utilization

Rye is used both for making bread and as animal feed. Rye bread, although nutritious and, to some people, palatable, is not comparable with wheaten bread as regards crumb quality and bold appearance of the loaf; as living standard rise, the consumption of rye bread falls while that of wheaten bread rises. The production of rye exceeded that of wheat in Germany FR from 1939 until 1957; thereafter, production declined, and between 1958 and 1978 a larger amount of rye has been used for animal feed than for human food. Since 1979, however, usage for feed has

decreased, so that by 1988, 54% of the total was used for human consumption and only 41% for animal feed. In Eastern Germany, production of rye exceeded that of wheat until 1965/66, but since 1966 production of wheat has greatly exceeded that of rye. In 1989, Germany DR produced 4.2 million t of wheat but only 2.0 million t of rye. Outside Europe, rye is used mainly for animal feed (cf Ch 15). A small amount is used for distilling (cf. p. 230). The domestic utilization of rye in various countries is shown in Table 1.8.

Human consumption

In some countries, notably those of eastern Europe and Scandinavia, rye forms a significant part of the diet. Data for human consumption of milled rye products, average 1984–1986, in those countries in which average consumption exceeded 2 kg/head/year are given in Table 1.9.

Triticale

Area, production, yield

Triticale (*Triticosecale*) was first grown commercially in the U.S.A. in 1970. Rosner, one of the best know varieties, is a cross between durum wheat and rye. The planted area of triticale in the U.S.A. in 1971 was about 80,000 ha. An octaploid triticale (a bread wheat/rye hybrid) was reportedly grown on 26,000 ha in China in 1977. By 1989, the world area under triticale had increased to 1.6 million ha, of which China and Poland contributed 37.5% each, France 8.8% and Australia 6.8%.

World production of triticale steadily increased from 1.2 million t in 1982 to 3.1 million t in 1987, and to 4.2 million t in 1989. Poland's contribution to the total world production has dramatically increased from 3% in 1985 to 47% in 1988 and in 1989. In 1989, China contributed 23.6% of the total world production, France 13.9%, Spain and Australia 4% each. Average world yield of triticale matched that of rye (20 q/ha) in 1984, but has subsequently exceeded it: triticale 25 q/ha, rye 21 q/ha in 1987: triticale 26.5 q/ha, rye 20.6

TABLE 1.8
Domestic Utilization of Rye

Country	Year	Total domestic usage (thousand t)	Human food	Industrial usage*	Feed	Seed	Source of data
			Percentage of total consumption				
World	1984/86	33,072	33.4	14.3	43.9	0.4	2
EC	1989/90	2603	45.4	3.4	45.2	6.0	1
Bel./Lux.	"	33	39.4	—	54.5	6.1	1
Denmark	"	260	38.5	—	53.8	7.7	1
France	"	212	13.7	—	84.9	1.4	1
Germany, FR	"	1479	58.2	5.7	32.3	3.8	1
Greece	"	18	27.8	—	55.5	16.7	1
Ireland	"	—	—	—	—	—	1
Italy	"	27	7.4	—	85.2	7.4	1
Netherlands	"	69	79.7	—	17.4	2.9	1
Portugal	"	95	—	—	57.9	42.1	1
Spain	"	361	23.5	0.8	67.9	7.8	1
U.K.	"	50	62.0	—	36.0	2.0	1
Austria	1984/86	299	54.5	3.0	38.1	4.3	2
Canada	"	280	4.6	17.9	67.5	10.0	2
Sweden	"	185	68.4	4.8	22.0	4.8	2
U.S.A.	"	556	15.5	9.3	54.9	20.3	2
China	"	1100	87.3	4.9	2.0	5.7	2
Germany DR	"	2575	52.1	5.5	38.9	3.6	2
Japan	"	292	—	—	100.0	—	2
Turkey	"	350	66.0	10.3	12.8	10.9	2
Czechoslovakia	"	647	61.7	3.6	32.4	2.3	2
Poland	"	8111	29.8	7.6	54.8	7.8	2
Soviet Union	"	14,898	28.3	23.7	38.0	10.0	2

* Includes waste.
Sources: (1) EC Commission Documents, via H.G.C.A. Cereal Statistics (1991a); (2) F.A.O. Food Balance Sheets (1990).

TABLE 1.9
Human Consumption of Rye Products in Certain Countries, 1984–86 Average (kg/head/year)

Korea, DPR	3.4	Finland	21.2	Portugal	4.1
Korea, Rep.	2.6	Germany DR	47.5	Spain	2.0
Turkey	4.7	Germany FR	14.4	Sweden	15.3
Albania	2.5	Hungary	3.4	Switzerland	2.6
Austria	21.6	Iceland	5.6	Yugoslavia	2.8
Czechoslovakia	25.7	Norway	8.7	Former Soviet Union	15.2
Denmark	20.2	Poland	64.9		

Source: Food and Agriculture Organisation. F.A.O. Food Balance Sheets 1984–86, Rome, 1990.

q/ha in 1989. Average yields of triticale (q/ha) in 1989 were: 52 in Germany FR, 42 in France and Switzerland, 33 in Poland, 24–25 in Italy, Portugal and Spain, 15–17 in Australia, China and Hungary.

principal exporters being Belgium/Luxembourg (49.8% of the total), France (29.4%), Germany FR (7.7%) and Spain (9.7%).

Crop movement

Exports of triticale in 1988 amounted to only 1715 tonnes (0.05% of the world crop), the

Utilization

The main use for triticale will probably be as a feed crop.

Rice

Cultivation

The rice crop (*Oryza sativa*) is grown in the tropics where rain and sunshine are abundant, and in temperate regions. *O. sativa indica* is confined to the tropics, while *O. sativa japonica* is grown mainly in temperate regions. Although typically a cereal of the swamp, rice can be grown either on dry land or under water. The common practice of flooding the paddy fields has been adopted as a means of irrigation and also as a means of controlling weeds.

The Malayan word 'padi' means 'rice on the straw', but the anglicized form of the word, 'paddy', refers both to the water-covered fields in which rice is grown and also to the grain as harvested, viz. with attached husk or hull.

In much of Asia and Africa, rice is grown on hilly land without irrigation. In some Asian countries where irrigation is practised, two crops of rice are grown per year. The main crop is grown in the wet season, the subsidiary crop with irrigation in the dry season. Yields are lower in the main crop than in the subsidiary crop because of the lack of sunshine.

There are varieties of rice adapted to a wide range of environmental conditions: it can be grown in hot, wet climates, but equally in the foothills of the Alps, up to 1220 m in the Andes of Peru, 1830 m in the Philippines, and 3050 m in India. This wide adaptability of the rice plant is the explanation of its importance as a food crop.

The U.S.A. produces three types of rice: long-grain, medium-grain, and short-grain. The long-grain rice, comprising about 60% of the total rice crop, is grown in Texas, Louisiana, Arkansas, Mississippi and Missouri, while the medium-grain (about 30% of the total crop) and short-grain (about 10%) are grown in California (Webb, 1985). In these States the requirements of the rice crop are — level land with an impervious soil and abundant water for irrigation. Rice is a highly mechanized crop in the U.S.A., where planting, fertilizer treatment and weeding are all carried out on a large scale by means of aircraft. The crop is harvested by combine-harvesters. In con-trast, the rice grown in the major producing countries, amounting to 90% of the world crop, is managed entirely without mechanization.

Area, production, yield

The world production of rice is commensurate with that of wheat. In 1958/59 both crops yielded about 250 million tonnes worldwide; since 1961 the world production of paddy rice has been 82–108% of that of wheat. This production is achieved on an area equivalent to 60–65% of that of wheat, because the yield of rice is so much the greater (paddy rice 34.7 q/ha, wheat 23.6 q/ha, 1989 world averages). World production of rice has shown a steady increase from 1960 to 1989: 30% increase in the 1960s, 30% in the 1970s, 27% in the 1980s. Rice is the basic food for more than half of the world's population, and provides up to 80% of the food intake in some countries.

In 1989 the area under rice was 54 million ha in India/Pakistan/Bangladesh and 32 million ha in China. Other major producers contributing to the total world area of 146 million ha in 1989 were Indonesia (10.2 Mha), Thailand (10.2 Mha), Vietnam (5.8 Mha), Brazil (5.3 Mha), Burma (4.7 Mha), Philippines (3.4 Mha) and Japan (2.1 Mha).

China is by far the biggest producer of rice. The estimated production in 1990/91 (including that of Taiwan Province) was 182 million tonnes of rough (paddy) rice, out of an estimated world total production of 511 million tonnes. Other major producers of rice in 1989 (with estimated production in million tonnes of paddy) were India 109, Indonesia 44, Bangladesh 26, Thailand 20, Vietnam 17, Burma (Myanmar) 14, Korea (DPR and Republic) 14, Japan 13 and Brazil 10.

The yield of rice varies widely according to the method of cultivation. In general it is high in subtropical regions where the variety *japonica* is grown; contributory factors are the intensive cultivation practised in some areas in these regions, and the fact that *japonica* rice gives increased yield when heavily fertilized. The highest yields have been obtained in Australia, where the estimated yield in 1989 was 79 q/ha, most of the crop being grown in New South

TABLE 1.10
Domestic Utilization of Rice*

Country	Year	Total domestic usage (thousand t)	Percentage of total consumption			
			Human food	Industrial usage†	Feed	Seed
World	1984/86	365,112	88.0	7.0	1.8	3.2
EC						
Bel./Lux.	"	103	63.1	24.3	12.6	—
Denmark	"	16	93.3	—	6.7	—
France	"	330	73.3	0.6	25.5	0.6
Germany, FR	"	165	89.7	7.9	2.4	—
Greece	"	67	89.5	7.5	—	3.0
Ireland	"	5	100.0	—	—	—
Italy	"	351	82.3	4.0	4.6	9.1
Netherlands	"	63	100.0	—	—	—
Portugal	"	197	98.5	—	—	1.5
Spain	"	268	96.6	0.4	—	3.0
U.K.	"	238	72.7	18.5	8.8	—
Australia	"	99	72.8	13.1	—	14.1
Bangladesh	"	18,505	88.5	8.0	—	3.5
Brazil	"	7768	84.3	11.5	—	4.2
Canada	"	132	96.2	3.8	—	—
China (and Taiwan)	"	135,289	89.1	6.0	2.3	2.6
Egypt	"	1769	88.2	5.9	—	5.9
India	"	70,187	88.2	5.4	0.4	6.0
Indonesia	"	29,019	88.8	8.0	2.1	1.1
Japan	"	10,678	90.9	7.6	0.6	0.9
Former Soviet Union	"	2558	93.1	2.5	—	4.4
Thailand	"	11,372	77.1	12.1	6.7	4.1
U.S.A.	"	2087	52.0	43.6	—	4.4
Vietnam	"	12,290	86.5	8.9	2.1	2.5

* Dehulled, milled and broken rice as milled rice equivalent.
† Including waste.
  Source: F.A.O. Food Balance Sheets, 1984–86, Rome, 1990.

Wales. Paddy yields of over 70 q/ha were obtained in Korea DPR in 1989; of over 60 q/ha in Japan, Korea Rep., U.S.A., Italy, Egypt and Puerto Rico; of 55.5 q/ha in China, and of over 50 q/ha in Spain, France, Greece, Peru and Uruguay. In 1989 the average yield was only 25–27 q/ha in India, Pakistan and Bangladesh.

Crop movements

The total world exports of paddy rice in 1989 were only 11.5 million tonnes, or 2.3% of the total world production of 506.9 million tonnes. Thailand was the principal supplier (52% of the total exports in 1988), followed by the United States (26%), Vietnam (12%) and Pakistan (7%). Italy and Spain were large importers of paddy

rice in 1988, taking 45% of the total imports. Other major importers were China, India, Iraq, Iran and Malaysia.

Utilization

Rice is used mainly for human food. In Japan rice is used to brew a type of beer called saké (cf. p. 228). Data for the domestic utilization of rice according to purpose are shown in Table 1.10.

Human consumption

Figures for the consumption of rice in certain countries are shown in Table 1.11. Consumption of rice exceeds 100 kg/head/year in most Asian countries and also in some African countries

TABLE 1.11
Human Consumption of Brown or Milled Rice in Certain Countries
(kg/head/year, 1984–86 Average)

Bangladesh	162	Indonesia	155	Mauritius	84
Brunei	95	Japan	80	Nepal	109
Burma	238	Kampuchea	204	Philippines	115
China	116	Korea DPR	164	Sierra Leone	114
Gambia	84	Korea Rep.	148	Sri Lanka	126
Guinea-Bissau	116	Laos	216	Suriname	112
Guyana	181	Liberia	130	Thailand	171
Hong Kong	71	Madagascar	150	Vietnam	178
India	81	Malaysia	118		

Source: Food and Agriculture Organisation, Food Balance Sheets 1984–86, Rome, 1990.

— Liberia, Madagascar, Guinea-Bissau, Sierra Leone — and some South American countries — Guyana, Suriname. Consumption of milled rice does not exceed 7 kg/head/year in European countries (except Portugal: 19 kg/head/year, 1984–1986 average); it was 4.5 kg/head/year in the U.S.A. in 1984–1986.

Rice provided approximately 8000 kJ of energy per day per cap. in Burma in 1984–1986, more than 6000 kJ/day/cap. in Kampuchea, Korea DPR, Laos, Thailand and Vietnam, and more than 4000 kJ/day/cap. in Liberia, Madagascar, Bangladesh, Indonesia, Korea Rep., Malaysia, China, Guyana, Philippines, Sri Lanka and Guinea-Bissau.

Maize

Cultivation

Maize (corn, in the U.S.A.; *Zea mays* L.) is cultivated in regions that experience periods of at least 90 days of frost-free conditions; the highest yields are obtained when the crop matures in 130–140 days. The crop needs temperatures of 10°–45°C and rainfall of 25–500 cm/an.; it can be grown in the tropics and the temperate regions (although mostly between the latitudes of 30° and 47°), and at altitudes from sea level to 12,000 ft, suitable types being available for these varying conditions.

Area, production, yield

In 1989, out of a total world area of 129 million ha sown to maize, the U.S.A. accounted for 20%, China for 16%, Brazil for 10%, Mexico for 5%, and India for 4.5%. Nevertheless, the U.S.A. produced 40% of the world's total maize tonnage in 1989 (and 43% in 1990/91) because of the use of high-yielding strains of hybrid maize (cf. p. 99).

Total world production of maize in 1990/91 was 471 million tonnes, the principal producing countries (with their share of the total) being the U.S.A. (43.2%), China (17.8%), Brazil (5.2%), eastern Europe (5.2%), EC countries (4.9%), the former Soviet Union (3.0%), Mexico (2.3%) and South Africa (1.8%).

Yields in the U.S.A. increased from 14 q/ha in 1934–1938 to 69 q/ha in 1981 and to 75.6 q/ha in 1987, but even higher yields have been recorded elsewhere: 100 q/ha in New Zealand in 1989; 96 q/ha in Austria, 1989; 93 q/ha in Greece in 1984 (87 q/ha in 1990); 92 q/ha in Switzerland in 1988 (76 q/ha in 1989); 75 q/ha in Germany FR and 80 q/ha in Italy in 1989. Other countries with yields in excess of 50 q/ha in 1989 were Belgium, Canada, France, Korea DPR and Egypt. Elsewhere, yields may be much lower: 10 q/ha in India in 1987 (13 q/ha in 1989); 12 q/ha in the Philippines (1989); 17 q/ha in Mexico (1989); 21 q/ha in Brazil (1989).

Crop movements

The proportion of the total world crop of maize that entered into world trade was 16.8% in 1988. In 1937–1940 the U.S.A. exported about 1.5 million tonnes of maize per annum, but by 1980 the figure had increased gradually to 63 Mt/an, falling back to 27 Mt in 1986, and rising again to 60 Mt in 1989/90.

The other principal exporting countries in 1989/90 were China (3.2 Mt), Argentina (3.0 Mt) and South Africa (2.9 Mt), contributing to the world total of 70 Mt of exports.

In 1989/90 the former Soviet Union was the biggest importer of maize (16.9 Mt), followed by Japan (16.1 Mt), Korea Rep. (6.2 Mt) and Taiwan (5.3 Mt). Imports of maize to Western Europe from the U.S.A. were 15.3 Mt in 1981/82, but declined to 6.6 Mt in 1984/85, partly on account of the imposition of an import levy by the European Community, and the EC's policy of encouraging the production of barley and feed wheat in substitution for imported grain. Imports of maize by the U.K. (from all sources) had been about 3 Mt in 1980, but fell gradually to 1.3 Mt in 1988. Between 1980/81 and 1984/85 imports of maize from the U.S.A. to Eastern Europe fell from 6.7 to 0.7 Mt, whereas those to the former Soviet Union increased from 4.9 to 14.9 Mt (Leath and Hill, 1987).

Field damage to maize

Abnormally dry conditions, such as the 1983 drought in the U.S.A., are conducive to the growth of the fungus *Aspergillus flavus* and the production of the mycotoxin aflatoxin. In the U.S.A. maize grain containing more than 20 µg/ kg of aflatoxin may not be shipped across State boundaries.

The distribution of aflatoxin within the grain is such that, on dry milling, the aflatoxin is concentrated 2–3– fold in the milling by-products — screenings, germ, hominy feed — whilst the main products contain aflatoxin at only 12–30% of the level in the original grain. Thus, for example, maize grain containing 51 µg/kg of aflatoxin could yield grits with 6 µg/kg and flour with 15 µg/kg (Romer, 1984).

On the other hand, abnormally wet conditions during growth promote the development of *Fusarium*, which produces the mycotoxins zearalenone, trichothecene and deoxynivalenol (vomitoxin), which may continue to be produced during storage of contaminated grain. The problems with the *Fusarium* mycotoxins mostly concern animals and animal feed (Romer, 1984).

Utilization

Maize is used for animal feeding, for human consumption, and for the manufacture of starch, syrup and sugar, industrial spirit and whisky (cf. pp. 314 and 230). The products of milling (cf. p. 136) include maize grits, meal, flour (and derived products), protein (gluten feed) and corn steep liquor. The ready-to-eat breakfast cereal 'corn flakes' is made from maize grits (cf. p. 248).

Data for the total domestic utilization of maize in certain countries and for its use for human food, industrial processing ('wet milling') and other uses, and for seed are shown in Table 1.12.

The U.S.A. is by far the largest domestic user of maize. In 1984/86, 138.2 Mt were used domestically, of which 79% was for animal feed and 0.3% for seed. In 1984, 26.9 Mt of maize were processed in the U.S.A., of which 60.8% was used for wet processed products, 22.6% for alcohol and distilled spirits, 13.4% for dry milled products (corn meal, flour, hominy grits, brewers' grits, flakes) and 3.2% for making breakfast cereals (USDA, via Leath and Hill, 1987). Products of wet milling include starch, sugars and corn oil.

Human consumption

Maize formed the staple diet of the early native American civilizations — Aztecs, Mayas, Incas — and it often forms the staple diet in present-day Latin American countries and in parts of Africa. Paraguay, Romania and Albania also have high human consumption of maize. Data for the human consumption of maize in countries in which consumption exceeds 40 kg/head/year are shown in Table 1.13. In Western European countries, Australia, New Zealand and Canada consumption does not exceed 10 kg/head/year. In the U.S.A. it was 8.4 kg/head/year in 1984/86.

Miscellaneous cereals

A group of small-seeded cereals and forage grasses used for food or feed includes the sorghums

TABLE 1.12
Domestic Utilization of Maize in Certain Countries

Country	Year	Total domestic usage (thousand t)	Human food	Industrial usage*	Feed	Seed	Source of data
			Percentage of total consumption				
World	1984/86	440,960	20.7	13.9	63.9	1.5	2
EC	1989/90	27,639	9.6	12.7	77.0	0.7	1
Bel./Lux.	"	746	3.4	60.3	36.3	—	1
Denmark	"	88	22.7	9.1	68.2	—	1
France	"	5909	6.0	5.0	87.4	1.6	1
Germany FR	"	2682	25.0	13.8	59.5	1.7	1
Greece	"	1925	0.5	3.1	96.1	0.3	1
Ireland	"	60	—	—	100.0	—	1
Italy	"	7182	2.4	8.6	88.7	0.3	1
Netherlands	"	1600	3.1	40.0	56.3	0.6	1
Portugal	"	1150	17.4	4.3	77.5	0.8	1
Spain	"	4578	0.8	15.5	83.3	0.4	1
U.K.	"	1719	64.5	17.3	18.2	—	1
Australia	1984/86	201	29.8	2.0	67.7	0.5	2
Austria	"	1565	1.1	4.6	93.6	0.7	2
Canada	"	7033	1.3	18.9	79.4	0.4	2
China	"	65,306	38.7	6.8	53.1	1.4	2
Japan	"	13,660	18.5	1.5	80.0	—	2
Former Soviet Union	"	25,422	0.4	17.6	73.4	8.6	2
U.S.A.	"	138,184	1.5	19.2	79.0	0.3	2
Yugoslavia	"	11,294	5.1	10.7	83.2	1.0	2

* Including waste.

Sources: (1) EC Commission Documents, via H.G.C.A. Cereal Statistics (1991a); (2) F.A.O. Food Balance Sheets, 1984–86, F.A.O., Rome, 1990.

TABLE 1.13
Human Consumption of Maize in Certain Countries
(kg/head/year, 1984–86 averages)

Africa				America, Central	
Benin	60	South Africa	110	El Salvador	83
Botswana	80	Swaziland	112	Guatemala	106
Cape Verde	80	Tanzania	73	Honduras	94
Egypt	57	Zambia	168	Mexico	120
Kenya	124	Zimbabwe	118	Nicaragua	58
Lesotho	124	America, South		Europe	
Malawi	170	Paraguay	51	Albania	71
Namibia	72	Venezuela	41	Romania	46
				Asia	
				Philippines	49

Source: FAO, Food Balance Sheets 1984–86, Rome 1990.

and millets, and is sometimes known as 'miscellaneous cereals' or 'coarse grains'. Crop improvement and increased usage, however, justify the separate consideration of sorghum and the millets, although statistical data, e.g. crop movements, are generally available only for the group as a whole.

Sorghum

Cultivation

The crop is grown in latitudes below 45° in all continents; in the U.S.A. it is grown in the Great Plains area, chiefly in Texas, where it is the most important crop, and in Kansas. The most favourable mean temperature for the crop is 27°C (80°F) and, although it does well in semi-arid conditions, it repays irrigation. The crop is not troubled by serious pests or diseases, and has the advantage that it can be sown late, in case other crops fail (Matz, 1969).

Area, production, yield

The world area under sorghum in 1989 was 44.4 million ha, and the world production was 59.2 million tonnes, giving a world average yield of 13.3 q/ha. In 1989, the largest area under sorghum was in India, with 36% of the total area, but producing only 20.4% of the total world crop. The U.S.A., with 10% of the world area, produced 26.5% of the world crop, and China produced 10% of the world crop. Since 1980, yields of over 60 q/ha have been obtained in Italy (62.8 q/ha in 1986) and Hungary (61.2 q/ha in 1986), and yields of over 50 q/ha in France (58.5 q/ha in 1987), China, Taiwan Province (54.3 q/ha in 1989) and Spain (52.5 q/ha in 1989). Yields were lower in the principal producing countries: 34.8 q/ha in the U.S.A., 31.9 q/ha in China (excluding Taiwan Province), and 7.6 q/ha in India, all in 1989.

Production of sorghum in the U.S.A. has increased considerably since 1940, reaching 28.5 million tonnes in 1985, when the average yield was 41.9 q/ha. The increase has been due to a change over from other crops (cotton, wheat, maize) because of U.S. government agricultural programmes, and to the availability of sorghum hybrids (cf. p. 49 (Ch.4)) which give 20–40% higher yields than types previously available. Other factors contributing to increased productivity are multiple-row planting equipment, and improved tillage and cultivation machinery. The types now grown are suitable for combine harvesting. However, production of sorghum in the U.S.A. had fallen back to 14.3 million tonnes by 1990/91.

Crop movements

Of the total world production of 59 million tonnes of sorghum in 1989/90, 9.0 Mt, or 15%,

entered world trade. The principal exporters were the U.S.A. with 7.3 Mt (81.1% of all exports) and Argentina, with 1.2 Mt (13.3%). Imports were taken mainly by Japan (3.9 Mt; 44% of total world imports) and Mexico (3.0 Mt; 33.3%).

Utilization

Sorghum is the staple food in many parts of Africa, Asia, Central America and the Arab countries of the Middle East, and also serves as the main source of beverages in some countries. About 300 million people rely on sorghum for their sustenance.

Of the total world sorghum production of 66 million tonnes per annum in 1984–1986, about 35% was used for human food, 56.6% for animal feed, 1.3% for seed and 6.9% for processing and other uses. In Australia, Japan, the U.S.A., Mexico, Argentina and Europe, 98–99.5% of the sorghum consumed was used for animal feed, 90–95% in the former Soviet Union and Venezuela, in 1984–1986. By contrast, 80–94.5% of the sorghum consumed in Ethiopia, the Sudan and India, and 78% in Nigeria, was used for human food in 1984–1986.

Waxy sorghum

Waxy sorghum varieties (cf. p. 100) served as sources of starch in 1942 and subsequently to replace imported tapioca when supplies of the latter from the Netherlands Indies were cut off during World War II. These varieties have starch with physical properties similar to those of cassava, from which tapioca is prepared.

Data for the total domestic utilization of sorghum in certain countries, and its use for human food, animal feed, seed, and processing and other uses, average for 1984–1986, are shown in Table 1.14.

In the U.S.A., 92 thousand tonnes were used in 1984/86 for purposes other than human food, animal feed or seed. These include:

1. Wet milling, to make starch and its derivatives, with edible oil and gluten feed as by-products (cf. p. 267).
2. Dry milling, to make a low protein flour which

TABLE 1.14
Domestic Utilization of Sorghum in Certain Countries (Average, 1984–86)

Country	Total domestic usage (thousand t)	Percentage of total consumption			
		Human food	Processing and other uses	Animal feed	Seed
World	65,957	35.1	6.9	56.6	1.3
Argentina	2659	—	<1	99.5	<1
Australia	405	—	—	98.3	1.7
China	6141	66.2	11.5	21.2	1.1
Ethiopia	1097	94.5	5.0	—	<1
Europe	1099	—	<1	98.6	1.2
India	10,156	88.4	6.9	1.2	3.5
Japan	4383	1.0	—	99.0	—
Mexico	7789	—	2.0	97.7	<1
Nigeria	4918	78.1	17.6	2.5	1.8
Soviet Union	1326	—	10.0	89.7	<1
Sudan	1849	80.7	12.5	4.4	2.3
U.S.A	14,938	—	<1	99.1	<1
Venezuela	1254	—	5.0	94.3	<1

Source of data: Food and Agriculture Organisation, Food Balance Sheets 1984–86, Rome, 1990.

is used for adhesives and in oil-well drilling muds (cf. p. 313).

3. The fermentation industry, for brewing, distilling and the manufacture of industrial alcohol (Martin and MacMasters, 1951; cf. p. 226).

Human consumption

The level of human consumption of sorghum and millet in those countries in which consumption of these two cereals together exceeds 30 kg/head/year (82 g/head/day) is shown in Table 1.15. Consumption is between 20 and 30 kg/head/year in numerous other African countries and in the Yemen (Arab Republic and Democratic Republic).

The Millets

The name 'millet' is applied to numerous small-seeded grasses which originated in Asia or Africa and are widely grown in these continents.

Area, production, yield

The world area occupied by millet in 1989 was 36.7 million ha, of which 53% was in Asia,

TABLE 1.15
Human Consumption of Sorghum and Millet in Certain Countries, 1984/86 Average (kg/head/year)

Country	Sorghum	Millet
Africa		
Botswana	36.5	0.9
Cameroon	0.5	36.0
Chad	—	64.6
Ethiopia	23.8	3.5
Gambia	—	61.2
Mali	—	121.5
Mauretania	6.7	28.1
Niger	45.7	162.5
Nigeria	40.3	28.8
Senegal	9.8	83.5
Somalia	31.5	—
Burkino Faso	73.1	56.1
Burundi	31.0	2.0
Uganda	14.2	23.0
Sudan	69.2	11.9
Togo	36.7	16.5
Asia		
China	3.9	4.7
India	11.8	10.2
Yemen, Arab Republic	54.9	—
Yemen, Democratic Republic	—	36.1

Source: Food and Agriculture Organisation, Food Balance Sheets 1984–86, Rome, 1990.

principally India (43.5%) and China (7.2%), and 35% in African countries, principally Nigeria (9.3%), Niger (8.7%), Sudan (3.5%), Burkino Faso (3.0%), Senegal (2.7%) and Mali (2.7%).

The former Soviet Union accounted for 7.6% of the total area.

Production in 1989 was 29.5 million tonnes for the whole world, with Asia contributing 54.6% and Africa 31.6%. The biggest producers (with percentages of the world total) were: India (32.2%), China (19.3%), Nigeria (11.9%). The former Soviet Union produced 10.2% of the world total, the millets, particularly proso and foxtail, being much more important crops than sorghum in that country.

World average yield of millet in 1989 was 8 q/ha with much higher yields in some countries. The yield averaged 37.5 q/ha in Egypt, 21.5 q/ha in China, 20.8 q/ha in Saudi Arabia, and 17.6 q/ha in Japan. Yields of 48.3 q/ha were recorded in France in 1985 and 36.4 q/ha in Spain in 1987. In many African countries, however, the average yield is low, e.g. 2.3 q/ha in the Sudan in 1989, because the crop is often grown in marginal areas under adverse conditions of unfertile soil, heat, and limited rainfall. In such conditions millets nevertheless often do better than other cereals. Higher yields would be obtained in good conditions.

Types of millet

There are at least twelve distinct botanical species (representing ten different genera) described as millets (the specific names are listed in Ch. 2, Table 2.3, p. 47), but of these only six species comprise the bulk of the world production of millets. Of the total world production of about 30 million tonnes of millets per annum, the contribution of these six in 1981–1985 (in percent of the total) was pearl millet 45%, foxtail 19%, proso 17%, finger millet 13%, teff 4%, and fonio 1%, leaving about 1% for all other types, including barnyard, little and kodo millets.

Pearl millet is a hardy plant capable of yielding a crop where most other grain cereals would fail. It is grown extensively as a food crop in subtropical regions of Asia and Africa, and is well suited to conditions of limited moisture and low fertility. In west Africa, pearl millet is grown in the north, where rainfall is less than 76 cm per an., while sorghum replaces millet in the wetter south.

Proso is grown chiefly in the former Soviet Union, Manchuria and China. Throughout the former Soviet Union proso is a staple food, eaten as a thick porridge called kasha.

Foxtail millet is an important food in China and other Asiatic countries. It is grown for fodder in the U.S.A.

Finger millet (ragi) is an important food grain in southern Asia and parts of Africa where it is able to withstand high temperatures. In northern India finger millet replaces rice as the principal food crop.

TABLE 1.16
Estimated World Distribution of Types of Millett, 1981–85, Average*
(% of World Total Production†)

Type of millet	Developed countries		Africa			Asia	South America	World
	U.S.S.R.	Others‡	W	N and Cent	E and S			
Pearl	—	—	21.3	2.4	1.0	20.2	—	44.9
Finger	—	—	—	—	2.9	9.8	—	12.7
Proso	7.8	0.5	—	—	—	8.1	0.6	17.0
Foxtail	—	0.1	—	—	—	19.4	—	19.5
Teff	—	—	—	—	3.6	—	—	3.6
Fonio	—	—	1.0	—	—	—	—	1.0
Others§	—	—	—	—	—	1.3	—	1.3
All	7.8	0.6	22.3	2.4	7.5	58.8	0.6	100.0

* Source: Official and FAO estimates, based on country information.
† World production was 29.7 million tonnes, 1981–85, annual average.
‡ Australia, U.S.A., Europe.
§ Barnyard, Little and Kodo millets.

Teff is confined largely to the highlands of Ethiopia, where it grows up to 2700 m. It is a very small seeded grass.

Fonio is grown only in west Africa.

The distribution of these six types of millet among the various regions of the world in 1981–1985 (the latest period for which data are available) is shown in Table 1.16.

Crop movements

Most of the millet crop is used in the country of production: in 1988, out of a world production of 29.6 million tonnes, only 0.21 million tonnes (0.7%) is recorded as entering world trade. The principal exporters were Argentina (31.9% of total exports), Australia (21.9%) and the U.S.A. (21.4%), with smaller contributions from the Netherlands, China and Hungary. Imports were received mainly by Japan (15.7% of the total imports), Germany FR (14.3%), the Netherlands (13.8%), Niger (9.0%) and the U.K. (8.1%).

Utilization

About 400 million people rely on millet for their sustenance. It has been estimated that out of a total world production of 29 million tonnes of millet in 1984–1986, some 74.7% was used for human food, 10.8% for animal feed, 2.7% for seed, and 11.8% for processing and other uses. The percentage of domestic utilization used for human food, in 1984–1986, was 75% in Africa, 83% in Asia.

Human consumption

The level of human consumption of millet in certain countries in 1984–1986 is shown in Table 1.15 (p. 25). The highest consumption level recorded was 162.5 kg/head/year in Niger.

References

AMOS, A. J. (1973) Ergot — recent work reduces risks. *Milling* **155**: 26.

ANONYMOUS (1991) World grain and feed trade review. *World Grain 1991*, Nov/Dec, pp. 7–14

BRITISH PATENT SPECIFICATION Number 523,116 (1940) Heat (or steam) treatment of buggy wheat.

COMMONWEALTH SECRETARIAT, *Grain Bull.*, a monthly publication, London (ceased publication after December, 1976).

COMMONWEALTH SECRETARIAT, *Grain Crops*, an annual publication, London (ceased publication after No. 15, 1973).

FOOD AND AGRICULTURE ORGANIZATION (1990) *Production Year Book*. F.A.O. Rome.

FOOD AND AGRICULTURE ORGANIZATION (1988) *Trade Year Book*. F.A.O. Rome.

FOOD AND AGRICULTURE ORGANIZATION (1990) *Food Balance Sheets, 1984–1986*. F.A.O. Rome.

HARRISON, K. R., DOARKS, P. F. and GREER, E. N. (1969) Detection of heat damage in dried wheat. *Milling* **151** (7): 37.

HOME-GROWN CEREALS AUTHORITY (1991a) *Cereal Statistics* H.M.S.O., London.

HOME-GROWN CEREALS AUTHORITY (1991b) *Weekly Digest* **18** (14th Oct).

KENT, N. L. (1969) Thrips in home-grown wheat. *Milling* **152** (1): 22.

LEATH, M. N. and HILL, L. D. (1987) Economics, production, marketing and utilization of corn. In: *Corn: Chemistry and Technology*, WATSON S. A. and RAMSTAD, P. E. (Eds) Amer. Ass. Cereal Chemists Inc. St Paul, MN. U.S.A.

MARTIN, D. J. and STEWART, B. G. (1991) Contrasting dough surface properties of selected wheats. *Cereal Foods World* **36** (6): 502–504.

MARTIN, J. H. and MacMASTERS, M. M. (1951) Industrial uses for grain sorghum. *U.S. Department of Agriculture, Yearbook on Agriculture (1951)* p. 349.

MATZ, S. A. (1969) *Cereal Science* Avi Publ. Co. Inc., Westport, Conn., U.S.A.

MINISTRY OF AGRICULTURE, FISHERIES AND FOOD (1990) *Domestic Food Consumption and Expenditure 1989*, Annual Report of National Food Survey Committee, H.M.S.O., London.

PERCIVAL, J. (1921) *The Wheat Plant*. Duckworth, London (Reprinted 1975).

ROMER, T. (1984) Mycotoxins in corn and corn milling products. *Cereal Foods World* **29**: 459–462.

STARZYCKI, S. (1976) Diseases, pests and physioloy of rye. In: *Rye, Production, Chemistry and Technology*, BUSHUK, W. (Ed.) Ch. 3, 27–61. American Association of Cereal Chemists. Inc., St. Paul MN. U.S.A.

WEBB, B. D. (1985) Criteria of rice quality in the United States. In: *Rice: Chemistry and Technology*, pp. 403–442, JULIANO, B. O. (Ed.) Amer. Ass. Cereal Chemists Inc., St. Paul, MN. U.S.A.

Further Reading

ADRIAN, J. and JACQOT, R. (1964) *Sorghum and the Millets in Human and Animal Feeding*. Centre recherches sur la Nutrition du C.N.R.S, Bellevue (Seine et Oise), Vigot Frères (Eds) Paris.

ANDRES, C. (1980) Corn — a most versatile grain. *Food Processing*, May 1978.

BARGER, G. (1931) *Ergot and Ergotism*. Gurney and Jackson, London.

BOVE, F. J. (1970) *The Story of Ergot*. S. Karger, Basel and New York.

BROWN, L. R. (1972) The Green Revolution and world protein supplies. *PAG Bull.* **2** (2): 25.

BUSHUK, W. (Ed) (1976) *Rye, Production, Chemistry and Technology.* Amer. Assoc. of Cereal Chemsts. St. Paul, MN. U.S.A.

FULLER, J. G. (1969) *The Day of St Anthony's Fire.* Hutchinson, London.

GRIST, D. H. (1959) *Rice.* 3rd edn Longmans, London.

HOME-GROWN CEREALS AUTHORITY (1989) Oats market developments. *H-GCA Weekly Digest,* 18th Sept.

HOUSE, L. R. and RACHIE, K. O. (1969) Millets, their production and utilization. *Cereal Sci. Today* **14**: 92.

HULSE, J. H., LAING, E. M. and PEARSON, O. E. (1980) *Sorghum and the Millets.* Academic Press, London.

JULANO, B. O. (Ed) (1985) *Rice: Chemistry and Technology.* Amer. Assoc. of Cereal Chemists. St. Paul MN. U.S.A.

LORENZ, K. and HOSENEY, R. C. (1979) Ergot on cereal grain. *CRC Crit. Rev. Fd Sci. Nutr.* **11** (4): 311.

MORTON, I. D. (Ed.) (1987) *Cereals in a European Context.* Ellis Horwood, London.

ROONEY, L. W. and CLARK, L. E. (1968) Biochemistry and processing of sorghum grain. *Cereal Sci. Today* **13**: 258.

SIMMONS, I. G. (1989) *Changing the Face of the Earth — Culture, Environment, History.* Basil Blackwell, NY.

2
Botanical Aspects of Cereals

Grasses

Cereals are cultivated grasses that grow throughout the temperate and tropical regions of the world. As members of the Gramineae (or grass family) they share the following characteristics, but these are developed to different degrees in the various members:

Vegetative features of grasses

1. Conspicuous nodes in the stem.
2. A single leaf at each node.
3. Leaves in two opposite ranks.
4. Leaves consist of sheath and blade.
5. Tendency to form branches at nodes and adventitious roots at the bases of nodes.
6. Lower branches may take root and develop into stems as tillers.

Variation in vegetative features among species may be illustrated by reference to maize and wheat. In wheat branches occur only at the base of the main stem or culm, to produce tillers (Fig. 2.1) (Percival, 1921). While all tillers have the capacity to bear ears, the later formed ones may not actually do so; this habit is characteristic of most cereals.

In maize branches occur higher on the main stems and they are much shorter as the internodes do not extend (Fig. 2.2).

Leaf bases are very close together and the leaves consist almost entirely of blades which surround the inflorescence, and the shortness of its stalk leads to branches that are almost entirely inflorescence. At the tip of the main culm there is also an inflorescence as in wheat, but maize is unique

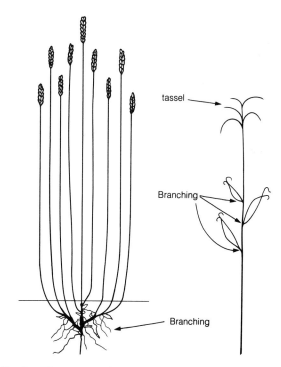

FIG 2.1 The pattern of branching in the wheat plant.

FIG 2.2 The pattern of branching in the maize plant

among cereals in that, on the branches, only the female organs develop in the florets and on the main culm only male organs develop. The adventitious roots that develop at the base of the main stem provide support for the aerial parts of the plant (Ennos, 1991); in maize they are called prop-roots and they are particularly well developed as is appropriate to the large and heavy nature of the aerial parts.

29

Reproductive features of grasses

1. All stems and branches normally form terminal inflorescences.
2. Flowers are produced in spikelets.
3. Each flower is enclosed between two bracts, the lemma and palea (pales or flowering glumes).
4. At the base of each spikelet are two glumes (empty or sterile glumes).

All cereal inflorescences are branched structures but the type of branching varies. The loose spreading structure found in oats is known as a panicle (Fig. 2.3).

The main axis of the panicle, the peduncle, bears several extended branches on which the spikelets are attached through short stalks or pedicels. Within the spikelet florets alternate (Fig. 2.4); the two closest to the base are similar in size but florets become progressively smaller towards the tip.

Each floret (Fig. 2.5) contains the female organs, a carpel containing a single ovule, with its stigma; and the male parts, three stamens, each consisting of filament and anther. Pollen released from the anthers, which split when ripe, is transferred by wind to the receptive stigma on another plant. The elaborate feathery style has an extensive sticky surface well suited to intercepting wind-borne pollen. Before the anthers mature, they remain enclosed between the pales but at the time of flowering or 'anthesis' the pales are forced open by the expansion of organs called lodicules at their base (lodicules swell as a result of an influx of water). The filaments of the stamens rapidly extend, projecting the opening anthers outside the pales, allowing the pollen to be shed onto the wind.

Rice inflorescences are also panicles but spikelets contain only one floret. Glumes are mostly insignificant small scales. Rice florets are unlike those of other cereals in having six stamens. (Fig. 2.6)

In sorghum the situation is complex: inflorescences are panicles but they may be compact or open (Hulse *et al.*, 1980). Spikelets occur in pairs, one is sessile and the other borne on a short pedicel. The sessile spikelet contains two florets,

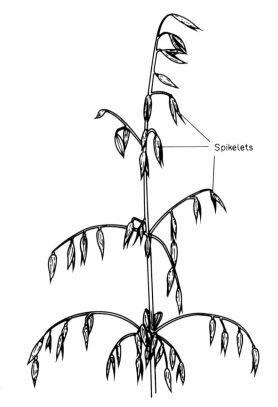

FIG 2.3 The oat panicle. Reproduced from Poehlman (1987) by courtesy of Avi Publishers, New York.

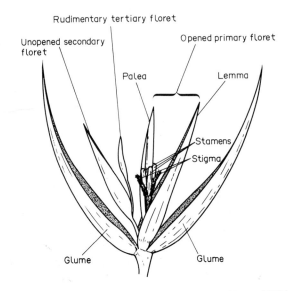

FIG 2.4 Spikelet of oat. Reproduced from Poehlman (1987) by courtesy of Avi Publishers, New York.

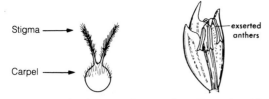

FIG 2.5 Reproductive organs in an oat floret. Reproduced from Poehlman (1987) by courtesy of Avi Publishers, New York.

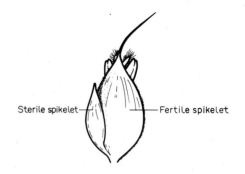

FIG 2.7 A pair of spikelets of sorghum. Reproduced from Poehlman (1987) by courtesy of Avi Publishers, New York.

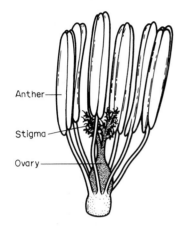

FIG 2.6 Reproductive organs of a rice floret. Note the six stamens present. Reproduced from Poehlman (1987) courtesy of Avi Publishers, New York.

one perfect and fertile and the other sterile. The pedicelled spikelet is either sterile or develops male organs only (Fig. 2.7).

In barley the type of inflorescence is a spike (Fig. 2.8).

It is more compact than a panicle, the spikelets being attached to the main axis or rachis by much shorter rachillas. The rachis is flattened and adopts a zigzag form, spikelets occur in groups of three alternately on the rachis. In six-rowed types all spikelets develop to maturity and each bears a grain in its single floret. In two-rowed types only the central spikelet in each three develops in this way; the others are sterile. The glumes are very small but the pales fully surround the grain and remain closely adherent to it even after threshing. The lemma tapers to a long awn which does break off during threshing. Many variants of the barley spike are illustrated by Briggs (1978).

The inflorescences of wheat, rye and triticale

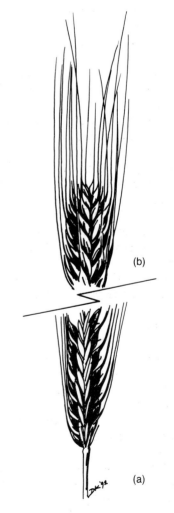

FIG 2.8 Spikes of barley, showing: *a.* the two-rowed and *b.* the six-rowed forms.

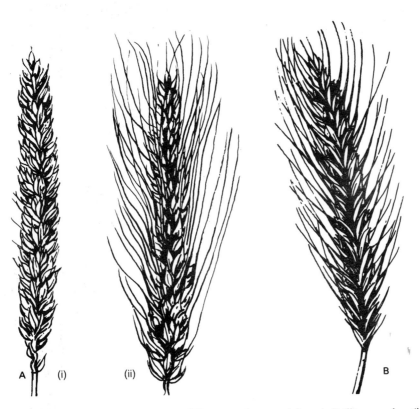

FIG 2.9 Spikes of *A*. wheat and *B*. rye. Wheat may be awned (bearded) (ii) or awnless (i).

are also spikes with spikelets alternating on a rachis, each spikelet however contains up to six florets (Fig. 2.9).

It is unusual for all six florets in a spikelet to be fertile and those at the extremes of the inflorescence may bear only one or even no fertile florets. Variants of wheat spikes are illustrated by Peterson (1965). As in oats the grains in the two basal florets are the largest. Those in the centre of the spike are larger than those at the extremes (Bremner and Rawson, 1978). The variation in size occurs as a function of the ability of each grain to compete for nutrients but also of the period of development, the earliest to flower being those in the basal florets of the central spikelets (Fig. 2.10).

Millets are an extremely diverse group, there

are many different species belonging to several different tribes (see Fig. 2.25) and no generalizations about their inflorescences are possible (details of most are given in Hulse *et al.*, 1980.) Pearl millet has a spike which may be anything between a few centimeters to a metre long. It is densely packed with groups of 2–5 spikelets, surrounded by 30–40 bristles (Fig. 2.11). Florets may be bisexual or male only.

In maize the male spikes occurring at the top of the culm bear spikelets in pairs, one being sessile, and the other pedicellate. Both types contain two florets, each with three anthers. The entire male inflorescence is known as the tassel (Fig. 2.2). On the female inflorescences the spikelets again carry two florets but only one is fertile, the upper functions while the lower one

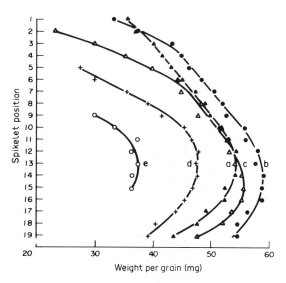

FIG 2.10 Typical profiles of wheat grain weights within a spike. The grain in the basal floret is designated *a*. and represented by ▲. Other grains are lettered progressively towards the tip of the spikelet, and represented by ●, △, +, and ○. (Bremner and Rawson, 1978)

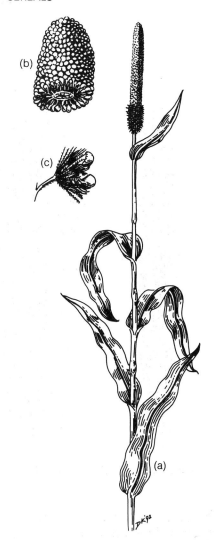

FIG 2.11 Pearl millet: *a*. spike, *b*. part of spike enlarged, *c*. spikelet.

aborts (Fig. 2.12). Each fertile floret contains a single ovary; its style is not of the feathery type typical of most cereals but a long threadlike structure covered in fine hairs which entrap wind-borne pollen. A single ear may contain 800 fertile florets so the same number of stigmas or 'silks' is present.

The ear or cob is wrapped by modified leaf sheaths forming husks or shucks and the silks emerge together from the distal open end of the protective husk (Fig. 2.13).

Perhaps the protection afforded by the husk obviates the need for enclosure of the reproductive structures by bracts and pales. These are insignificant in maize and as a result grains are not separated one from another on the cob. In some cases their mutual pressure imposes an angular form on them. While most cereals are dependent to some degree on cultivation for their survival, maize has the ultimate dependence on man since there is no mechanism for dispersal of its seeds remaining. The concentration of the sexual organs on separate spikes encourages cross-pollination, which is the norm for maize. Its

outbreeding habit fits it admirably for F1 hybrid production, whereby yields have been increased dramatically through the heterosis or hybrid vigour which results.

Life cycle of cereals

Although other parts of cereal plants have value, particularly in providing feed and bedding for livestock, the ripe fruit or grains are economically by far the most valuable parts of the plant.

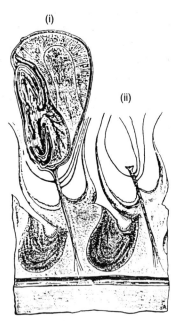

FIG 2.12 Radial section of a maize cob, showing (i) a perfect (fertile) and (ii) a rudimentary (empty) floret. Based on A. L. and K. B. Winton (1932).

Nevertheless the quality of the grain is dependent upon the condition of the plant; diseases that affect the leaves, roots and stem can reduce the photosynthetic area, the ability of the plant to take up water and nutrients from the soil, and the ability of the plant to stand. Agricultural scientists have, through experimentation, determined the optimal times for field treatments such as fertilizer, herbicide and pesticide applications. For communicating this information it has been found convenient to define stages of plant growth and several scales have been devized. Those that exist for wheat, barley and oats have been compared by Landes and Porter (1989). The decimal scale of Zadoks *et al.* (1974) is illustrated here (Fig. 2.14).

Scales usually start at the time of seed germination but a life cycle is, by definition, a continuously repeating sequence of events and, as such, has no absolute beginning or end.

The beginning of each new generation occurs when pollination is effected. As in all plants this results when pollen produced in the anther contacts the stigma on the carpel of another, or even

FIG 2.13 Cob of maize, showing the protective husk.

the same floret. Once on the stigma, pollen grains have a mechanism whereby a pollen tube is produced. The tube progresses towards the micropyle and, having effected access by this route, it allows nuclei from the pollen grain to pass into the ovule and fuse with nuclei present there. The primary fusion is of the sperm nucleus with the egg nucleus. The product is a cell, the successive divisions of which produce the embryo. A further set of fusions, however, produces the first endosperm nucleus. Three, not two, nuclei are involved, one from the pollen and two polar nuclei from the ovule. All endosperm cells are ultimately derived from this first endosperm cell and each inherits chromosomes from three nuclei rather than the more usual two. Endosperm cells thus have one and a half times as many chromosomes as cells elsewhere in the plant. The details of the development of endosperm, embryo and other grain tissues of different cereals are described in relevant texts (Kiesselbach, 1980; Bushuk,1976; Watson, 1987; Hulse *et al.*, 1980; Percival, 1921; Evers and Bechtel, 1988; Palmer, 1989; Hoshikawa, 1967).

An important phase in the life cycle, from the point of view of grain quality, is germination. This occurs when ripe grain is subjected to damping to an adequate moisture content at an appropriate temperature. In primitive grains the appropriate conditions are those to be expected in the natural habitat of the species at the beginning of the growing season, but breeding and cultivation over many years have diminished this relationship.

The processing requirements in respect of

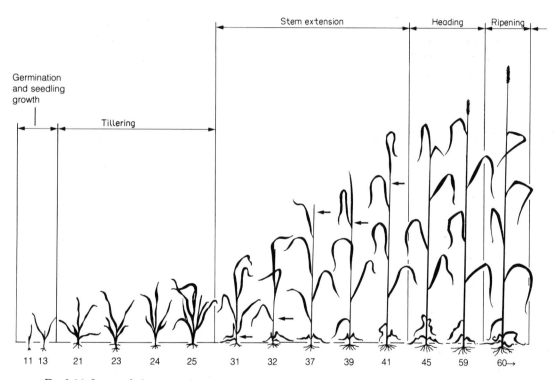

FIG 2.14 Stages of plant growth corresponding to the phases defined in Zadoks' scale. Major phases are represented by higher order numbers thus: 0. germination, 10. seedling growth, 20. tillering, 30. stem elongation, 40. booting, 50. inflorescence emergence, 60. anthesis, 70. milk development, 80. dough development, 90. ripening.

Within major phases additional lower order numbers indicate events of lesser importance. The descriptions corresponding to the numbers on the horizontal axis are: 11, first leaf unfolded; 13, 3 leaves unfolded; 21, main shoot and 1 tiller; 23, main shoot and 3 tillers; 24, main shoot and 4 tillers; 25, main shoot and 5 tillers; 31, first node detectable; 32, second node detectable; 37, flag (last) leaf just visible; 39, flag leaf ligule/collar just visible; 41, flag leaf sheath extending; 45, boots* swollen; 59, emergence of inflorescence completed; 60–95 anthesis to full ripeness.

* the term 'boot' refers to the swollen sheath of the last leaf, when the inflorescence within causes it to expand. The inflorescence is said to be 'in boot'.

germination vary with uses. In grains destined for malting the requirement is for ready and vigorous germination as soon after harvest as possible but this must be combined with resistance to sprouting, or premature germination prior to harvest. Resistance to sprouting is, in fact, desirable in all cereals, irrespective of their intended use, because the growth of the embryonic axis is accompanied by the production of hydrolytic enzymes which render stored nutrients in the endosperm soluble, thus reducing the amount of starch and protein harvested. Additionally the presence of high germination enzyme levels in cereals intended for flour production gives rise to excessive hydrolysis during processing. Bread-making flours are particularly sensitive to such high enzyme levels as processing conditions are well suited to enzyme-catalyzed hydrolysis.

Germination is a complex syndrome, the details of which are not fully understood. The important events are shown in the flow diagram (Fig. 2.15) below.

Grain anatomy

The basic structural form of cereal caryopses is surprisingly consistent, to the extent that a 'generalized' cereal grain can be described (Fig. 2.16).

Although frequently referred to as seeds, cereal grains are in botanical terms *fruits*. The fruits of grasses are classified as caryopses (singular:

caryopsis), which is a type of achene. All achenes are dry (rather than fleshy like many common fruits). All fruits, whether dry or fleshy, typically contain one or more seeds. In the case of caryopses the number of fruits contained is one, and the seed accounts for the greater part of the entire fruit when mature. It comprises:

1. Embryonic axis;
2. Scutellum;
3. Endosperm;
4. Nucellus;
5. Testa or seedcoat.

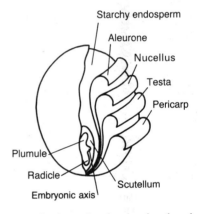

FIG 2.16 Generalized cereal grain, showing the relationships among the tissues. The proportions that they contribute, in individual cereals, are shown in Table 2.1.

Embryo

The embryonic axis and the scutellum together constitute the embryo. The embryonic axis is the plant of the next generation. It consists of primordial roots and shoot with leaf initials. It is connected to and couched in the shield-like scutellum, which lies between it and the endosperm. There is some confusion about the terminology of the embryo as the term 'germ' is also used by cereal chemists to describe part or all of the embryo. If the botanical description is adopted as above and 'germ' reserved for the embryo-rich fraction produced during milling, then there can be no confusion.

The scutellum behaves as a secretory and

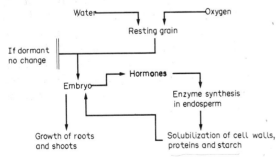

FIG 2.15 The main events involved in germination of a seed.

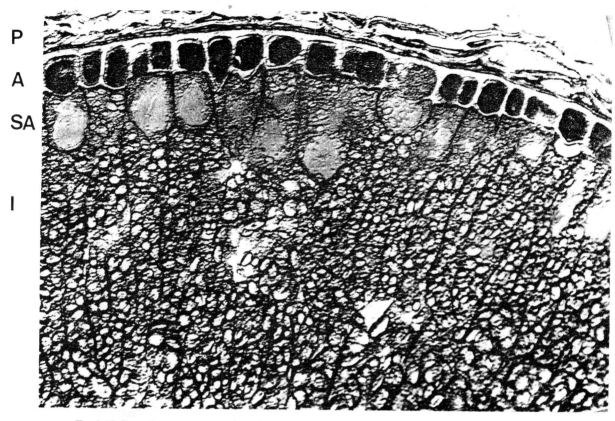

FIG 2.17 Part of a transverse section of a grain of Hard Red Winter wheat, 14.4% protein content, showing concentration of protein in subaleurone endosperm. Protein concentration diminshes towards the central parts of the grain in all cereals. P, pericarp; A, aleurone layer; SA, subaleurone endosperm; I, inner endosperm. (Reproduced from N. L. Kent, *Cereal Chem.* 1966, **43**: 585, by courtesy of the Editor.)

absorptive organ, serving the requirements of the embryonic axis when germination occurs. It consists mainly of parenchymatous cells, each containing nucleus, dense cytoplasm and oil bodies or spherosomes. The layer of cells adjacent to the starchy endosperm consists of an epithelium of elongated columnar cells arranged as a pallisade. Cells are joined only near their bases.

Exchange of water and solutes between scutellum and starchy endosperm is extremely rapid. Secretion of hormones and enzymes and absorption of solubilized nutrients occurs across this boundary during germination. The embryonic axis is well supplied with conducting tissues of a simple type and some conducting tissues are also present in the scutellum (Swift and O'Brien, 1970).

Although the fusions of nuclei, occurring during sexual fertilization, and leading to the formation of endosperm and embryo respectively, take place approximately at the same time, the development of the embryo tissue, by cell division, is relatively delayed. When the embryo does enlarge, it compresses the adjacent starchy endosperm tissue giving rise to a few layers of crushed, empty cells the contents of which have either been resorbed or have failed to develop. The crushed cells are described variously as the cementing, depleted or fibrous layer.

Endosperm

The endosperm is the largest tissue of the grain. It comprises two components that are clearly distinguished. The majority, a central mass described as starchy endosperm, consists of cells packed with nutrients that can be mobilized to support growth of the embryonic axis at the onset of germination. Nutrients are stored in insoluble form, the major component being the carbohydrate starch. Next in order of abundance is protein. In all cereals there is an inverse gradient involving these two components, the protein percentage per unit mass of endosperm tissue increasing towards the periphery (Fig. 2.17).

Cell size also diminishes towards the outside and this is accompanied by increasing cell wall thickness. The walls of the starchy endosperm of wheat are composed mainly of arabinoxylans, while in barley and oats $(1\rightarrow3)$ and $(1\rightarrow4)$ β-D glucans predominate. Cellulose contributes little to cereal endosperm walls except in the case of rice (see p. 64).

Surrounding the starchy endosperm is the other endosperm tissue, the aleurone, consisting of one to three layers of thick-walled cells with dense contents and prominent nuclei.

The number of layers present is characteristic of the cereal species, wheat, rye, oats, maize and sorghum having one and barley and rice having three. Unlike the tissue they surround, aleurone cells contain no starch but they have a high protein content and they are rich in lipid. They are extremely important in both grain development, during which they divide to produce starchy endosperm cells, and germination, when in most species they are a site of synthesis of hydrolytic enzymes responsible for solubilizing the reserves.

The balance between aleurone and scutellum in the latter role varies among species. Both tissues synthesize the enzymes in response to hormones including giberellic acid, transmitted from the embryonic axis (via the scutellum in the case of the aleurone). A further function of aleurone cells in some millets and sorghum is the transfer of metabolites into the starchy endosperm during grain maturation. This activity is deduced from the knobbly, irregular thickenings on their walls often associated with such transfers (Rost and Lersten, 1970).

At maturity the starchy endosperm dies but aleurone cells continue to respire, albeit at a very slow rate, for long periods. The aleurone tissue covers the outer surface of the embryo but its cells in this area may become separated and degenerate. Their ability to respire and to produce enzymes on germination is in some doubt (Briggs, 1978).

Seed coats

Surrounding the endosperm and embryo lie the remains of the nucellus, the body within the ovule in which the cavity known as the embryo sac develops. Following fertilization the embryo and endosperm expand at the expense of the nucellus, which is broken down except for a few remnants of tissue and a single layer of squashed empty cells from the nucellar epidermis. Epidermal cells in many higher plants secrete a cuticle and a cuticle is present on the outer surface of the nucellar epidermis of many cereals.

The outermost tissue of the seed is the testa or seed coat (the nucellar epidermis is also regarded as a seed coat but its origin is different from that of the testa which develops from the integuments). The testa may consist of one or two cellular layers. In some varieties of sorghum a testa is absent altogether. Where two layers are present the long axes of their elongated cells lie at approximately 90° to each other. Frequently the testa accumulates corky substances in its cells during grain ripening and this may confer colour on the grain and certainly reduces the permeability of the testa. A cuticle, thicker than that of the nucellar epidermis, is typical, and this also plays a role in regulating water and gaseous exchange.

Both testa and nucellus are tissues which once formed part of the ovule of the mother plant. They are thus of an earlier generation than the endosperm and embryo which they surround and to which they closely adhere.

Pericarp

The pericarp (or fruitcoat) is a multilayered structure consisting of several complete and incomplete layers. In all cereal grains the pericarp is dry at maturity consisting of largely empty cells. During development it serves to protect and support the growing endosperm and embryo. At this time chloroplasts are present in the innermost layers and starch accumulates as small granules in the central layers. By maturity all starch has disappeared and the cells in which it was present are largely squashed or broken down. An exception to this is provided by some types of sorghum in which at least some of the cells and some of the starch granules persist.

The innermost layer of the pericarp is the inner epidermis or epicarp. In many cereals this is an incomplete layer. Its cells are elongate and thus termed 'tube cells'. They are sometimes squashed flat in mature grains. Outside this layer lies the 'cross cell layer'. Unique to grasses, this layer takes its name from the fact that the long axes of its elongate cells lie at right angles to the grain's long axis. They are arranged side by side, in rows, in wheat, rye and barley. It is the cross cells and tube cells that contain chloroplasts in the immature green grain. The 'mesocarp', outside the cross cells, is not found as a true layer in mature grains except in sorghums. It is in this pericarp tissue that starch accumulates during grain maturation, but in most cases becomes reabsorbed before maturity.

The outermost layer of the pericarp and indeed of the caryopsis is the outer epidermis or 'epicarp'; it is one cell thick. It is adherent to the 'hypodermis' which may be virtually absent, as in oats or some millets, one or two layers thick, as in wheat, rye, sorghums and pearl millet or several layers thick, as in maize. The outer epidermis has a cuticle which controls water relations in growing grains but generally becomes leaky on drying (Radley, 1976). Hairs or trichomes are present at the nonembryo end of wheat, rye, barley, triticale and oats. They are collectively known as the 'brush' and they have a high silicon content. Trichomes have a spiral sculptured surface that

TABLE 2.1
Proportions of Parts of Cereal Grains (%)

Cereal	Hull	Pericarp and testa	Aleurone	Starchy endosperm	Embryo — Embryonic axis	Scutellum
Wheat						
Thatcher	—	8.2	6.7	81.5	1.6	2.0
Vilmorin 27	—	8.0	7.0	82.5	1.0	1.5
Argentinian	—	9.5	6.4	81.4	1.3	1.4
Egyptian		7.4	6.7	84.1	1.3	1.5
Barley						
Whole grain	13	2.9	4.8	76.2	1.7	1.3
Caryopsis	—	3.3	5.5	87.6	1.9	1.5
Oats						
Whole grain	25	9.0		63.0	1.2	1.6
Caryopsis (groat)	—	12.0		84.0	1.6	2.1
Rye	—	10.0		86.5	1.8	1.7
Rice						
Whole grain	20	4.8		73.0	2.2	
Caryopsis						
Indian	—	7.0		90.7	0.9	1.4
Egyptian	—	5.0		91.7	3.3	
Sorghum	—	7.9		82.3	9.8	
Maize						
Flint	—	6.5	2.2	79.6	1.1	10.6
Sweet	—	5.1	3.3	76.4	2.0	13.2
Dent	—	6	82		12	
Proso millet	16	3	6.0	70	5	

Sources as in Kent (1983).

differs among cereals allowing their origin to be determined when they are found detached from the grain (Bennett and Parry, 1981).

The tissues described above are present in cereal grains even if they are of the naked or free threshing type. In some cereal grains, such as oats, barley, rice and some millets, the pales are closely adherent and are thus not removed by threshing. These are therefore part of the grain as traded and their additional contribution to grain mass has to be borne in mind when comparing the relative proportions of nutrients in different species (Table 2.1).

Grain characteristics of individual cereals

In spite of structural similarities, there are wide variations among cereal grains in size and shape. Comparisons in size and form are shown diagrammatically in Fig. 2.18, and mass differences are given in Table 2.2.

Wheat

Wheat exhibits few differences from the generalized structure described (Fig. 2.16). Its most striking morphological characteristic is a crease or elongated re-entrant region parallel to its long axis, and on the ventral side — that is the opposite

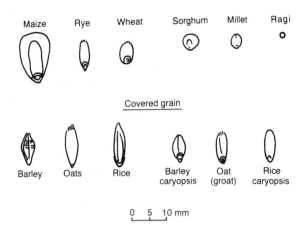

FIG 2.18 Grains of cereals showing comparative sizes and shapes. The caryopses of the three husked grains (barley, oats, rice) are shown with and without their surrounding pales.

side to the embryo. At the inner margin of the crease there lies between the testa and the endosperm, a column of nucellar tissue (Fig. 2.19).

Whereas elsewhere in the grain the nucellar epidermis is all that remains of that tissue, several layers of parenchymatous cells persist to maturity in the crease. They carry the remains of vascular tissue by which nutrients were transported into the developing grain and it is likely that their final passage into the endosperm was *via* the nucellar cells.

TABLE 2.2
Dimensions and Weight per 1000 Grains of the Cereals

Cereal	Dimensions		Weight per 1000 (grains)	
	Length (mm)	Width (mm)	Average (g)	Range (g)
Millets				
Teff				0.14–0.2
Proso			6	
Pearl	2	1.0–2.5	7	5–10
Rye	4.5–10	1.5–3.5	21	15–40
Sorghum	3–5	2–5	28	8–50
Rice	5–10	1.5–5	27	
Oats	6–13	1.0–4.5	32	
Triticale			36	28–45
Barley	8–14	1.0–4.5	35	32–36
Wheat	5–8	2.5–4.5	37	{ CWRS:27 English:48
Maize	8–17	5–15	324	150–600

Source: Kent (1983).

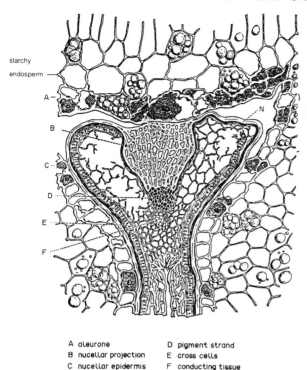

starchy
endosperm —

A —

B

C

D

E

F

N

A aleurone   D pigment strand
B nucellar projection   E cross cells
C nucellar epidermis   F conducting tissue

FIG 2.19 Transverse section of a wheat grain showing the tissues of the crease region. Modified from A. L. and K. B. Winton (1932).

Outside the nucellar projection (as the column is called) there is a discontinuity in the testa, in which lies a bundle of corky cells forming the pigment strand. Its name is derived from the dark colour which it exhibits in red wheats. It is not highly pigmented in white wheats. Between the nucellar projection and the endosperm lies a void known as the endosperm cavity.

Wheat aleurone tissue is but one cell thick, the cells being approximately cubic with 50 μm sides. Some remnants of the mesocarp remain as intermediate cells. The disintegration of most of the mesocarp leaves a space between cross cells and outer epidermis, which is thus only loosely attached except in the crease where the mesocarp persists. The tube cells are separated by many large gaps except where they cover the tips of the grain.

The cellular nature of the tissues is shown in Fig. 2.20 (although an old publication *The Stucture and Composition of Foods* by Winton and Winton (1932), from which this illustration is taken, pro-

vides the best collection of detailed descriptions of cereal tissues).

Among cereals wheat has a relatively high protein content in the starchy endosperm (say 8–16%). Two distinct populations of starch granules are present, two thirds of the starch mass being contributed by large lenticular granules between 8 and 30 μm and one third by near-spherical granules of less than 8 μm diameter (see Ch. 3).

Rye

Rye grains are more slender and pointed than wheat grains but they also have a crease and indeed share many of the features described for wheat. The beadlike appearance of the cell walls of the pericarp is less distinct than in wheat. Rye grains may exibit a blue-green cast due to pigment present in the aleurone cell contents. Two populations of starch granules are present as in wheat; the larger granules, seen under the microscope, often display an internal crack.

Barley

Barley grains mostly have a hull of adherent pales which is removable only with difficulty. It amounts to about 13% of the grain (by weight) on average, the proportion ranging from 7 to 25% according to type, variety, grain size and latitude where the barley is grown. Winter barleys have more hull than spring types: six-row (12.5%) more than two-row (10.4%). The proportion of hull increases as the latitude of cultivation approaches the equator, e.g. 7–8% in Sweden, 8–9% in France, 13–14% in Tunisia. Large and heavy grains have proportionately less hull than small, lightweight grains.

Grains are generally larger and more pointed than wheat though they are not less broad. They have a ventral crease which is shallower than those of wheat and rye and its presence is obscured by the adherent palea. Cross cells are in a double layer and their walls do not appear beaded. Only one cellular testa layer is present. Two to four (mostly three) aleurone layers are present, cells being smaller than those in wheat; about 30 μm in each direction. Blue colour may be present due

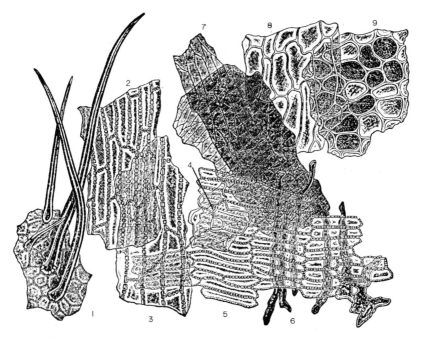

FIG 2.20 Bran tissues of wheat in surface view. 1–6 are pericarp components. 1. Fragment of epidermis with brush hairs; 2. epidermis on body of grain; 3. hypodermis; 4. intermediate cells; 5. cross cells, 6. tube cells; 7. outer and inner testa layers; 8. nucellar epidermis. 9. aleurone cells. Modified from A. L. and K. B. Winton (1932).

to anthocyanidin pigmentation. In the starchy endosperm two populations of starch granules are present in most types, though in some mutants, exploited for their chemically different starch, only one population may be present.

Oats

The oat grain is characterized by pales that are not removed during threshing. As they do not adhere to the groat (the name describing the actual caryopsis of the oat) within, they can be removed mechanically. Oats are traded with the husk in place. In this condition they have an extremely elongated appearance and even with the husk removed groats are long and narrow. Within a spikelet the two grains present differ in form, the lower being convex on the contact face while the upper grain has a concave face (Fig. 2.21).

The groat's contribution to the entire grain mass varies from 65 to 81% in cleaned British-grown oats (average 75%). Differences are due to

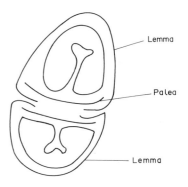

FIG 2.21 Diagrammatic transverse section through an oat spikelet showing the relationship between the grains within, accounting for their different shapes.

both variety and environment. The groat contribution tends to be higher in winter-sown than in spring-sown types, in Scottish-grown than in English-grown samples, and in small, third grains than in the large, first (main) grains of varieties with three grains per spikelet (Hutchinson *et al.*, 1952).

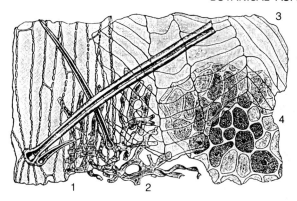

FIG 2.22 Bran coats of oat in surface view. 1. Outer epidermis; 2. hyperdermis; 3. testa; 4. aleurone. Modified from A. L. and K. B. Winton (1932).

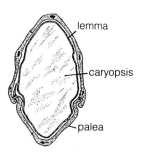

FIG 2.23 Transverse section through a rice grain showing the 'locking' mechanism of the lemma and palea. Modified from A. L. and K. B. Winton (1932).

Naked oats *Avena nuda* L. is a type of oat which readily loses its husk during threshing, thus obviating any need for a special dehulling stage in milling. Naked oats have high protein and oil contents, and a high energy value.

Only two pericarp layers can be distinguished in oats, an epidermis with many trichomes on the outer surface (unlike the Triticeae fruits, these are not restricted to the non-embryo end) and a hypodermis, consisting of an irregular, branching collection of worm-like cells with long axes lying in all directions (Fig. 2.22).

The testa comprises a single cellular layer with cuticle. In cross section the nucellar epidermis can be seen as a thin colourless membrane, its cellular structure cannot be discerned, and a cuticle separates it from the testa.

In the endosperm there is a single aleurone layer. The cell walls are not thick, as they are in wheat and rye. Conversely, the starchy endosperm cells have thicker walls than in wheat. Starch granules are polydelphous (compound) consisting of many tiny granuli which fit together to form a spherical structure. Possibly eighty or more granuli up to 10 μm diameter constitute a single compound granule. Individual free granuli are also present in the spaces among the aggregates. Endosperm cells of the oat have a relatively high oil content.

Rice

The pales of rice are removed from the grain only with difficulty, as they are locked together by a 'rib and groove' mechanism (Fig. 2.23).

Once the pales are removed, the outer epidermis of the pericarp is revealed as the outer layer of the caryopsis. Unlike the other small grained cereals described above, rice does not have a crease. It is laterally compressed and the surface is longitudinally indented where broader ribbed regions of the pales restricted expansion during development. Distinctively, in all except one of the tissue layers (the tube cells) surrounding the endosperm, the cells are elongated transversely (in other grains only the cross cells are elongated in this direction). Cells of the epidermis have wavy walls making them quite distinct from those of the hypodermis, with their flat profiled walls. Cross cells have many intercellular spaces and the cells are worm-like; similar to, but lying at right angles to, the tube cells. The testa has one cellular layer, with an external cuticle. Cells of the nucellar epidermis are similar to those of the testa but the walls have a beaded appearance in section. Aleurone cells are similar to those of oats, but the number of layers varies around the grain from one to three.

Starch granules in the starchy endosperm cells are similar to those of oats. Unusually, the embryo of rice is not firmly attached to the endosperm.

The proportion of husk in the rice grain averages about 20%. Varieties of rice are classified according to grain weight, length and shape — which is described as round, medium or long, and defined by their aspect ratio (length to breadth). In shape, grains of the *indica* type are short, broad and thick, with a round cross-section; grains of *japonica* rice are long, narrow and slightly flattened in shape.

Maize (Dent corn)

Although there are many types of maize and their morphology and anatomy vary (Watson 1987), it is possible only to describe one type here; dent corn is the most abundantly grown and this explains its selection. The maize grain is the largest of cereal grains. The basal part (embryo end) is narrow, the apex broad. The embryonic axis and scutellum are relatively large.

The pericarp is thicker and more robust than that of the smaller grains. It is known as the hull and the part of the hull overlying the embryo is known as the tip-cap. An epidermis is present as the outermost layer, no hairs are present. Beneath the epidermis lie up to 12 layers of hypodermis, which appear increasingly compressed towards the inside. Both tube cells and cross cells are present, cross cells occuring in at least two layers. Cell outlines are extremely irregular and there are many spaces among the anastomosing cells. No cellular testa layer is present but a cuticular skin persists to maturity. The same applies to the nucellus.

In spite of the great size of the endosperm of maize, individual aleurone cells are small, comparable to those of oats and rice. One layer of them is present. In blue varieties it is the aleurone cells that provide the colouration. In the starchy endosperm many small starch granules (average 10 μm) occur. Protein (zein) also occurs in tiny granular form. Horny endosperm occurs as a deep cap surrounding a central core of floury endosperm. The designation 'dent' results from the indentation in the distal end of the grain which contracts on drying. The dent is not found in other types of maize such as flint maize, popcorn, and sweetcorn. The most significant differences among maize types lie in the endosperm character and shape. The covering layers are similar but other types may have fewer hypodermis layers than dent corn.

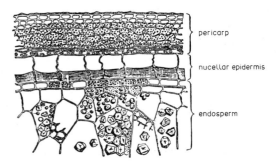

FIG 2.24 Section through a sorghum grain. Modified from A. L. and K. B. Winton (1932).

Sorghum

The following description applies essentially to the type of grain sorghum which is grown in the U.S.A. but variations found in other types are also noted. Grains are near-spherical with a relatively large embryo. A single outer epidermis of the pericarp surrounds the grain. Within it are two or three hypodermis layers. The mesocarp is one of two unusual features of the sorghum grain; it consists of parenchymatous cells that still contain starch at maturity. The cells are not crushed as in other cereals (Fig. 2.24).

The starch granules are up to 6 μm diameter; smaller than those in the endosperm. Cross cells do not form a complete layer, they are elongated vermiform cells aligned in parallel but separated by large spaces. Tube cells on the other hand are numerous and closer packed. The second unusual feature of sorghum anatomy is the absence of the testa. The nucellar layer is, however, well developed; in fact it is the most conspicuous of all the bran layers and may be up to 50 μm thick and coloured yellow or brown. There are different opinions among authors as to the identity of the seed coat layers. Some refer to a testa rather than a nucellar epidermis. There is general agreement that it may be clear or coloured and also that it may be incomplete.

Aleurone cells are similar in size and appearance to those of maize as are the inner endosperm cells and the starch granules that they contain. Peripheral endosperm cells however are less clearly distinguished from the aleurone cells to which they are adjacent.

In some sorghums pigmentation occurs; all tissues may be coloured, but not all together. The starchy endosperm may be colourless or yellow, the aleurone may or may not contain pigmentation. The pericarp and nucellar layers may contain variable amounts of tannins: polyphenolic compounds responsible for red colouration. Types with large proportions of tannins in the pericarp are known as 'bird proof' or 'bird repellant' because, it is assumed, of the unpalatability of the polyphenols. They are certainly attacked less by birds than are white types (cf. Ch. 4).

An African sorghum 'kaffir corn' does not have the usual conspicuous nucellus described above. The hypodermis is more robust due to thicker cell walls. Both white and red varieties are cultivated. Further information may be found in Hulse *et al.* (1980), Winton and Winton (1932) and Hoseney and Variano-Marston (1980).

The Millets

Not all types can be described here as there are many and not all are well documented (Hulse *et al.*, 1980). Some details are given of the most widely grown: pearl millet.

The grain is about one third the size of sorghum grain, the epidermis/hypodermis combination being thicker than that of sorghum; cross cells and tube cells are spongy. No cellular seed coat is present but a membranous cuticle is. A single layer of aleurone cells surrounds a starchy endosperm with horny and floury regions. The aleurone cells have conspicuous knobbly thickening. Similar cells have been noted in foxtail millet and assigned a transfer function.

Taxonomy

The names commonly applied to individual species are sometimes confusing as many different names may be used for the same plant. Worse still, the same name may be applied to more than one plant or its fruit. Theoretically the application of systematic nomenclature should remove all

such confusion and, indeed, in practice it undoubtedly clarifies communications. However, experts are not agreed on all details and several systems of scientific nomenclature exist. Attempts have been made to reach a concensus and the International Standards Organization has published a list of the agreed names. These are given in Table 2.3 which, in addition to systematic and common names, includes the number of chromosomes typical of each species.

Classification

The primary objective of plant classification is the grouping of plants and populations into recognizable units with reasonably well defined boundaries and stable names. Modern taxonomists strive to establish a phylogenetic arrangement of the taxa, based on known or presumed genetic relationships.

Living organisms are classified in a hierarchical system in which descending groupings indicate progressively closer relationships. The lowest taxonomic level to which all cereals belong is the *Family*. A Family may be divided into *sub-families*, each of which is further divided into *tribes*. Within a tribe there may be several *genera* and there may be several *species* within a genus. The species is the highest level at which routine natural breeding among members would be expected.

Within a species there may be several *cultivars*, which, if accepted by an appropriate authority, may be recognized as commercial varieties. At the species level binomial designation applies, the first part of the name being that of the *genus*, for example '*Hordeum*'. Addition of the *trivial* name '*vulgare*' completes the *specific* name: *Hordeum vulgare*. (In the case of barley the two-rowed and four-rowed types are distinguished at the '*convar*' level — this lies between species and cultivar in the taxonomic hierarchy.) It is customary to print specific names in italics or to underline them. Designation of taxonomic status is somewhat arbitrary. Those competent taxonomists responsible for establishment of species are credited by their name being suffixed to the species name, either in full or in shortened form. The most frequent

TABLE 2.3
Taxonomic Details of Cereals

Cereal type	Systematic name (Chromosomes)	Common name
Wheat	*Triticum aestivum* L. emend Fiori et Paol (2n = 42)	wheat, breadwheat
	Triticum aestivum ssp *compactum* Hosteanum (2n = 42)	club wheat
	Triticum dicoccum Schrank (2n = 28)	emmer wheat
	Triticum durum Desfontaines (2n = 28)	durum wheat
	Triticum monococcum L. (2n = 14)	small spelt, einkorn
	Triticum polonicum (2n = 28)	diamond wheat, Polish wheat
	Triticum spelta L. (2n = 42)	spelt wheat, dinkel
	Triticum sphaerococcum Percival (2n = 42)	shot wheat, Indian dwarf wheat
	Triticum turgidum L. (2n = 28)	English wheat, rivet wheat, poulard wheat
Triticale	*Triticosecale* ssp Whitmark (2n = 42 or 56)	triticale (hexaploid or octaploid)
Rye	*Secale cereale* L. (2n = 14)	rye
Barley	*Hordeum vulgare* L. sensuo lato (2n = 14)	barley
	Hordeum vulgare convar *distichon* (2n = 14)	two-rowed barley
	Hordeum vulgare convar *hexastichon* (2n = 14)	six-rowed barley
Oats	*Avena sativa* L. (2n = 42)	common oats, white oats
	Avena byzantina Karl Koch (2n = 42)	Algerian oats, red oats
	Avena nuda L. (2n = 42)	naked oats, hull-less oats
Rice	*Oryza sativa* L. (2n = 24)	rice
	Oryza rufipogon (Griffiths)* (2n = 24)	red rice
Maize	*Zea mays* L. (2n = 20)	maize, corn
	Zea mays convar *amylacea* (2n = 20)	soft maize, flour maize
	Zea mays convar *ceratina* (2n = 20)	waxy maize
	Zea mays convar *everta* (2n = 20)	popcorn
	Zea mays convar *indentata* (2n = 20)	dent maize
	Zea mays convar *indurata* (2n = 20)	flint maize
	Zea mays convar *saccharata* (2n = 20)	sweet corn, sugar maize
	Zea mays convar *tunicata* (2n = 20)	pod maize

* May be classed as a millet.

TABLE 2.3

Continued

Cereal type	Systematic name (Chromosomes)	Common name
Sorghum	*Sorghum bicolor* (L.) Moench ($2n = 20$)	sorghum, guinea-corn, great millet, Egyptian millet, kaffir corn, dari, milo, waxy milo, jowar, cholain, milo-maize, kaoliang, feteritas, shallu, broomcorn
Millets (alphabetical)	*Brachiaria deflexa* (Schumacher) Hubert	fonio
	Coix lacryma-jobi L. ($2n = 20$)	adlay, Job's tears
	Digitaria exdis (Kippist) Stapf. ($2n = 54$)	acha, fonio, hungry rice
	Digitaria iburua Stapf. ($2n = 54$)	black fonio
	Echinochloa crus-galli (L.) P. Beauvoir var *frumentacea* (Roxburgh) ($2n = 36$ or 54) W. F. Wight	Japanese millet, barnyard m., Billion dollar grass, white panicum, cockspur grass
	Eleusine coracana (L.) Gaertner ($2n = 36$)	finger m., ragi, Indian m., birdsfoot m., African m., marica
	Eragrostis tef (Zuccagni) Trotter ($2n = 40$)	teff, teffgrass
	Panicum miliaceum L. ($2n = 36$ or 72)	common m., Australian white m., Chinese red m., hog m., Moroccan yellow m., plate m., proso (cheena) m., Turkish yellow m., U.S.A. proso m. (cheena), U.S.A. red m., corn m., samai, broomcorn
	Panicum miliare Lamarck ($2n = 36$)	little m.
	Paspalum scorbiculatum L.	kodo m., ditch m., kodra, kodon, varagu, scrobic
	Pennisetum glaucum (L.) R. Brown ($2n = 14$)	pearl m., inyati m., babala seed, bulrush m., cattail, pale rajeen grass, bajra, African m., candle m., cumbu
	Setaria italica (L.) P. Bauvois	foxtail m., German m., Australian m., Chinese yellow m., seed of Anjou sprays, seed of Burgundy sprays, seed of Italian sprays, Hungarian m., Italian m.

name suffixed is Linnaeus or L., crediting the Swedish biologist who devized the system.

Several classifications of the Gramineae family exist. One is shown in Table 2.3. In Fig. 2.25 a suggested family-tree is shown. The diagram also indicates the photosynthetic pathway adopted by members of some of the groupings. The C3 is typical of temperate plants, and the C4 is appropriate to tropical plants.

Breeding

The process of fertilization described earlier in this chapter is typical of outbreeding species. Some of the most important species of cereal do not conform to this pattern however. Although other mechanisms may also be involved, one certain barrier to cross-pollination is the failure of anthers to emerge from the pales before shedding

Family	Subfamily	Tribe	Genus

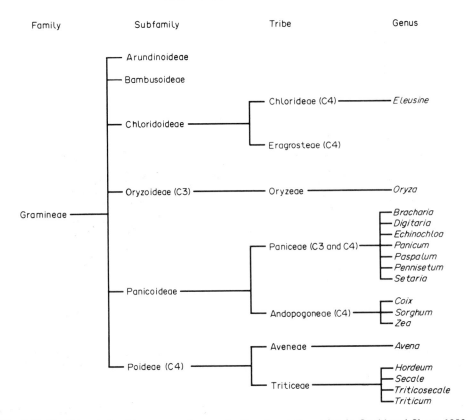

FIG 2.25 Possible relationships among the cereals. Based on information in Gould and Shaw, 1983.

their pollen. Pollen thus falls on the stigma within the same floret, leading to self-fertilization.

Barley is about 99.5% self-pollinated as may be inferred from the strong adhesion of its pales. Rice, oats and wheat are 97–98%, and sorghum about 94% self-pollinated. Finger millet and foxtail millet also have a high degree of self pollination. Rye, on the other hand, is almost entirely cross-pollinated and indeed may be self sterile; maize engages in about 95% crossing and pearl millet about 80% crossing.

Plant breeders seek to achieve improved varieties by selection and selective crossing. Techniques used are becoming more sophisticated: they have progressed from simple selection to complex crossing programmes and use of biotechnology and induced mutation. Strategies are necessarily different in self-pollinated types, in which true breeding lines can be easily established, and out-crossing types where this is difficult and

may even lead to progressive deterioration of stocks (Fig. 2.26).

Because new genetic material is not naturally introduced into self-pollinating plants, they are homozygous. That is, the chromosomes derived from pollen bear genes which not only control the same characters as those derived from the ovum but also influence them in the same way. Continued self-pollination in successive generations would not therefore produce variation. Variation can thus be introduced by crossing with a compatible plant with different genes or by mutation — a change to genes induced by chemical, radiation or other influence.

Following a cross, heterozygosity is reduced by one half with each successive self-fertilization. Complete homozygosity is theoretically unattainable but a practical state of uniformity is normally reached after five to eight generations of self-fertilizing.

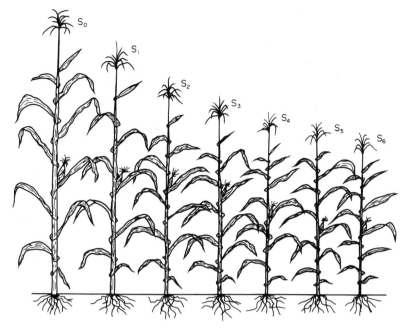

FIG 2.26 Reduction in vigour in maize with successive generations of inbreeding. S_0 represents the original inbred plant and S_{1-6} successive inbred generations. Reproduced from Poehlman (1987), by courtesy of Avi Publications, New York.

Hybridization

The crossing to produce heterozygosity and selection of a pure line from an advanced line following the cross is known as hybridization. It should not be confused with the breeding of *hybrids* in which the F1 generation is grown commercially. The *F1 generation* is the first filial generation or that produced immediately from the cross. This method is used most effectively with species such as maize and sorghum which naturally outcross. The simplest method of production is to grow the F1 of a cross beween inbred lines, but more sophisticated techniques are now frequently used. These include modified single or three-way crosses, double crosses, multiple crosses, backcrosses, synthetic varieties and composites. Explanations of these terms can be found in specialized texts such as that of Poehlman (1987). Whereas continued inbreeding of outbreeding species leads to loss of vigour, F1 hybrids produced from inbred lines exhibit heterosis, or hybrid vigour. This is the increase in size or vigour of a hybrid over the average of

its parents, this being referred to as the mid-parent value. In practice, hybrid vigour is the increase in size and vigour over the better parent as this is the only achievement of interest to growers or processors.

When growing hybrid maize, the farmer generally obtains fresh hybrid seed each year from growers who specialize in its production. The specialized grower chooses an isolated field on which he grows two inbred lines of maize in alternate strips, one row of male parent plants to about four rows of female parent plants. At the appropriate time, the female plants are detasselled, and are subsequently fertilized by pollen from the adjacent male parent plants. Seed is later collected from the female parent only. An alternative method of producing hybrid maize is by use of male sterility, which may be genetic or cytoplasmic.

Before 1970, over 90% of maize seed in the U.S.A. produced in Texas utilized male-sterile cytoplasm, but after 1970, when the fungus *Helminthosporium maydis* race T caused much damage, cytoplasmic male-sterility was not used for a few years. It subsequently made a return but the detasselling method is most frequently used.

As both male and female reproductive parts occur in the same spikelets of sorghum, production of hybrid sorghum seed awaited development of male-sterile types. Hybrid sorghum was first grown commercially in 1956. F1 hybrids are commercially successful in millets also.

Attempts have been made to produce F1 hybrids from inbreeding cereals and this has been achieved, but not yet commercially, in wheat, oats and barley. In China there are reports of successful commercial rice F1 hybrids (Poehlman 1987).

Intergeneric hybrids

Genes capable of coding for all plant characters are accommodated on a relatively small number of chromosomes. In wheat the minimum number is 7. Somatic (non-reproductive) cells are diploid, that is, they contain two sets of chromosomes; they thus have two genes for each character. One gene of each pair may be dominant and it is this that is expressed. Other degrees of co-operation are also possible. In some plants there are two or three or even four pairs of chromosomes per cell instead of the usual one pair. Those with only one pair are termed euploid, while those with more are polyploid. Polyploid plants may arise as a result of duplication of the inherent chromosomes to form *autopolyploids*, or by combining chromosome complements (genomes) from two or more species, leading to *allopolyploids*. Polyploids of both types can occur in nature or they can be induced by breeders. They are larger than either parent and generally exhibit greater vigour also. An example of a natural allopolyploid is bread wheat, *Triticum aestivum*. It is hexaploid, with six of each type of chromosome in each somatic cell. The three different genomes (each of 2×7 chromosomes) came from three different species of grass, none of which had all the characteristics of the resulting species. Other wheats have fewer genomes, for example *T. durum* is tetraploid and *T. monococcum* is diploid. Oats and some millets are further examples of natural allopolyploid (hexaploid) species grown commercially. For the ploidy of other cereals see Table 2.3.

An example of an allopolyploid produced by design is triticale (*Triticosecale*), a hybrid of wheat and rye which contains complete genomes of both parents. Triticales have been produced independently with several wheat parents (rye is always the pollen parent). Both hexaploid triticales, produced from a tetraploid wheat and diploid rye, and octoploid triticales, with a hexaploid wheat parent, have been produced. The philosophy behind triticale breeding has been the combination of the hardiness of rye with the grain quality of wheat (cf. Ch. 4.) The most successful triticales have been secondary triticales, those resulting from a cross between primary triticale (rye × wheat) and wheat. Other cereal intergeneric hybrids have been bred experimentally but none has reached commercial scale production. The future of triticale as a widely grown commercial cereal remains uncertain.

Breeding objectives

To the grower the most important characters of a cultivar may be related to the responses of the whole plant to environmental pressures, while to the processor: the miller, the maltster or the starch manufacturer it is the grain characteristics that are most important. Poehlman (1987) lists objectives for a number of cereals, acknowledging that the emphasis may vary from one part of the world to another.

The primary objectives for all species are yield potential, yield stability (including: optimum maturity date, resistance to lodging and shattering, tolerance to drought and soil stress, and resistance to disease and pests) and grain quality. The factors involved in yield stability are to some extent linked, for example resistance to lodging (the tendency of standing crops to bend at or near ground level, thus collapsing) might involve the height of the plant, the size and weight of the ear in relation to culm strength and resistance to stem rot pathogens and to insects such as the corn borer.

In some cereals additional objectives include the production of hull-less varieties of oats and barley and the adaptation to mechanized harvesting of sorghum and millets. In barley, low yield remains a problem and varieties with smooth awns are also desirable. Some winter-sown cereals such as wheat, oats and barley provide herbage for stock grazing on vegetative parts of immature plants in the autumn, and good fodder properties may thus be desirable. Fodder aspects also feature in breeding programmes for millets and sorghum.

Agronomic characters of importance to breeding programmes include the seasonal habit. Winter-sown crops of a given cereal produce greater yields than those sown in Spring. Spring-sown types need to grow rapidly in order to complete the necessary phases of growth in a relatively short period. They must not require a period of cold treatment known as stratification when it promotes germination or vernalization when it is necessary to promote production of reproductive structures in the first flowering season (some cereals are naturally perennials although they are treated as annuals in agriculture). A more ambitious objective may be the breeding of crops with special qualities previously not associated with the species concerned. An example of the successful achievement of this objective is the breeding of maize with high amylose or high amylopectin starch, or high lysine maize. Breeding objectives related to processing are discussed for individual cereals in Ch. 4.

References

BENNET, D. M. and PARRY, D. W. (1981) Electron-probe microanalysis studies of silicon in the epicarp hairs of the caryopses of *Hordeum sativum* Jess, *Avena sativa* L., *Secale cereale* L. and *Triticum aestivum* L. *Ann. Bot.* **48**: 645–654.

BREMNER, P. M. and RAWSON, H. M. (1978) The weights of individual grains of the wheat ear in relation to their growth potential, the supply of assimilate and interaction between grains. *Aust. J. Plant Physiol.* **5**: 61–72.

BRIGGS, D. E. (1978) *Barley*. Chapman and Hall, London.

BRIGGS, D. E. and MACDONALD, J. (1978) Patterns of modification in malting barley. *J. Inst. Brew.* **89**: 260–273.

BUSHUK, W. (1976) *Rye, Production, Chemistry and Technology*. Amer. Assoc. of Cereal Chemists Inc. St Paul MN. U.S.A.

ENNOS, A. R. (1991) The mechanics of anchorage in wheat *Triticum aestivum* L. 2. Anchorage of mature wheat against lodging. *J. exp. Bot.* **42**: 1607–1613.

EVERS, A. D. and BECHTEL, D. B. (1988) Microscopic structure of the wheat grain. In: *Wheat: Chemistry and Technology*, POMERANZ, Y. (Ed.) Amer. Assoc. of Cereal Chemists Inc, St. Paul MN. U.S.A.

FULCHER, R. G. (1986) Morphological and chemical organization of the oat kernel. In: *Oats: Chemistry and Technology*, WEBSTER, F. H. (Ed.) Amer. Assoc. of Cereal Chemists Inc. St Paul MN. U.S.A.

GOULD, F. W. and SHAW, R. B. (1983) *Grass Systematics*. Texas A and M University Press.

HOSENEY, R. C. and VARIANO-MARSTON, E. (1980) Pearl millet: its chemistry and utilization. In: *Cereals for Food and Beverages*, INGLETT, G. E. and MUNCK, L. (Eds.) Academic Press Inc. NY. U.S.A.

HOSHIKAWA, K. (1967) Studies on the development of rice 2. Process of endosperm tissue formation with special reference to the enlargement of cells, and 3. Observations on the cell division. *Proc. Crop Sci. Soc. Japan.* **36**: 203–216.

HUBBARD, C. E. (1954) *Grasses*. Penguin Books Ltd. Harmondsworth, Middx.

HULSE, J. H., LAING, E. M. and PEARSON, O. E. (1980) *Sorghum and the Millets: Their Composition and Nutritive Value*. Academic Press Inc. London.

HUTCHINSON, J. B., KENT, N. L. and MARTIN, H. F. (1952) The kernel content of oats. *J. natn. Inst. agric. Bot.* **6**: 149.

I.S.O.5526 (1986) Cereals, pulses and other food grains — Nomenclature. *International Standards Organization.* (Dual numbered as B.S.I. 6860 1987).

JULIANO, B. O. and BECHTEL, D. B. (1985) The rice grain and its gross composition. In: *Rice: Chemistry and Technology*, JULIANO, B. O. (Ed.) Amer. Assoc. of Cereal Chemists Inc. St Paul MN. U.S.A.

KENT, N. L. (1966) Subaleurone endosperm of high protein content. *Cereal Chem.*, **43**: 585.

KENT, N. L. (1966) *Technology of Cereals*, 1st edn, Pergamon Press Ltd, Oxford.

KENT, N. L. (1983) *Technology of Cereals*, 3rd edn, Pergamon Press Ltd, Oxford.

KIESSELBACH, T. A. (1980) *The Structure and Reproduction of Corn*. University of Nebraska Press. (Reprint of 1949 edn).

LANDES, A. and PORTER, J. R. (1989) Comparison of scales used for categorizing the development of wheat, barley, rye and oats. *Ann. appl. Bot.* **115**: 343–360.

NICHOLSON, B. E., HARRISON, S. G. MASEFIELD, G. B. and WALLACE, M. (1969) *The Oxford Book of Food Plants*. Oxford University Press.

PALMER, G. H. (1989) Cereals in malting and brewing. Ch. 3. In: *Cereal Science and Technology*, PALMER, G. H. (Ed.) pp. 61–242, Aberdeen University Press, Aberdeen.

PERCIVAL, J. (1921) *The Wheat Plant*. Duckworth and Co. London. Reprinted 1975.

PETERSON, R. F. (1965) *Wheat: Botany, Cultivation and Utilization*. Leonard Hill Books, London.

POEHLMAN, J. M. (1987) *Breeding Field Crops*. 3rd edn. AVI Publishing Co. Ltd Westport, Connecticut. U.S.A.

RADLEY, M. (1976) The development of wheat grains in relation to endogenous growth substances. *J. exp. Bot.* **27**: 1009–1021.

REXEN, F. and MUNCK, L. (1984) *Cereal Crops for Industrial use in Europe*. The Commission of the European Communities.

ROST, T. C. and LERSTEN, N. R. (1970) Transfer aleurone cells in *Setaria lutescans* (Gramineae) *Protoplasma*. **71**: 403–408.

SWIFT, J. G. and O'BRIEN, T. P. (1970) Vascularization of the scutellum of wheat. *Aust. J. Bot.* **18**: 45–53.

WANOUS, M. K. (1990) Origin, taxonomy and ploidy of the millets and minor cereals. *Plant Varieties and Seeds.* **3**: 99–112.

WATSON, S. A. (1987) Structure and composition. In: *Corn: Chemistry and Technology*, WATSON, S. A. and RAMSTAT, P. E. (Eds.) Amer. Assoc. of Cereal Chemists Inc, St. Paul, MN. U.S.A.

WINTON, A. L. and WINTON, K. B. (1932) *The Structure and Composition of Foods. Vol. 1. Cereals, Starch, Oil seeds, Nuts, Oils, Forage Plants*. Chapman and Hall. London.

ZADOKS, J. C., CHANG, T. T. and KONZAAK, C. F. (1974) A decimal code for the growth stages of cereals. *Weed Res.* **14**: 415–421.

ZEE, S.-Y. and O'BRIEN, T. P. (1970) Studies on the ontogeny of the pigment strand in the caryopsis of wheat. *Aust. J. Bot.* **23**: 107–110.

Further Reading

ANON. (1989) *Biological Nomenclature.* Institute of Biology, London.

BEWLEY, J. D. and BLACK, M. (1978) *Physiology and Biochemistry of Seeds in relation to Germination. Vol. 1. Development, Germination and Growth.* Springer-Verlag. Berlin.

BEWLEY, J. D. and BLACK, M. (1982) *Physiology and Biochemistry of Seeds in relation to Germination. Vol. 2. Viability, Dormancy and Environmental Control.* Springer-Verlag. Berlin.

BLAKENEY, A. B. (1992) Developing rice varieties with different textures and tastes. *Chemy Aust.* Sept. 475–476.

BUSHUK, W. and LARTER, E. N. (1986) Triticale: production, chemistry and technology. *Adv. Cereal Sci. Tech.* **8**: 115–158.

MAUSETH, J. D. (1988) *Plant Anatomy.* Benjamin/Cummings Publ. Co. Inc. Menlo Park, CA. U.S.A.

POMERANZ, Y. (Ed.) (1989) *Wheat is Unique.* Amer. Assoc. of Cereal Chemists Inc. St. Paul MN. U.S.A.

RINGLUND, K., MOSLETH, E. and MARES, D. J. (Eds.) *Fifth International Symposium on Pre-Harvest Sprouting in Cereals.* Westview Press, Boulder, Colorado. U.S.A.

3

Chemical Components

Carbohydrates

The terminology surrounding carbohydrates frequently serves to confuse rather than to clarify. Archaic and modern conventions are often inter-mixed and definitions of some terms are inconsistent with their use. Even the term carbohydrate itself is not entirely valid. It originated in the belief that naturally occurring compounds of this class could be represented formally as hydrates of carbon. i.e. $C_x(H_2O)_y$. This definition is too rigid however as the important deoxy sugars like rhamnose, the uronic acids and compounds such as ascorbic acid would be excluded and acetic acid and phloroglucinol would qualify for inclusion. Nevertheless the term carbohydrate remains to describe those polyhydroxy compounds which reduce Fehlings solution either before or after hydrolysis with mineral acids (Percival, 1962).

It is customary to classify carbohydrates according to their degree of polymerization. Thus: monosaccharides (1 unit), oligosaccharides (2–20 units) and polysaccharides (>20 units).

Monosaccharides are the simplest carbohydrates; most of them are sugars. Monosaccharides may have 3–8 carbon atoms but only those with 5 carbons (pentoses) and 6 carbons (hexoses) are common. Both pentoses and hexoses exist in a number of isomeric forms, they may be polyhydroxyaldehydes or polyhydroxyketones. Structurally, they occur in ring form which may be six-membered (pyranose form) or five-membered (furanose form).

In mature cereal grains the monomers are of little importance in their own right but, as components of polymers, they are of extreme importance, both in their contribution to the structural and storage components of the grain, and to the behaviour of grains and their products during processing. In this context the most important monosaccharide, because of its abundance, is the six-carbon polyhydroxyaldehyde:

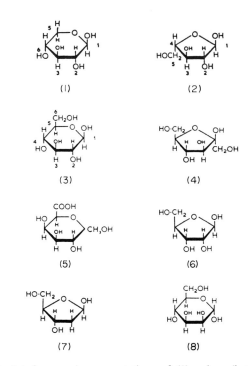

(1)

(2)

(3)

(4)

(5)

(6)

(7)

(8)

FIG 3.1 Structural representations of (1) xylose (beta-D-xylopyranose), (2) arabinose (alpha-L-arabinofuranose), (3) glucose (beta-D-glucopyranose), (4) fructose (beta-D-fructo-furanose, (5) D-galacturonic acid, (6) ribose (beta-D-ribofuranose), (7) deoxyribose (beta-D-deoxyribofuranose), and (8) mannose (alpha-D-mannopyranose).

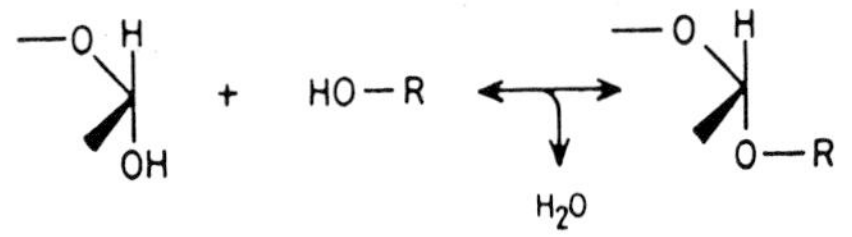

FIG 3.2 Formation of the glycosidic link.

D-glucose. It is the monomeric unit of starch, cellulose and beta-D-glucans.

The most important pentoses are the poly-hydroxyaldehydes D-xylose and L-arabinose, because of their contribution to cell wall polymers. The structures of these compounds and of some other monosaccharides found in cereals are shown in Fig. 3.1.

The most abundant derivatives of monosaccharides are those in which the reducing group forms a glycosidic link with the hydroxyl group of another organic compound (as in Fig. 3.2), frequently another molecule of the same species or another monosaccharide. Sugar molecules may be joined to form short or long chains by a series of glycosidic links, thus producing oligosaccharides or polysaccharides.

Oligosaccharides

The smallest oligosaccharide, the disaccharide, comprises two sugar molecules joined by a glycosidic link. Although this may appear to be a simple association it is capable of considerable variation according to the configuration of the glycosidic link and the position of the hydroxyl group involved in the bonding. Three important variants among disaccharides involving only α-D-glucopyranose are shown in Fig. 3.3.

In these compounds the reducing group of only one of the monosaccharide molecules is involved in the glycosidic link and the reducing group of the other remains functional.

In sucrose, another important disaccharide found in plants, fructose and glucose residues are joined through the reducing groups of both; hence their reducing properties are lost. Sucrose is readily hydrolyzed under mildly acid conditions, or enzymically, to yield its component monomers which of course again behave as reducing sugars. Sucrose is the main carbon compound involved in translocating photosynthate to the grain. It features prominently during development rather than in the mature grain because it

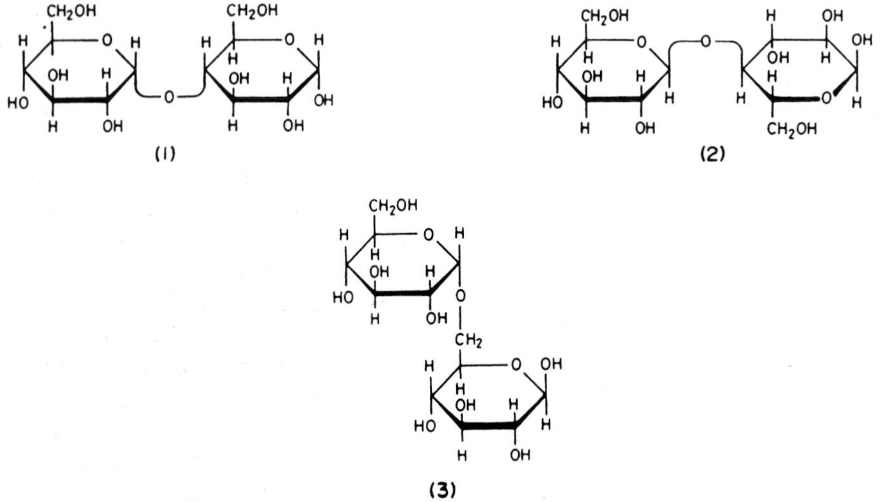

FIG 3.3 Structural conformation of (1) maltose (α-D-glucopyronosyl-(1→4)-α-D-glucopyranose), (2) cellobiose (β-D-glucopyranosyl-(1→4)-α-D-glucopyranose), (3) isomaltose (α-D-glucopyranosyl-(1→6)-β-D-glucopyranose).

is converted during maturation, to structural and longer-term storage carbohydrates such as starch. In sweet corn the sucrose content is higher by a factor of 2–4 throughout grain development than in other types of maize at a similar stage, as the rate of conversion is slower (Boyer and Shannon, 1983).

Literature values for sugars in cereals vary with methods of analysis and with varieties examined and in consequence tables which bring together results of different authors can be misleading. Henry recently analyzed two varieties of each of six cereal species. All results were obtained by the same methods and are thus comparable. Values for free glucose and total (including that in sucrose and trisaccharides) are given in Table 3.1.

TABLE 3.1
Total Soluble Glucose and Fructose in Two Varieties of Each of Six Cereals*

	Barley	Oat	Rice	Rye	Triticale	Wheat
Glucose	0.17	0.12	0.14	0.21	0.25	0.11
	0.09	0.13	0.19	0.29	0.21	0.11
Fructose	2.31	1.01	0.84	5.79	3.22	1.73
	1.98	1.00	0.75	5.11	3.05	2.46

* Data from Henry, 1985.

Free sugars are not distributed uniformly throughout the grain. The distribution in the maize grain is shown in Table 3.2.

The embryo has the highest concentration of free sugars in other cereals also. This is reflected in the distribution among mill fractions, as illustrated with respect to rice in Table 3.3.

TABLE 3.2
Proportions of Free Sugars in the Anatomical Fractions of the Maize Grain*

Grain part	% of dry matter
Endosperm	0.5–0.8
Embryo	10.0–12.5
Pericarp	0.2–0.4
Tip cap	1.6
Whole grain	1.61–2.22

* Data from Watson, 1987.

TABLE 3.3
Proportions of Soluble Sugars in Mill Fractions of Rice*

Mill fraction	% of dry matter
Rough	0.5–1–2
Brown	0.7–1.3
Milled	0.22–0.45
Hull	0.6
Bran	5.5–6.9
Embryo	8–12

* Data from Juliano and Bechtel, 1985.

Polysaccharides

Oligomers and polymers in which glucose residues are linked by glycosidic bonds are known as glucans. The starch polymers, amylose and amylopectin, are glucans in which the α-$(1{\rightarrow}4)$-link, as in maltose (Fig. 3.2), features. Additionally, in amylopectin the α-$(1{\rightarrow}6)$-link, as in isomaltose (Fig. 3.3) occurs, giving rise to branch points. The same linkages are present in the other main storage carbohydrate found in sweet corn. The product is known as phytoglycogen, it is highly branched with α-$(1{\rightarrow}4)$ unit chain lengths of 10–14 glucose residues and outer chains of 6–30 units (Marshall and Whelan, 1974). Unlike the true starch polymers phytoglycogen is largely soluble in water and as a result the soluble saccharides of sweet corn contribute about 12% of the total grain dry weight. The starch polymers are discussed at greater length in a later section of this chapter.

In cellulose the β-$(1{\rightarrow}4)$ form of linkage, as present in cellobiose (Fig. 3.3) occurs. β-Links are also involved in the other important cell wall components, $(1{\rightarrow}3,1{\rightarrow}4)$-$\beta$-D-glucan. These polymers contribute about a quarter of the cell walls of wheat aleurone but they are particularly important in oat and barley grains, in the starchy endosperm of which they may contribute as much as 70% (Fincher and Stone, 1986). With water they form viscous gums and contribute significantly to dietary fibre. Both the ratio of $(1{\rightarrow}3)$ to $(1{\rightarrow}4)$ links and the number of similar bonds in an uninterrupted sequence differ between the species. Extraction and analysis of the mixed linkage compounds are particularly difficult in the presence of such large excesses of α-glucan (Wood, 1986).

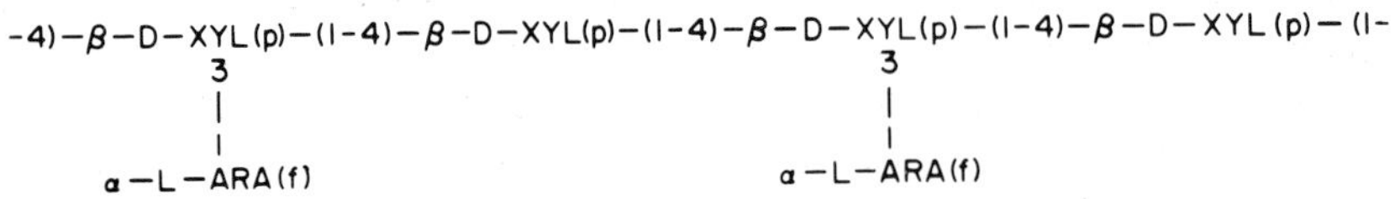

-4)-β-D-XYL(p)-(1-4)-β-D-XYL(p)-(1-4)-β-D-XYL(p)-(1-4)-β-D-XYL(p)-(1-
                3   3
                | |
                | |
           α-L-ARA(f)   α-L-ARA(f)

FIG 3.4 Structure of arabinoxylan of wheat aleurone and starchy endosperm cell walls. p, represents the pyranose or 6–membered ring form; f, represents the furanose or 5–membered ring form.

Pentosans

While glucans are polymers of a single sugar species the common pentosans (polymers of pentose sugars) comprise two or more different species, each in a different isomeric form. Thus arabinoxylans, found in endosperm walls of wheat and other cereals, have a xylanopyranosyl backbone to which are attached single arabinofuranosyl residues (Fig. 3.4).

Starch

Starch is the most abundant carbohydrate in all cereal grains, constituting about 64% of the dry matter of the entire wheat grain (about 70% of the endosperm), about 73% of the dry matter of the dent maize grain and 62% of the proso millet grain. It occurs as discrete granules of up to 30 μm diameter and characteristic of the species in shape.

Starch granules are solid, optically clear bodies which appear white when seen as a bulk powder because of light scattering at the starch–air interface. They have a refractive index of about 1.5. Specific gravity depends upon moisture content but it is about 1.5. The mysteries of granule structure, development and behaviour have exercized the minds of scientists for hundreds of years and continue to do so. Granules from different species differ in their properties and there is even variation in form among granules from the same storage organ. Shape is determined in part by the way that new starch is added to existing granules, in part by physicochemical conditions existing during the period of growth and in part by composition.

Composition

The main way in which composition varies is in the relative proportions of the two macromolecular species of which granules consist (Fig. 3.5).

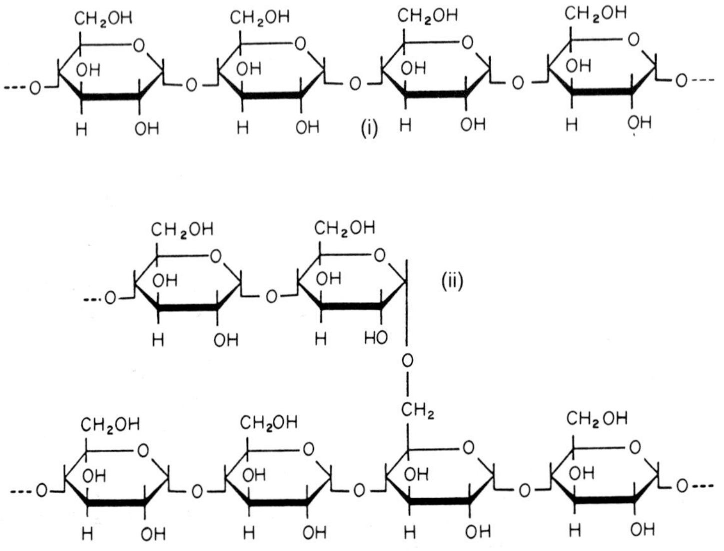

FIG 3.5 Structural representation of amylose (i) and amylopectin (ii).

Amylose comprises linear chains of (1→4) linked α-D-glucopyranosyl residues. Amylopectin has, in addition, (1→6) tri-O-substituted residues acting as branch points. Amylose has between 1000 and 4400 residues, giving it a molecular weight between 1.6×10^5 and 7.1×10^5. In solution amylose molecules adopt a helical form and may associate with organic acids, alcohols or, more importantly, lipids to form complexes in which a saturated fatty acid chain occupies the core of the helix. Binding of polyiodide ions in the core in the same way is responsible for the characteristic blue coloration of starch by iodine.

The average length of amylopectin branches is 17–26 residues. As their reducing groups are involved in bonding, only one is exposed. The molecule is generally considered to consist of 3 types of chain (Fig. 3.6):

A chains — side chains linked only via their reducing ends to the rest of the molecule.

B chains — those to which A chains are attached.

C chains — chains which carry the only reducing group of the molecule.

The amylose contents of most cereal starches lie between 20 and 35%, but mutants have been used in breeding programmes to produce cultivars with abnormally high or low amylose contents. It is in diploid species such as maize and barley that such breeding has been most successful as polyploid species are more conservative, with single mutations having less chance of expression (cf. Ch. 2). High amylopectin types are generally described as waxy as the appearance of the endosperms of the first mutants discovered had suggested a waxy composition. Waxy maize cultivars have up to 98% amylopectin (100% according to some references). High amylose maize starches consist of up to 80% amylose.

Granular form

Although some variation exists within species, there are many characteristic features by which

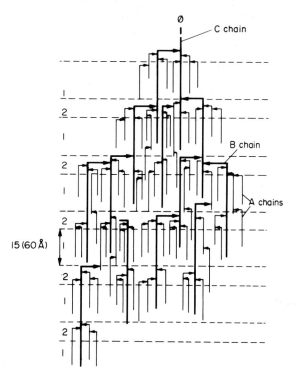

FIG 3.6 Structure of (potato) amylopectin proposed by Robin *et al.* (1974). Bands marked 1 are considered to be crystalline while alternating 2 bands are amorphous. *Reproduced by courtesy of American Association of Cereal Chemists.*

TABLE 3.4
*Characteristics of Starch Granules of Cereals**

Cereal	Shape and diameter (μm)	Features
Wheat	Large, lenticular: 15–30	Characteristic equatorial groove
	Small, spherical: 1–10	Angular where closely packed
Triticale	Large, lenticular, 1–30	As wheat
	Small, spherical: 1–10	
Rye	Large lenticular: 10–40	As wheat, often displaying radial cracks. Visible hilum
Barley	Small, spherical: 2–10	As wheat
	Large, lenticular: 10–30	
	Small, spherical: 1–5	
Oats	Compound, ovoid: up to 60	Comprising up to 80 granuli
	Simple, angular: 2–10	
Rice	Compound granules	Comprising up to 150 angular granuli: 2–10 μ
Maize	Spherical:	In floury endosperm
	Angular:	In flinty endosperm
	Both types 2–30; average 10	
Sorghum	Spherical/angular: 16–20; average 15	As maize
Millet, pearl	Spherical/angular: 4–12; average 7	As maize

* Based on Kent, 1983.

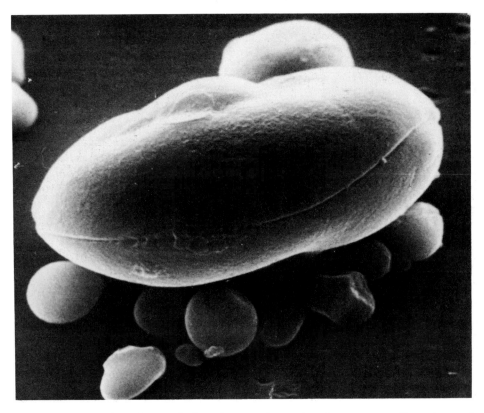

FIG 3.7 Scanning electon micrograph of one large starch granule and numerous small starch granules of wheat. The large granule shows the equatorial groove. From A.D. Evers, *Stärke*, 1971, **23**: 157. Copyright by Leica U.K., Reproduced with permission of the Editor of *Die Stärke*.

an experienced microscopist can identify the source, either from observation of an aqueous suspension at room temperature or with the additional help of observed changes when the suspension is heated, leading to gelatinization at a temperature characteristic of the species and type (Snyder, 1984). The characteristic blue staining reaction with iodine/potassium iodide solution does not occur with waxy granules, which contain virtually no amylose, they stain brownish red to yellow. It is characteristic for amylose percentage to increase during endosperm development, consequently staining reactions change during growth.

Granules of cereals from the Triticeae tribe (see Ch. 2) are of two distinct types. The larger ones are biconvex while the smaller ones are nearly spherical (Fig. 3.7). Granules from most other cereals are similar in shape to the smaller population of Triticeae granules, but rice and oats have some compound granules in which many granuli fit together to produce large ovoid wholes. Shapes of high-amylose granules are varied and may be related to their individual composition. The later developers tend to be filamentous, some resembling strings of beads. Characteristics of starch granules from cereals are shown in Table 3.4.

Within the endosperm of a species small differences in granule shape may arise as a result of packing conditions. These can be seen in grains as mealy and vitreous (or horny) regions. In mealy regions, packing is loose and granules adopt what appears to be their natural form. In horny regions close packing causes granules to become multi-faceted as a result of mutual pressure. Small indentations can also arise from other

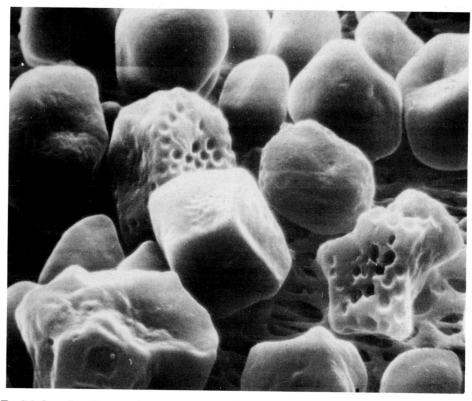

FIG 3.8 Scanning electron micrograph of maize starch granules of spherical and angular types. Some angular granules show indentations due to pressure from protein bodies.

endosperm constituents such as protein bodies. (Fig. 3.8).

Pitting on the surface can be caused by enzymic hydrolysis and it is possible to find such granules in some cereal grains in which germination has begun or in which insect damage has occurred. There is no evidence that these two physical modifications to granule form change the chemical properties of the granules.

As granules are transparent some manifestations of internal structure can be detected, even if their significance cannot be fully appreciated. One such internal feature is the hilum exhibited by granules of some species. It is a small airspace, considered to represent the point of initiation around which growth occurred (Hall and Sayre, 1969). This assumes that granules grow by deposition of new starch material on the outer surface of existing granules, and indeed this has been demonstrated by detection of radioactively labelled precursors incorporated into growing granules (Badenhuizen, 1969). Such a system of growth allows for the change in shape that occurs in starches of the Triticeae, by preferential deposition on some parts of the surface. As a result they change from tiny spheres to larger lentil shaped granules (Evers, 1971).

Some structures not evident in untreated granules can be revealed or exaggerated by treatment with weak acid or amylolytic enzymes. In cereal starches a lamellate structure results from removal of more susceptible layers and persistence of more resistant layers. Layers may be spaced progressively more closely towards the outside. The number of rings appears to coincide with the number of days for which a granule grows (Buttrose, 1962). Lamellae cannot be revealed in granules from plants grown under conditions of continuous illumination (Evers, 1979).

Size distributions

The literature contains many tables of granule size ranges and size distributions of granules from different botanical sources. While such tables are useful guides they do not all accord in detail and some fail to indicate the nature of the distribution. For example the bimodal distribution of the Triticeae is not always indicated although this is an important characteristic by which the source of a starch may be recognized. In wheats the proportional relationship between large biconvex and small spherical granules is fairly constant (approx 70% large granules w/w), and this is the same for rye and triticale.

In barley there is a wider variation, in part due to the existence of more mutant types (Goering *et al.*, 1973). Among 29 cultivars, small granules accounted for between 6% and 30% of the total starch mass.

Granule organization

Under crossed polarizers starch granules exhibit birefringence in the form of a maltese cross. This indicates a high degree of order within the structure. The positive sign of the birefringence suggests that molecules are organized in a radial direction (French, 1984). Amylomaize starch exhibits only weak birefringence of an unusual type (Gallant and Bouchet, 1986).

Starch granules exhibit X-ray patterns, indicating a degree of crystallinity. Cereal starches give an A pattern, tuber, stem and amylomaize starches give a B pattern and bean and root starches a C pattern. The C pattern is considered to be a mixture of A and B. It is accepted that the crystallinity is due to the amylopectin as it is shown by waxy granules. Furthermore, amylose can be leached from normal granules without affecting the X-ray pattern. The A and B patterns are thought to indicate crystals formed by double helices in amylopectin. The double helices occur in the outer chains of amylopectin molecules, where they form regions or clusters. The crystalline parts of starch granules are responsible for many of the physical characteristics of the granules' structure and behaviour. Nevertheless they involve less than half the total starch present. Some 70% is amorphous; this comprises all the amylose but must also include much of the amylopectin. The evidence of biochemical studies and electron microscopy has pointed to the existence of structures with a periodicity of 5–10 nm, reflecting the alternating crystalline and amorphous zones of amylopectin.

Granule surface and minor components

The distribution of amylose and amylopectin molecules in one starch granule was estimated by French (1984): for one spherical granule 15 μm in diameter, with a mass of 2.65×10^{-9} g there would be about 2.5×10^9 molecules of amylose (D.P = 1000, 25% of total starch) and 7.4×10^7 molecules of amylopectin (D.P. = 100,000, 75% of starch). If the molecular chains are perpendicular to the surface of the granule there would be about 14×10^8 molecular chains terminating at the surface. Of these, 3.5×10^8 would be amylose molecules and the rest would be amylopectin chains.

Surface characteristics of granules are also affected by the minor components of starches. Bowler *et al.* (1985) reviewed developments in work on these although they point out that this is an under-researched area. Non-starch materials present in commercial starch granules can arise from two sources. They may be inherent components of the granules in their natural condition or they may arise as deposits of material solubilized or dispersed during the process by which the starch is separated.

The main non-starch components of starch granules are protein and lipid. Amounts vary with starch type: in maize 0.35% of protein ($N \times 6.25$) is present on average. Slightly more is present in wheat starch (0.4%). The most significant proteins in terms of their recognized effects on starch behaviour are amylolytic enzymes bound to the surface. Even traces of *alpha*-amylase can have drastic effects on pasting properties through hydrolyzing starch polymers at temperatures up to the enzymes' inactivation temperatures.

SDS PAGE (sodium dodecyl sulphate, polyacrylamide gel elecrophoresis) showed surface proteins of wheat starch to have molecular masses of under 50 K while integral proteins were over 50 K. Altogether ten polypeptides have been separated between 5 K and 149 K. The major 59 K polypeptide is probably the enzyme responsible for amylose synthesis. It has been shown to be concentrated in concentric shells within granules. Two other polypeptides of 77 K and 86 K are likely to be involved in amylopectin synthesis. Perhaps the most interesting of the surface proteins is that in the 15 K band. This has been found in greater concentration on starches from cereals with soft endosperm than on those from cereals with hard endosperm. The protein has been called 'friabilin', because of its association with a friable endosperm (cf Ch. 4) (Greenwell and Schofield, 1989).

Phosphorus is another important minor constituent of cereal starches. It occurs as a component of lysophospholipids. They consist of 70% lysophosphatidyl choline, 20% lysophosphatidyl ethanolamine and 10% lysophosphatidyl glycerol. The proportion of lysophospholipids to free fatty acids varies with species: in wheat, rye, triticale and barley over 90% occurs as lysophospholipids, in rice and oats 70% and in millets and sorghum 55%. In maize 60% occurs as free fatty acids (Morrison, 1985).

Removal of lipids from cereal starches reduces the temperatures of gelatinization-related changes and increases peak viscosity of pastes. In other words they become more like the lipid-free potato starch.

Technological importance of starch

Much of the considerable importance of starch in foods depends upon its nutritional properties; it is a major source of energy for humans and for domestic herbivorous and omnivorous animals. In the human diet it is usually consumed in a cooked form wherein it confers attractive textural qualities to recipe formulations. These can vary from those of gravies and sauces, custards and pie fillings to pasta, breads, cakes and biscuits (cookies). Much of the variation in texture depends upon the degree of gelatinization, which in turn depends upon the temperature, and the amount of water available during cooking. Digestibility in the intestines of single-stomached animals is also increased by gelatinization.

Gelatinization

This is a phenomenon manifested as several changes in properties, including granule swelling and progressive loss of organized structure (detected as loss of birefringence and crystallinity), increased permeability to water and dissolved substances (including dyes), increased leaching of starch components, increased viscosity of the aqueous suspension and increased susceptibility to enzymic digestion.

At room temperature starch granules are not totally impermeable to water, in fact water uptake can be detected microscopically by a small increase in diameter. The swelling is reversible and the wetting and drying can be cycled repeatedly without permanent change. If the temperature of a suspension of starch in excess water is raised progressively, a condition is reached, around 60°C, at which irreversible swelling begins, and continues with increasing temperature. The change is endothermic and can be quantified by thermal analysis techniques.

Typical heats of gelatinization in J per g of dry starch are: wheat 19.7, maize 18.0, waxy maize 19.7 and high amylose maize 31.79 (Maurice et al., 1983). Swelling involves increased uptake of water and can thus lead to increased viscosity by reducing the mobile phase surrounding the granules; accompanying leaching of starch polymers into this phase can further increase viscosity. The swelling behaviour of starch heated in water is often followed using a continuous automatic viscometer, such as the Brabender Amylograph (Shuey and Tipples, 1980). Upon heating a slurry of 7–10% starch w/w in water at a constant rate of 1°–5°C per min, starch eventually gelatinizes and begins to thicken the mixture. The temperature at which a rise in consistency is shown is called the pasting temperature. The curve then generally rises to a peak, called the peak viscosity. When the temperature reaches 95°C, that temperature is maintained for 10–30 min and stirring is continued to determine the shear stability of

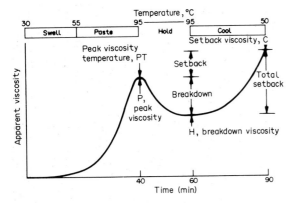

FIG 3.9 Chart showing characteristics recorded by the Brabender Amylograph.

the starch. Finally the paste is cooled to 30°C and the increase in consistency is called set-back. (Fig. 3.9)

Retrogradation (see also Ch. 8)

Suspensions of gelatinized granules containing more than 3% starch form a viscous or semi-solid starch paste which, on cooling, sets to a gel. Three dimensional gel networks are formed from the amylose-containing starches by a mechanism known as 'entanglement'. The relatively long amylose molecules that escape from the swollen granules into the continuous phase become entangled at a concentration of 1–1.5% in water. On cooling the entangled molecules lose translational motion, and the water is trapped in the network. Crystallites begin to form eventually at junction zones in the swollen discontinuous phase, causing the gel slowly to increase in rigidity (Osman, 1967). When starch gels are held for prolonged periods, retrogradation sets in. As applied to starch this means a return from a solvated, dispersed, amorphous state to an insoluble, aggregated or crystalline condition. Retrogradation is due largely to crystallization of amylose, which is much more rapid than that of amylopectin. It is responsible for hardening of cooked rice and shrinkage and syneresis of starch gels and possibly firming of bread. Although regarded as crystalline, retrograded gels are susceptible to amylolysis, however a fraction has

recently been found that is resistant to enzyme attack. Known as resistant starch, it behaves as dietary fibre and is most abundant in autoclaved amylomaize starch suspensions (Berry, 1988).

Starch damage (see Chs 6 and 8)

Granule damage of a particular type alters the properties of starch in some ways similar to gelatinization. Defining the exact type of damage is difficult and this accounts for the continued use of the general term. The essential characteristics associated with damaged starch are somewhat similar to gelatinized granules, but there are differences also. Thus mechanical damage results in:

1. increased capacity to absorb water, from 0.5–fold starch dry mass when intact to 3–4–fold when damaged (gelatinized granules absorb as much as 20–fold);
2. increased susceptibility to amylolysis;
3. loss of organized structure manifested as loss of X-ray pattern, birefringence, differential scanning calorimetry gelatinization endotherm;
4. reduced paste viscosity;
5. increased solubility, leading to leaching of mainly amylopectin. (In gelatinized granules, amylose is preferentially leached (Craig and Stark, 1984).)

At a molecular level the disorganization of granules appears to be accompanied by fragmentation of amylopectin molecules during damage whereas gelatinization achieves loss of organization without either polymer being reduced in size.

Controlling starch damage level during milling of wheat flour is important as it affects the amount of water needed to make a dough of the required consistency (see Ch. 7) (Evers and Stevens, 1985).

Cell walls

The older literature describes the components of cereal grain cell walls as pentosans and hemicelluloses. Pentosans are defined earlier in this chapter, but hemicelluloses are more difficult to define and indeed the term is even now only used loosely. Hemicelluloses were originally assumed to be low molecular weight (and therefore more soluble) precursors of cellulose. Coultate (1989) writes that the name as applied now covers the xylans, the mannans and the glucomannans, and the galactans and the arabinoxylans, however others use the name to include β-glucans also (Hoseney, 1986).

Cell walls are important in several contexts.

1. As a structural framework with which the grain is organized.
2. As a physical boundary to access by enzymes produced outside the cell.
3. As as source of energy in ruminants and of dietary fibre in single stomached animals, including man.
4. They or their derivatives affect processing of raw or cooked cereal products.

Cell walls of different cereals have some common components but composition is not consistent among species. Cellulose is one component present in all cell walls, it is the material of the simplest and the youngest structures. In most cases additional carbohydrates of varying complexity are deposited as a matrix, and some protein also becomes included. Lignin is a common component of secondary thickening in the pericarp of all cereal grains. It is found in the pales but this is relevant to processing only in those grains of which they remain a part after threshing (i.e. oats, barley and rice). The walls of nucellus and seedcoat (see Ch. 2) are generally unlignified; they may contain some corky material. The pigment strand, which is continuous with the seedcoat in grains where a crease is present, is lignified and later becomes encrusted with a material of unknown composition. Similar unidentified material encrusts testa cell walls on their inner surfaces (Zee and O'Brien, 1970). The more precise composition of cell walls has been reviewed by Fincher and Stone (1986). Walls of cereal endosperm (aleurone and starchy-endosperm) consist predominantly of arabinoxylans and (1→3,1→4)-β-glucans, with smaller amounts of cellulose, heteromannans, protein and esterified phenolic acids. They are unlignified and contain little, if any, pectin and xyloglucan, or hydroxyproline-rich glycoprotein, all of which

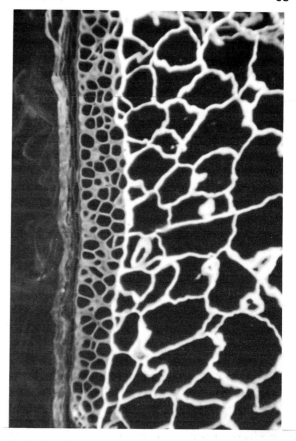

FIG 3.10 Cell walls of barley endosperm fluorescing as a result of staining with calcofluor White MR. The bright fluorescence of the starchy endosperm cell walls contrasts with that of the walls in the triple aleurone layer (cf. Ch 2) (Previously unpublished photograph kindly supplied by Dr S. Shea Miller, Centre for Food and Animal Research, Agriculture Canada, Ottawa.)

are common components of other primary cell walls.

Walls with high β-D-glucan content, such as endosperm cell walls in oats and barley, can be identified under the fluorescence microscope, because of a specific precipitation reaction with Calcofluor White MR (new) (Fig. 3.10). The blue fluorescence is intense and excitation by a wide range of wavelengths is possible (Fulcher and Wong, 1980).

Rice is exceptional in containing significant proportions of pectin and xyloglucan, together with small amounts of hydoxyproline-rich protein. The cellulose content of rice cell walls is also unusually high (25–30%), and mannose-containing sugars in some may contribute as much as 15%.

A significant non-carbohydrate molecule intimately associated with arabinoxylans in wheat and other cereal cell walls is ferulic acid, a phenol carbonic acid very abundant in plant products. It is esterified to the primary alcoholic group of the arabinose side chain (Amado and Neukom, 1985). The formation of diferulic acid cross-links is at least partly responsible for the gelation of aqueous flour extracts or solutions of cereal pentosans in the presence of oxidizing agents. Ferulic acid exhibits intensive blue fluorescence when irradiated with light of 365 nm. The reaction is particularly marked in aleurone cell walls and thus can be used for identifying these in ground cereal products under the fluorescence microscope.

Fractionation into water soluble and insoluble pentosans by various protocols is common analytical practice as it has been found to distinguish different functional properties. Thus it is the water-soluble pentosans (mainly arabinoxylans) of wheat that have a very high water absorbing capacity. They are linear molecules while those of the insoluble fraction are highly branched. The backbone of arabinoxylans consists of D-glucan units linked by β-(1→4) glycosidic bonds. Single α-L-arabinofuranose residues are attached randomly to the xylan and cause the water solubility of the arabinoxylans. As much as 23% of water in a bread dough may be associated with pentosans (Bushuk, 1966). It has been suggested (Hoseney, 1984) that pentosans reduce the rate of CO_2 diffusion through the dough, behaving in this way similarly to gluten.

Fibre

Extraction of individual cell wall components is complex and unsuitable for routine analysis. Nevertheless an estimate of cell wall content is often required, particularly in relation to nutritional attributes of a product. Analytical procedures have been devized to determine undigestible material as 'fibre', but not all experts are agreed as to which chemical entities should be included.

The term is frequently qualified to reflect the method of analysis employed because different methods produce different values (it should also be noted that some methods are themselves inconsistent). The following types of fibre may be encountered, the definitions are based on a glossary by Southgate et al. (1986).

Crude fibre — The residue left after boiling the defatted food in dilute alkali and then in dilute acid. The method recovers 50–80% of cellulose, 10–50% of lignin and 20% of hemicellulose. Results are inconsistent.

Acid detergent fibre (ADF) — The cellulose plus lignin in a sample; it is measured as the residue after extracting the food with a hot dilute H_2SO_4 solution of the detergent cetyl trimethylammonium bromide (CTAB).

Neutral detergent fibre (NDF) — The residue left after extraction with a hot neutral solution of sodium dodecyl sulphate (SDS) also known as sodium lauryl sulphate. It is designed to divide the dry matter of feeds very nearly into those which are nutritionally available by the normal digestive process and those which depend on microbial fermentation for their availability.

Dietary fibre — All the polymers of plants that cannot be digested by the endogenous secretions of the human digestive tract.

The last definition differs from those that precede it in that it is not based on the method by which it is determined. It represents the value that the analytical methods seek to achieve.

Other terms are also in use, such as '*unavailable carbohydrate*' and '*plantix*', which depart from the indication that only fibrous material (i.e. occurring as fibres) is included. Instead it suggests a matrix of plant materials. It will probably be some time before a consensus is achieved because of lack of agreement on whether a functional or compositional definition is more appropriate. In the meantime methods that distinguish several classes of indigestible material will be the most useful. That of Southgate et al. (1986) distinguishes among cellulose, non-cellulosic polysaccharides and lignin. That of Asp et al. (1983) distinguishes soluble and insoluble fibre. Insoluble components include galacto- and gluco-mannans, cellulose and lignin, and the soluble class includes galacturonans (pectins), (1→3,1→4)-β-glucans and arabinoxylans.

Proteins

Although an enormous range of proteins exists in nature they are all composed of the same relatively simple units: amino acids. The diversity of proteins comes about because the amino acids are arranged in different sequences and those sequences are of different lengths. There are only twenty amino acids commonly found in proteins. Cereal proteins are important in human and animal nutrition, they provide the unique gas-retaining qualities in wheat flour doughs and bread, but in all organisms proteins are present which function as enzymes. Within the growing plant the genetic code is interpreted through the synthesis and activation of enzymes, providing the means by which characteristics of individual species are expressed. When seen in the context of this function it is perhaps easier to appreciate the subtlety of the differences in behaviour that can be achieved among what, at first sight, appear to be molecules of relatively simple construction. The subtle functional differences are possible because of the diversity of the properties of the amino acids and the relationships in which they are capable of engaging with other amino acids or even with lipid, carbohydrate and other molecules. Additional variation comes about as a result of the environment in which a protein finds itself. A change in pH, temperature or ionic strength can lead to a single protein species behaving in different ways.

Structure

All amino acids have in common the presence of an alpha-amino group (NH_2) and a carboxyl group (-COOH). It is through the condensation of these groups that neighbouring amino acids are joined by a peptide bond, as in Fig. 3.11.

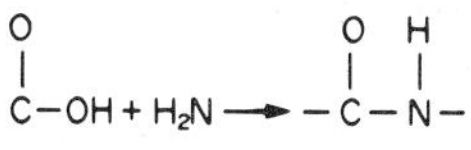

$$\underset{\text{C-OH}}{\overset{\text{O}}{|}} + H_2N \longrightarrow \underset{\text{C-N-}}{\overset{\text{O H}}{|\ |}}$$

FIG 3.11 The peptide bond.

A sequence of a large number of units linked by peptide bonds is called a polypeptide.

The differences among amino acids lie in the side-chains attached to the carbon atom lying between their carboxyl and amino groups. Side chains may be classified according to their capacity for interacting with other amino acids by different mechanisms. The types of interaction and the amino acids capable of engaging in them are listed in Table 3.5.

TABLE 3.5

Grouping of Amino Acid Residues According to their Capacity for Interacting Within and Between Protein Chains

Type of interaction	Amino acid
(1) Covalent — disulphide bonding Dissociated by oxidizing and reducing agents, e.g. performic acid; 2-mercaptoethanol	Cysteine/cystine
(2) Neutral — hydrogen bonding Dissociated by strong H-bonding agents, e.g. urea, dimethyl formamide	Asparagine Glutamine Threonine Serine Cysteine
(3) Neutral — hydrophobic interaction Dissociated by ionic and non-ionic detergents, e.g. sodium salts of long chain fatty acids	Tyrosine Tryptophan Phenylalanine Proline Methionine Leucine Isoleucine Valine Alanine★ Glycine★
(4) Electrostatic — acid hydrophilic — basic hydrophilic Dissociated by acid, alkali, or salt solutions	Aspartic acid Lysine Arginine Histidine Glutamic acid

★ Amino acids with short, aliphatic side-chains show very little hydrophobicity. Both glycine and alanine are readily soluble in water.

Table from Simmonds, 1989.

The order in which amino acids occur in a polypeptide defines its 'primary structure'. Because of the range of interactions that can occur among the side-chains, different sequences are capable of different interactions giving rise to a secondary structure. The units of secondary structure in turn react to give rise to the tertiary structure which defines the three-dimensional conformation adopted as a result of side-chain interactions. The secondary and tertiary structures of a protein change in response to the environment but the primary structure remains unaltered unless its length is reduced by hydrolysis.

All the interactions listed in Table 3.5 can contribute to tertiary structure but the most stable types are the covalent disulphide bonds formed by oxidation of sulphydryl group on interacting cysteine/cystine residues.

Such bonds also occur between cysteine/cystine residues on different polypeptides giving rise to a stable structure involving more than one polypeptide. Inter-peptide links can thus produce in a protein a fourth or *quaternary* level of structure.

Disulphide bonds are stronger than non-covalent bonds but they are nevertheless capable of entering into interchange reactions with substances containing free sulphydryl groups. Such reactions are of great importance in dough formation.

Cereal proteins

The complexity of cereal proteins is enormous and the determination of the structure of gluten — the protein complex responsible for the dough-forming capacity of wheat flour — has been described as one of the most formidable problems ever faced by the protein chemist (Wrigley and Bietz, 1988). To simplify their studies cereal chemists have sought to separate the proteins into fractions that have similarities in behaviour, composition and structure. As protein studies have proceeded and knowledge has accumulated, the validity of earlier criteria of classification has been, and continues to be, challenged.

One of the most significant means of classifying plant proteins is that which Osborne (1907) made on the basis of solubility. Water soluble proteins were described as 'albumins', saline soluble as 'globulins', aqueous alcohol soluble as 'prolamins' and those that remained insoluble as 'glutelins'.

There are differences in amino acid composition between proteins in the Osborne classes (see Ch. 14) but there is also heterogeneity within each class and this may be as significant as between class differences. Newer analytical methods have shown that the solubility classes overlap considerably. The distinction formalized by Osborne that remains unquestionably valid is that between albumins and globulins on the one hand and prolamins and glutelins on the other. In composition there is a marked difference due mainly to the extremely high content of proline and glutamine in the less soluble fractions (the name 'prolamine' reflects this characteristic). An extremely low lysine content is also characteristic of insoluble cereal proteins.

Soluble proteins

These are found in starchy endosperm, aleurone and embryo tissues of cereals. They account for approximately 20% of the total protein of the grain. Albumins are usually more prevalent than globulins. The amino acid composition of soluble proteins is similar to that of proteins found in most unspecialized plant cells suggesting that they include those that constitute the cytoplasm found in most cells. They are a complex mixture including:

1. metabolic enzymes;
2. hydrolytic enzymes;
3. enzyme inhibitors;
4. phytohaemaglutenins (proteins that clot red blood cells).

Globulins may also arise as storage proteins, occurring in protein bodies, particularly in oat and rice endosperm. In other cereals, storage proteins arising in protein bodies are exclusively of the insoluble types (Payne and Rhodes, 1982).

The number of individual proteins in the soluble categories is large. By two-dimensional electrophoresis 160 components have been separated in aqueous extracts from wheat endosperm and a different pattern of 140 components have been separated from the 0.5 M NaCl extracts (Lei and Reeck, 1986).

Enzymes

Enzymes may be considered in the context of the stage of the grain's life cycle. Thus, most enzyme activity during maturation is concerned with synthesis, particularly the synthesis of storage products. Some hydrolytic enzymes involved in breakdown of starch and protein stored in the pericarp are found before maturity and may persist (Fretzdorf and Weipert, 1990).

In the mature grain the enzyme levels are relatively low if the grain is sound and dry. If damaged, as in milling, lipids become exposed to lipase. This is particularly relevant to oats; and to germ and bran fractions of other grains.

On adequate damping, germination begins and enzymes concerned with solubilization are produced. Cell walls are hydrolyzed, permitting greater access to storage products by enzymes that catalyze hydrolysis of starch and protein (see Ch. 2).

Technologically the highest profile enzyme is *alpha*-amylase, as large quantities are essential in successful malting and brewing and small quantities are necessary in breadmaking. Excessive *alpha*-amylase in milling wheats is disastrous, leading to dextrin production during baking, making the crumb sticky. Polyphenol oxidases can lead to production of dark specks in stored flour products. Other classes of enzymes of technological importance, found in cereals are *beta*-amylases, proteases, *beta*-glucanases, lipases, lipoxygenase and phytase.

Amylases

Both *alpha*- and *beta*-amylases are α-$(1\rightarrow4)$-D-glucanases; by definition catalyzing the hydrolysis of the same bonds within starch molecules. Their action is synergistic because *beta*-amylase gains greater access to the substrate through the activity of *alpha*-amylase. As this last observation implies, their modes of action are quite different: *alpha*-amylase is endo-acting while *beta*-amylase is exo-acting (Fig. 3.12).

Exo-enzymes catalyze removal of successive low molecular weight products from the non-reducing chain-end, the product removed through *beta*-amylase activity is maltose due to the hydrolysis of alternate α-$(1\rightarrow4)$-glycosidic bonds. *beta*-Amylase is inactive on granular starch but is capable of rapid action when the substrate is in solution. As the exo-action produces a large number of small sugars with reducing power, the reducing

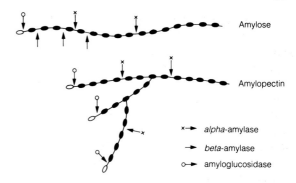

FIG 3.12 Diagrammatic representation of hydrolytic cleavage catalyzed by *alpha*-amylase, *beta*-amylase amd amyloglucosidase respectively. From D. H. Simmonds (1989). Reproduced by courtesy of CSIRO.

capacity of the solution increases rapidly. When the substrate is amylose the iodine staining reaction is reduced only slowly as the chain lengths, on which iodine binding depends, are slowly reduced.

By contrast endo-action of *alpha*-amylolysis through random fragmentation reduces iodine staining relatively quickly in relation to the increase in reducing power. A further consequence of the rapid reduction in molecular size resulting from *alpha*-amylolysis is a marked reduction in viscosity of a starch suspension. This is exploited in laboratory tests for the enzyme. Assaying for *beta*-amylase is more difficult because the rate of maltose production is influenced by the presence of *alpha*-amylase, and the enzymes are almost always present together. Even in well washed starch they are absorbed on the granule surface (Bowler *et al.*, 1985).

Grain quality is more influenced by the *alpha*-enzyme as its amount is more variable according to the condition of the grain. *beta*-Amylase is present in resting grain and increases only a few fold on germination through release of a bound form.

Alpha-amylase is actually synthesized during germination and activity increases progressively, as germination proceeds, by several hundred fold. In different cereals the site of synthesis of *alpha*-amylase varies; in wheat, rye and barley it occurs first in the scutellum and later in the aleurone, in maize only the scutellum is involved. Several isoenzymes of the *alpha*-amylase type exist in most cereals, they fall into two groups depending upon their isoelectric points. The Triticeae cereals contain two groups while sorghum, millet, maize, oats and rice have only one (Kruger and Reed, 1988).

Even the combined action of *alpha-* and *beta-*amylases cannot completely digest solubilized starch. Neither of them can catalyze hydrolysis of α-(1→6)-bonds and hence branch points remain intact. Also, those α-(1→4) bonds close to branch points resist hydrolysis. Hence only about 85% of starch is converted to sugars. In order to increase yield of sugars in commercial processes, debranching enzymes may be used. Amyloglucosidase from fungal sources is a popular expedient, it catalyzes hydrolysis of both α-(1→4)- and α-(1→6)-bonds leaving glucose as the ultimate product. Some brewing processes permit the use of this enzyme and saké (see Ch. 9) production is dependent upon it.

β-Glucanases

These enzymes assume greatest importance in processing of barley in which β-glucans contribute 70% of cell walls. There are two endo-β-glucanases in barley malt, both synthesized during germination. Each catalyzes hydrolysis of β-(1→4) linkages adjacent to β-(1→3) links, ultimately producing a mixture of oligosaccharides containing three or four glucosyl units (Woodward and Fincher, 1982). The two isoenzymes are synthesized in different sites, I in the scutellum and II in the aleurone. Before being susceptible to these enzymes it is thought that another enzyme, β-glucan solubilase renders the substrate soluble (Bamforth and Quain, 1989).

Proteolytic enzymes

Although proteolytic enzymes may be important technologically in baking, their significance is usually masked by the greater effects of *alpha-*amylase. In brewing their role is better understood. Both endo-peptidases and exo-enzymes (the carboxypeptidases which catalyze cleavage of single amino acids from the carboxyl terminus) are present.

Proteolysis increases access by amylases to starch granules as well as producing nitrogenous nutrients, for the growing embryo in nature and for yeast during fermentation for beer production.

Lipid modifying enzymes

Enzymes of two types are important in catalyzing breakdown of lipids: lipase and lipoxidase. Both are capable of causing rancidity in cereals; thus both hydrolytic and oxidative rancidity are recognized. Lipoxidase can only catalyze degradation of free fatty acids and monoglycerides and therefore follows lipolysis. Lipolysis proceeds slowly in the dry state; enzymic oxidation occurs rapidly on wetting.

The problem of rancidity is potentially greatest in oats which have a high oil content (4–11%, average 7%). Maize also has a relatively high oil content because of its large embryo (about 4.4%), brown rice contains about 3% but other cereals contain only 1.5–2%. Problems caused by hydrolysis catalyzed by lipase are prevented in the case of processed oats by 'stabilization', a steaming process which inactivates the enzyme (cf. Ch. 6). In other cereals that are milled, potential storage problems can be avoided in starchy endosperm, if it is separated from other grain parts where enzyme and substrate are concentrated. This is common practice in the cases of sorghum and maize grits, in which the embryo presents the greatest hazard, and in wheat and rice, in which the aleurone layer also has a high lipid content. In wheat, lipase activities in the embryo and aleurone layer are 10–20-fold that of the endosperm (Kruger and Reed, 1988). The storage lives of bran, germ and wholemeal flour are considerably less than that of white flour for this reason (see Ch. 7).

As well as true lipases, esterases are also present in cereals and in most studies the two classes have not been distinguished. Like other hydrolases, they are synthesized during the early stages of germination, although oats are exceptional in having a high lipase activity in resting grain.

Lipases catalyze hydrolysis of triglycerides to produce diglycerides and free fatty acids, diglycerides to give monoglycerides and free fatty acids;

and monoglycerides to give glycerol and free fatty acids. The unsaturated fatty acids can be converted to hydroperoxides which, in turn, are changed to hydroxy acids by lipoxygenase, lipoperoxidase and other enzymes, as well as by non-enzymic processes (Youngs, 1986).

Lipoxygenase is an effective bleaching agent; a coupled oxidation destroys the yellow pigments in wheat endosperm. Cosmetically, this is beneficial in bread doughs but undesirable in pasta products in which the yellow colour is valued (Hoseney, 1986).

Phytase

Phytase catalyzes hydrolysis of phytic acid (inositol hexaphosphoric acid) to inositol and free orthophosphate. In wheat its activity increased six-fold on germination and more activity was found in hard wheats than in soft (Kruger and Reed, 1988). In oats the activity is much lower than in wheat, rye and triticale (Lockhart and Hurt, 1986).

In rice the phytin level dropped from 2.67 to 1.48 mg/g of dry mass after one day of germination, then to 0.44 mg/g after five days. Phytase activity levelled off after seven days (Juliano, 1985).

Phenol oxidases

In the mature wheat grain several polyphenol oxidases are present in the starchy endosperm, they are more concentrated in the bran. On germination an increase, including new isoenzyme synthesis, occurs, mainly in the coleoptile and roots. Monophenolase also increases. Durum wheat has less activity than other wheat types (Kruger and Reed, 1988).

Catalase and peroxidase

Catalase and peroxidase are haemoproteins. Peroxidase is involved in the degradation of aromatic amines and phenols by hydrogen peroxidase. Its activity is greater in wheat than in other cereals. Catalase catalyzes degradation of hydrogen peroxide to water and oxygen. Its

physiological function is not understood but it increases during germination (Kruger and Reed, 1988).

Insoluble proteins

The state of knowledge of many insoluble cereal proteins has now advanced even to complete sequencing of their amino acids. This is true of prolamins of maize which are known as zeins, since they come from *Zea*. Barley prolamins are hordeins, rye prolamins are secalins and oat prolamins are avelins. A different basis for nomenclature is applied to the naming of wheat prolamins which are called gliadins.

The cereal prolamins have been reviewed by Shewry and Tatham (1990). On the basis of sequencing, four major groups of zeins have been defined. The groups differ in their amino acid content as well as the sequence in which they occur. As prolamins they are by definition rich in proline and glutamine, and low in lysine and tryptophan. The groups are designated α, β, γ and δ. The β- and δ-groups are relatively rich in methionine and the δ-group is also rich in cysteine and histidine.

α-Zeins

The predominant group is the α-group, contributing 75–80% of the total insoluble fraction of zein. By electrophoresis the apparent molecular weights of the two major α-zeins are 19,800 and 22,000. They can be separated by isoelectric focussing into a series of monomers and oligomers, though some of the latter can be extracted only after reduction of the S–S bonds by which the monomers are held together. It is frequently found in peptide sequences that the domain in the centre is quite different from those at the C- and N-terminal parts. In α-zeins the C-terminal domain consists of 10 amino acids in a unique sequence, similarly the N-terminus has a unique sequence of 36–37 residues in which one or two cysteine residues are present. The central domain comprises repetitive blocks of 20 residues that are rich in leucine and alanine. The tertiary structure of α-zeins, divined from circular dichroism, suggests a high content of α-helix and low β-sheet content.

β-Zeins

β-Zeins contribute 10–15% of total prolamin. They are rich in methionine and cysteine and can be extracted only in the presence of a reducing agent indicating mutual association through disulphide bonds. No sequences are repeated and all differ from those in α-zeins. The tertiary structure consists mainly of β-sheet, and aperiodic structure (β-turns and random coil).

γ and δ-Zeins

γ-Zeins account for 5–10% of the total prolamin. Like β-zeins, they require the presence of a reducing agent for extraction. Eight hexapeptide sequences in the central domain are flanked by unique N- and C-terminal regions. The repeat sequences are very hydrophilic, rendering the proteins very soluble when reduced.

δ-Zeins also require reduction before extraction, no sequences are repeated in the central region but between 17 and 29 methionine residues occur here.

Other tropical cereals

Although only the prolamins of maize among the tropical cereals have been studied extensively, available evidence indicates that sorghum, pearl millet and Job's tears contain essentially similar groups.

Temperate cereals

Under the classical nomenclature the necessity to reduce disulphide bonds would define β-, γ- and δ-zeins as glutelins. Indeed to regard all insoluble cereal proteins as prolamins is not universally accepted among protein chemists. Such a classification can be extended to temperate cereals but at the present time the traditional Osborne classification is more widespread especially in wheat proteins where the functional aspects are particularly important. From two dimensional electrophoresis studies it has been established that up to 20 different polypeptides are found in glutenins (the glutelins of wheat — see p. 69). An even greater number — up to 50 — may be found in gliadins. One argument advanced for distinguishing between wheat prolamins and glutelins is the different physical properties of the two classes when hydrated: gliadins behave as a viscous liquid and glutenins as a cohesive solid. Although both influence gluten behaviour, it is the larger polymeric glutenins that wield the greater influence.

One of the most attractive theories concerning the relationship between glutenin structure and function is the linear gluten hypothesis (Fig. 3.13).

It envisages a series of polymeric subunits joined head to tail by interchain disulphide bonds. The essential features of the subunits are terminal α-helices and central regions of many β-turns (β-turns also occur in the body tissue protein elastin, they are capable of much extension under tension and can return to their former folded condition on

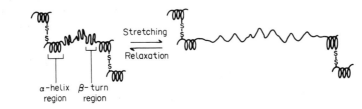

FIG 3.13 Schematic representation of a polypeptide subunit of glutenin within a linear concatenation. The subunits are joined head-to-tail via S–S bonds to form polymers with molecular weights of up to several million. The subunits are considered to have a conformation that may be stretched when tension is applied to the polymers, but when the tension is released the native conformation is regained through elastic recoil. The N- and C- terminal ends of some high molecular weight subunits, where interchain S–S bonds are located, are now thought to be alpha-helix rich domains, whereas the central domains are thought to be rich in repetitive beta-turn structures. The presence of repetitive beta-turn structures may result in a beta-spiral structure, which may confer elasticity. From D. J. Schofield (1986). *Reproduced by courtesy of The Royal Society of Chemistry, London.*

subsequent relaxation of the tension). In general the sulphur-containing cysteine residues occur in the α-helical regions, so that the disulphide bonds form between these regions in adjacent polypeptides. β-Turns thus remain unencumbered by interchain bonds that might otherwise restrict their extension. Molecular weights of glutenins are upward of 10^5.

The unusually high content of the amino acids asparagine and glutamine found in gluten proteins may be significant in providing stability of gluten, through their tendency to become involved in hydrogen bonding. Hydrophobic and electrostatic reactions associated with other amino acid side chains also contribute.

The relative importance of glutenins and gliadins varies in wheats from different parts of the world. In Australian and Italian wheats gliadin variations have the strongest association with bread quality. In European wheats high molecular weight glutenin subunits with apparent molecular weight of 90–150 K are paramount in determining quality. Each wheat possesses a complement of 3–5 types, and a variety of individual subunits (allelic forms) may represent each type, giving rise to variation in baking properties. Gliadins are thought to behave as plasticizers, the proportional relationship between them and glutenins is an important

factor. Too low a gliadin content leads to inhibition of bubble expansion while the reverse results in excessive expansion and collapse.

Gliadin complements are characteristic of individual cultivars and these, revealed through polyacrylamide gel electrophoresis (PAGE), are exploited in establishing the varietal identity of wheat cultivars (Fig. 3.14) and for detecting adulteration of *T. durum* products with *T. aestivum* additions.

While this technique may be useful in other species also, it has not been developed to the same degree as in wheat. An even more sensitive method of identifying protein components is high performance liquid chromatography (HPLC). It is faster, and capable of greater resolution than PAGE. Its widespread use is limited by its greater expense and demands for technical expertise.

High lysine mutants

To produce cereals with better balanced proteins, from a nutritional point of view, breeders have exploited mutants with high lysine and high arginine contents. It is the storage proteins that are deficient in these amino acids so the mutants selected frequently achieve the improved balance through a deficiency in storage proteins (Hoseney and Variano-Marston, 1980). In the 'opaque' varieties of maize high lysine content is associated with 'opaque' (o_2) and 'floury' (fl_2) genes being double recessive, and the consequent inhibition of zein synthesis (Watson, 1987) (cf. Ch. 4). Thus the 'high lysine' varieties of maize, barley, sorghum and pearl millets have lower yields than their conventional counterparts.

Lipids

Lipids have been defined as those substances which are:

1. insoluble in water,
2. soluble in organic solvents such as chloroform, ether or benzene,
3. contain long chain hydrocarbon groups in their molecules, and
4. are present in or derived from living organisms (Kates, 1972).

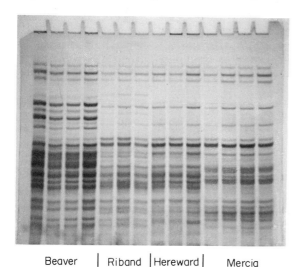

Beaver | Riband | Hereward | Mercia

FIG 3.14 PAGE electrophoretogram showing distinctive gliadin patterns of four U.K. wheat varieties. *Courtesy of FMBRA, Chorleywood, England.*

This covers a wide range of compounds including long-chain hydrocarbons, alcohols, aldehydes and fatty acids, and derivatives such as glycerides, wax esters, phospholipids, glycolipids and sulpholipids. Also included are substances which are usually considered as belonging to other classes of compounds, for example the 'fat soluble' vitamins A, D, E and K, and their derivatives, as well as carotenoids and sterols and their fatty acid esters (Kates, 1972).

The terms lipid, fat and oil are often used loosely, but, applied strictly 'lipids' include all the above while only triglycerides (triacylglycerols) are described as fats and oils. Fats are solid at room temperature while oils are liquid. Although many fats and oils originate in living organisms (where they function as a means of storing energy) this is not a feature of their definition as it is for lipids (see (d) above).

Nomenclature

With many series of compounds several conventions by which they are named, coexist. The earlier 'trivial' names may have been chosen to reflect the original source or other arbitrary connection. They provide no indication of the structure of the molecules. As knowledge increases and more compounds of the series are identified, the need for a systematic system of names, and the means of achieving it increase. Such is the case with lipids and a systematic convention for their nomenclature was recommended by the International Union of Pure and Applied Chemists (IUPAC) (Sober, 1968).

Fatty acids

Fatty acids present in cereal lipids mainly consist of a long hydrocarbon chain covalently linked to a carboxylic acid group (Fig. 3.15).

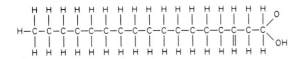

FIG 3.15 Generalized structure of a fatty acid.

A fatty acid in which all bonds are single is said to be saturated. In the absence of two adjacent hydrogens, a double bond is formed and the resultant fatty acid is described as unsaturated. Where more than one double bond is present the term polyunsaturated is applied. The systematic description of the compound depends on where double bonds are substituted. If the remaining hydrogens are on the same side of the chain, the conformation is called 'cis-'. If on different sides a 'trans-' double bond exists (Fig. 3.16).

FIG 3.16 *Cis-* and *trans*-configurations.

As well as a systematic nomenclature, a shorthand way of desribing the fatty acid may be used. Thus *cis*-9-octadecenoic acid has a shorthand description C18:1.9*cis*, indicating 18 carbons (octadec-), a double bond (-en-) in the *cis* form between the ninth and tenth position, counting from the functional-group-carbon.

Acylglycerols (glycerides)

Glycerides are compounds formed by esterification of the tertiary alcohol glycerol, and one to three fatty acids. Esterification involves removal of the elements of water and replacing the hydrogen of hydroxy groups of glycerol with the acyl group *RCO*. The residue of a fatty acid forming the ester is an acyl group (acyl = carboxylic radicle *RCO* where *R* is aliphatic). Hence the systematic name for glycerides is acylglycerols. Glycerol has three hydroxyl groups capable of ester formation and, depending on the number esterified, the resulting compounds may be mono-acylglycerols, di-acylglycerols or tri-acylglycerols. Plants usually store lipids as tri-acylglycerols, and cereal grains conform to this plant characteristic. The highest tri-acylglycerol levels occur in aleurone and scutellar tissue, but there are appreciable quantities in cereal embryonic axes and in the endosperm of oats (Morrison, 1983). They are the main lipid stored in all cereal endosperm, and in wheat the endosperm contributes about 12% of the total in the grain. Mono- and di-acylglycerols occur only in small quantities as intermediates in the biosynthesis of tri-acylglycerols or products of their breakdown.

Phosphoglycerides (Phospholipids)

The principal phosphoglycerides in cereal grains are phosphatidyl-choline, phosphatidyl-ethanolamine, phosphatidyl-inositol, N-acylphosphatidyl-ethanolamine and its monoacyl (lyso) derivative. The monoacylphospholipids: lysophosphatidyl-choline, lysophosphatidyl-ethanolamine and lyso-phosphatidyl-glycerol are the major internal starch lipids. Monoacylphospholipids are also formed from diacylphospholipids by enzymic hydrolysis (Morrison, 1983).

The structures of diacylphosphoglycerides are shown in Fig. 3.17, in which R, R′ and R″ are acyl groups.

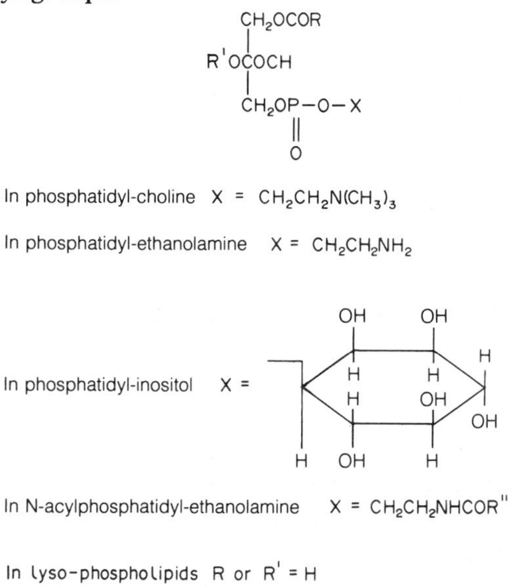

$$CH_2OCOR$$
$$R'OCOCH$$
$$CH_2OP-O-X$$
$$\parallel$$
$$O$$

In phosphatidyl-choline $X = CH_2CH_2N(CH_3)_3$

In phosphatidyl-ethanolamine $X = CH_2CH_2NH_2$

In phosphatidyl-inositol   X =

In N-acylphosphatidyl-ethanolamine $X = CH_2CH_2NHCOR''$

In lyso-phospholipids R or R′ = H

FIG 3.17 General formulae of diacylphosphoglycerides.

Glycosyl-glycerides (glycolipids)

Monoglycosyl-diglyceride and diglycosyl-diglyceride are the major glycolipids, with some mono- and diglycosyl-monoglycerides. Triglycosyl-diglyceride and tetraglycosyl-glycerides have also been reported. In wheat and most other cereals, the sugar is mainly galactose, sometimes with small amounts of glucose or none. Other minor glycolipids include sterylglycosides (Morrison 1983).

The structure of gycosyl-diglycerides is shown in Fig. 3.18, where R and R′ are acyl groups and S is a sugar (mono-saccharide–tetra-saccharide).

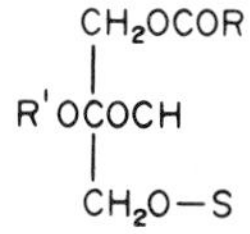

$$CH_2OCOR$$
$$R'OCOCH$$
$$CH_2O-S$$

FIG 3.18 General formula of glycosyl-diglycerides. In glycosyl-monoglycerides R or R′ = H.

Mineral matter

About 95% of the minerals in the actual fruits of cereals (i.e. the grain without adherent pales in the case of husked types) consists of phosphates and sulphates of potassium, magnesium and calcium. The potassium phosphate is probably present in wheat mainly in the form of KH_2PO_4 and K_2HPO_4. Some of the phosphorus is present as phytic acid. Important minor elements are iron, manganese and zinc, present at a level of 1–5 mg/100g, and copper, about 0.5 mg/100g. Besides these, a large number of other elements are present in trace quantities. Representative data from the literature are collected in Table 3.6. The content of mineral matter in the husk of barley, oats and rice is higher than that in the caryopses, and the ash is particularly rich in silica (Table 3.7). Also see Ch. 14.

Vitamins

The distribution and nutritional significance of vitamins in cereals are discussed in Ch. 14.

TABLE 3.6
Mineral Composition of Cereal Grains (mg/100g d.b.)

| Element | Wheat | Barley | Oats | | Rye | Triticale | Rice | | |
			Whole grain	Groat			Paddy	Brown	White
Main									
Ca	48	52	94	58	49	37	15	22	12
Cl	61	137	82	73	36	—	15	—	19
K	441	534	450	376	524	485	216	257	100
Mg	152	145	138	118	138	147	118	187	31
Na	4	49	28	24	10	9	30	8	6
P	387	356	385	414	428	487	260	315	116
S	176	240	178	200	165	—	—	—	88
Si	10	420	639	28	6	—	2047	70	10
Minor									
Cu	0.6	0.7	0.5	0.4	0.7	0.8	0.4	0.4	0.2
Fe	4.6	4.6	6.2	4.3	4.4	6.5	2.8	1.9	0.9
Mn	4.0	2.0	4.9	4.0	2.5	4.2	2.2	2.4	1.2
Zn	3.3	3.1	3.0	5.1	2.0	3.3	1.8	1.8	1.0
Trace									
Ag	0.05	0.005	—	—	—	—	0.02	—	—
Al	0.4	0.67	0.6	0.6	0.56	—	0.9	—	—
As	0.01	0.01	0.03	—	0.01	—	0.007	—	—
B	0.4	0.2	0.16	0.08	0.3	—	0.14	—	—
Ba	0.7	0.5	0.4	0.008	—	—	1.2	—	—
Br	0.4	0.55	0.3	—	0.19	3.3	0.1	—	—
Cd	0.01	0.009	0.02	—	0.001	—	—	—	0.005
Co	0.005	0.004	0.006	0.02	0.01	—	0.007	0.007	0.006
Cr	0.01	0.01	0.01	—	—	—	0.06	—	0.003
F	0.11	0.15	0.04	0.04	0.1	—	0.07	—	0.04
Hg	0.005	0.003	—	—	—	—	0.001	—	—
I	0.008	0.007	0.007	0.06	0.004	—	—	0.002	0.002
Li	0.05	—	0.05	—	0.017	—	0.5	—	—
Mo	0.04	0.04	0.04	—	0.03	—	0.07	—	—
Ni	0.03	0.02	0.15	—	0.18	—	0.08	0.1	0.02
Pb	0.08	0.07	0.08	—	0.02	—	0.003	—	—
Rb	—	—	—	—	—	0.3	0.4	—	—
Sb	—	—	—	—	—	—	0.05	—	—
Sc	—	—	—	—	—	—	0.005	—	—
Se	0.05*	0.21	0.2	0.01	0.23	—	0.01	0.04	0.03
Sn	0.3	0.065	0.21	—	0.19	—	0.03	—	0.03
Sr	0.1	0.2	0.21	—	—	0.5	0.02	—	—
Ti	0.15	0.1	0.2	—	0.08	—	1.4	—	—
V	0.007	0.005	0.1	—	—	—	0.01	—	—
Zr	—	—	—	—	—	—	0.007	—	—
Ash%	1.9	3.1	2.9	2.1	2.2	2.1	7.2	1.8	0.6

| Element | | Maize | Sorghum | Millets | | | | |
				Pearl	Foxtail	Proso	Kodo	Finger
Main								
Ca		20	30	36	29	13	37	352
Cl		55	52	32	42	21	13	51
K		342	277	454	273	177	165	400
Mg		143	148	149	131	101	128	180
Na		40	11	11	5	7	5	16
P		294	305	379	320	221	245	323
S		145	116	168	192	178	156	184
Si		—	200	—	—	—	—	—

TABLE 3.6
Continued

Element	Maize	Sorghum	Millets Pearl	Foxtail	Proso	Kodo	Finger
Minor							
Cu	0.4	1.0	0.5	0.7	0.5	1.0	0.6
Fe	3.1	7.0	11.0	9.0	9.0	3.0	4.5
Mn	0.6	2.6	1.5	2.0	2.0	—	1.9
Zn	2.0	3.0	2.5	2.0	2.0	—	1.5
Trace							
Ag	—	<0.005	<0.005	—	—	—	0.4
Al	0.057	1.8	1.7	—	—	—	—
As	0.03	—	0.01	—	—	—	<0.05
B	0.3	0.13	0.19	—	—	—	2.2
Ba	3.0	0.08	0.04	—	—	—	—
Br	0.26	0.14	0.38	—	—	—	—
Cd	0.012	—	—	—	—	—	<0.01
Co	0.008	<0.05	0.05	—	—	—	0.02
Cr	0.004	0.05	0.03	—	—	—	—
F	0.04	—	—	—	—	—	—
I	0.2	—	0.0016	—	—	—	0.2
Li	0.005	0.07	0.01	—	—	—	0.2
Mo	0.03	0.2	0.07	—	—	—	0.02
Ni	0.04	0.3	0.11	—	—	—	0.6
Pb	0.01	0.11	0.02	—	—	—	0.2
Rb	0.3	0.12	0.34	—	—	—	—
Sc	0.01	—	—	—	—	—	0.006
Sn	0.01	0.07	0.004	—	—	—	3.3
Sr	0.02	0.18	0.03	—	—	—	0.03
Ti	0.17	0.1	0.02	—	—	—	0.04
V	0.01	0.05	<0.01	—	—	—	—
Zr	0.02	—	—	—	—	—	—
Ash %	1.7	1.7	2.4	3.7	2.2	3.0	2.2

N.B. A dash in the Table indicates that no reliable information has been found.
Sources as in Kent, 1983.
* Level found in wheat growing in normal soils. Much higher values, e.g. up to 6 mg 100g, are found in wheat growing in seleniferous soils.

TABLE 3.7
*Ash and Silicia in the Husk of Cereal Grains**

Material	Yield of ash (%)	Silica in ash (%)
Barley husk	6.0	65.8
Oat husk	5.2	68.0
Rice husk	22.6	95.8

* Sources as in Kent, 1983.

References

ALDRICK, A. J. (1991) The nature, sources, importance and uses of cereal dietary fibre. *H-GCA Review no.22.* Home-Grown Cereals Authority. London.

AMADO, R. and NEUKOM, H. (1985) Minor constituents of wheat flour: The pentosans. In: *New Approaches to Research on Cereal Carbohydrates*, HILL, R. D. and MUNCK, L. (Eds) Elsevier, Amsterdam.

AMAN, P. and NEWMAN, C. W. (1986) Chemical composition of some different types of barley grown in Montana, U.S.A. *J. Cereal Sci.* 4: 133–141.

ASP, N.-G., JOHANSSON, C. G., HALLMER, H. and UILJESTROM, M. (1983) Rapid enzymic assay of insoluble and soluble dietary fibre. *J. Agric. Food Chem.* 31: 476–482.

BADENHUIZEN, N. P. (1969) *The Biogenesis of Starch Granules in Higher Plants.* AppletonCrofts NY. U.S.A.

BAMFORTH, C. W. and QUAIN, D. E. (1989) Enzymes in brewing and distilling. In: *Cereal Science and Technology*, PALMER, G. H. (Ed.) Aberdeen University Press.

BECKER, R. and HANNERS, G. D. (1991) Carbohydrate composition of cereal grains. In: *Handbook of Cereal Science and Technology*, LORENZ, K. and KULP, K. (Eds) Marcel Decker Inc. NY. U.S.A.

BERRY, C. S. (1988) Resistant starch — a controversial component of 'dietary fibre' *BNF Nutr. Bull.* 54: 141–152.

BOCK, M. A. (1991) Minor constituents of cereals. In: *Handbook of Cereal Science and Technology*, pp. 555–594, LORENZ, K. J. and KULP, K. (Eds) Marcel Decker Inc. NY.

BOWLER, P., TOWERSEY, P. J., WAIGHT, S. G. and GALLIARD, T. (1985) Minor components of wheat starch and their technological significance. In: *New Approaches to Research on Cereal Carbohydrates*, pp. 71–79, HILL, R. D. and MUNCK, L. (Eds) Elsevier, Amsterdam.

BOYER, C. D. and SHANNON, J. C. (1983) The use of endosperm genes for sweet corn improvement. In: *Plant Breeding Reviews*, Vol. 1., pp. 139–161, JANICK, J. (Ed.) Avi Publishing Co. Westport, CT. U.S.A.

BUSHUK, W. (1966), Distribution of water in dough and bread. *Bakers' Digest* **40** (October): 38–40.

BUTTROSE, M. S. (1962) The influence of environment on the shell structure of starch granules. *J. Cell Biol.* **14**: 159–167.

COULTATE, T. P. (1989) *Food, The Chemistry of its components.* 2nd edn. Roy. Soc. of Chem. Letchworth.

CRAIG, S. A. S, and STARK, J. R. (1984) The effect of physical damage on the molecular structure of wheat starch. *Carbohydr. Res.* **125**: 117–125.

EVERS, A. D. (1971) Scanning electron microscopy of wheat starch. 3. Granule development in endosperm. *Stärke* **23**: 157–162.

EVERS, A. D. (1979) Cereal starches and proteins. In: *Food Microscopy*, VAUGHAN, J. G. (Ed.) Academic Press. London.

EVERS, A. D. and STEVENS, D. J. (1985) Starch damage. *Adv. Cereal Sci. Technol.* **7**: 321–349.

FINCHER, G. B., and STONE, B. A. (1986) Cell walls and their components in cereal grain technology. *Adv. Cereal Sci. Technol.* **8**: 207–295.

FRENCH, D. (1984) Physical and chemical organization of starch granules. In: *Starch Chemistry and Technology*, WHISTLER, R. L., BEMILLER, J. N. and PASCHALL, E. F. (Eds) Academic Press Inc NY. U.S.A.

FRETZDORF, B. and WEIPERT, D. (1990) Enzyme activities in developing triticale compared to developing wheat and rye. In: *Fifth International Symposium on Pre-Harvest Sprouting in Cereals*, RINGLUND, K., MOSLETH, E. and MARES, D. J. (Eds) Westview Press Inc. Boulder. Colorado. U.S.A.

FULCHER, R. G. and WONG, S. I. (1980) Inside Cereals — a fluorescence microchemical view. In: *Cereals for Food and Beverages*, pp. 156–164, INGLETT, G. E. and MUNCK, L. (Eds) Academic Press. NY. U.S.A.

GALLANT, D. J. and BOUCHET, B. (1986) Ultrastructure of maize starch granules. *Food Microstruct.* **5**: 141–155.

GALLIARD, T. (1983) Enzymic degradation of cereal lipids. In: *Lipids in Cereal Technology*, pp. 111–147, BARNES, P. J. (Ed.) Academic Press Inc. London.

GOERING, K. J., FRITTS, D. H. and ESLICK, R. F. (1973) A study of starch granule size distribution in 29 barley varieties. *Stärke* **25**: 297–302.

GREENWELL, P. and SCHOFIELD, J. D. (1989) What makes cereals hard or soft? *Proc. SAAFOST. 10th Congress Cereal Sci. Symp.*, Vol. 2, Natal Technikon Printers, Durban R.S.A.

HALL, D. M. and SAYRE, J. G. (1969) A scanning electron microscopy study of starch. Root and tuber starches. *Textile Res. J.* **39**: 1044–1052.

HENRY, R. J. (1985) A comparison of the non-starch carbohydrates in cereal grains. *J. Sci. Food Agric.* **36**: 1243–1253.

HOSENEY, R. C. (1984) Functional properties of pentosans in baked foods. *Food Tech.* **38** (January): 114–117.

HOSENEY, R. C. (1986) *Principles of Cereal Science and Technology.* Amer. Assn. of Cereal Chemists Inc. St Paul MN. U.S.A.

HOSENEY, R. C. and VARIANO-MARSTON, E. (1980) Pearl millet: its chemistry and utilization. In: *Cereals for Food and Beverages*, INGLETT, G. E. and MUNCK, L. (Eds) Academic Press Inc. NY. U.S.A.

JOHNSON, L. A. (1991) Corn: Production, processing, and utilization. In: *Handbook of Cereal Science and Technology*, LORENZ, K. and KULP, K. (Eds) Marcel Decker Inc. NY. U.S.A.

JULIANO, B. O. (1985) Biochemical properties of rice. In: *Rice: Chemistry and Technology*, pp 175–206, JULIANO, B. O. (Ed.) Amer Ass. of Cereal Chemists Inc., St Paul, MN. U.S.A.

JULIANO, B. O and BECHTEL D. B. (1985) The rice grain and its gross composition. In: *Rice: Chemistry and Technology*, Ch. 2, pp. 17–57, JULIANO, B. O. (Ed.) Amer. Assoc. of Cereal Chemists Inc. St. Paul MN. U.S.A.

KATES, M. (1972) *Techniques of Lipidology: Isolation, Analysis and Identification of Lipids.* Elsevier Publ. Co. Inc. NY.

KENT, N. L. (1983) *Technology of Cereals*, 3rd edn. Pergamon Press, Oxford.

KRUGER, J. E. and REED, G. (1988) Enzymes and color. In: *Wheat: Chemistry and Technology*, pp. 441–500, POMERANZ, Y. (Ed.) Amer. Ass. of Cereal Chemists Inc. St. Paul, MN. U.S.A.

LEI M.-G and REECK, C. G. R. (1986) Two dimensional electrophoretic analysis of wheat kernel proteins. *Cereal Chem.* **63**: 111–116.

LOCKHART, H. B. and HURT, H. D. (1986) Nutrition of oats. In: *Oats: Chemistry and Technology*, pp. 297–308, WEBSTER, F. H. (Ed.) Amer. Ass. of Cereal Chemists Inc., St Paul, MN. U.S.A.

MARSHALL, W. R. and WHELAN, W. J. (1974) Multiple branching in glycogen and amylopectin. *Arch. Biochem. Biophys.* **161**: 234–238.

MAURICE, T. J., SLADE, L., SIRETT, R. R. and PAGE, C. M. (1983) Polysaccharide–water interactions — Thermal behaviour of rice starch. *3rd Int. Symp. on Properties of Water in Relation to Food Quality and Stability.* Baune France.

MORRISON, W. R. (1983) Acyl lipids in cereals. In: *Lipids in Cereal Chemistry*, pp. 11–32, BARNES, P. J. (Ed.) Academic Press Inc. London.

MORRISON, W. R. (1985) Lipids in cereal starches, In: *New Approaches to Research on Cereal Carbohydrates*, HILL, R. D. and MUNCK, L. (Eds.) Elsevier Science Publ. B. V. Amsterdam.

OSBORNE, T. B. (1907) The proteins of the wheat kernel. *Publ. 84. Carnegie Inst.* Washington. D.C.

OSMAN, E. M. (1967) Starch in the food industry. In: *Starch: Chemistry and Technology*, Vol. 2, WHISTLER, R. L. and PASCHAL, E. F. (Eds) Academic Press Inc. NY. U.S.A.

PAYNE, P. I. and RHODES, A. P. (1982) Cereal storage proteins: structure and role in agriculture and food technology. *Encycl. Plant Physiol.* New Ser. **14A**: 346–369.

PERCIVAL, E. G. V., (1962) *Structural Carbohydrate Chemistry*, 2nd edn, PERCIVAL, E. G. V., (Ed.) Garnet Miller Ltd London.

ROBIN, J. P., MERCIER, C., CHARBONNIERE, R. and GUILBOT, A. (1974) Lintnerized starches. Gel filtration and enzymatic studies of insoluble residues after prolonged acid treatment of potato starch. *Cereal Chem.* **51**: 389–406.

ROONEY, L. W. and SERNA-SALDIVAR, S. O. (1991) Sorghum. In: *Handbook of Cereal Science and Technology*, LORENZ, K. and KULP, K. (Eds) Marcel Decker Inc. NY. U.S.A.

SCHOFIELD, J. D. (1986) Flour proteins: structure and functionality in baked products. In: *Chemistry and Physics of Baking*, BLANSHARD, J. M. V., FRAZIER, P. J. and GALLIARD, T. (Eds.) Roy. Soc. of Chem. London.

SHEWRY, P. R. and TATHAM, A. S. (1990) The prolamin storage proteins of cereal seeds: structure and evolution. *Biochem. J.* **267**: 1–12.

SHUEY, W. C. and TIPPLES, K. H. (1980) *The Amylograph Handbook*. Amer. Assoc of Cereal Chemists Inc. St. Paul, MN. U.S.A.

SIMMONDS, D. H. (1989) *Wheat and Wheat Quality in Australia*. CSIRO, Australia.

SNYDER, E. M. (1984) Industrial microscopy of starches. In: *Starch: Chemistry and Technology*, WHISTLER, R. L., BEMILLER, J. N. AND PASCHALL, E. F. (Eds) Academic Press Inc. N.Y. U.S.A.

SOBER, Ed. (1968) *Handbook of Biochemistry*. CRC Press Inc Boca Raton. FL. U.S.A.

SOUTHGATE, D. A. T., SPILLER, G. A., WHITE, M. and McPHERSONKAY, R. (1986) Glossary of dietary fiber components. In: *Handbook of Dietary Fibre in Human Nutrition*, SPILLER, G. E. (Ed.) CRC Press Inc., Boca Raton. FL. U.S.A.

SOUTHGATE, D. A. T. (1986) The Southgate method of dietary fiber analysis. In: *Handbook of Dietary Fibre in Human Nutrition*, SPILLER, G. E. (Ed.) CRC Press, Inc. Boca Raton FL. U.S.A.

WATSON, S. A. (1987) Structure and composition. In: *Corn: Chemistry and Technology*, pp. 53–82, WATSON, S. A. and RAMSTAD, P. E. (Eds) Amer. Assoc. of Cereal Chemists Inc. St Paul MN. U.S.A.

WOOD, P. J. (1986) Oat β-glucan: structure, location, and properties. In: *Oats: Chemistry and Technology*, pp. 121–152, WEBSTER, F. H. (Ed.) Amer. Assoc. of Cereal Chemists Inc. St. Paul, MN. U.S.A.

WOODWARD, J. R. and FINCHER, G. B. (1982) Substrate specificities and kinetic properties of two (1–3, 1–4)-β-D-glucan hydrolases from germinating barley (*Hordeum vulgare*). *Carbohydrate Res.* **106**: 111–122.

WRIGLEY, C. W. and BIETZ, J. A. (1988) Proteins and amino acids. In: *Wheat: Chemistry and Technology*, 3rd edn, Vol. 1, pp. 159–275, POMERANZ, Y. (Ed.) Amer. Assoc. of Cereal Chemists Inc. St. Paul MN. U.S.A.

YOUNGS, V. L. (1986) Oat lipids and lipid related enzymes. In: *Oats: Chemistry and Technology* , pp. 205–226, WEBSTER, F. H. (Ed.) Amer. Assoc. of Cereal Chemists Inc. St. Paul, MN. U.S.A.

ZEE, S.-Y. and O'BRIEN, T. P. (1970) Studies on the ontogeny of the pigment strand in the caryopsis of wheat. *Aust. J. Biol. Sci.* **23**: 1153–1171.

Further Reading

ALAIS, C. and LINDEN, G. (1991) Cereals — Bread. In: *Food Biochemistry*, Ch. 10, pp. 119–129. Ellis Horwood.

ALEXANDER, R. J. and ZOBEL, H. F. (1988) *Developments in Carbohydrate Chemistry*. Amer. Assoc. of Cereal Chemists Inc. St. Paul MN. U.S.A.

BANKS, W. and GREENWOOD, C. T. (1975) *Starch and its Components*, Edinburgh Univ. Press. Edinburgh.

FOWDEN, L. and MIFLIN, B. J. (Eds.) (1983) *Seed Storage Proteins*. Roy. Soc. London.

GUNSTONE, F. D. (1992) *A Lipid Glossary*. The Oily Press Ltd, Ayr.

HOFFMANN, R. A., KAMERLING, J. P. and VLIEGENTHART, J. F. G. (1992) Structural features of a water-soluble arabinoxylan from the endosperm of wheat. *Carbohydrate Res.* **226** (2): 303–311.

IZYDORCZYK, M., BILADARIS, C. G. and BUSHUK, W. (1991) Physical properties of water-soluble pentosans from different wheat varieties. *Cereal Chem.* **68**: 145–150.

KRUGER, J. E., LINEBACK, D. and STAUFFER, C. E. (Eds.) (1987) *Enzymes and their Role in Cereal Technology*. Amer. Assoc. of Cereal Chemists Inc. St. Paul MN. U.S.A.

MAROUSIS, S. N., KARATHANOS, V. T. and SARACACOS, G. D. (1991) Effect of physical structure of starch materials on water diffusivity. *J. Fed. Process. Preserv.* **15**: 161–166.

MITCHELL, J. R. and LOCKWARD, D. A. (Eds.) *Functional Properties of Food Macromolecules*. Elsevier, Appl. Sci. Publ. London

RITTENBURG, J. H. (1990) *Development and Application of Immunoassay for Food Analysis*. Elsevier Science Publishers Ltd Barking.

4

Cereals of the World: Origin, Classification, Types, Quality

Wheat

Origin

The cultivation of wheat (*Triticum* spp.) reaches far back into history, and the crop was predominant in antiquity as a source of human food. It was cultivated particularly in Persia (Iran), Egypt, Greece and Europe. Numerous examples of ancient wheat have been unearthed in archaeological investigations; the grains are always carbonized, although in some cases the anatomical structure is well preserved.

Races and species of wheat

The many thousands of known species and varieties of the genus *Triticum* (wheat) can be grouped into three distinct races which have been derived from separate ancestors, and which differ in chromosome numbers. One classification of the races, with the probable wild types, chromosome numbers (2n) and the cultivated forms is shown in Table 4.1.

Einkorn, emmer and spelt are husked wheats, i.e. the lemma and palea form a husk which remains around the kernel after threshing. Emmer was used for human food in prehistoric times (cf. p. 95); there is archaeological evidence that it was grown about 5000 B.C. in Iraq. The principal wheats of commerce are varieties of the species *T. aestivum*, *T. durum* (cf. p. 86) and *T. compactum* (cf. p. 85).

Common or Bread Wheat, *T. aestivum* (hexaploid 2n = 42) is an allopolyploid; three genomes, each corresponding to a normal diploid set of chromosomes, are distinguishable and are known to have had separate origins in the past.

TABLE 4.1
Wild and Cultivated Wheat Types

Race	Wild type	2n	Cultivated forms	
			Species name	Common name
Small spelt	*T. aegilopoides*	14	*T. monococcum*	Einkorn
Emmer	*T. dicoccoides*	28	*T. dicoccum*	Emmer
			T. durum	Macaroni wheat (durum)
			T. polonicum	Polish
			T. turgidum	Rivet, Cone
Large spelt	} probably	42	*T. aestivum*	Bread wheat
Dinkel	} *T. monococcum*		*T. spelta*	Dinkel, Spelt
	× *T. speltoides*		*T. compactum*	Club
	× *Ae. squarrosa*		*T. sphaerococcum*	Indian Dwarf

78

The hexaploid wheats are believed to have arisen by hybridization of the diploid species *T. tauschii (Aegilops squarrosa)* with the tetraploid hybrid of *T. monococcum* and *T. speltoides (Ae. speltoides)*, two diploid species, with doubling of the chromosomes. Evidence points to the *spelta* group as being the oldest of the hexaploids, with the *aestivum* group having arisen from the *spelta* group by mutation of a single gene, and the *compactum* and *sphaerococcum* groups having arisen from *aestivum* similarly by the mutation of single genes. The evidence is presented by Quisenberry and Reitz (1967).

Cultivated varieties, which are of widely differing pedigree and are grown under varied conditions of soil and climate, show wide variations in characteristics.

The climatic features in countries where spring wheat is grown — maximum rainfall in spring and early summer, and maximum temperature in the mid- and late-summer — favour production of rapidly maturing grain with endosperm of vitreous texture and high protein content, traditionally suitable for breadmaking. Winter wheat, grown in a climate of relatively even temperature and rainfall, matures more slowly, producing a crop of higher yield and lower nitrogen content, better suited for biscuit and cake-making than for bread, although in the U.K., where winter wheat comprises about 96% of the total (cf. pp. 87), winter wheat is used for breadmaking. The yield of durum wheat (cf. p. 86), which is grown in drier areas, is lower than that of bread wheat.

The high-yielding Indian wheats, developed for use in the Green Revolution (cf. p. 6), make chapatties (cf. p. 270) of indifferent quality (Brown, 1972).

Wheat types

In a general way, wheats are classified according to (1) the texture of the endosperm, because this characteristic of the grain is related to the way the grain breaks down in milling, and (2) the protein content, because the properties of the flour and its suitability for various purposes are related to this characteristic.

Vitreous and mealy wheats

The endosperm texture may be vitreous (steely, flinty, glassy, horny) or mealy (starchy, chalky). Samples may be entirely vitreous or entirely mealy, or may consist of a mixture of vitreous and mealy grains, with one type predominating. Individual grains are generally completely vitreous or completely mealy, but grains which are partly vitreous and partly mealy ('piebald' or 'metadiné') are frequently encountered. The specific gravity of vitreous grains is generally higher than that of mealy grains: 1.422 for vitreous, 1.405 for mealy (Bailey, 1916).

The vitreous or mealy character is hereditary, but is also affected by environment. Thus, *T. aegilopoides, T. dicoccoides, T. monococcum* and *T. durum* are species with vitreous kernels, whereas *T. turgidum* and many varieties of *T. compactum* and *T. aestivum* are mealy (Percival, 1921). However, the vitreous/mealy character may be modified by cultural conditions. Mealiness is favoured by heavy rainfall, light sandy soils, and crowded planting, and is more dependent on these conditions than on the type of grain grown. Vitreousness can be induced by nitrogenous manuring or commercial fertilizing and is positively correlated with high protein content; mealiness is positively correlated with high grain-yielding capacity.

Vitreous kernels are translucent and appear bright against a strong light, whereas mealy kernels are opaque and appear dark under similar circumstances. The opacity of mealy kernels is an optical effect due to the presence of minute vacuous or air-filled fissures between and perhaps within the endosperm cells. The fissures form internal reflecting surfaces, preventing light transmission, and giving the endosperm a white appearance. Such fissures are absent from vitreous endosperm.

The development of mealiness seems to be connected with maturation, since immature grains of all wheat types are vitreous, and vitreous grains are found on plants that grow and ripen quickly: spring wheats, and those growing in dry continental climates. Mealy grains are characteristic of varieties that grow slowly and have a long maturation period.

Vitreous kernels sometimes acquire a mealy appearance after being conditioned in various ways, e.g. by repeated damping and drying, or by warm conditioning (cf. p. 122). The proportion of vitreous kernels in a sample is a characteristic used in grading some types of U.S. wheat (cf. p. 85).

Hard and soft wheats

Wheat types may also be classified as hard or soft, and as strong or weak (see below). Vitreous grains tend to be hard and strong, mealy grains to be soft and weak, but the association is not invariable.

'Hardness' and 'softness' are milling characteristics, relating to the way the endosperm breaks down. In hard wheats, fragmentation of the endosperm tends to occur along the lines of the cell boundaries, whereas the endosperm of soft wheat fractures in a random way. This phenomenon suggests a pattern of areas of mechanical strength and weakness in hard wheats, but fairly uniform mechanical weakness in soft wheat. One view is that 'hardness' is related to the degree of adhesion between starch granules and the surrounding protein, viz. that differences in endosperm texture must be related to differences in the nature of the interface between starch granules and the protein matrix in which they are embedded (Barlow et al., 1973; Simmonds, 1974). The interface was shown to be rich in water-extractable proteins, although no specific biochemical component that might control the adhesion between starch granules surface and protein matrix was identified (Simmonds et al., 1973).

Well-washed prime starch separated from wheat endosperm contains some 0.15–0.2% by weight of protein — 'starch granule protein' (SGP) — and the SGP comprises about 1% of the total protein of the grain. Part of the SGP is located on the surface of the granules, the remainder is an integral part of the granule structure. Characterization of the SGP by polyacrylamide gel electrophoresis in the presence of sodium dodecyl sulphate has shown that the surface SGP comprized five polypeptides of low relative molecular mass (M_r) from 5 to 30 k, while the integral SGP also comprised five polypeptides, but of higher M_r: 59–149 k. It has also been found that one particular polypeptide of the surface SGP, with M_r 15 k, is strongly present in all soft wheats examined, but only weakly present in hard (T. aestivum) wheats, and completely absent from the very hard durum (T. durum) wheats. Hardness/softness of wheat endosperm is known to be genetically controlled (Berg, 1947) by a gene, Ha, which is located on the 5D chromosome, and it was subsequently discovered that the gene controlling the formation of the M_r 15 k protein is also situated on the 5D chromosome, and is either identical with the Ha gene, or is located quite close to it. The M_r 15 k protein has been named 'friabilin' and it appears to act as a 'non-stick' agent, since it is strongly present in soft-endosperm wheats that fragment easily and at random, but is only weakly present in, or entirely absent from, hard wheats that fragment only with difficulty, and generally not along the interface between starch granule and surrounding protein matrix (Schofield and Greenwell, 1987; Greenwell, 1987; Greenwell and Schofield, 1989).

Hard wheats yield coarse, gritty flour, free-flowing and easily sifted, consisting of regular-shaped particles, many of which are whole endosperm cells, singly or in groups. Soft wheats give very fine flour consisting of irregular-shaped fragments of endosperm cells (including a proportion of quite small cellular fragments and free starch granules), with some flattened particles, which become entangled and adhere together, sift with difficulty, and tend to clog the apertures of sieves (cf. p. 144). The degree of mechanical damage to starch granules produced during milling is greater for hard wheats than for soft (cf. pp. 62 and 149).

Hardness affects the ease of detachment of the endosperm from the bran. In hard wheats the endosperm cells come away more cleanly and tend to remain intact, whereas in soft wheats the sub-aleurone endosperm cells tend to fragment, a portion coming away while the rest remains attached to the bran.

The granularity of flour gives a measure of the relative hardness of the wheat, the proportion of the flour passing through a fine flour silk (when milled under standard conditions) decreasing with increasing hardness. Greer (1949) found that the percentage of the total flour passing through a No. 16 standard silk (aperture width: 0.09 mm) under standard conditions was 49–56% for four related varieties of hard English wheat, whereas it was 63–71% for ten unrelated varieties of soft English wheat. The granularity of flour can also be expressed as the Particle Size Index, as determined by means of an Alpine air-jet sieve. Ease of sifting, however, is affected by other factors besides hardness of endosperm, e.g. moisture content (cf. p. 152).

The principal wheats of the world are arranged according to their degree of hardness as follows:

extra hard: Durum, some Algerian, Indian;

hard: CWRS (Manitoba), American HRS, Australian Prime Hard;

medium: Plate, Russian, some Australian, American HRW, some European;

soft: Some European, some Australian, American SRW, American Soft White.

Protein content

The protein content of wheat varies over a wide range (6–21%) and is influenced less by heredity than by edaphic factors — soil and climatic conditions — prevailing at the place of growth, and by fertilizer treatment. Ranges of protein content encountered among samples of various wheat types are shown in Table 4.2.

TABLE 4.2
Protein Content Ranges of Wheat Types*

Wheat type	Approximate protein range (%)
HRS (United States)	11.5–18
Durum	10–16.5
Plate (Argentina)	10–16
CWRS (Manitoba)	9–18
HRW (United States)	9–14.5
Russian	9–14.5
Australian	8–13.5
English	8–13
Other European	8–11.5
SRW (United States)	8–11
White (United States)	8–10.5

* Sources: Schruben (1979); Kent-Jones and Amos (1947).

Protein content *per se* is not a factor determining milling quality, except in so far as the protein content tends to be higher in vitreous than in mealy wheats, and vitreousness is often associated with hardness and good milling quality. Samples of the English soft wheat varieties Riband or Galahad may have high protein content and a large proportion of vitreous grains and yet mill as soft wheats; on the other hand, a low-protein, predominantly mealy-grained sample of the hard varieties Hereward or Mercia will mill as a hard wheat. The protein content of the endosperm — its quality and its chemical structure — is, however, a most important characteristic in determining baking quality (cf. p. 66).

Strong and weak wheats

Wheats yielding flour which has the ability to produce bread of large loaf volume, good crumb texture, and good keeping properties (cf. pp. 174 and 192) generally have a high protein content and are called 'strong', whereas those yielding flour from which only a small loaf with coarse open crumb texture can be made, and which are characterized by low protein content, are called 'weak'. The flour from weak wheats is ideal for biscuits (cookies) and cakes, although unsuitable for breadmaking unless blended with stronger flour.

Flour from strong wheats is able to carry a proportion of weak flour, i.e. the loaf maintains its large volume and good crumb structure even when a proportion of weak flour is blended with it; it is also able to absorb and retain a large quantity of water.

The main types of wheat are classified according to their baking strength as follows:

strong: CWRS (Manitoba), American HRS, Russian Spring, some Australian;

medium: American HRW, Plate, S.E. European, Australian Prime Hard;

weak: N.W. European, American SRW, American Soft White, Australian Soft.

Hardness (milling character) and strength (baking character) are inherited separately and independently (Berg, 1947). Hence, it should be possible, through breeding, to combine good milling quality with, for example, the type of gluten associated with weak wheats, to produce a good milling biscuit wheat. The varieties Slejpner, Haven and Maris Huntsman are hard but possess no particular baking strength; Minaret and Flanders are soft wheats which are generally acceptable for breadmaking.

Grain size and shape

The maximum yield of white flour obtainable from wheat in milling is ultimately dependent upon the endosperm content, and the latter is affected by the size and shape of the grain, and by the thickness of the bran.

The specific (bushel) weight (bu wt) measurement (test wt per bu in the U.S.A.; hectolitre wt or natural wt in Europe) estimates the weight of a fixed volume of grain, and gives a rough indication of kernel size and shape. Wheats of high bu wt are usually considered to mill the more readily and to yield more flour. However, these measurements can be misleading, as soft mealy wheats often have high bu wt. Moisture content also affects bu wt.

Shellenberger (1961) found that the volumetric bran content is lower in large than in small grains, viz. 14.1% and 14.6%, respectively, from samples of the same types of wheat, showing the economic importance of large kernel size.

World wheats

U.K.-grown and imported wheats cover a wide range in quality and characteristics. In an endeavour to produce flour of regular quality, the British flour miller makes a 'milling grist' by blending together various types of wheat so that particular properties lacking in one component of the grist may be provided by another.

In recent years, the sources of imported wheat used by U.K. flour millers have been more restricted than formerly, and comprise principally the EC countries (mainly France) and Canada. Of the total wheat milled in the U.K. in 1978–1980 (5 million t per an.), 56% was home-grown, 7% was other EC wheat, and 37% imported, non-

EC (mainly Canadian). Proportionately more 'imported, non-EC' wheat is used for breadmaking and less (possibly none) for making biscuits and cakes, for household flour, and for flour for other purposes. Thus, in 1978–1980 the average composition of the grist used for breadmaking in the U.K. was 41% home-grown, 9% other EC, and 50% imported, non-EC.

However, since 1981 there has been a considerable reduction in the quantity of imported wheat used for breadmaking (for reasons explained in Ch. 8, p. 193). Thus, in 1990/91, while the total wheat milled in the U.K. comprised 87% home-grown, 6% other EC, and 7% imported non-EC, the average composition of the breadmaking grist was approximately 79% home-grown, 10% other EC, and 11% imported non-EC (see also Fig. 8.1).

America

The principal wheat-producing countries of the American continent are Canada, U.S.A. and Argentina.

Canada

The wheat grown in western Canada is segregated into five classes according to season of sowing (spring or winter), grain colour (red or white), grain texture (hard or soft), and species (*T. aestivum* or *T. durum*). Spring-sown types predominate — over 95% of the Canadian wheat crop is spring sown — and there is more red than white, and more hard than soft. The classes are further described as follows.

Canada Western Red Spring (CWRS)

CWRS wheat — formerly known as 'Manitoba' — is grown in the provinces of Manitoba, Saskatchewan and Alberta. It is a high protein wheat of excellent milling and baking qualities. It can be used alone, or in blends with lower-protein wheat, to produce hearth breads, noodles, flat breads and steam breads. CWRS wheat is marketed in three grades, of which the primary grade characteristics are shown in Table 4.3. The minimum content of hard, vitreous kernels is 65% for

TABLE 4.3
Grade Characteristics of Canada Western Red Spring Wheat and Canada Utility Wheat, 1991★

Grade	Minimum test weight		Maximum limits (%) of			
			Foreign material		Wheat of other classes	
			Other than		Contrasting	
	lb/bu	kg/hl	cereal grains	Total	classes	Total
Canada Western Red Spring						
No. 1 CWRS	60.1	75.0	0.2	0.75	1.0	3.0
No. 2 CWRS	57.7	72.0	0.3	1.5	3.0	6.0
No. 3 CWRS	55.3	69.0	0.5	3.5	5.0	10.0
Canada Western Utility						
No. 1 CWU	60.1	75.0	0.3	2.0	3.0	10.0
No. 2 CWU	57.7	72.0	0.5	4.0	5.0	20.0
CW Feed	no min.		1.0	10.0	no limit, but not more than 10% amber durum	

★ Source: Official Grain Grading Guide, 1991 edition. Canadian Grain Commission, Winnipeg.

No. 1 CWRS, and 35% for No. 2 CWRS. No minimum is prescribed for No. 3 CWRS.

The protein content of CWRS wheat ranges 9–18%, but Nos 1 and 2 CWRS can be supplied at guaranteed minimum protein levels of 11.5%, 12.5%, 13.5% and 14.5% (on 13.5% m.c. basis). The average yield in 1989 was 18 q/ha (22 q/ha in 1986), the yield and protein content tending to be related inversely. The moisture content at harvest is usually 11–13%, and for inclusion in the straight grades the moisture content must not exceed 14.5%. Wheat with 14.6–17.0% m.c. is graded 'tough'; with 17.1% m.c. or over, 'damp'.

Marquis is a variety of CWRS wheat which was grown extensively for a long period. It was bred by crossing Red Fife, a good milling variety which was liable to frost damage, with Hard Red Calcutta, an Indian early-ripening variety. Marquis inherited the good milling and early ripening characters of its parents. However, Marquis was susceptible to rust (cf. p. 7) and has been largely replaced by other varieties, such as Thatcher, bred from the double cross (Marquis × Iumillo) × (Marquis × Kanred), and Selkirk, bred from (McMurachy × Exchange) × Redman. The variety Selkirk is equal to Marquis in breadmaking quality and, in addition, is resistant to stem rust race 15B. In 1958 Selkirk comprised over 80% of the spring wheat area in Manitoba and 28% of that grown in Saskatchewan, but by 1969 it was being grown on only 6% of the wheat area in western Canada.

The grading (Table 4.3) includes a maximum limit for 'wheat of other classes'. Until recently, 'wheat of other classes' meant 'wheat of classes or varieties not equal to Marquis', but Marquis was replaced as the standard variety by Neepawa in 1987. Neepawa was bred from (Thatcher × Frontana) × (Thatcher × Kenya Farmer) × (Thatcher × Frontana-Thatcher), and is superior to Marquis in resistance to stem rust. The variety of CWRS most widely grown at present is Katepwa: it is resistant to stem rust (*Puccinia graminis tritici*) and common root rot, and moderately resistant to leaf rust (*Puccinia triticina*) and loose smut, though moderately susceptible to speckled leaf disease (*Septoria*).

The Export Standards for the CWRS grades, shown in Table 4.4, are somewhat stricter than the primary grade characteristics, shown in Table 4.3.

Early frosts may reduce the yield of grain and lower its milling quality by increasing the proportion of small shrivelled grains with low endosperm content, and adversely affect baking quality, because the milled flour is of high maltose content and produces a flowy dough.

Canada Western Utility (CWU) wheat and Feed wheat

Red Spring Wheat which does not attain the required standards for Nos 1–3 CWRS may be

TABLE 4.4
Export Standards For Canadian Wheats, 1991–1992*

	Minimum test weight		Foreign material		Wheat of other classes	
			Other than		Contrasting	
Grade	1b/bu	kg/hl	cereal grains	Total	classes	Total
Canada Western Red Spring						
No. 1 CWRS	62.5	78.0	0.2	0.4	0.3	1.5
No. 2 CWRS	61.7	77.0	0.2	0.75	1.5	3.0
No. 3 CWRS	60.1	75.0	0.2	1.25	2.5	5.0
Canada Western Red Winter						
No. 1 CWRW	62.5	78.0	0.2	1.0	1.0	3.0
No. 2 CWRW	59.3	74.0	0.2	2.0	2.0	6.0
No. 3 CWRW	55.3	69.0	0.2	3.0	3.0	10.0
Canada Western Soft White Spring						
No. 1 CWSWS	62.5	78.0	0.2	0.75	—	1.5
No. 2 CWSWS	60.9	76.0	0.2	1.0	—	3.0
No. 3 CWSWS	60.1	75.0	0.2	1.5	—	5.0
Canada Prairie Spring						
No. 1 CPS	61.7	77.0	0.2	0.75	3.0	5.0
No. 2 CPS	60.1	75.0	0.2	1.5	5.0	10.0
Canada Western Utility						
No. 1 CWU	62.5	78.0	0.2	1.0	3.0	5.0
No. 2 CWU	60.9	76.0	0.2	2.0	5.0	10.0
Canada Western Amber Durum						
No. 1 CWAD	64.2	80.0	0.2	0.5	2.0	3.0
No. 2 CWAD	63.8	79.5	0.2	0.8	2.5	5.0
No. 3 CWAD	62.5	78.0	0.2	1.0	3.5	7.0
No. 4 CWAD	56.9	71.0	0.5	3.0	10.0	15.0
Canada Western Feed Wheat						
	59.3	74.0	0.5	5.0	—	—

* Source: Grains from Western Canada, 1991–1992. The Canadian Wheat Board, Winnipeg.

graded as Nos 1 or 2 Canada Western Utility wheat or as Feed Wheat, the grade characteristics for which are shown in Table 4.3, and the Export Standards in Table 4.4. Canada Western Feed Wheat may contain up to 10% of Amber Durum and up to 5% of heat-damaged grains. CWU wheat is a medium protein red wheat with hard kernel characteristics. It can be used to make pasta or, when blended with other wheat, to make bread. Acceptable reference varieties for CWU wheat are Glenlea, Wildcat and Bluesky. The CW Feed Wheat, as its name implies, is suitable, alone or blended with other grains, for feeding to animals.

Canada Western Red Winter (CWRW)

CWRW wheat is grown in southern Alberta. It is a medium protein, strong wheat with hard

kernel characteristics, suitable for the production of French-style hearth breads and of noodles, flat breads and steam breads. Export Standards are listed in Table 4.4.

Canada Western Soft White Spring (CWSWS)

CWSWS wheat is grown under irrigation in the southern regions of western Ontario and in British Colombia. The protein content of the top grades normally ranges from 9.0 to 10.0% (14% m.c. basis), and the flour is suitable for making cakes, biscuits and crackers, and also, either alone or in blends, for making flat bread and noodles. See Table 4.4.

Canada Prairie Spring (CPS)

CPS wheat is a relatively new class of Canadian wheat, now available as CPS White wheat and

CPS Red wheat. It is a semi-hard wheat with 12–13% protein content (13.5% m.c. basis) of excellent milling quality, suitable for making French-type breads and, in blends, for making noodles, steam breads, pan breads, crackers and related products. See Table 4.4.

Canada Western Amber Durum (CWAD)

CWAD wheat, a tetraploid (2n = 28), has a particularly hard grain and is milled to provide semolina (cf. p. 154) for making pasta products (see Ch. 10). The flour is unsuitable for bread-making. CWAD wheat is graded into four grades according to test weight, content of foreign material, other cereal grains and seeds, content of wheat of other classes and varieties of durum not equal to Hercules, and content of immature kernels (see Table 4.4).

The wheat known as Canada Eastern is low in protein content (about 9%): it is suitable for high ratio cake flour (cf. p. 178) and for biscuits, when mixed with more extensible wheats. Being low in diastatic power, it is also suitable for sausage rusk, which requires flour of low maltose content and high absorbency.

United States

Five principal types of wheat are grown in the U.S.A.: their names, the proportion that each contributed to the total crop in 1991/92, and their bu wt ranges are shown in Table 4.5.

TABLE 4.5
*Characteristics of U.S. Wheat Types**

Type	Proportion (1991/92 crop) %	Bushel wt† (lb)	Hectolitre wt† (kg)
Hard Red Winter (HRW)	44	62–64	77.3–79.8
Soft Red Winter (SRW)	17	60–64	74.8–79.8
Hard Red Spring (HRS)	23	63–64	78.6–79.8
White	11	61–63	76.1–78.6
Durum	5	63–64	78.6–79.8

* Data from Dahl (1962), Johnson (1962), Shellenberger (1961), *Milling and Baking News* (19 Feb. 1980), U.S. Wheat Reviews, HGCA (1991).
† 1lb/bu = 1.247 kg/hl.

Hard Red Winter (HRW)

HRW wheat is grown in Texas, Oklahoma, Kansas, Colorado, Nebraska, Montana, South Dakota and California. HRW wheat is used for making yeasted bread and hard rolls. Grading of HRW wheat, according to content of dark, hard and vitreous kernels, as formerly practised, was officially discontinued in December, 1979.

Soft Red Winter (SRW)

SRW wheat is grown in Missouri, Illinois, Ohio, Indiana, Arkansas and Michigan States. That grown east of the Great Plains region is called Red Winter; the remainder is Western Red. SRW wheat mills well but is a weak wheat, low in protein content. The flour is used for biscuits (cookies) and crackers, cakes and pastries.

Hard Red Spring (HRS)

HRS wheat is grown in Minnesota, North Dakota, Montana and South Dakota. The milling quality is only slightly inferior to that of CWRS, and the protein content is comparable. HRS wheat is graded according to content of dark, hard and vitreous kernels into Dark Northern Spring (75% or more), Northern Spring (25–75%) and Red Spring (less than 25%). HRS wheat is used for quality yeasted bread and rolls.

White wheat

This includes hard and soft types. Hard White must contain 75% or more of hard kernels. Soft White, also known as Pacific White in Britain, contains less than 75% of hard kernels. White wheat is grown in the west coast States and in Michigan and New York States. The flour from White Wheat is unsuitable for breadmaking, but is ideal for biscuits (cookies), crackers, cakes and pastries. The Soft White is similar to Canada Eastern White wheat, but slightly stronger and higher in protein content. The diastatic power tends to be low.

Club wheat

Club wheat or *Triticum compactum*, is a hexaploid white wheat grown principally in Washington

State, U.S.A. The flour milled from it is similar in characteristics to that milled from white wheat of *T. aestivum*, and is used for making biscuits (cookies).

Durum wheat

Durum wheat or *Triticum durum*, is a tetraploid wheat grown in North Dakota (85% of the crop in 1990) and California, and is used for making pasta products (see Ch. 10). Hard Amber Durum has 75% or more of hard and vitreous kernels of amber colour; Amber Durum has 60–75%; Durum has less than 60%.

U.S. wheat grading

The main classes of wheat are graded according to test wt per bu, content of damaged and shrunken kernels, foreign material and wheat of other classes, as shown in Table 4.6. Parcels which do not fall within the limits of grades U.S. Nos. 1–5, or are of low quality in certain respects, and durum wheat of over 16% m.c., are designated 'sample grade'. 'Ergoty wheat' contains more than 0.3% of ergot (cf. p. 15). 'Smutty wheat' contains more than 30 smut balls per 250g (cf. p. 8). 'Garlicky wheat' contains more than 2 green garlic bulbils per kg. 'Infested wheat' is wheat infested with live weevils or other insects injurious to stored grain (cf. p. 106). 'Treated wheat' has been scoured, limed, washed, sulphured, or treated in such manner that the true quality is not reflected by either the numerical grade or sample grade designation alone (*Federal Register*, 30 June, 1987).

The (U.S.) Food and Drug administration announced in July 1977 that grain is subject to seizure if it contains 32 or more insect-damaged kernels per 100 g or 9 mg or more rodent pellets and/or fragments of rodent excreta pellets per kg. Wheat so adulterated may be used for animal feed but not for human food (*Federal Register*, 12 July, 1977).

Protein testing

The U.S. Federal Grain Inspection Service authorized the official protein testing of HRW and HRS wheats under the Grain Standards Act on 1 May 1978 and for all other classes of U.S. wheat except mixed and unclassed from 1 May 1980. Protein content will be determined either by a near-infra red reflectance (NIR) method or by the Kjeldahl method (*U.S. Wheat Review*, May 1980).

U.S. wheat surplus

The surplus of wheat left over at the end of the season in the U.S.A. when the new crop is coming in is known as the 'carry over'. Since 1952 the carry over has greatly increased, and by 1962 amounted to 37 million tonnes, considerably more than one entire U.S. wheat harvest (31.3 million t in 1962). Subsequently the carry over deceased, partly because of wheat-area restriction, and partly because of increased exports to

TABLE 4.6
Grade Characteristics of U.S. Wheat*

Grade	Minimum test wt				Maximum limits (%) of						
	HRS, White Club		Other classes		Heat-damaged kernels	Damaged kernels (total)	Foreign material	Shrunken and broken kernels	Defects (total)	Wheat of other classes	
	kg/hl	lb/bu	kg/hl	lb/bu						Contrasting	Total
U.S. No. 1	72.3	58	74.8	60	0.2	2	0.4	3	3	1	3
U.S. No. 2	71.1	57	72.3	58	0.2	4	0.7	5	5	2	5
U.S. No. 3	68.6	55	70.0	56	0.5	7	1.3	8	8	3	10
U.S. No. 4	66.1	53	67.3	54	1.0	10	3.0	12	12	10	10
U.S. No. 5	62.3	60	63.6	51	3.0	15	5.0	20	20	10	10

* Source: U.S. Dept. Agric. (1991).
1 lb/bu = 1.247 kg/hl.

TABLE 4.7
Composition of U.S. Wheat Carry Over* (million bu)†

Year	HRW	SRW	HRS	Durum	White	Total
1976/77	605	72	250	92	93	1112
1979/80	451	39	317	78	89	974
1989/90	297	40	216	60	81	694
1991/92	151	46	117	50	50	414

* Sources: *Milling and Baking News*, 19 February 1980; 15 August, 1989; 17 December, 1991.
† 1 million bu of wheat = 27,223 metric tonnes of wheat.

countries, such as the former Soviet Union, India and China, which had suffered loss of crops. In million tonnes, the carry over was 22.3 for 1969, 9.2 for 1973/74, 32.0 for 1977/78, 18.9 for 1989/90 and 11.3 for 1991/92. Composition of the carry over for 1976/77, 1979/80, 1989/90 and 1991/92, by wheat type, is shown in Table 4.7.

Argentina

Argentina is the main producer of bread wheat in South America. The wheat is classified as Hard Red Winter, and is known as 'Plate' wheat. The grain is hard, red and semi-vitreous, small, thin and elongated in shape. In quality, the wheat is strong, with about 12% protein content, but the gluten has limited extensibility, and the wheat is low in diastatic power, and suitable only as a filler in breadmaking grists.

The types of Plate wheat are named after the ports of shipment: Rosafé, grown in the north around Rosario, and Santa Fé are shipped from Rosario; Baril, grown in the central area, from Buenos Aires; Barusso, grown in the south, from Bahia Blanca. Barusso is a softer wheat, with lower protein content, than the other types.

Australia

Wheat is grown in the relatively high rainfall areas of New South Wales, Victoria, South Australia and Queensland. There are two classes: hard and soft wheats. The hard type is of medium strength, and is suitable as a filler in breadmaking grists. The soft type is weak, and the flour is good for biscuits and pastry production. Australian wheat exports are mostly of the soft type and are marketed mainly in the Mid- and Far-East for

making noodles and ethnic breads. The grains are large, brittle and slightly elongated, with thin bran of white or yellow colour. The moisture content is seldom above 11%, and the grain is millable up to 16% m.c. Nos 1 and 2 Prime Hard wheat are traded with guaranteed protein levels of 13.0%, 14.0% and 15.0% (natural m.c. basis). Attempts to raise the protein content by undersowing with clover have produced promising results.

Former Soviet Union

Good-quality Hard Red Spring, Hard Red Winter and Durum wheats are produced in the former Soviet Union. The grains of the HRS and HRW wheats are small, red, hard and horny, and the wheat may contain frosted grains and grains damaged by wheat bug (cf. p. 9). The big climatic variation within the former Soviet Union gives rise to a wide range of wheat quality from very strong to medium strong. Average Russian wheat is weaker than CWRS in baking strength and is suitable as a filler-wheat. The protein averages 12%.

Britain

Winter and spring types are grown, although winter wheat comprised about 97% of the total in 1991. Average yield of wheat in the U.K. in 1991 was 72 q/ha (H-GCA, 1991). The bran colour is generally red. Much of the grain is harvested at relatively high moisture contents (16–20%) and needs drying. Varieties grown in 1991 that were suitable for inclusion in breadmaking grists were Avalon, Camp Remy, Mercia and Urban (winter wheats); Alexandria, Axona and Tonic (spring wheats), all of which have hard endosperm texture. Varieties most favoured for use in biscuit grists were the winter varieties Apollo, Beaver, Brock, Galahad, Hornet, Longbow, Norman and Riband, all of soft endosperm texture. In 1989, the varieties Mercia and Avalon made up about 37% of total seed sales, Galahad, Apollo and Brock about 23% of the total.

The protein content may range from 8% to 13%, according to locality, within one season. At any one locality, the average range of protein contents among varieties would be about 2%, the higher-yielding varieties having the lower protein contents.

Alpha-Amylase activity of wheat is not a factor determining milling quality. Nevertheless, flour milled from wheat with excessively high activity of alpha-amylase may not be suitable for baking, and therefore the flour miller generally applies the Falling Number (FN) test (formerly known as the Hagberg Falling Number test; cf. p. 183) to his wheat as a guide to the suitability of the milled flour for further processing. He would reject parcels giving FN of less than 180 (based on a 7 g sample) for milling into flour for breadmaking and for cream crackers, or less than 140 for flour for biscuits.

Grading

There is no official grading scheme for wheat in Britain. Wheat is described as 'millable' or 'non-millable', but the large variation in quality within the former class is not recognized by any official system of price support. However, since 1971, if supplies of high quality wheat are short, the larger milling combines have offered financial premiums for high quality wheat of selected varieties, the level of such premiums varying according to supply and demand, but probably averaging 10% over the price of 'other milling wheat' for wheat of 11% protein content (on 14% m.c. basis), or 7.5% premium for 10.5–10.9% protein content, provided the moisture content does not exceed 16% and the FN is 200 or more.

A voluntary grading scheme, or marketing guide, for U.K.-grown milling wheat was inaugurated by the Home-Grown Cereals Authority in January 1975. The 1991 edition of the H-GCA's 'Marketing Guide' specifies the following requirements for Milling wheat:

— freedom from objectionable smells, pest infestation, discoloured grains, ergot and other injurious material;
— not overheated during drying or storage;
— moisture content not exceeding 15% or 16% (according to contract);
— content of admixture and screenings not exceeding 2%;
— pesticide residues within the limits prescribed by U.K. or EC legislation.

In addition, the Marketing Guide classifies U.K.-grown wheat varieties as 'favoured for breadmaking', 'others — hard' (which may be incorporated in some grists or used for animal feed), and 'others — soft' (which can be suitable for biscuit-making, used in other grists, or for animal feed). The Guide recommends a minimum protein content of 11% (on 14% m.c. basis) for breadmaking wheat, although 10.5% protein content may be acceptable in some years, and a Falling Number of 250 or more (on 7 g sample). For biscuit-making, the Guide specifies a soft-milling wheat with a low capacity for water absorption, a protein content of 9–10%, and a Falling number of 140 or more.

The requirements specified for Standard Feed Wheat by the H-GCA are as follows: the wheat shall be suitable for use as animal feed, be of typical colour, free from objectionable smell and free from living pests; 16% max m.c.; 68 kg/hl min. sp.wt; 0.05% max. ergot content and 3% max. miscellaneous impurities; 5% max. other cereals, and 12% max. total impurities (which include, besides those already mentioned, sprouted, broken, shrivelled grains, and grains damaged by heat or insects).

The National Institute of Agricultural Botany (headquarters at Cambridge) undertakes thorough examination of new wheat varieties and issues recommendations to British farmers. The milling and baking qualities of contemporary varieties are assessed by the Flour Milling and Baking Research Association: a section of their 1991 edition of 'Classification of Home-Grown Wheat Varieties' (by Stewart, Susan Salmon and Lindley) is shown in Fig. 4.1.

Variety	Endosperm texture	Flour yield	Flour colour	Water absorption	Bread-making value	Biscuit value	Country of origin	Parentage
WINTER WHEATS								
Admiral*	Soft C	C	C	C	(-)	B	UK	Mithras x Hobbit — Hedgehog line
Apollo	Soft D	D	C	C	C	A	Ger	Maris Beacon x Kronjuwel
Apostle	Hard B	C	A	A	B	D	UK	(Alcedo x Avalon) x Moulin
Avalon	Hard B	B	C	A	B	D	UK	Maris Ploughman x Bilbo
Axial*	Soft C	B	A	C	C	C	Fr	[(Talent x Maris Beacon) x Arminda] x Festival
Beaver	Soft D	C	D	D	D	A	UK	(Hedgehog x Norman) x Moulin
Boxer	Hard B	C	B	A	B	C	UK	Griffin x RPB 181/70D
Brigand	Soft D	C	C	C	D	B	UK	Maris Huntsman x Bilbo
Brimstone	Hard B	A	B	B	B	C	UK	(54/218 x Hobbit 30/2) x Hustler
Brock	Soft C	B	B	C	D	A	UK	Hobbit 30/2 x Talent
Crest	Soft C	C	A	C	D	B	UK	F_1 x [5878 x (Corin x Stuart)]
Dean	Hard B	B	B	A	B	D	UK	Disponent x Norman
Estica*	Hard B	C	B	B	B	(-)	Neth	Arminda x Virtue
Fenman	Soft C	B	C	C	D	B	UK	[(Maris Ranger x Durin) x Maris Beacon] x Hobbit 'sib'
Flanders	Soft D	B	A	C	B	C	Fr	Champlein x FD281/348
Flint	Hard A	C	C	B	D	D	UK	F_1 x [5878 x (Corin x Stuart)]
Foreman	Soft C	B	C	D	D	C	UK	Norman 'sib' x Disponent
Fortress	Hard B	B	C	B	C	D	UK	Wizard x F_1 3575
Fresco	Hard B	A	A	A	C	D	UK	Moulin x Monopol
Galahad	Soft C	C	C	C	D	C	UK	(Joss Cambier x Durin) x Hobbit 'sib'
Gawain	Soft D	C	C	C	D	C	UK	Durin derivative x Brigand
Haven	Hard B	B	D	B	D	C	UK	(Hedgehog x Norman) x Moulin
Hereward*	Hard B	A	A	B	B	D	UK	Norman 'sib' x Disponent
Hornet	Soft D	C	C	C	D	B	UK	Norman x Hedgehog
Huntsman	- see Maris Huntsman							
Hussar*	Hard B	B	C	(-)	(-)	(-)	UK	Squadron x Rendezvous
Kador	Hard	A	B	B	C	D	Fr	(Champlein x Cappelle-Desprez) x B21
Longbow	Soft C	B	B	C	D	B	UK	[(Maris Ranger x Durin) x Maris Beacon] x Hobbit 'sib'
Mandate	Soft C	D	C	D	D	C	UK	Norman x Hedgehog
Maris Huntsman	Hard B	D	C	C	D	C	UK	[(CI 12633 x Cappelle-Desprez) x Hybrid 46] x Professeur Marchal
Maris Widgeon	Hard	A	B	B	A	D	UK	Holdfast x Cappelle-Desprez
Mercia	Hard B	B	A	B	B	D	UK	(Talent x Virtue) x Flanders
Norman	Soft C	B	C	C	D	B	UK	[(Maris Ranger x Durin) x Maris Beacon] x Hobbit 'sib'
Parade	Hard B	C	B	A	B	D	UK	Granta x Marksman
Pastiche*	Hard B	A	A	A	B	D	UK	Jena x Norman
Rendezvous	Hard B	B	B	A	B	D	UK	(VPM x Hobbit 'sib') x Virtue
Riband	Soft C	B	B	D	D	B	UK	Norman x (Maris Huntsman x TW161)
Sitka	Soft C	C	B	(-)	(-)	(-)	UK	Flambeau x TJ8989/4
Sleipner	Hard B	B	C	B	C	C	Swe	(W20102 x CB149) x Maris Huntsman x Bilbo
Soldier	Hard B	B	C	B	(-)	D	UK	Squadron x Rendezvous
Soleil	Hard A	B	A	A	C	D	Fr	LD 339 (Jass x Top)
Stag	Soft C	C	B	(-)	(-)	(-)	UK	Galahad x Hammer
Talon*	Hard B	B	A	A	B	D	Ger	[Huntsman x Seva x (NS 732 x Huntsman)]
Tara*	Soft C	A	C	D	D	C	UK	(Clement x Marksman) x Brock
Torfrida*	Hard B	B	C	A	B	D	UK	Rendezvous x (Moulin x Mercia)
Urban	Hard A	A	B	A	B	D	Ger	Diplomat x Kranich
Virtue	Hard B	B	C	B	D	D	UK	Maris Huntsman x Durin
Wasp	Soft C	C	B	C	(-)	B	UK	Galahad x Boxer
Widgeon	- see Maris Widgeon							
SPRING WHEATS								
Alexandria	Hard B	C	C	A	B	D	Neth	Bastion x Mironovskaja
Axona	Hard A	B	C	A	B	C	Neth	HPG 522/66 x Maris Dove
Baldus*	Hard A	A	A	B	B	(-)	Neth	Sicco x [(N66 x MGH6537) x Kolibri]
Canon	Hard A	A	B	A	B	D	Swe	[(Sicco x 125022) x Sappo] x Kadett
Harlequin	Hard A	A	A	B	C	(-)	UK	TW275 x TAS 706/19/2
Isis	Hard A	C	B	A	B	(-)	Fr	69-C-10 x 235-25
Tonic	Hard A	D	C	A	B	D	UK	RPB 87-73 x RPB 94-73
Touchstone	Hard A	B	B	A	B	(-)	UK	Norman x Musket
Troy*	Hard A	A	A	B	B	(-)	UK	TAS 894/5/3 x Sicco

nb: The classifications are based on mean results from at least five sites over several years; individual results may vary widely.

(-)   indicates that the variety was not tested for that character

* Trials information on these varieties is not yet complete

FIG 4.1 Classification of British-grown wheat varieties according to their value for milling and baking purposes, by B. A. Stewart, Susan E. Salmon and J. A. Lindley (Flour Milling and Baking Research Association, 1991) reproduced by kind permission of the FMBRA. The characteristics in the second to seventh columns are graded on a four-point scale, A–D, A indicating superior, D inferior, qualities. In the fifth column, A indicates greater, D smaller, water absorption by the flour.

Utilization

About 3% of British-grown wheat is used for seed and about 3% for industrial purposes; the remainder is shared almost equally between flour-milling and animal feeding requirements. Approximately 5 million tonnes of wheat are milled annually in the U.K. Utilization of British-grown and imported wheat milled in the U.K. in 1990/91, according to purpose, was approximately as shown in Table 4.8.

TABLE 4.8
Wheat Milled in the U.K. 1990/91* (percentages of Total Wheat Milled)

Purpose	British grown	Imported EC	Imported Other	Total
Bread	50	6	7	63
Biscuits	15	—	—	15
Cakes	2	—	—	2
Household flour	6	—	—	6
Other	14	—	—	14
All	87	6	7	100

* Sources: MAFF, NABIM, H-GCA.

Other Western European wheat

Wheat from France, Germany, Belgium and the Baltic States is similar to U.K.-grown wheat, but is harvested somewhat drier (15–17% m.c.) and is a little stronger.

France

Production of wheat in France averaged 9 million tonnes per annum during 1950/59, the harvested area averaging 4.2 million ha, giving an average annual yield of 21.4 q/ha. By 1989, production had increased to 32 million tonnes on an area only a little larger, 5.0 million ha, because yield had increased to 63.7 q/ha. Total storage capacity in 1978 was 34.6 million tonnes, of which 31% was on farms and 69% with co-operatives and merchants.

A system for classifying and grading French wheat was introduced in 1969. The price of wheat for domestic consumption is fixed according to its class and grade. There are four classes, for which the minimum requirements are as shown in Table 4.9.

TABLE 4.9
Minimum Quality Requirements for French Wheat Classes

Class	Protein* (%)	Zeleny Sedimentation test†
I	13	38
II	12	28
III	11	18
IV	10	13

* Dry basis.
† See p. 183.

Within each class, the wheat is graded into one of three grades. The quality requirements for Grades 1 and 2 are shown in Table 4.10. Grain which does not conform to these requirements, but reaches the required standards for intervention (see below), is 'hors grade', i.e. outside grade.

Utilization

Of the 11 million tonnes of wheat used domestically in 1987/88 in France, about 9% was used for seed, about 1% for industrial purposes, and the remainder equally divided between human foods and animal feed (45% each).

EC grain standards

Wheat grown in the countries comprising the European Community (EC) may be offered into Intervention (i.e. purchase by a national intervention authority). The intervention authority is committed to buy, in unlimited quantities, any lots of the cereal offered which meet the minimum quality and quantity requirements.

A three-tier intervention system operates in the U.K., wheat of breadmaking quality being purchased at a 'reference price' which is fixed 5–7.5% above the intervention price for feed wheat. Breadmaking wheat and feed wheat offered into intervention in the U.K. must be of sound, fair and marketable quality (defined as being of typical colour of the grain, free from abnormal smell and free from live pests (including mites)), and must meet the quality standards set out in Table 4.11. Breadmaking wheat is subdivided into Premium Wheat and Common Wheat according

TABLE 4.10
Quality Requirements for French Wheat Grades

	Minimum limits		Maximum limits		
Grade	Hectolitre weight (kg/hl)	Hagberg Falling Number	Moisture content (%)	Broken kernels (%)	Total impurities (%)
1	76	200	15.5	4	2
2	75	150	16.0	5	3

TABLE 4.11
U.K. Standards for Wheat, Barley and Rye for Intervention, 1990–1991*

	Breadmaking wheat		Feed wheat	Rye	Barley
	Premium	Common			
Minimum quantity, t	100	100	100	100	100
Maximum moisture content, %	14.5	14.5	14.5	14.5	14.5
Minimum specific wt, kg/hl	72	72	72	68	63
Maximum total impurities, %	10	10	12	12	12
of which:					
Broken grains	5	5	5	5	5
Grain impurities,	10	10	12	5	12
of which:					
Other cereals and grains					
damaged by pests	5	5	5	—	5
Grain overheated during drying	0.5	0.5	3	3	3
Sprouted grains	6	6	6	6	6
Miscellaneous impurities,	3	3	3	3	3
of which:					
Noxious seeds	0.1	0.1	0.1	0.1	0.1
Damaged by spontaneous overheating	0.05	0.05			
Ergot	0.05	0.05	0.05	0.05	
Minimum protein content, % (d.b.)	14.0	11.5	N/A		
Minimum Zeleny test	35	20	N/A		
Minimum Hagberg Falling Number	240	200	N/A		
EC dough machinability test	Pass	Pass	N/A		

* Source: Intervention Board, 1990.
N/A not applicable.

to quality, as measured by protein content, Zeleny sedimentation test and Hagberg Falling Number. The Intervention buying-in price for Common Wheat is about 5% above that for Feed Wheat; that of Premium Wheat about 7% above that for Feed Wheat (Intervention Board, 1990).

South-eastern Europe

Wheat from Bulgaria and Romania is somewhat harder and stronger than that from the west of Europe and can be used as a filler in breadmaking grists.

World wheat stocks

Between 1952 and 1970 a large surplus of wheat gradually built up in wheat-exporting countries until, by 1970, 64 million tonnes (one-fifth of the annual world production at that time) were available for export. Since 1970, stocks have rapidly diminished, and by mid 1973 there was an acute shortage; prices, in consequence, rose steeply. The average market price for wheat in England and Wales, which was about £23 per tonne in 1970, rose to about £58 per tonne in 1973, and has continued to rise, reaching about £100 per tonne in 1980, and £126 per tonne in 1990/91 (Common Breadmaking Wheat:

Intervention buying-in price) (Intervention Board, 1990).

The causes of the shortage in the 1970s were:

— cut back on planted areas of wheat, particularly in Canada and the U.S.A.;
— increased demand for wheat for animal feed to meet increased meat consumption in developed countries;
— disappointingly slow implementation of the potentialities of the so-called 'Green Revolution' in Asia — the use of new high-yielding varieties (cf. p. 6);
— a series of natural disasters of which the most far reaching was the Russian crop failure in 1972. The former Soviet Union bought 18 million tonnes of wheat (30% of world stocks) to make good its shortfall. In the same year, China and India also bought large quantities of wheat.

Wheat condition and quality — for the miller

Condition

For the miller, wheat in good condition is:

1. *Of good appearance*: the grains are normal in colour (not discoloured) and bright; unweathered, free from fungal and bacterial diseases, unsprouted, and free from musty odour.
2. *Undamaged*: the grains are not mechanically damaged by the thresher, by insect infestation or by rodent attack, and have not been damaged by over-heating during drying;
3. *Clean*: the grain is free from admixture with an abnormal quantity of chaff, straw, stones, soil, or with weed seeds and grains of other cereals, or of other types or varieties of wheat; the grains should be completely free from admixture with ergot, garlic or wild onion, bunt, rodent excreta and insects;
4. *Fit for storage*: the moisture content should not exceed 16% for immediate milling, or 15% if the wheat is to go into storage (but see p. 104).

Tests for grain *condition* therefore examine the appearance and odour of the grain, measure the degree of contamination with impurities, estimate the moisture content, and investigate whether the grain has suffered heat damage to the protein during farm drying. Tests for heat damage include estimates of the amount of soluble, i.e. undenatured, protein (Every, 1987), the germination test, and the tetrazolium test (cf. p. 113).

Quality

Besides these four aspects of *condition* — which are dependent mainly upon the agricultural history of the wheat before the miller received it — the miller also wants the wheat to be of good *milling quality*, that is, to perform well on the mill: to give an adequate yield of flour, to process easily, and to yield a product of satisfactory quality.

The quality of wheat on the mill is measured by the yield and purity of the flour obtained from it. Relatively more flour of lower mineral content (ash yield) and lower grade colour (GC: cf. p. 184) is obtained from good milling wheats, when properly conditioned (cf. Ch. 5) and milled under standard conditions, than from poor milling wheats.

Purity of the flour (white flour) means relative freedom from admixture with particles of bran. Bran is dark coloured whereas the endosperm is white: the Grade Colour (GC) of the flour is an index of the degree of bran contamination (cf. p. 184). Another measure of bran contamination is the yield of ash obtained upon incineration, as bran yields more ash than endosperm does (cf. pp. 184 and 170).

The yield and purity of white flour are dependent upon the way the endosperm separates from the bran when the wheat is ground; on the toughness of the bran, i.e. its resistance to fragmentation; and on the friability of the endosperm and the ease with which the flour is sifted. All these characteristics are related to grain texture and to the type of wheat (cf. p. 80).

Tests which measure baking strength and which give indication of the suitability of the flour for various purposes are described later (p. 184).

Barley

Origin

The cultivation of barley (*Hordeum* spp; cf. p. 11) reaches far back into history; it was known to the ancient Egyptians, and grains of six-row barley have been discovered in Egypt dating from pre-dynastic and early dynastic periods. Barley is mentioned in the Book of Exodus in connection with the ten plagues.

Barley was used as a bread grain by the ancient Greeks and Romans. Hunter (1928) illustrates Greek coins dating from 413 to 50 B.C. which incorporate ears or grains of barley into their design. Barley was the general food of the Roman gladiators who were known as the *hordearii* (Percival, 1921). Calcined remains of cakes made from coarsely ground grain of barley and *Triticum monococcum*, dating from the Stone Age, have been found in Switzerland.

Bread made from barley and rye flour formed the staple diet of the country peasants and the poorer people of England in the 15th century (cf. p. 95), while nobles ate wheaten bread. As wheat and oats became more generally available, and with the cultivation of potatoes, barley ceased to be used for breadmaking. Barley is still a staple grain, however, in the Near East.

Classification

There are three main types of barley: (a) hulled, six-row; (b) hulled, two-row; and (c) hull-less, all of which have been regarded as sub-groups of one species, *Hordeum sativum*, or *H. vulgare*. The characteristics of hulled barley are described in Ch. 2, p. 41.

Hull-less barley

Hull-less or naked types of barley are cultivated extensively in southeast Asia. The yield of grain is lower than that of the hulled types, and the spikelets have a tendency to shed grain when ripe, thus further reducing the yield. Hull-less types have weaker straw and are more liable to lodge than are the hulled types. The absence of hull makes them unsuitable for malting (cf. p. 219), but they are useful for food, having a higher digestibility (94%) than the hulled types (83%). In 1971, 0.36 million tonnes of hulled barley and 0.14 million tonnes of naked barley were grown in Japan, and approximately 0.17 million tonnes of hulled and 0.14 million tonnes of hull-less were used for food.

'Waxy' barley

A type of barley in which nearly all the starch is in the form of amylopectin (cf. p. 57), and known as 'waxy' barley, is known. It was of interest as a possible replacement for tapioca starch, but has not been a commercial success.

High amylose barley

A high amylose barley, in which about 40% of the starch is in the form of amylose (as compared with a usual figure of 25–27% in most cereal genotypes), is also known: it would be useful in the manufacture of malt whisky. The malt enzymes digest amylose to glucose completely (cf. p. 67) but digestion of amylopectin is incomplete, being hindered near the cross linkages: the yield of alcohol in the yeast fermentation of malted barley is thus related to the amylose content of the starch. The high amylose cultivar *Glacier* is being hybridized with other barley cultivars in order to introduce the high amylose factor into more suitable cultivars.

U.S. grades of barley

Barley, in the U.S.A., is defined as grain that, before removal of dockage, consists of 50% or more of kernels of cultivated barley (*Hordeum vulgare* L. or *H. distichum* L.) and not more than 10% of other grains for which standards have been established under the United States Grain Standards Act. The term 'barley' as used in these standards does not include hull-less barley (*Federal Register*, 1985).

Barley grown in the U.S.A. is divided into three subclasses, as follows:

TABLE 4.12
U.S. Grades of Barley *

	Minimum limits of			Maximum limits of							
	Test	Suitable malting	Sound	Damaged Kernels		Wild	Foreign	Other	Skinned and broken	Thin	Black
Grade No.	wt (1 lb/bu)†	type (%)	barley (%)	Total‡ (%)	Heat (%)	oats (%)	material (%)	grains (%)	kernels (%)	barley (%)	barley (%)
Sub-class Six-rowed Malting barley and Six-rowed Blue Malting barley											
U.S. No. 1	47	95	97	2	0.2		1	2	4	7	0.5
U.S. No. 2	45	95	94	3	0.2		2	3	6	10	1
U.S. No. 3	43	95	90	4	0.2		3	5	8	15	2
Sub-class Two-rowed Malting barley											
U.S. No. 1 choice	50	97	98		0.2	1	0.5		5	5	0.5
U.S. No. 1	48	97	98		0.2	1	0.5		7	7	0.5
U.S. No. 2	48	95	96		0.2	2	1		10	10	1
U.S. No. 3	48	95	93		0.2	3	2		10	10	2
Sub-class Six-rowed barley, Two-rowed barley and the class Barley											
U.S. No. 1	47		97	2	0.2		1		4	10	0.5
U.S. No. 2	45		94	4	0.3		2		8	15	1
U.S. No. 3	43		90	6	0.5		3		12	25	2
U.S. No. 4	40		85	8	1		4		18	35	5
U.S. No. 5	36		75	10	3		5		28	75	10

* Source: U.S. Dept. Agric. (1989).
† 1 lb/bu = 1.247 kg/hl.
‡ Includes heat-damaged kernels.

1. *Malting barley*: six-row or two-row barley, which has 90% or more of the kernels with white aleurone layer, and which is not semi-steely in the mass;
2. *Blue malting barley*: six-row barley, which has 90% or more of the kernels with blue aleurone layer, but otherwise as Malting barley;
3. *Barley*: barley that does not meet the requirements of the other sub-classes. Each sub-class is graded into numerical grades. The grade requirements are shown in Table 4.12.

The term 'U.S. Sample Grade' is applied to barley which does not meet the requirements of the numbered grades, which contains a quantity of smut so great that one or more of the grade requirements cannot be determined accurately, or contains more than seven stones or more than two crotolaria seeds (*Crotolaria* spp.) per kg, or has a musty, sour or commercially objectionable foreign odour except smut or garlic odour, or contains the seeds of wild brome grasses, or is heating or otherwise of distinctly low quality.

U.K. standards for intervention

Barley offered into Intervention in the U.K. (i.e. offered for purchase by the government; cf. p. 90) must attain the standards shown in Table 4.11.

In 1988/89 in the U.K. the winter varieties Igri and Magic provided 22% of the total barley seed certifications (*H-GCA Cereal Statistics*, 1989).

Oats

Origin

Two species of oats, *Avena sativa* L. and *A. byzantina* C. Koch, both hexaploids, are economically important since they provide the majority of present-day cultivated varieties. In Europe, types of *A. sativa* predominate. It is believed (Coffman, 1946) that a wild hexaploid species, *A. sterilis* L., is the progenitor of *A. fatua* L. (a weed popularly known as the 'wild oat') and of *A. byzantina* (red oat), and that *A. sativa* (white oat) has been derived from the latter. In his studies in 1916, the Russian botanist Vavilov discovered plants of *A. sativa* in patches of Emmer wheat (*Triticum dicoccum*; cf. p. 78) in areas of the Middle East where oats are not cultivated, and concluded that the spread of common cultivated types of *A. sativa* from their original centre, probably in the Mediterranean basin, had been determined largely by the spread of Emmer wheat, the oats being carried as a weed as the Emmer spread northwards. Eventually, in a harsher climate, the oats came to dominate the wheat (Jones, 1956). The first oat crops seem to have been grown in Europe at about 1000 B.C. Oats as a crop reached America in A.D. 1602.

Naked oats, *A. nuda* L. (hexaploid) is a type of oats which readily loses its husk during threshing. It is believed to have been derived from the tetraploid species *A. barbata*. Naked oats has considerable potential for food and feed use, since the grains require no dehulling and have high protein, oil and energy values. However, naked oats are more susceptible than ordinary hulled oats to mechanical damage during threshing, and research is aimed at the breeding of types more resistant to threshing damage. *A. nuda* is grown principally in India, Tibet and China, and it is currently being grown experimentally in Wales.

U.S. grading of oats

In the U.S.A., 'oats' is defined as grain consisting of 50% or more of cultivated oats (*A. sativa* and/or *A. byzantina*), containing not more than

TABLE 4.13
U.S. Grades of Oats*

| Grade No. | Minimum limits of | | Maximum limits of | | |
	Test wt (1b/bu)†	Sound cultivated oats (%)	Heat damaged kernels (%)	Foreign material (%)	Wild oats (%)
U.S. No. 1	36	97	0.1	2	2
U.S. No. 2	33	94	0.3	3	3
U.S. No. 3	30	90	1.0	4	5
U.S. No. 4	27	80	3.0	5	10

* Source: U.S. Dept. Agric. (1987).
† 1 lb/bu = 1.247 kg/hl.

25% of wild oats and other grains for which standards have been established. Oats are classed as White, Red, Grey, Black or Mixed; each class is graded into four numerical grades and Sample grade, as shown in Table 4.13.

The term 'Sample grade' is applied to oats which do not meet the requirements of grades 1–4, or which contain more than 16% m.c., or which contain more than 0.2% (by wt) of stones, or are musty, sour, or heating, or which contain objectionable foreign odour except that of smut or garlic, or which are otherwise of distinctly low quality.

Rye

Origin

Rye, *Secale cereale*, was domesticated in about the 4th century B.C. in Germany, later in southern Europe. According to Vavilov (1926), cultivated rye has been derived from the rye grass that occurred as a weed in wheat and barley crops.

In Roman times the chief cereal crop in the south of Britain was probably wheat, but rye was introduced by Teutonic invaders who used it for making bread, and it was grown in East Anglia.

During the Middle Ages the poorer people in England ate bread made from rye, or from a rye/wheat mixture known as 'maslin' or 'meslin', or from barley and rye (cf. p. 93). In 1764, according to Ashley (1928), bread made in the north of England contained 30% of rye, that in Wales 40%. At this time, rye was still an important crop in the north of the country and, in addition, was regularly imported from Germany and Poland, where it was plentifully grown.

The native rye grown in the U.K. until about 1945 was long-strawed, gave low yields, and had high (10–13%) protein content. Since then, new varieties have been introduced from Sweden and Germany; they are short-strawed and suitable for combine-harvesting, equal to wheat in yielding ability on good land, and low (7–8%) in protein content. Tetra Petkus, for example, is a tetraploid variety (i.e. with double the normal number of chromosomes) produced in Germany by treating Petkus rye with colchicine. The grain yield is

TABLE 4.14
*U.S. Grades of Rye**

| Grade No. | Minimum test wt (lb/bu)† | Maximum limits for | | | | |
| | | Damaged kernels (rye and other grains) | | Foreign material | | |
		Total (%)	Heat damaged (%)	Total (%)	Other than wheat (%)	Thin rye (%)
U.S. No. 1	56	2	0.2	3	1	10
U.S. No. 2	54	4	0.2	6	2	15
U.S. No. 3	52	7	0.5	10	4	25
U.S. No. 4	49	15	3.0	10	6	—

* Source: U.S. Dept. Agric. (1987).
† 1 lb/bu = 1.247 kg/hl.

reported to be 115% of that of Petkus, and the variety is credited with superior winter hardiness, test wt (bushel wt; cf. p. 82) and agronomic characteristics. The open or cross pollination of rye increases the difficulty of keeping the strain pure, and therefore new seed must be raised in isolation for this purpose.

U.S. grades of rye

Only one class of rye is recognized: it is defined as any crop which, before removal of dockage (cf. p. 93), consists of 50% or more of rye and not more than 10% of other grains for which standards have been established under provisions of the U.S. Grain Standards Act.

The grades of rye recognized in the U.S.A., and their limiting characteristics, are shown in Table 4.14. For all grades, the maximum permitted content of ergot (cf. p. 15) is 0.3%. Sample grade is defined in the same way as for Sample grade barley (cf. p. 94).

Triticale

Origin

Triticale — an abbreviation of 'Triticosecale' — is a polyploid hybrid cereal derived from a cross between wheat (*Triticum*) and rye (*Secale*). It had its beginnings in the late 19th century in Europe, but intensive work began only in 1954, at the University of Manitoba. The objectives in making the cross were to combine the grain quality, productivity and disease resistance of wheat with the vigour and hardiness of rye.

There are two types of triticale: durum wheat × rye hybrid is an hexaploid (42 chromosomes), and the usual kind of triticale; bread wheat × rye hybrid is an octoploid (56 chromosomes).

Early varieties of triticale did not offer much promise for milling and baking; they resembled their rye parent more than their wheat parent, particularly in a tendency towards undesirably high activity of *alpha*-amylase. Moreover, the early types were poor yielding, and generally had shrivelled kernels. Grain yields of more recently produced types are much increased, but triticale is not yet equal to wheat on the basis of yield. Thus, average yields in 1989 in Poland were: 33.3 q/ha for triticale, 37.7 q/ha for wheat, and in Germany: 52.0 q/ha for triticale, 62.1 q/ha for wheat. Nevertheless, the world average yield of triticale in 1989, 26.6 q/ha, exceeded that of wheat, 23.6 q/ha. Further breeding will be necessary to eliminate defects which include susceptibility to lodging, low tillering capacity, endosperm shrivelling, sprouting in the ear, susceptibility to ergot (cf. p. 15) and lack of adaptability. Development of types insensitive to day length will permit the crops to be grown in lower latitudes. Types with short, stiff straw have been developed recently by introducing dwarfing genes from short-strawed wheats.

TABLE 4.15
U.S. Grades of Triticale*

| Grade | Minimum test wt (lb/bu)† | Damaged kernels | | Maximum limits of | | | |
| | | | | Foreign material | | | |
		Heat damaged (%)	Total‡ (%)	Other than wheat or rye (%)	Total§ (%)	Shrunken and broken kernels (%)	Defects (total) (%)
U.S. No. 1	48	0.2	2	1	2	5	5
U.S. No. 2	45	0.2	4	2	4	8	8
U.S. No. 3	43	0.5	8	3	7	12	12
U.S. No. 4	41	3.0	15	4	10	20	20

* Source: U.S. Dept. Agric. (1987).
† 1 lb/bu = 1.247 kg/hl.
‡ includes heat-damaged kernels.
§ Includes material other than wheat or rye.

U.S. grades of triticale

Triticale, in the U.S.A., is defined as grain that, before removal of dockage, consists of 50% or more of triticale (*Triticosecale* Wittmack) and not more than 10% of other grains for which standards have been established under the United States Grain Standards Act, and that, after removal of dockage, contains 50% or more of whole triticale (*Federal Register*, 1986).

Triticale, in the U.S.A., is graded into four numerical grades, of which the limiting characteristics are as shown in Table 4.15.

Rice

Origin

Some authorities have traced the origin of rice, *Oryza sativa*, to a plant grown in India in 3000 B.C., but the first mention of rice in history was in 2800 B.C. when a Chinese emperor proclaimed the establishment of a ceremonial ordinance for the planting of rice. Rice culture gradually spread westwards from northeastern Asia and was introduced to southern Europe in medieval times by the invading Saracens.

Classification

Rice provides up to 75% of the dietary energy and protein for 2.5 billion people in Asia. In south Asia, rice provides 68% of the total energy contribution, wheat 10% and maize 2.5%; rice also provides 69% of the total dietary protein (Juliano, 1990, quoting CGIAR,TAC).

The peoples of various countries have varying preferences for types of rice: round-grain rice is preferred in Japan, Korea and Puerto-Rico, possibly because the cooked grains are more adherent, whereas long-grained rice (e.g. Patna rice) is preferred in the U.S.A. People in most countries prefer white rice, but in India and Pakistan the preference is for red, purple or blue strains.

Classification of types of rice according to cooking properties on the basis of the final gelatinization temperature has been suggested (Juliano *et al.*, 1964). Grains of *indica* type soften rapidly on cooking and become mushy, whereas the cooked grains of *japonica* are non-sticky and well separated.

Technological problems associated with the Green Revolution concern quality of the new varieties of rice. For example, the high yielding dwarf varieties of rice contain starch of high amylose content (cf. p. 6) which tends to lower the cooking quality (Brown, 1972).

Ordinary rice is non-glutinous and has vitreous endosperm. Another type, known as glutinous, sweet or waxy rice (*Oryza sativa* L. *glutinosa*) has a chalky, opaque endosperm, the cut surface of which has the appearance of paraffin wax. The starch of ordinary rice consists of amylose and amylopectin (cf. p. 57) and gives a blue colour

with iodine. The starch from glutinous rice, however, gives a reddish-brown colour with iodine;
it contains 0.8–1.3% of amylose The amylose
content of milled rice may be classified as waxy,
1–2%; intermediate, 20–25% and high, >25%
(Juliano, 1979). Flour produced from waxy rice
is used as a thickener in processing foodstuffs as
it withstands the freeze/thaw cycling well; it is
also used for puddings and cakes.

U.S. grades of rice

Rough rice (paddy), brown rice and milled rice
are classified as long-, medium- or short-grain
types according to length/width ratio of the
grains, as shown in Table 4.16. Six numerical
grades of rough rice (paddy) are recognized, with
requirements as shown in Table 4.17. 'Sample
grade' is defined in a similar way to sample grade
oats (cf. p. 95), except that the limiting m.c. is
14%. Grades and grade requirements are also set
out for Brown Rice and Milled Rice.

Maize

Origin

Maize or Indian corn (*Zea mays*, L.) is undoubtedly a plant of American origin, since there

TABLE 4.16
U.S. Classification of Rice (Paddy) According to Length/Width Ratio*

Rice form	Length/width ratio		
	Long-grain	Medium-grain	Short-grain
Rough	3.4+	2.3–3.3	2.2 and less
Brown	3.1+	2.1–3.0	2.0 and less
Milled	3.0+	2.0–2.9	1.9 and less

* Source: Rice Inspection Handbook. Federal Grain
Inspection Service, U.S. Dept. Agric., Washington, D.C.,
1982.

is no evidence — archaeological, linguistic, pic-
torial, or historical — of the existence of maize
in any part of the Old World before 1492, whereas
cobs and ears of maize are frequently encountered
among archaeological remains in North and
South America. It is believed that maize originated in Mexico, whence it spread northwards to
Canada and southwards to Argentina, and has
subsequently been taken to Africa, India, Australia
and the warmer parts of Europe (Watson and
Ramstad, 1987).

Types

The principal types of maize are listed in Table
2.3 (p. 46).
Maize grains may be white, yellow or reddish

TABLE 4.17
U.S. Grades and Grade Requirements for Rough Rice* (Long-, Medium- and Short-grain types)

	Maximum limits of							
	Seeds and heat-damaged grains			Red rice, Damaged grains (singly or combined) (%)	Chalky grains			
Grade	Total No./500g	Heat damaged and objectionable (singly or combined) No./500g	Heat damaged No./500g		long-grained (%)	medium or short grained (%)	Other types (%)	Colour requirements
U.S. No. 1	4	3	1	0.5	1.0	2.0	1.0	white; creamy
U.S. No. 2	7	5	2	1.5	2.0	4.0	2.0	slightly grey
U.S. No. 3	10	8	5	2.5	4.0	6.0	3.0	light grey
U.S. No. 4	27	22	15	4.0	6.0	8.0	5.0	grey or slightly rosy
U.S. No. 5	37	32	25	6.0	10.0	10.0	10.0	dark grey or rosy
U.S. No. 6	75	75	75	15.0	15.0	15.0	10.0	dark grey or rosy

* Source: U.S. Dept. Agric., 1991.

in colour. White is preferred in western Europe, but yellow is preferred for poultry feeds; the reddish type is favoured in Japan.

The texture of the endosperm of maize is variable, according to the type and the region of the endosperm. The crown region of the endosperm (at the opposite end from the germ), which is light in colour, contains loosely packed starch granules with little protein, whereas the horny region (towards the base), which is more intensely coloured in yellow varieties, has smaller starch granules which are embedded in sheets of proteinaceous material.

The yellow pigment in yellow varieties is zeaxanthin, which contains 200–900 µg of beta-carotene per 100 g. The oil and protein contents of the horny endosperm are more than double those in the crown region. In the more floury types of maize the crown region predominates; the endosperm is all soft with hardly any vitreous endosperm. In some of these floury types the crown region contracts during maturation, producing a noticeable indentation. Such types, varieties of *Z.mays indentata*, are called dent maize (corn). Varieties of *Z.mays indurata*, in which the horny region predominates, are called flint maize (corn). The starches of dent corn and flint corn comprise about 73% of amylopectin and 27% of amylose.

The term 'corn' is frequently applied to the predominant local cereal, viz. wheat in England, oats in Scotland. In the U.S.A., 'corn' means specifically maize.

The starch of waxy corn (*Z.mays ceratina*) is 100% amylopectin. A high amylose type of maize is also known: its endosperm contains a mutant amylose-extender (*ae*), which increases the amylose content of the starch to 50–80%. Waxy and high-amylose types of maize are used for wet-milling (cf. p. 264); waxy maize is also used for cattle feed.

Hybrid maize

Hybrid lines of maize have been developed that give 15–20%, sometimes up to 50%, higher yields than those of inbred lines, the use of which has been almost superseded in the U.S.A. Thus,

hybrid corn comprised less than 1% of the total crop in the U.S.A. in 1933, but by 1945 it accounted for about 90% (Watson and Ramstad, 1987).

Opaque-2 maize

Maize mutants known as opaque-2 and floury-2 contain the genes o_2 and fl_2 which confer high lysine content; floury-2 also has high methionine content. The kernels of opaque-2 maize are soft, chalky and non-transparent, and contain very little hard, vitreous, horny endosperm. They are also more liable to damage by kernel rot, insects, rodents and harvesting machinery. Opaque-2 is claimed to have 10–12% protein content (cf. 8% for normal maize), 0.49% lysine content (cf. 0.24% in normal maize), a higher niacin content and a lower leucine content than in normal maize. In feeding trials with adult humans using opaque-2 no improvement was obtained from individual supplementation of the diets with lysine, tryptophan or methionine, indicating that the amino acid composition of opaque-2 is well balanced for humans.

The yield of early types of opaque-2 maize was only 85–90% of that of normal maize, but has been improved during the 1980s through selection. However, the use of opaque-2 in the U.S.A., though nutritionally desirable, is restricted because the yield is still less than that of normal dent maize.

U.S. grades of maize

In the U.S.A., maize is classified as White, Yellow or Mixed, and is further qualified as Flint or Flint and Dent. Five numerical grades are recognized, with characteristics as shown in Table 4.18. 'Sample grade' is defined in a similar way to sample grade oats (cf. p. 95).

Sorghum

Origin

Sorghum probably originated in North Africa about 3000 B.C. and was certainly being cultivated in Egypt in 2200 B.C., as recorded in

TABLE 4.18
U.S. Grades of Maize *

Grade No.	Minimum test wt (lb/bu)†	Maximum limits of			
		Moisture content (%)	Broken corn and foreign material‡ (%)	Damaged kernels	
				Heat damaged (%)	Total (%)
U.S. No. 1	56	14.0	2.0	0.1	3.0
U.S. No. 2	54	15.5	3.0	0.2	5.0
U.S. No. 3	52	17.5	4.0	0.5	7.0
U.S. No. 4	49	20.0	5.0	1.0	10.0
U.S. No. 5	46	23.0	7.0	3.0	15.0

* Source: U.S. Dept. Agric (1987)
† 1 lb/bu = 1.247 kg/hl.
‡ Passing through a 12/64 in. (4.76 mm) round hole sieve.

wall paintings of that period. Thence it spread throughout Africa and to India and the Middle East. It reached China and America more recently.

Classification

There are four main classes of sorghum that have been bred for particular qualities: grain sorghum for kernel quality and size; sweet sorghums for stem sugar content and forage quality; broom corns for length of panicle branches and suitability of the panicle for use as brooms and brushes; grassy sorghums for forage. All the cultivated sorghums are generally known as *Sorghum vulgare* Pers., although the correct binomial terminology for these types is *Sorghum bicolor* (Linn.) Moench.

Grain Sorghum

Grain sorghum is a coarse grass which bears loose panicles containing up to 2000 seeds per panicle. It is an important crop and the chief food grain in parts of Africa, Asia, India/Pakistan, and China, where it forms a large part of the human diet.

Types of cultivated grain sorghum include kaffir corn, milo and durra (Africa), feteritas (Sudan), shallu, jowar, cholum and 'Indian millet' (India), and kaoliang (China).

A waxy sorghum is known, in which the starch is composed almost entirely of amylopectin.

A sugary type of sorghum, sugary milo, is low in starch (31.5%) but contains 28.5% of a water-soluble polysaccharide resembling phytoglycogen from sweet maize.

Bird-proof Sorghum

About one-third of the grain sorghum in South Africa is 'bird-proof' sorghum. Polyphenols or tannins in the nucellar layer in this type of sorghum are distasteful to birds and give the crops some protection. Nevertheless, the presence of tannins reduces protein digestibility; the available methionine (by *Streptococcus zymogenes* assay) was inversely related to tannin content (Hewitt and Ford, 1982). As tannins are simply inherited, they can be eliminated by breeding. Tannins in existing high-tannin varieties can be inactivated, e.g. by addition of absorbents such as polyethyleneglycol. The C.S.I.R. in South Africa has developed a process which neutralizes the polyphenols in the intact grain without affecting significantly its physiological properties.

U.S. Grades of Sorghum

Four grades of grain sorghum are recognized in the U.S.A., with characteristics as shown in Table 4.19. 'Sample grade' is defined in the same way as for sample grade oats (cf. p. 95).

Millet

Origin and Use

The millets (cf. p. 25) have been used for food and for brewing since prehistoric times in Asia, Africa and Europe. Swiss Stone-age dwellers grew proso; finger millet (ragi) was cultivated from

TABLE 4.19
U.S. Grades of Grain Sorghum

Grade No.	Minimum test wt (lb/bu)†	Moisture content (%)	Maximum limits of			
			Damaged kernels		Broken kernels‡ (%)	Foreign material (%)
			Heat damaged (%)	Total (%)		
U.S. No. 1	57	13	0.2	2	3	1
U.S. No. 2	55	14	0.5	5	5	2
U.S. No. 3	53	15	1.0	10	7	3
U.S. No. 4	51	18	3.0	15	9	4

* Source: U.S. Dept. Agric. (1991).
† 1 lb/bu = 1.247 kg/hl.
‡ Passing through a 5/64 in. (2.0 mm) triangular-hole sieve.

early times in India and Central Africa; pearl millet has similarly been used for thousands of years in Africa and the Near East. Foxtail millet is probably indigenous in China, although some authorities believe that it originated in India. For further information about the various types of millet, see Ch. 1 (p. 25).

References

ASHLEY, W. (1928) *The Bread of our Forefathers: An Enquiry in Economic History*, Clarendon Press, Oxford.

BAILEY, C. H. (1916) The relation of certain physical characteristics of the wheat kernel to milling quality. *J. agric Sci.* **7**: 432.

BARLOW, K. K., BUTTROSE, M. S., SIMMONDS, D. H. and VESK, M. (1973) The nature of the starch–protein interface in wheat endosperm. *Cereal Chem.* **50**: 443–454.

BERG, S. O. (1947) Is the degree of grittiness of wheat flour mainly a varietal character? *Cereal Chem.* **24**: 274.

BROWN, L. R. (1972) The Green Revolution and world protein supplies. *PAG Bull.* **2**(2): 25.

CANADIAN GRAIN COMMISSION (1991) *Official Grain Grading Guide, 1991*. Canadian Grain Commission Winnipeg.

CANADIAN WHEAT BOARD (1992) *Grains from Western Canada, 1991–1992*. Canadian Wheat Board, Winnipeg.

CODEX ALIMENTARIUS COMMISSION (1991) Joint FAO/WHO Food Standards Programme. ALINORM 91/29. Food and Agriculture Organization, Rome.

COFFMANN (1946) Quoted by Jones, E. T. (1956).

DAHL, R. P. (1962) Classes of wheat and the surplus problem. *Northwest Miller* **266**(8): 30.

EVERY, D. (1987) A simple, four minute protein solubility test for heat damage in wheat. *J. Cereal Sci.* **6**: 225–236.

FOOD AND AGRICULTURE ORGANIZATION OF THE UNITED NATIONS (1980) *Production Year Book 1980*. F.A.O. Rome.

GREENWELL, P. (1987) Wheat starch granule proteins and their technological significance. In: *Proc 37th Australian Cereal Chemistry Conference*, Melbourne, Royal Australian Chemical Institute.

GREENWELL, P., and SCHOFIELD, J. D. (1989) The chemical basis of grain hardness and softness. *Proc. ICC Symposium on Wheat End-Use Properties, Lahti, Finland* pp. 59–72, University of Helsinki, Finland.

GREER, E. N. (1949) A milling character of home-grown wheat. *J. agric Sci.* **39**: 125.

HEWITT, D. and FORD, J. E. (1982) Influence of tannins on the protein nutritional quality of food grains. *Proc. Nutr. Soc.* **41**: 7.

HOME-GROWN CEREALS AUTHORITY (1989) Quality and quantity requirements for Intervention. In: *Cereal Statistics—1989*, H-GCA, London.

HOME-GROWN CEREALS AUTHORITY (1991) Wheat for flour production. A marketing guide. *H-GCA Weekly Bull.* **25** (10 June).

HOME-GROWN CEREALS AUTHORITY (1991) U.S. wheat situation. *H-GCA Weekly Digest*, **18** (14 Oct.).

HOME-GROWN CEREALS AUTHORITY (1992) *Weekly Digest* **18** (13 Jan).

HUNTER, H. (1928) *The Barley Crop*. Crosby Lockwood, London.

INTERVENTION BOARD (1990) Support buying of cereals. *Leaflet* No. CD15. Intervention Board, Reading, England.

JOHNSON, J. A. (1962) *Wheat and Flour Quality*. Kansas State University, Mimeo Publication.

JONES, E. T. (1956) The origin, breeding and selection of oats. *Agric Rev.* June.

JULIANO, B. O., BAUTISTA, G. M. LUGAY, J. C. and REYES, A. C. (1964) Studies on the physico-chemical properties of rice. *Agric. Fd. Chem.* **12**: 131.

JULIANO, B. O. (1979) The chemical basis of rice grain quality. In: *Proc. Workshop on Chemical Aspects of Rice Grain Quality*, pp. 69–90. Int. Rice Res. Inst., Los Banos Laguna, Philippines.

JULIANO, B. O. (1990) Rice grain quality: problems and challenges. *Cereal Fds Wld* **35**(2): 245–253.

KENT-JONES, D. W. and AMOS, A. J. (1947) *Modern Cereal Chemistry*, 4th edn. Northern Publ. Co. Ltd, Liverpool.

MICKUS, R. R. (1959) Rice (*Oryza sativa*). *Cereal Sci. Today*, **41**: 138.

NATIONAL ASSOCIATION OF BRITISH AND IRISH MILLERS LTD, *Facts about The U.K. Milling Industry* (annually). NABIM, London.

NATIONAL INSTITUTE OF AGRICULTURAL BOTANY, *Farmers' Leaflets* (annually). N.I.A.B. Cambridge.

PERCIVAL, J. (1921) *The Wheat Plant*. Duckworth, London. (Reprinted 1975).

QUISENBERRY, K. S. and REITZ, L. P. (Eds.) (1967) *Wheat and Wheat Improvement*. Amer. Soc. Agron. Inc. Monograph No. 13 Amer. Soc. Agron. Inc., Madison, Wis, U.S.A.

RACHIE, K. O. (1975) *The Millets — Importance, Utilization and Outlook*. International Crops Research Institute for Semi-Arid Tropics, Begumpet, Hyderabad, India.

SCHOFIELD, J. D. and GREENWELL, P. (1987) Wheat starch granule proteins and their technological significance. In: *Cereals in a European Context*, pp. 407–420, MORTON, I. D. (Ed.) Ellis Horwood.

SCHRUBEN, L. W. (1979) Principles of wheat protein pricing. In: *Wheat Protein Conference*, Agric. Res. Manual ARM9, Sci. and Ed. Admin, U.S.D.A.

SHELLENBERGER, J. A. (1961) World wide review of milling evaluation of wheats. *Ass. Oper. Millers Bull.* 2620.

SIMMONDS, D. H. (1974) Chemical basis of hardness and vitreosity in the wheat kernel. *Bakers' Digest.* 48(5): 16.

SIMMONDS, D. H., BARLOW, K. K. and WRIGLEY, C. W. (1973) The biochemical basis of grain hardness in wheat. *Cereal Chem.* 50: 553–562.

STEWART, B. A., SALMON, SUSAN, E. and LINDLEY, J. A. (1991) *Milling and Baking Quality of Home-Grown Wheat Varieties*, 24th edn. Flour Milling and Baking Research Association, Chorleywood Herts, England.

UNITED STATES DEPARTMENT OF AGRICULTURE (1987) Official United States Standards for grain. *Fed. Reg.* 52(125): 24414–24442.

UNITED STATES DEPARTMENT OF AGRICULTURE (1991) Official United States Standards for grain. *Fed. Reg.* 56(63): 13424; 56(126): 29911.

UNITED STATES DEPARTMENT OF AGRICULTURE (1991) United States Standards for rice. *Fed. Reg.* 56(67): 14214.

VAVILOV, N. (1926) Studies on the origin of cultivated plants. *Bull. appl. Bot. Plant Breeding (Leningrad)* 16: 139.

WATSON, S. A. and RAMSTAD, P. E. (1987) *Corn: Chemistry and Technology*. Amer. Assoc. of Cereal Chemists Inc. St Paul MN. U.S.A.

WEBB, B. D. (1985) Criteria of rice quality in the United States. In: *Rice: Chemistry and Technology*, pp. 403–442, JULIANO, B. O. (Ed.). Amer. Assoc. of Cereal Chemists Inc. St Paul, MN. U.S.A.

Further Reading

ANDRES, C. (1980) Corn — a most versatile grain. *Food Processing*. May: 78.

BRIGGLE, L. W. (1979) Triticale. *Crop Sci.* 2: 197.

BRITTON, D. K. (1969) *Cereals in the United Kingdom: Production, Marketing and Utilization*. Pergamon Press, Oxford.

BUSHUK, W. (1976) *Rye: Production, Chemistry and Technology*. Amer. Assoc. Cereal Chemists Inc., St. Paul, MN. U.S.A.

BUSHUK, W. (1978) Wheat grading systems of the major exporting countries: Argentina, Australia, Canada, France and U.S.A. *Proc. 10th Nat. Conf. on Wheat Utilization Research*, Tucson Arizona, Nov. 1977, U.S.D.A., ARM-W.4.

CASEY, P. and LORENZ, K. (1977) Millet — functional and nutritional properties. *Bakers' Digest* Feb: 45.

COFFMAN, F. A. (1961) *Oats and oat improvement*. Amer. Soc. Agron., Madison, Wis. U.S.A.

EVANS, L. T. and PEACOCK, W. J. (1981) *Wheat Science: Today and Tomorrow*. Cambridge University Press.

GRIST, D. H. (1959) *Rice* 3rd edn. Longmans, London.

HOSENEY, R. C. (1986) *Principles of Cereal Science and Technology*. Amer. Assoc. of Cereal Chemists Inc. St. Paul, MN. U.S.A.

HOUSE, L. R. and RACHIE, K. O. (1969) Millets, their production and utilization. *Cereal Sci. Today.* 14: 92.

HULSE, J. H. LAING, E. M. and PEARSON, O. E. (1980) *Sorghum and the Millets*. Academic Press, London.

JULIANO, B. O. (Ed.) (1985) *Rice: Chemistry and Technology*. Amer. Assoc. of Cereal Chemists Inc. St. Paul, MN. U.S.A.

NATIONAL INSTITUTE OF AGRICULTURAL BOTANY. Varieties of cereals. *Farmers Leaflet* No. 8, N.I.A.B., Cambridge.

PERTEN, H. (1977) Specific characteristics of millet and sorghum milling. In: *Proc. Symposium on Sorghum and Millets for Human Food*, DENDY, D. A. V. (Ed.). Tropical Products Institute London.

POMERANZ, Y. (1971) Functional characteristics of triticale, a man-made cereal. *Wallerstein Lab. Commun.* 34: 175.

POMERANZ, Y. (Ed.) (1989) *Wheat is Unique: Structure, Composition, Processing, End-Use Properties, and Products*. Amer. Assoc. of Cereal Chemists Inc. St. Paul MN. U.S.A.

TSEN, C. C. (Ed.) (1974) *Triticale: First Man-made Cereal*. Amer. Assoc. of Cereal Chemists Inc. St. Paul MN. U.S.A.

WATSON, S. A. (Ed.) (1987) *Corn: Chemistry and Technology*. Amer. Assoc. of Cereal Chemists Inc. St. Paul, MN. U.S.A.

WEBSTER, F. H. (Ed.) (1986) *Oats: Chemistry and Technology*. Amer. Assoc. of Cereal Chemists Inc. St. Paul, MN. U.S.A.

5

Storage and Pre-processing

Storage

Compared with many other fruits, cereal grains are extremely amenable to storage, principally as their moisture content at harvest is relatively low and their composition is such that biodeterioration is slow. Harvesting is seasonal but the need for fresh cereal products is continuous. The least requirement for storage, therefore, is for the period between harvests. Under appropriate conditions this can easily be met and indeed storage for many years without serious loss of quality is possible. Even in biblical times long periods of storage were apparently achieved.

In spite of the diversity of cereal grain types and the ambient conditions throughout the cereal producing and consuming world the hazards of storage are fundamentally similar although the relative difficulties involved in their avoidance vary with location. Successful storage methods range from primitive to highly sophisticated.

The hazards besetting cereal grain storage are:

— excessive moisture content;
— excessive temperature;
— microbial infestation;
— insect and arachnid infestation;
— rodent predation;
— bird predation;
— biochemical deterioration;
— mechanical damage through handling.

The complexity of storage problems results from the matrix of interactions of the various hazards. They are considered separately in the text but their combined effects, some of which

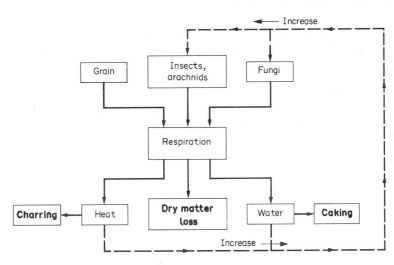

FIG 5.1 Schematic diagram of the cumulative effects of grain storage at moisture contents above the safe levels. The effects on stored grain itself are shown in **bold type**.

103

are shown in Fig. 5.1, should be borne in mind throughout.

Moisture content and storage temperature

Moisture content is expressed as a percentage of the grains' wet weight. The safe moisture contents for storage vary according to the type of cereal but it is widely assumed that they are equivalent to the equilibrium moisture content of the respective grains at 75% RH and 25°C (Table 5.1).

TABLE 5.1
Equilibrium Moisture Contents of Grains at 75% RH and 25°C

Cereal Type	Moisture %	
Barley	14.3	(25°–28°C)
Maize	14.3	
Oats	13.4	
Rice	14.0	
Rye	14.9	(25°–28°C)
Sorghum	15.3	(″)
Wheat		
durum	14.1	(″)
red	14.7	(″)
white	15.0	(″)

Based on values in Bushuk and Lee, 1978.

In temperate regions the moisture contents at which grain is stored are closer to those described as wet rather than dry. The significance of moisture contents cannot be considered alone as the deleterious effects of excessive dampness are affected critically by ambient temperature and the composition of the surrounding atmosphere. The increase in relative humidity of the interseed atmosphere with temperature, is slight. It amounts to about 0.6–0.7% moisture increase for each 10°C drop in temperature.

The relationship between moisture content and temperature as they affect storage of wheat is shown in Fig. 5.2.

The relationship depicted takes account only of the maintenance of grain quality as assessed by grain viability. The relationship is also important, however, through its effects on infesting organisms, as Fig. 5.3 shows.

The values used in Figs 5.2 and 5.3 refer

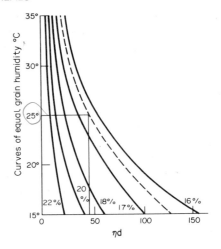

FIG 5.2 Potential storage time (number of days) of wheat grain as a function of temperature (°C) and moisture content (%), the germination rate maintained being 70%. From Guilbot, (1963) *Producteur Agricole Francais*, Suppl. mai ITCF, Paris.

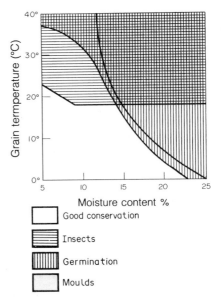

FIG 5.3 Risks to which stored cereal grain is exposed as a function of grain temperature in °C and moisture content in %. From Burges, H. D. and Burrel, M. J. (1964) *J. Sci Food Agric.* 1: 32–50.

to sound clean samples. Broken grains are almost always present to some extent as a result of damage during harvesting or transferring to stores. In broken grains endogenous enzymes and their substrates, kept separate in the whole grain, can achieve contact and lead to necrotic deterioration. Further, the most nutritious elements of the grain, endosperm and embryo, are exposed to moisture, micro-organisms and animal pests whereas in the whole grain they are protected by fruit coat, seed coats and possibly husks. Impurities can also reduce storage time in that weeds present in the crop ripen and dry at a different rate from the crop itself. Hence, still green plant material — with a relatively high moisture content — can carry excessive moisture into store even when mixed with dry grain.

Changes during storage in the grains themselves

Respiration

In a natural atmosphere gaseous exchange will occur in a stored cereal crop. This is due to respiration and it involves a depletion in atmospheric oxygen and an increase in carbon dioxide with the liberation of water, and energy (as heat). Respiration rates measured in a store normally include a major contribution from micro-organisms that are invariably present at harvest; nevertheless even ripe dry grain, suitable for storing, contains living tissues in which respiration takes place, albeit at a very slow rate. The aleurone and embryo are the tissues involved and, like other organisms present, their rate of respiration increases with moisture content and temperature. Respiration is a means of releasing energy from stored nutrients (mainly carbohydrates) and a consequence of long storage is a loss of weight. Under conditions unfavourable to respiration this may, however, be insignificant and under any circumstances it is likely to be of little consequence in relation to other storage hazards. Respiration can be reduced by artificially depleting the oxygen in the atmosphere.

Germination

Germination of grain is an essential and natural phase in the development of a new generation of plant. It involves the initiation of growth of the embryo into a plantlet. Roots develop from the radicle, and leaves and stem develop from the plumule (see Ch. 2). Hydrolytic enzymes are released into the starchy endosperm, and these catalyze the breakdown of stored nutrients into a soluble form available to the developing plantlet.

The conditions required for germination are also conducive to other, more serious, hazards such as excessive mould growth. They would rarely occur throughout a well managed store but could develop in pockets due to moisture migration. Deterioration results from loss of weight due to enzyme activity and a loss of quality resulting from excessive enzyme activity in the products of processing. These problems would apply even if the germination process were terminated through turning the grain. Having germinated in store the grain would also be useless for seed purposes as the process cannot be restarted.

Microbial infestation

Fungal spores and mycelia, bacteria and yeasts are present on the surfaces of all cereal crops. During storage they respire and, given adequate moisture, temperature and oxygen, they grow and reproduce, causing serious deterioration in grains.

A distinction may be drawn between those that attack developing and mature grain in the field and those that arise during storage. Field fungi thrive in a relative humidity (RH) of 90–100% while storage fungi require 70–90% RH. Several investigations have shown that a RH of 75% is required for germination of fungal spores (Pomeranz, 1974).

Storage fungi are predominately of the genera *Aspergillus*, of which there are five or six groups, and *Penicillium*, the species of which are more clearly defined. Some of the more common storage fungi and the minimum relative humidity in which they can thrive are listed in Table 5.2.

As with other spoilage agents dependent upon a minimum moisture content, fungi may be a problem even when the overall moisture content in the store is below the safe level. This can result from air movements leading to moisture migration. Warm air moving to a cooler area will give up moisture to grains, thus remaining

TABLE 5.2
Approximate Minimum Equilibruim Relative Humidity for Growth of Common Storage Fungi

Mould	RH(%) limit
Aspergillus halophiticus	68*
A. restictus group	70*
A. glaucus group	73*
A. chevalieri	71†
A. repens	71†
A. candidus group	80*
A. candidus	75†
A. ochraseus group	80*
A. flavus group	85*
A. flavus	78†
A. nidulans	78†
A. fumigatus	82†
Penicillium spp.	80–90*
P. cyclopium	82†
P. martensii	79†
P. islandicum	83†

* Christensen, C.M. and Kaufmann, H.H. (1974) Microflora. In: *Storage of Cereal Grain and their Products.* Christensen, C. M., (Ed.) Amer. Assoc. of Cereal Chemists Inc. St. Paul, MN. U.S.A.
† Ayerst, G. (1969). The effects of moisture and temperature on growth and spore germination in some fungi. *J. Stored Prod. Res.* 5: 127–141.
Table adapted from Bothast, R. J. (1978).

in equilibrium with them. Unless temperature gradients are extreme the exchanges occur in the vapour phase; nevertheless, variations in moisture content up to 10% within a store are possible.

If mould growth continues in the presence of oxygen, fungal respiration increases, producing more heat and water. If the moisture content is allowed to rise to 30% a succession of progressively heat-tolerant micro-organisms arises. Above 40°C mesophilic organisms give way to thermophiles.

Thermophilic fungi die at 60°C and the process is kept going by spore forming bacteria and thermophilic yeasts up to 70°C.

In recent years attention has been given to the toxic products of fungi such as *Aspergillus flavus* and *Fusarium moniliforme* which produce aflatoxin and zearalenone (see Ch. 14).

Insects and arachnids

Insects that infest stored grains belong to the beetle or moth orders: they include those capable of attacking whole grain (primary pests) and those that feed on grain already attacked by other pests (secondary pests). All arachnid pests belong to the order Acarina (mites) and include primary and secondary pests. Most of the common insects and mites are cosmopolitan species found throughout the world where grain is harvested and stored (Storey, 1987). Insects and mites can be easily distinguished as arachnids have eight legs and insects, in their most conspicuous form, have six. Reference to the most conspicuous form is necessary as some insects (including those that infest grain) develop through a series of metamorphic forms. There are four stages: the egg, the larva, the pupa and the adult or imago. Although some female insects lay eggs without mating having occurred, this is less usual than true sexual reproduction, and this and dispersion are the two principal functions of the adult. Large numbers of eggs are produced and these are very small. Those of primary pests may be deposited by the female imago inside grains, in holes bored for the purpose prior to the egg laying. Under suitable conditions eggs hatch and from each a larva

TABLE 5.3
Alphabetical List of Primary Insect and Arachnid Pests

Systematic name	Common name	Family
Acarus siro. L	Grain (or flour) mite	Acaridae
Cryptolestes ferrugineus Stephens	Rust red grain beetle	Cucujidae
Rhyzopertha dominica F.	Lesser grain borer	Bostrichidae
Sitophilus granarius L.	Grain weevil	Curculionidae
Stiophilus oryzea L.	Rice weevil	"
Sitophilus zeamais Motschulsky	Maize weevil	"
Sitotroga cerealella Olivier	Angoumois grain moth	Gelechiidae

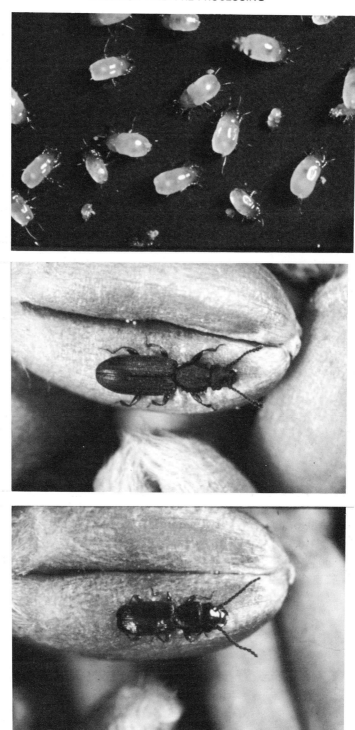

FIG 5.4 *Acarus siro*, the flour mite (top), *Sitophilus granarius*, the grain weevil, and *Cryptolestes ferrugineus*, the rust-red grain beetle (bottom). Crown Copyright Central Science Laboratory 1993.

TABLE 5.4
Alphabetical List of Most Important Secondary Insect Pests

Systematic Name	Common Name	Family
Anagasta kuehniella Zella*	Mediterranean flour moth	Phycitidae
Gadra cautella Walker	Almond moth	"
Cryptolestes pusillus Schonherr	Flat grain beetle	Cucujidae
Cryptolestes turacus Grouv	Flour mill beetle	"
Ephestia elutella Hübner	Tobacco moth	Phycitidae
Oryzaephilus surinamensis L.	Saw-toothed grain beetle	Cucujidea
Oryzaephilus mercator Fauv.	Merchant grain beetle	"
Plodia interpunctella Hübner	Indian meal moth	Phycitidae
Tenebriodes mauritanicus L.	Cadelle	Ostomatidae
Tribolium castaneum Herbst.	Red flour beetle	Tenebrionidae
Tribolium confusum Duval	Confused flour beetle	"
Trogoderma granarium Everto	Khapra beetle	Dermestidae

★ Formerly *Ephestia kuehniella*.

emerges. The larva is the form most damaging to the stored crop as it feeds voraciously. In consequence it grows rapidly, passing through a series of moults during which its soft cuticle is shed, thus facilitating further growth. Finally pupation occurs; the pupa, chrysalis or cocoon does not eat and appears inactive. However, changes continue and the final metamorphosis leads to the emergence of the adult form. The life cycle of mites is simpler as eggs hatch into nymphs which resemble the adult form, although there are only six legs present at this stage. By a series of four moults the adult form is achieved.

The time taken for development of both insects and mites is influenced by temperature, the greater the temperature the more rapid the development up to the maximum tolerated by the species.

The primary pests — those attacking whole grains — are given in Table 5.3.

The three most damaging of these pests in the U.K. are shown in Fig. 5.4.

The most important secondary insect pests — those feeding only on damaged or previously attacked grains — are given in Table 5.4.

The saw-toothed grain beetle is shown in Fig. 5.5.

FIG 5.5 *Oryzaephilus surinamensis*. Saw-toothed grain beetle. Crown Copyright. Central Science Laboratory 1993.

FIG 5.6 X-ray photograph of wheat grains, two uninfested (*bottom left*), the others showing cavities caused by insect infestation. An insect is visible within one of the cavities. (Part of a picture in *J. Photograph. Sci.*, 1954, 2: 113; reproduced by courtesy of Prof. G. A. G. Mitchell and the Editor of *Journal of Photographic Science*.)

Those insects listed in the tables are considered major pests. They are particularly well adapted to life in the grain bin and are responsible for most of the insect damage to stored grain and cereal products. Minor pests occur mainly in stores in which grain has started to deteriorate due to other causes, while incidental pests include those that arrive by chance and need not even be able to feed on grains. For further information on minor and incidental insect pests specialist works such as Christensen (1974) should be consulted.

Among the major primary pests five species develop inside grains. Weevils (grain, rice and maize) lay eggs inside while lesser grain borers and Angoumois grain moths deposit eggs outside but their newly hatched larvae promptly tunnel into grains. The presence of the insect and the damage it causes may not be evident from outside even though only a hollow bran coat may remain. Detection by means of soft X-rays is possible (Fig. 5.6).

Damage caused by insects and mites

Serious grain losses due to consumption of grain by insects and mites occurs only after prolonged storage under suitably warm conditions. They are most serious in hot climates. Other problems caused by insects include creation of hot-spots around insect populations where metabolic activity leads to local heating. Moisture movements and condensation in cooler areas results in caking, and encourages fungal infestation (see Fig. 5.1).

Introduction of insects and mites from wheat stores to flour mills can cause serious deterioration in the products. Mite excreta taints flour with a minty smell and hairs from the animals' bodies can cause skin and lung disorders in workers handling infected flour. Silk from the larvae of the Mediterranean flour moth webs together causing agglomeration of grains and blockages in handling and processing equipment. In tropical countries termites can weaken the structure of a store, leading to its collapse.

Vertebrate pests

The principal vertebrate pests in cereal stores are rodents and birds. In many countries the three main rodent species involved are:

Rattus norvegicus — the Norway, common or brown rat;
Rattus rattus   — the roof, ships or black rat;
Mus musculus   — the house mouse.

Apart from consuming grains, particularly the embryo of maize, rodents cause spoilage through their excretions which contain micro-organisms pathogenic to man. These include salmonellosis, murine typhus, rat-bite fever and Weil's disease. Rodents also damage stores' structural elements, containers, water pipes and electric cables.

In well-managed stores access by rodents is denied and good housekeeping practice, such as removal of grain spillages, maintenance of uncluttered surroundings and regular inspections, prevent problems. The same is true of birds. These are serious pests only when access is easy, as for example in hot countries where grain may be left to dry in the sun. Damage to drains and blockage of pipes by nests can give rise to secondary storage problems through promoting local dampness in some stores.

Design of storage facilities

The requirements of long term safe storage are protection against dampness caused by weather or other sources, micro-organisms, destructively high temperature, insects, rodents and birds, objectionable odours and contaminants and unauthorized disturbance. Clearly the simplest stores such as piles on the ground, unprotected, are suitable for short periods only. Other simple stores,

however, have several advantageous features. Thus underground stores provide protection from temperature fluctuations, the most successful simple ones being found in hot dry regions. They are filled, to leave little air space, and sealed, to approach the concept of hermetic storage under which insects and moulds rapidly use up oxygen, giving rise to high CO_2 content of the intergrain atmosphere. In more humid regions ventilation is desirable as the crop may have to be stored before reaching a safe moisture level. Such a system is suited to cob maize rather than threshed grains, as adequate space for air movement within the store is essential. Clearly the requirements of ventilation and exclusion of insects are not immediately compatible and hence careful design is essential.

Storage of maize as cobs is practised now largely by small scale growers producing for the requirements of the local community. It was at one time adopted more widely even in highly commercial practice, much as small grain cereals were stored unthrashed in ricks.

In the commercial context stores are needed for three purposes:

1. Holding stocks on the farm prior to sale.
2. Centralization before distribution or processing during the year following harvest.
3. Storage of annual surpluses over a longer period.

Farm stores may consist of any available space that will keep out the elements. The facilities for protection against mould and pests are very variable. Stores range from small wooden enclosures in the barn, to round steel bins holding 25–80 tonnes, to silos of larger capacities. Good on-farm storage facilities allow farmers to choose the time to sell, to receive the best prices.

The degree of centralization depends upon the marketing regime within the country of production. In North America, Country elevators and Terminal elevators with storage capacity up to 500,000 tonnes exist. The country elevators provide a local staging *en route* to terminal elevators which include high-capacity equipment for cleaning, drying and conditioning of grain. The term 'elevator' is applied to the entire facility although it refers literally to the mechanism (normally belt and bucket) by which grains are raised to a level from which they can be deposited into the large capacity silos invariably found on the sites. Elevators are associated with good transport facilities by road, rail, water or all three. Many are capable of loading grain into vessels at a rate of 2,750 t/h.

It is sometimes necessary to provide storage for grain beyond the normal capacity of an elevator facility or elsewhere. In such conditions a relatively inexpensive expedient is the flat store. This is little more than a cover for a pile of dry grain adopting its natural form as poured. Such a form is described by the angle of repose. In the case of wheat the angle is 27° to the horizontal, hence flat stores have roofs close to this angle. Very temporary stores may make use of inflatable covers.

Flat stores are easy to fill but, as they have flat floors, removal of stocks is more difficult, usually requiring the use of mechanical shovels. In contrast, silos usually have a floor formed like a conical hopper whose walls make an angle greater than 27° to the horizontal. Piles created by grains falling freely from a central spout are not uniform as whole grains tend to roll from the apex down the sloping surfaces. Small impurities and broken grains roll less readily and thus become trapped in the central core of the pile. Such a core is described as the spoutline. As the interstices can amount to 30% of the occupied space, fines in the spoutline can reach that level. Because air circulation and hence heat loss is prevented, the spoutline can be associated with early deterioration through overheating. The diameter of the spoutline is proportional to the width of the bin.

Also in contrast to tall tower-like stores, flat stores require little strength in the side walls. In a silo much of the pressure of the column of grain is borne not by the floor but by the side walls. This is because each grain rests on several grains below it so that some of the weight is distributed laterally until it reaches the walls and, by friction, rests on them. In all stores some settling occurs and this varies according to the cereal type. Wheat is relatively dense and settling may be only 6% of volume but oats may pack as much as 28%.

Settling is a continuous process arising in part from the collapse of hulls, brush hairs, embryo tips etc.

Control of pests and spoilage of grains in store

Deterioration in store is less likely if care is taken to ensure that the grain is in a suitable condition for storing. Criteria for the latter include a suitably low moisture content, a low mould count and freedom from insects. Wheat containing live insects can be sterilized by passage through an entoleter (Fig. 5.7), run at about 1450 rev/min. (BP 965267 recommends speeds of 3500 rev/min for conditioned wheat, 1700 rev/min for dry wheat.) Hollow grains and insects may be broken up and can be removed by subsequent aspiration.

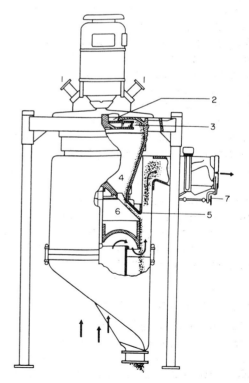

FIG 5.7 Diagrammatic section through an Entoleter Aspirator. 1. Grain inlet: 2, 3, impeller; 4, scouring cone; 5, grain discharge over 6, cone; 7, valve controlling air flow. Arrows indicate path taken by air. (Reproduced from *Milling*, 1969, Oct 10, by courtesy of the Editor.)

The store itself should provide protection from weather (particularly wet) and intrusion by insects and rodents. High temperatures are undesirable and variation should be reduced to a minimum as this can lead to local accumulation of moisture. All spoilage agents depend upon respiration and hence a depletion of oxygen inhibits their proliferation and activity. To achieve this it is necessary to provide a seal around the grain and a minimal headspace. In a sealed store oxygen depletion can be achieved by natural or artificial means. Natural depletion results from respiration which in most organisms consumes oxygen and produces carbon dioxide. Artificial atmosphere control comes about by flushing of interstitial and head spaces with a gas other than oxygen, usually nitrogen or CO_2 as these are relatively inexpensive. Complete removal of oxygen is not possible. Experiments carried out in artificial conditions in the U.K. showed that baking properties of wheats were maintained for eighteen years in low oxygen conditions. At ambient temperatures germinative energy was seriously reduced and although this reduction was prevented by storage at 5°C this was the only advantage of low temperature in addition to oxygen depletion recorded (Pixton, 1980).

Sealed conditions are unusual and prevention of spoilage in many cases depends upon careful maintaining of the stored grains' condition, and prophylactic treatments with chemicals. Fortunately, nearly all threats to grain quality cause temperature rises and monitoring of temperature, through incorporation of thermocouples, can reveal a great deal about condition.

Forced ventilation can reduce temperatures but it may be necessary to remove the cause by use of chemical treatments. Such treatments are relevant primarily when the problem is caused by insects. Because of the possible persistence of pesticides on cereals, their use in stores is increasingly becoming regarded as a last resort. In most countries strict codes of practice apply to their use. In the U.K. the legislation is contained in the Food and Environment Protection Act 1985, the Control of Pesticides Regulations 1986 and HSE Guidance Notes (E440/85) on Occupational Exposure Limits.

Pesticides used to control insects, during storage of cereal grains, are of two types. Those that are designed as a respiratory poison, and are hence applied as gas or volatile liquid, are described as *fumigants*. Those designed to kill by contact or ingestion are described as *insecticides*. They may be applied in liquid or solid form.

Of the gaseous fumigants, methyl bromide and phosphine (PH_3) are the main examples. Examples of 'liquid' fumigants are mixtures of 1,2 dichloroethane and tetrachloromethane: although the most effective fumigant is methyl bromide, this gas does not penetrate bulk grain well and the use of a carrier gas such as tetrachloromethane is an alternative to the fan-assisted circulatory systems required if methyl bromide is used alone. Few stores have the necessary fans.

The period of treatment required depends upon the susceptibility of the species of insects present to the fumigant. For example a three day exposure to phosphine may eliminate the saw-toothed grain beetle but six days at low temperature may be needed to kill the grain weevil.

'Liquid' fumigants penetrate bulks well. The proportions need to be adjusted to suit the depth of the grain stored. Up to three metres deep a 3:1 mixture of 1,2 dichloroethane:tetrachloromethane is suitable but for penetration to a depth of 50 m equal proportions are needed. Fumigation requires the stores to be sealed to prevent escape of the toxic fumes.

Pesticide residues

Some of the various types of pesticide (herbicides, fungicides, insecticides and rodenticides) used in the field or in storage, may persist in grains being processed or indeed into foods as consumed. In the U.K. the *maximum residue limits* (MRLs) permitted are mostly defined in EC Directive 86/232 and amendment 88/298, which came into force on 29 July 1988. Additional MRLs came into force in December 1988 referring to pesticides that have been refused approval in the U.K. but are used elsewhere; or those that have been consistently found in U.K. monitoring, where the limits provide a check that good agricultural practice is being observed (see Table 5.5).

TABLE 5.5
The Maximum Residue Limits for Pesticides (mg/kg) in Cereals Excluding Rice (UK)

Aldrin and Dieldrin	0.01	Captafol	0.05
Carbaryl	0.5	Carbendazim	0.5
Carbon disulphide	0.1	Carbon tetrachloride	0.1
Chlordane	0.02	*Chlorpyrifos methyl*	10
DDT (total)	0.05	Diazinon	0.05
Dichlorvos	2	Endosulfan	0.1*
Endrin	0.01	Ethylene dibromide	0.05
Entrimfos	10	*Fenitrothion*	10
Hexachlorobenzene	0.01	α and	
γ-Hexachloro cyclohexane	0.1	β-Hexachlorocyclohexane	0.02
Hydrogen cyanide	15	Heptachlor	0.01
Inorganic bromide	50	Hydrogen phosphide	0.1
Mercury compounds	0.02	*Malathion*	8
Methyl bromide	0.1	*Methacrifos*	10
Pirimiphos-methyl	10	Phosphamidon	0.05
Trichlorphon	0.1	Pyrethrins	3

* 0.2 for maize.
Italic type indicates limits are set by Good Agricultural Practice. Others are set by detection limit.
Source: Osborne *et al.* (1988).

Monitoring of residues of fumigants applied at levels specified by manufacturers in an experiment, revealed that only traces remained in cooked products. They were associated mostly with the bran fractions.

Insecticide residues of laboratory-milled wheat grains, treated at manufacturers' recommended rates, were four times as concentrated in bran and fine wheatfeed as in white flour. In commercial samples the germ contained five times as much as in the white flour. The milling process removed only about 10% of the residue on the whole grain, however only 50–70% of that in white flour survived bread baking.

Overall, the results suggested that insecticides applied at recommended rates are unlikely to lead to residues above the MRLs of Codex Alimentarius (Osborne *et al.*, 1988).

In the U.K. less than one quarter of stored grain is treated with pesticides (MAFF, 1991, H-GCA, 1991). As part of a surveillance programme, surveys have been carried out, since 1987, by the Flour Milling and Baking Research Association, in association with the National Association of British and Irish Millers, on samples representing those purchased by all flour millers. Residues have been low throughout but a decline

in residue level was also noted. Of the 1340 samples tested between April 1990 and March 1992, none approached the MRL for organophosphorus or other classes of pesticide.

Preprocessing treatments

Before processing of cereals it is necessary to carry out certain treatments on them. In cases where grains undergo a period of storage before use, some or all the treatments may take place before entering the store.

The treatments include *drying, cleaning, grading, conditioning* and *blending*. Of these, drying is likely to be carried out on the farm but the remainder are more likely to be performed at the mill or the elevator or receival as appropriate to the marketing system in use.

Drying

Purchasing-contracts typically stipulate an acceptable range of moisture contents corresponding to that required for safe storage. These may be considerably lower than the moisture content at harvest and some means of drying must be used.

At its simplest, and in suitable climates, drying can consist of spreading the grain to dry in the sun assisted by frequent turning to expose all grains. Such treatment is unlikely to cause damage to grain beyond that which any handling imposes. More sophisticated drying methods employing artificially heated air or surfaces have inherent dangers that must be avoided if grain quality is to be maintained. In this context quality may be defined in several ways but the most fundamental is retention of the ability to germinate. This test is particularly relevant to malting barley as the malting process requires grain to be viable. In other cereals viability provides an indication that other quality factors have not diminished during drying. In milling-wheat, the most damaging change is denaturation of gluten and the consequent deterioration in baking quality. Aside from a germination test other, more rapid, tests include those using 2,3,5-tri-phenyltetrazolium chloride. A 0.2% aqueous solution applied to longitudinally bisected grains for 2 h in the dark produces, in viable grains, a red coloration around the embryo. The colour results from the activity of the enzyme dehydrogenase, present in respiring grains but inactivated by uncontrolled drying. A more recently introduced test depends on the reduced solubility of heat-damaged proteins in sodium chloride solution (Every, 1987).

The temperature at which damage occurs depends upon the initial moisture content of the grain. Recommendations for safe drying conditions are based on two types of experiment: those in which grains are held at constant moisture during heating in closed containers and those in which temperature is maintained at a constant value. Neither truly represents the situation in all dryers and few if any experiments have been carried out on a commercial scale. Nevertheless the recommendations are a useful guide and the sensible practice is to use conditions, as far as practicable, on the safe side of the critical values. The relationships between initial moisture contents and safe drying temperatures for wheat are shown in Fig. 5.8.

Similar conditions are suitable for seed grain and malting barley (Nellist, 1978). For grains being fed to stock, temperatures of 82°–104°C may

be used as denaturation of proteins is not disadvantageous and increased digestibility of starch may arise with positive advantage.

In drying rice an additional hazard arises as fissuring increases, giving rise to more broken grains. Since intact grains are valued in milled rice considerable care has to be taken not to remove moisture too rapidly. Although it is customary to harvest while the grain is green with a high moisture content it may undergo a period of drying in the field while still in the ear. This applies in the less industrialized growing areas and demands adequate temperatures but care must be taken to avoid too rapid a loss of moisture. Techniques for shading in windrowed crops include overlapping ears of one row with the stems and leaves of the next. In more industrialized production areas, drying is effected in specially designed driers. Grains are dried as paddy or rough rice, that is with the husk still surrounding them. Control of moisture loss is achieved through the use of a multi-pass system. Installations may be of the mixing or non-mixing type. The principle of both involves downward flow between two screens. On one side of the screen warm air is generated and passes through the curtain of moving grains, grain nearer to the warm air source tends to become warmer than the more remote grains and the introduction of a mixing mechanism minimizes the effects of this gradient. The diagrams in Fig. 5.9, show the differences.

Many types of hot air drier exist, they are frequently classified as batch or continuous but they may also be distinguished on the basis of the direction of airflow in relation of that of the grain (Nellist, 1978). Thus in the *Cross flow* type, air flows across the path of the grains. The layer of grain adjacent to the air inlet is soon dried and its temperature rapidly approaches that of the inlet air. The grain on the exhaust side remains at or near the wet bulb temperature of the drying air for most of the drying period and its final temperature will normally be much less than the inlet air temperatures. Because of the large moisture gradient, mixing after drying is essential. Cross flow driers are the most abundant.

In *Concurrent flow* driers both grain and air flow in the same direction. Thus hot air meets cool

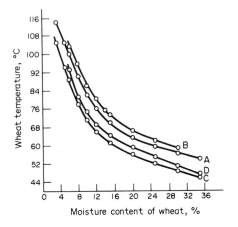

FIG 5.8 Relationship between wheat moisture content and maximum safe drying temperature. Curves *A* and *B* correspond to nil germination after 60 min and 24 min heat treatment respectively; curves *C* and *D* correspond to start of damage after 60 min and 24 min treatment, respectively. (From J. B. Hutchinson, *J. Soc. Chem. Ind.* 1944, **63**: 104, reproduced by courtesy of the Society of Chemical Industry.)

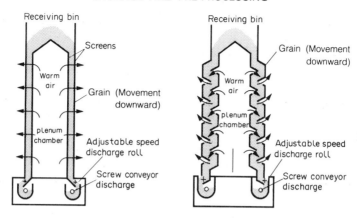

FIG 5.9 Non-mixing (left) and mixing (right) type columnar driers. (Wasserman and Calderwood, 1972).

grain, but the heat and moisture transfer that occurs ensures that grain temperature does not rise to the inlet air temperature and that the air temperature falls rapidly. For the final phase of drying, the air and grain are almost at the same temperature. Advantages of this design are that high drying-air temperatures can be used to give high initial drying rates without overheating the grain and the period during which the grain and air are in temperature equilibrium is thought to help temper the grains and relieve stress cracks. In *Counterflow* driers air travels in the opposite direction to the grain and the dried grain temperature approaches that of the inlet air. The air tends to exhaust near to saturation, and drying is therefore efficient. The temperature of the hot air at inlet and the final grain temperature are almost the same (Nellist, 1978).

Separations

Cleaning may occur before storage, or shortly before processing, or both. Cleaning before storage has the advantage that it helps to minimize deterioration in store which is aggravated by the presence of broken grains, dust and green plant parts. Receival at harvest time however is often accompanied by pressure to deal with grains rapidly and hence time for cleaning may not be available. Even grains cleaned before storage may require cleaning again before processing to remove undesirable elements resulting from the storage itself. Impurities, together with damaged and shrunken and broken grains, are collectively known as 'screenings' in the U.K., 'Bezatz' on the continent of Europe or 'dockage' in the U.S.A. Use of one term in the following includes all.

Impurities

Frequently encountered components of screenings may be classified according to their composition thus:

Vegetable matter — weed seeds, grains of other cereals, plant residues such as straw, chaff and sticks.
— fungal impurities — bunt balls, ergot.
Animal matter — rodent excreta and hairs, insects and insect frass, mites.
Mineral matter — mud, dust, stones, sand, metal objects, nails, nuts etc.
Other — string and twine. Miscellaneous rubbish.

Purity of cereal samples can be improved by cleaning and by separating. Cleaning involves the removal of material adhering to the surface of grains, while separation involves the removal of freely assorting material. Considerable ingenuity has been demonstrated in the design of devices

for eliminating impurities by both methods and
as a result a wide range of machines is produced
by a number of manufacturers. Wide though the
range is, the principles involved are few. Thus
cleaning depends upon *abrasion, attrition* or *impact*,
and separation upon discrimination by *size, shape,
specific gravity, composition* and *texture*. No single
machine can perform all the necessary operations
and it is customary for parcels of cereal to pass
through a series of operations based on the above
principles.

Metals

Because of the potential danger to be caused
to machinery by hard objects, these are usually
among the first to be removed. Thus metals are
removed by devices capable of detecting their
composition, installed in spouting through which
the grains are directed early in the process.
Ferrous metals are attracted by strong *magnets*.
Their removal from the magnetic surface by grain
flow cannot be avoided so magnets are designed
to revolve allowing removed metal objects to fall
into a reservoir out of the grain stream. Non-
ferrous metals are detected by the interruption
which they cause to an electric field when passing
through a *metal detector*. The interruption rapidly
activates a mechanism which temporarily diverts
the stream containing the offending item.

Destoners

Stones, sand and string also constitute potentially
damaging impurities and hence these are also
removed early. A single *separator* is capable of
removing these and other less hazardous materials
such as straw and some seeds of other cereals.
Size and shape are the distinguishing criteria
here, as the process is essentially one of sieving
through a screen coarser than the grain (to remove
string, straw and larger objects such as stones and
large grains of different species) and over a screen
finer than the required grains, through which
broken grains, small seeds and sand can pass.
Specific gravity separations may be made on
the selected material leaving the separator. The
grains leave the machine as a curtain and this is
subjected to an upward stream of forced air which
lifts light material such as chaff, while the denser
particles fall under gravity. The aspirated air
can flow to an expansion chamber where solid
particles are deposited; air can then be recycled.

For more rigorous removal of stones specialized *dry stoning machines* may be used. Stoners
have an inclined vibrating deck through which a
uniform current of air flows upward. Feed is
directed on to the lower end of the deck and
spreads and mixes in response to the vibration.
On reaching the region of the bed where the
fluidizing effect of air is experienced, the grains
are lifted beyond the range of influence of the
vibration. They fall under gravity and are discharged at the lower end of the deck. Stones are
too dense to be lifted by the air stream and
continue to the top of the deck where they are
discharged.

Unwanted species

Elimination of grains of other cereals and weed
seeds surviving the above treatments is achieved
by *disc separators* (see Fig. 5.10). Discrimination
is on the basis of length, the width being of no
consequence to this operation. A series of discs
are mounted on a single horizontal axle in a trough
partly filled with grain. The axle is driven causing
the discs to rotate through the bed of grain. Each
disc has, on both surfaces, a series of indentations
arranged concentrically. The indentations behave
as pockets into which grains of the required depth
or smaller can fit. They are thus lifted out of the
grain mass. As pockets pass the high point of the
rotation they fall out of the pockets into channels
adjacent to the disc faces by which they are
conducted into a discharge in common with
channels from the other discs. Material failing to
be picked up in pockets is driven by worm from
the feed end to the discharge end of the machine.
A gate at the discharge end can be adjusted to
control the amount of time spent in the trough.
The more impurities the shorter the time needed
for sorting.

By use of several machines, each with different
sized pockets in its discs, a number of separations
may be made in sequence, with the required

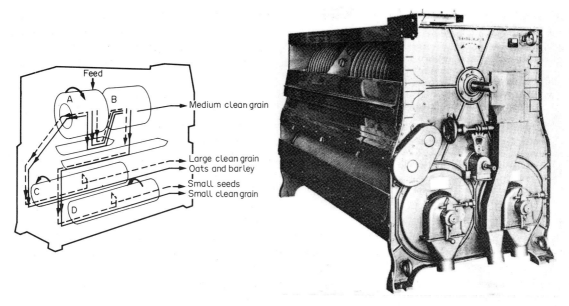

FIG 5.10 Carter-Day Disc Separator for length separation of impurities from wheat, rye and other cereals grains. Exterior view of machines, *right*; diagram of operation, *left*. (by courtesy of the Carter-Day Company, Minneapolis.)

cereal behaving as the selected material when eliminating larger impurities, or the rejected material when eliminating small impurities. In some machines several separations can be made within a single unit.

The same principle of separation in pockets on a rotating device is used in *trieur cylinders*. In these the interior surface of the cylinder has the pockets in it. The capacity of disc machines is greater than that of cylinders of the same diameter but the selectivity of cylinders is said to be better. The loading of discs or cylinders can be reduced if concentrators are used upstream of them. *Concentrators*, otherwise known as *gravity selectors* or *combinators* (Fig. 5.11), serve to effect a preliminary streaming of stocks so that the individual streams can be subjected to treatment appropriate to the impurities they are most likely to contain. Thus a light fraction would be treated to remove small seeds and grains of unwanted small cereals by use of discs or cylinders. The denser stream would not need to pass through these machines but may be routed to the more appropriate destoner.

The concentrator operates on a similar principle

to the destoner. Stocks are stratified according to their specific gravity by the oscillating action of the sloping deck, the lighter stocks are conveyed on a current of air while the heavier material is directed by friction on the oscillating deck. The routes of the two fractions can thus be separately ordered. Dust is removed by the air flow that provides the air cushion.

Aspiration

The rate at which a particle falls in still air is the resultant of the speed imparted to it by the force of gravity, balanced by the resistance to free fall offered by the air. The rate of fall, or 'terminal velocity', depends on the weight of the particle and its surface-area:volume ratio. Compact spherical or cubical particles thus have a higher terminal velocity than diffuse or flake-like particles. Instead of allowing the particles to fall in still air, it is more usual to employ an ascending air current into which the stock is introduced. The velocity of the air current can be regulated so that particles of high terminal velocity fall, while those of low terminal velocity are lifted. Utilizing this principle,

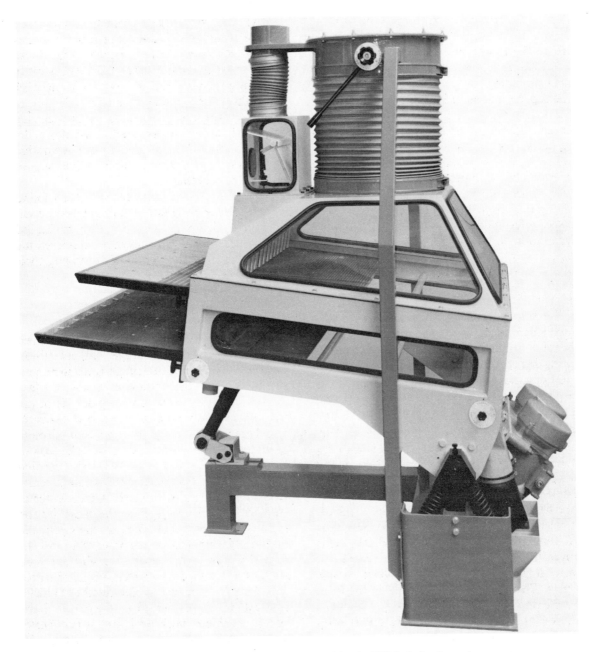

FIG 5.11 Gravity selector. (Reproduced by courtesy of Satake UK Ltd, Stockport.)

particles of chaff, straw, small seeds, etc., having terminal velocity lower than that of wheat, can be separated from wheat in an aspirator, in which an air current is directed through a thin falling curtain of stock. In a *duo-separator* the stock is aspirated twice, permitting a more critical separation to be made. In this type of machine, the lifted particles are separated from the air current by a type of cyclone, and the cleaned air is recycled to the intake side of the machine. Such closed-circuit aspirators save energy and minimize effluent problems.

The separations described thus far are related to the separation of the required species from other cereal species, weed seeds and other materials quite unlike the grains of the required species. There are occasions when it is required to make a separation of two types of grain of the required species. This usually means shrivelled grains or those that have been hollowed as a result of insect infestation. They are removed mainly by aspiration or other methods depending on specific gravity. In oats the following types are present:

1. Double (bosom) oats.
2. Pin oats.
3. Light oats.
4. Other types of oats.

In double oats the hull of the primary groat envelopes a normal groat, plus a second complete grain. Both groats are usually of poor size. Pin oats are thin, short and contain little or no groats, light oats consist of a husk which encloses only a rudimentary grain. Other types are twins, discoloured green and hull-less. (Deane and Commers, 1986).

While indented discs and/or cylinders separate on the basis of length, it is sometimes required, particularly with oats, to eliminate narrow grains by a separation on the basis of width. This involves the use of screens with elongated but narrow slots, through which only reject grains can pass. The screens are commonly in the form of a *perforated rotating cylinder* through which the bulk flows horizontally. Undersize grains drop through into a hopper. Small round holes recessed into the cylinder walls are used as an alternative to slots. The recess upends undersize grains directing them through the hole.

Where size and shape are similar and only small differences in specific gravity may exist, sophisticated gravity separators are required. An example of this type of machine is the *Paddy separator* (Fig. 5.12). In this a table mounted on rubber tyres is made to pivot around its horizontal axis. Compartments on the table are formed by a series of zig-zags arranged at right angles to the direction of motion. Grains accumulate and stratify within the channels and the motion of the table causes those light grains which rise to the top, to travel towards the upper side of the channel flanks, while the heavier stocks pass down the table. As might be expected with a separation based on small differences careful adjustment of variables is necessary.

Gravity separations have recently been found capable of separating satisfactory grains of wheat from sprouting grains. The sprouting, having led to some mobilization of starch, has rendered the affected grains lighter while not changing their overall size.

Cleaning

Cleaning, in contrast to separating, is performed on *scourers*. As their name suggests, these subject grains to an abrasive treatment designed to remove dirt adhering to the surface of grains. Surface layers such as beeswing (outer epidermis of the pericarp of wheat) may also be removed. Scourers propel grain within a chamber by means of rotors to which are fixed beaters or pins. The axis of the rotor may be horizontal or vertical according to type. In the horizontal type, beaters rotate within a cylindrical wire mesh. Grains enter at one end and are cast against the mesh by the inclined rotor beaters; cleaning is achieved by the friction of grains against each other or against the mesh. Dust removed passes through the screen and falls into hoppers by which it is discharged.

Aspiration following scouring removes particles loosened but not removed from the grains. Some vertical machines work on a screen and beater system but others combine a scouring action with aspiration.

FIG 5.12 Paddy separator. (Reproduced by courtesy of Satake UK Ltd. Stockport.)

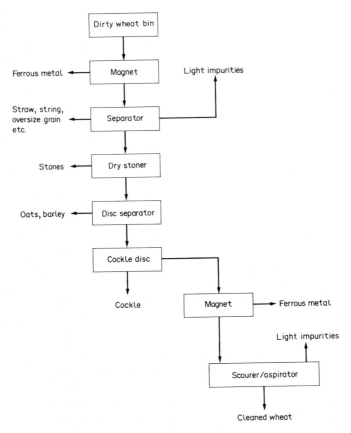

FIG 5.13 Schematic diagram of a possible flourmill screen room.

Durum wheats are subjected to particularly vigorous treatment with beaters. In addition to cleaning, the beaters also eliminate much of the embryos, considerably facilitating the cleaning of semolina produced and reducing its ash yield.

Scourers have replaced washers in the cleaning programmes of wheat in most countries on account of:

1. Problems of pollution control concerning the discharge of effluent.
2. Problems with microbiological control (mainly bacterial).
3. High costs of operating machinery and of water.

Washing may still be practised in the former Soviet Union, where ultrasonic vibrators were used to assist the cleaning process (Demidov and Kochetova, 1966). Wet cleaning is also in use in the United States for removing surface dirt from maize before milling (Alexander, 1987).

Screenroom operation

In practice, the grain passes in succession through a series of machines working on the various principles described. No single machine can remove all the impurities, but all the machines, considered as a unit, remove practically all the impurities for the loss of very little of the required grain.

A possible cleaning flow for wheat is shown in Fig. 5.13.

The efficiency of operation of screenroom machinery depends on machine design, feed rate and proportion of *cut-off* (reject fraction). As the feed rate is reduced, interference between particles decreases, and efficiency of separation increases. As the proportion of cut-off increases, the rejection of separable impurities becomes more certain.

In the conventional screenroom arrangement, it is customary to feed to each machine in succession, the entire feed, except for the screen-ings removed stage by stage. As the total quantity of screenings removed frequently amounts to only 1–1.5% of the feed by wt, every machine in the screenroom flow must have a capacity (i.e. be able to deal with a rate of feed) practically equal to that of the first machine.

Loop system

The feed rate to many of the machines can be reduced, and the efficiency of separation increased, by an arrangement known as the *loop* or *by-pass* system.

In the loop system, the first machine is set to reject a large cut-off (say 10% of the feed) containing all the separable impurities, together with a proportion of clean grain. The remaining 90% or so is accepted as clean, requiring no further treatment. The cut-off is retreated to recover clean wheat. As the cut-off amounts to only a fraction of the total feed, the feed rate in the retreatment machines can be much reduced, and the efficiency of separation improved. The loop system is frequently applied in the operation of disc separators and trieur cylinders, and is a feature of the Twin-Lift Aspirator, which embodies main aspiration of the whole feed, and re-aspiration of a substantial cut-off to recover clean grain.

The Bühler 'concentrator' operates on the loop system. A cut-off of about 25% by wt. is lifted by aspiration and specific gravity stratification: this contains various impurities but is stone-free, and goes to a gravity table or scourer for separation of clean wheat from the impurities. The remaining 75% is free of impurities other than stones and needs only to be de-stoned on a dry stoner or gravity table.

Individual components of a mixed grist to a flour mill (cf. p. 192) are cleaned separately before blending. Wheat types differ in grain size and shape, and in the types of impurities contained, and require individual treatment in the screenroom, particularly as regards choice of sieve sizes for the milling separator, and of indent sizes for the cylinders and discs.

Dust explosions

Dust is released whenever grain is moved: the atmosphere in grain silos and mills therefore tends to be dusty, leading to conditions under which dust explosions may occur. A suspension of dust in air, within certain limiting concentrations, may explode if a source of ignition above a certain limiting temperature is present. A series of dust explosions in U.S. grain elevators in 1978 caused the death of at least fifty persons, and led to the initiation of extensive programmes of research into the causes and avoidance of dust explosions.

Avoidance of dust explosions is directed to:

1. suppression of dust and
2. avoidance of sources of ignition.

Devices that minimize dust formation include light damping with water (about 1% by wt), dust-free intake nozzles, dust suppressors, and the 'Simporter' system — a method of mechanical grain handling in which the grain is conveyed 'sandwich fashion' between two belts held together by air pressure. By the use of inclined belts operating on this principle, grain can even be conveyed vertically.

Sources of ignition most frequently responsible for dust explosions in flour mills are welding and hand lamps. Other potential sources are flames, hot surfaces and bearings, spontaneous heating, electrical appliances, friction sparks, static electricity, magnets, bins and bucket elevators.

Damping

Damping, as a pretreatment for milling grists, is a long established practice. It probably originated as part of the cleaning process when washing featured as an important treatment. It was adopted as a treatment in its own right when the beneficial effects of washing followed by a period of lying were noted. In current practice the washing of most cereals for cleaning purposes has now been abandoned but conditioning — or tempering (both terms mean the controlled addition of moisture) remains an important stage of processing. Conditioning requires much less water than washing and, as the water is absorbed by the grains, there is no effluent problem. Neither is bacterial infestation usually a problem if lying times are short. If problems do arise they can be minimized by chlorination to levels above that of the water supplied.

The study of the effects of water on the physical properties of cereals, and hence their milling properties, is complex. The cells of the grain have several components, the physical properties of which are altered, each in its own way. Water also affects the adhesion between the components of the cells, the cells themselves and the various tissues of the grain.

Although applied to almost all cereals, conditioning is not a single process appropriate as a preparation for all milling treatments. Even different types of the same cereal undergoing the same treatment may receive different treatments or different degrees of the same treatment. Consequently each cereal type will be considered separately here.

Common or bread wheat

The effects of moisture content and distribution have received more attention in relation to wheat than any other cereal. Variation in endosperm texture demands different treatments; in general the amount of water added and the length of lying time increase in proportion to endosperm hardness.

The reasons for conditioning are:

1. to toughen the outer, non-starchy endosperm components of the grain so that large pieces survive the milling, and powdering of bran is avoided;
2. to mellow the endosperm to provide the required degree of fragmentation.

The term 'mellow' is much used to describe a desirable state of endosperm but a useful definition of its meaning is difficult to find. The changes induced by water penetration have been better described by Glenn et al. (1991) who found that endosperm strength decreased. 'Strength' as used by these authors has its strict materials-science meaning, i.e. a measure of the stress required to bring about fracture. The toughness (*senso stricto*) of the endosperm (manifested as the energy needed to bring about complete fracture) was found by the same authors to decrease with increasing moisture content, although this was less consistently observed than the increase in strength. The hardness of the endosperm is not the only factor determining the level of water addition. Clearly the grain moisture has an effect on the fractions separated during milling. It influences their surface properties which are particularly significant in relation to their behaviour on sieves if the covers are made from absorbent fibres such as nylon or (particularly) silk. Moisture can also be transferred to the fabric of the sieves themselves, possibly altering the frictional properties and even changing the tension and the effective aperture size. The ease with which starchy endosperm may be separated from bran is also influenced by moisture content and distribution. Both increase and decrease of difficulty of this separation with increasing moisture content have been imputed (although an increase is the more frequently claimed) but direct evidence is scarce and measurements must be difficult to make. It is generally accepted that, for the milling of high extraction rate flour, conditioning to a lower than normal moisture level is appropriate. Another factor influencing the degree of damping and post-damping lying is the target moisture content of the final products. Storage life of flour declines with increasing moisture content and most specifications impose a maximum moisture content — mainly to avoid buying water at flour prices. The relative costs of transport and the requirement to absorb adequate water in dough making also exert an influence however. In calculating the amount of water required to achieve the target moisture content of stocks, characteristics of the mill and its location have to be considered. These include:

1. the conveying system in use (pneumatic systems incur more moisture loss than elevator systems);
2. ambient temperature and humidity;
3. roll temperatures.

Losses of between 1 and 2.5% have been recorded due to evaporation of water from stocks.

Water penetration

Water is absorbed rapidly by capillarity into the empty pericarp cells. They are capable of taking up 80% of their weight, equivalent to about 8% (w/w) of total grain weight. Once wetted therefore, the bran can behave as a reservoir from which water can pass into the endosperm. The rate of passage is restricted by the presence of impermeable components of the testa and possibly the nucellar epidermis. Since these are thinnest in the non-adherent layers overlying the embryo this is the area through which most water gains access to the inner components of the grain. It advances towards the distal end of the endosperm, leading to near-uniform distribution within the whole grain. Whereas saturation of the pericarp takes only minutes, equilibration throughout the grain takes hours.

The exact time taken depends on several factors including:

1. the initial moisture content of the grain; damper wheats absorb more rapidly (Moss, 1977) as well as requiring less damping.
2. The degree of permeability of the testa — particularly in the embryo region. There is little information on this from conditioning studies, but studies on imbibition in relation to germination show this to vary (Wellington and Durham, 1961; King, 1984).
3. The mealiness of the endosperm — water permeates through mealy endosperm more quickly than through vitreous endosperm. The essential feature appears to be the greater proportion of air spaces in mealy tissues. As soft wheats are more mealy than hard, equilibration is achieved more rapidly in softer wheats (Stenvert and Kingswood, 1976). As the continuity of the protein matrix is influenced by total nitrogen content of the endosperm, moisture movement is slower in high nitrogen wheats than in low.
4. The temperature of the water — a 3-fold increase in penetration rate was recorded for

each 12°C rise, by Campbell and Jones (1955) within the range 20°–43°C, above 43°C smaller rate increases applied up to 60°C.

5. The uniformity of the water distribution among grains. Where less than 8% water (w/w) is added it is preferentially absorbed into the pericarp of the grains which it contacts first, leaving little for distribution to more remote grains. Equilibration among grains takes even longer than within a grain.

It is possible to exploit the above to achieve more rapid conditioning. Thus:

1. multistage damping provides a progressively higher starting moisture content for each addition;
2. abrasion of the grain surface can remove the impermeable layers;
3. damping causes expansion of the endosperm, but this is reversed as equilibrium is achieved or if drying occurs. Such changes induce stresses which cause minute stress cracks that increase effective mealiness;
4. warm (above ambient to 46°C), or hot (above 46°C) conditioning may be practised;
5. precision placement of water in a uniform fashion and/or vigorous mixing of grains during and immediately following damping may be employed.

Much research has been carried out on the length of time required to achieve equilibrium moisture conditions. With so many factors affecting the rate of penetration it is not surprising that results are variable. Estimates range from several hours to several days, with more recent work tending to favour the shorter periods. Such revision of ideas has influenced the manner in which conditioning of wheat is carried out (Hook *et al.*, 1984).

Conditioning practice

Both phases of conditioning (i.e. addition of water, and lying for a period of equilibration) require specially designed equipment, the design of which has evolved to suit changing circumstances and changing ideas about conditioning.

The amount of water required to be added to achieve the optimum moisture content of a particular type of wheat clearly depends upon the initial moisture content of the wheat. Wheat that travels long distances by sea tends to be traded at lower moisture content than wheat bought locally (water is expensive to transport!). If more than 3–5% needs to be added, it must be done in more than one stage. The amount capable of addition in one stage depends on the method of addition.

In preparing wheats for milling as a mixed grist, in which components are introduced into the mill at different moisture contents, the individual components are best conditioned separately and blended after lying. Although the time taken for water to penetrate into the grain is reduced as a function of temperature, the practice of conditioning at elevated temperatures has declined as a result of prohibitive costs and the introduction of alternative practices that allow reduced lying times. Details of warm and hot conditioning systems are provided in older textbooks (e.g. Lockwood, 1960; Smith, 1944). They will not be described here.

Principles underlying the design of modern conditioning systems are:

1. addition of accurately metered quantities of water;
2. uniform distribution of moisture.

Accurate water addition is achieved by controlling the water addition and flow rate of wheat, possibly using feed-forward and feed-back control systems. Uniform distribution is ensured by careful placement of water followed by rapid mixing. Most milling engineers now produce conditioning equipment with sophisticated control systems. The H_2O-Kay (Satake UK Ltd) system, consisting of the Kay-Ray controller and a Technovator mixer, employs moisture measurement by microwave attenuation meters. This is the only method by which the moisture of freshly damped wheat can be monitored, allowing corrections to be made to the levels of water addition on a feedback basis. The system has the additional facility for adding steam. A sampling stream is diverted from the main flow, through a cell in which its moisture content is measured. Additionally, flow

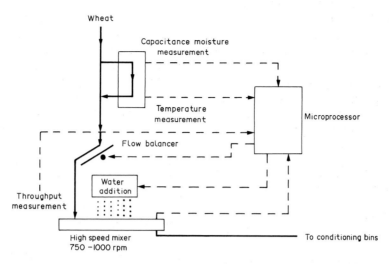

FIG 5.14 Schematic diagram showing damping control by the Bühler Intensive Dampener.

rate and specific weight are detected by gamma ray absorbence which is sensitive to mass of wheat in the measurement cell. The required amount of water is computed and added as the wheat-stream enters at the lower end of an inclined worm. Damped wheat is raised along the worm by adjustable paddles whose disposition is set to provide a transit time of 20 s through the worm. Tumbling and mixing for this period is chosen as grain-to-grain transfer of moisture is possible for this time, before being trapped in the pericarp. Measurements similar to those made prior to damping are made on a sampling stream taken from the flow leaving the chute. These measurements can influence the level of water addition. Up to 4.25% of water addition is claimed as the capability of this system.

Even more vigorous mixing is employed in the Bühler Intensive Dampener (Fig. 5.14), as a result of which it is said that additions up to 5% are possible. In this system wheat is mixed and conveyed horizontally, after damping, by blades of a rotor that turns at about 1000 rpm. The dampener is used with a control system in which a microprocessor calculates the required amount of water on the basis of incoming wheat moisture content, flow-rate, temperature, and target moisture content. Moisture measurement is by capacitance.

Durum wheat

The purpose of durum milling is to produce not flour but semolina (see Ch. 6). Hence the purpose of conditioning is solely to toughen the bran. Penetration of water into the endosperm is not required and in consequence lying times are short. Less than 4 h is typical and as much as 2% water may be added as a final addition only 20 min before milling.

Rye

Short lying times are typical of rye conditioning regimes as water penetrates more rapidly into rye grains than into wheat. Lying times depend on the efficiency of the damping system used but even with traditional methods of water addition, such as water wheel damping, periods of 6–15 h are typical. In the cold winter conditions of parts of North America, warm conditioning may be used to provide uniform milling conditions and to eliminate condensation in break sifters and spouts (Bushuk, 1976). In Western Europe a first temper lasting 5–6 h is typical, while in Eastern Europe a slightly longer first temper of 6–10 h may be followed by a second damping an hour or so before first break.

Rice

Rice is milled at 14% moisture content with no damping immediately preceding dehulling. For rice subjected to parboiling however an elaborate heat/water treatment is involved. Paddy rice has a hull of adherent pales (Fig. 2.23, p. 43) and water penetrates very slowly. Parboiling breaks the tight seal between the two pales, thus removing the main obstacle to entry. This was the original purpose of parboiling, which is an ancient tradition, originating in India and Pakistan. Other changes occur however, including an improvement in nutritional quality of the milled product. The hot water involved in treatment dissolves vitamins and minerals present in the hull and bran coats, and carries them into the endosperm. Conversely, rice oil migrates outwards during parboiling, reducing oil content of the endosperm and increasing it in the bran. Starch present in the endosperm becomes gelatinized, toughening the grain and reducing the amount of breakage during milling. Aleurone and scutellum adhere more closely so that more of each remains in the milled rice. Some discoloration of grains occurs but susceptibility to insect attack is reduced.

Barley

Conditioning for pearling consists of adjusting the moisture content of the grains to 15%, followed by a rest of 24 h.

Maize

The details of the conditioning protocol in preparation for degerming of maize depend upon the process in use: for the Beall degerminator, the moisture content is raised to 20–22%, while for entoleter or rolling processes an addition of only 2–3% to stored grain is appropriate. Following damping a rest of 1–2 h is usual but up to 24 h is possible.

Conditioning bins

In all storage bins used for cereal grains, sound design is important with regard to efficient empty-

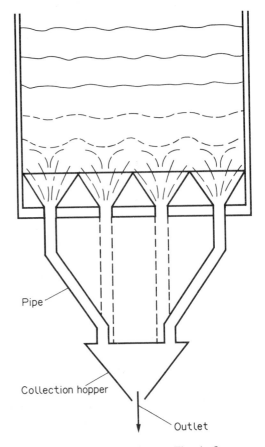

FIG 5.15 Manifold discharge from a First-in-first-out type conditioning bin. From: *Screenroom Operations 1*. Flour Milling Industry Correspondence Course. By courtesy of *Incorporated National Association of British and Irish Millers Ltd.*

ing. It is particularly important for conditioned grain. It is usual for a bin to be filled by deposition from a central spout: grains delivered to the centre tumble towards the edges but the pile remains deepest at a point below the delivery spout and a smooth conical profile with an 'angle of repose' characteristic of the particular sample remains at the top. As the bin empties, the profile changes because the grains in the centre fall more quickly than those at the edges. This leads to a reversal of the original surface profile, the centre becoming the lowest point. Such a means of emptying is disorderly, it leads to a mixture of grains and a sequence of removal different from that of filling. The sequence of emptying may not be important in general storage considerations but in conditioned grain, where lying time is controlled, it is important that the grains introduced into the bin first are also the first to leave. The difficulties in achieving this are compounded as damped grain flows less readily than dry and the tendency to non-sequential flow is exaggerated. Also the shortening of lying times associated with contemporary practice demands greater precision in controlling the lying period, emphasizing the need for sequential flow.

The expedient by which improvements have been made is the deconcentration of outlets. In new bins a matrix of funnel-shaped hoppers may be provided at the base, each leading to spouting that ducts to a common discharge hopper via a manifold (see Fig. 5.15).

Existing bins may be converted by imposing a cone-shaped baffle near the top of the discharge hopper. The cone may be point-up, surrounded by an annular space, directing flow to the periphery of the bin, or point-down with an annular space and an orifice in the centre also. In bins of square section pyramidal baffles replace the conical ones described for cylindrical bins.

References

ALEXANDER, R. J. (1987) Corn dry milling: processes, products, and applications. In: *Corn: Chemistry and Technology*, pp. 351–376, WATSON, S. A. (Ed.) Amer. assoc. of Cereal Chemists Inc. St. Paul, MN. U.S.A.

BOTHAST, R. J. (1978) Fungal deterioration and related phenomena in cereals, legumes and oilseeds. In: *Postharvest Biology and Biotechnology*, Ch. 8, pp. 210–243, HULTIN, H. O. and MILNER, M. (Eds.) Food and Nutrition Press Inc. Westport, Conn. U.S.A.

BUSHUK, W. (1976) *Rye, Production, Chemistry and Technology.* Amer. Assoc. of Cereal Chemists Inc. St. Paul, MN. U.S.A.

BUSHUK, W. and LEE, J. W. (1978) Biochemical and functional changes in cereals: storage and germination. In: *Postharvest Biology and Biotechnology*, pp. 1–33, HULTIN, H. O. and MILNER, M. (Eds.) Food and Nutrition Press Inc. Westport, Conn. U.S.A.

CAMPBELL, J. D. and JONES, C. R. (1955) The effect of temperature on the rate of penetration of moisture within damped wheat grains. *Cereal Chem.* 32: 132–139.

CHRISTENSEN, C. M. (1974) *Storage of Cereal Grains and Their Products.* Amer. Assoc. of Cereal Chemists Inc. St Paul, MN. U.S.A.

DEANE, D. and COMMERS, E. (1986) Oat cleaning and processing. In: *Oats: chemistry and Technology*, WEBSTER, F. H. (Ed.) Amer. Assoc of Cereal Chemists Inc., St. Paul, MN. U.S.A.

DEMIDOV, P. G. and KOCHETOVA, A. A. (1966) Effect of ultrasonic waves on the technological properties of grain. *Izv. Vyssh. Ucheb. Zaved., Pishch Tekhnol.* 1: 13.

EVERY, D. (1987) A simple four-minute protein-solubility test for heat-damage in wheat. *J. Cereal Sci.* 6: 225–236.

GLENN, G. M., YOUNCE, F. L. and PITTS, M. J. (1991) Fundamental physical properties characterizing hardness of wheat endosperm. *J. Cereal Sci.* 13: 179–194.

HOOK, S. C. W., BONE, G. T. and FEARN, T. (1984) The conditioning of wheat. A comparison of U.K. wheats milled at natural moisture content and after drying and conditioning to the same moisture content. *J. Sci. Food Agric.* 35: 591–596.

KING, R. W. (1984) Water uptake and pre-harvest sprouting damage in wheat: grain characteristics. *Aust. J. Agric. Res.* 35: 338–345.

LOCKWOOD, J. F. (1960) *Flour Milling.* 4th edn. Henry Simon Ltd. Stockport, Cheshire.

MOSS, R. (1977) The influence of endosperm structure, protein content and grain moisture on the rate of water penetration into wheat during conditioning. *J. Fd. Technol.* 12: 275–283.

NELLIST, M. E. (1978) *Safe Temperatures for Drying Grain.* Report No. 29, Nat. Inst. Agric Engng. (To Home Grown Cereals Authority.)

OSBORNE, B. G., FISHWICK, F. B., SCUDAMORE, K. A. and ROWLANDS, D. G. (1988) *The Occurrence and Detection of Pesticide Residues in UK Grain.* H-GCA Research Review No. 12. Home-Grown Cereals Authority, London.

PIXTON, S. W. (1980) Changes in quality of wheat during 18 years storage. In: *Controlled Atmosphere Storage of Grains*, pp. 301–310, SHEJBAL, J. (Ed.) Elsevier Scientific Publ. Co. NY. U.S.A.

POICHOTTE, J. L. (1980) Wheat storage. In: *Wheat*, pp. 82–84, HAFLIGER, E. (Ed.) Ciba Geigy Ltd, Basle, Switzerland.

POMERANZ, Y. (1974) Biochemical, functional, and nutritive changes during storage. In: *Storage of Cereal Grains and their Products*, pp. 56–114, CHRISTENSEN, C. M. (Ed.) Amer. Assoc of Cereal Chemists. St. Paul, MN. U.S.A.

ROBINSON, I. M. (1983) *Modern Concepts of the Theory and Practice of Conditioning and its Influence on Milling.* Gold medal thesis of Nat. Jt. Ind. Council for the Flour Milling Industry.

SMITH, L. (1944) *Flour Milling Technology.* 3rd. edn. Northern Publishing Co. Ltd Liverpool.

STENVERT, N. L. and KINGSWOOD, K. (1976) An autoradiographic demonstration of the penetration of water into wheat during tempering. *Cereal Chem.* 53: 141–149.

STOREY, C. L. (1987) Effects and control of insects affecting corn quality. In: *Corn: Chemistry and Technology*, pp. 185–200, WATSON, S. A. and RAMSTAD, P. E. (Eds.) Amer. Assoc. of Cereal Chemists Inc., St. Paul, MN. U.S.A.

WASSERMAN, T. and CALDERWOOD, D. L. (1972) Rough rice drying. In: *Rice: Chemistry and Technology*, pp. 140–165, HOUSTON, D. F. (Ed.) Amer. Assoc. of Cereal Chemists. Inc. St. Paul, MN. U.S.A.

WELLINGTON, P. S. and DURHAM, V. M. (1961) The effect of the covering layers on the uptake of water by the embryo of the wheat grain. *Ann. Bot.* 25: 185–196.

Further Reading

ANON. (1987) *Pests of Stored Wheat and Flour*. Module 2 in Workbook Series, National Association of British and Irish Millers.

BAKKER-ARKEMA F. W., BROOK, R. C. and LEREW, L. E. (1978) Cereal grain drying. *Adv. Cereal Sci. Technol.* 2: 1–77.

BHATTACHARYA, K. R. and ALI, S. Z. (1985) Changes in rice during parboiling, and properties of parboiled rice. *Adv. Cereal Sci. Technol.* 7: 105–159.

CORNWELL, P. B. *Pest Control in Buildings. A Guide to the Meaning of Terms*. Rentokill Ltd. E. Grinstead.

GORHAM, J. R. (Ed) (1991) *Ecology and Management of Food-Industry Pests*. FDA Technical Bulletin 4. Arlington VA. Assoc. of Official Analytical Chemists.

HOME-GROWN CEREALS AUTHORITY (1992) *Storage Crops* H-GCA London.

SAUER, D. B. (Ed.) (1992) *Storage of Cereal Grains and their Products*. Amer. Assoc. of Cereal Chemists Inc. St. Paul MN. U.S.A.

TKATCHUK, R., DEXTER, J. E. and TIPPLES, K. H. (1991) Removal of sprouted kernels from hard red spring wheat with a specific gravity table. *Cereal Chem.* 68: 390–395.

6

Dry Milling Technology

Introduction

The single term 'milling', applied in the context of cereals, covers a wide range of processes. In general they are methods of transforming whole grains into forms suitable for consumption or for conversion into consumable products. Milling processes do not themselves involve intentional heating, although in some cases, as in oat processing, a heating phase precedes the milling, and in maize dry milling a drying phase is included.

Characteristic (but not essential) features of milling processes are:

1. Separation of the botanical tissues of the grain (e.g. endosperm from pericarp, testa and embryo).
2. Reduction of the endosperm into flour or grits.

Some milling systems include both operations (e.g. white flour milling from wheat), while others involve only one (e.g. rice milling comprises only separation, and wholemeal wheat milling seeks only to reduce particle size).

Milling schemes are conveniently classified as wet or dry, but this indicates a difference in degree rather than an absolute distinction as water is used in almost all separations. Damping or 'tempering' features even in 'dry' milling; it is considered in detail in Ch. 5 as it is a pre-milling treatment. This chapter is concerned with so-called dry milling; wet milling processes are dealt with in Ch. 12. Emphasis is placed on preparation for human consumption but the term 'milling' also applies to production of animal feeds. A brief description of feed-milling is given in Ch. 15.

In milling processes involving both separation and size reduction the two operations may be carried out in two distinct phases, as in sorghum milling, or, to some extent, combined, as in bread- and soft-wheat milling, where the two processes continue throughout the multi-stage operation (although the emphasis changes as the process proceeds). In rice milling, two stages occur but neither seeks to fragment the endosperm. The first stage removes the husk and the second removes the bran.

Even the above distinctions are not absolute as processes are in use for decorticating wheat grains before reduction of size and some rice is milled into flour. It is clear that few generalizations can be made about cereals milling, most milling technologies depend upon a series of individual processes through which stocks pass in sequence.

Milling processes

The processes are of three types, they may:

1. Change the shape and size of the feedstock.
2. Separate fractions produced by (1)-type treatments.
3. Change the temperature and/or water content of the stocks.

The processes are described below.

Treatments that change shape and size

Abrasion

Effects depend upon severity, thus:

1. Surface abrasion is a relatively gentle process which removes all or part of the fruit coats

(pericarp) and possibly the embryo. Greater severity, or retreatment of already abraded stocks, removes part of the endosperm also. Grains are brought into contact with an abrasive surface, which may be of natural stone, carborundum, sculptured or perforated metal, or other material. Where perforated screens are used, these behave as sieves also, selectively permitting passage of particles. In the present context abrasion is used to include the similar process of attrition. There is a nice distinction between the two terms, depending on the roughness of the surfaces involved but it is insufficient to warrant separate consideration. When used specifically to remove the outer tissues of caryopses either process may be described as decortication.

2. Severe abrasion includes heavy or protracted grinding between surfaces such as those of a pestle and mortar and of grinding stones; it features therefore in many of the simple and historical methods used for preparing ground meals. Such grinding may reduce grains to a range of particle sizes including that of flour.

Roller milling

Whole grains or partially milled stocks are passed between rotating rollers for several reasons, including:

1. Grinding — a process in which grains are reduced to smaller particle size. All grinding stages are performed within rollermills, each of which is equipped with a pair of rolls. Rollstands usually consist of two mills back-to-back, within the same housing but operating independently and on different feedstocks. The rollers may be disposed diagonally, vertically or horizontally. They are aligned in parallel and rotate in opposite directions. One of the rolls rotates faster than the other so that a speed *differential* exists (see Fig. 6.1).

The rollers may be smooth surfaced or fluted, but when used in a grinding mode they rotate at different speeds.

On fluted rolls the profile of each flute

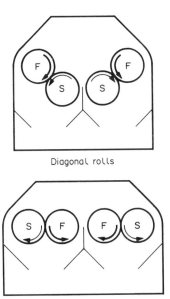

FIG 6.1 Disposition of rolls in roller mill stands. The fast roll of each pair is indicated by the broarder arrow.

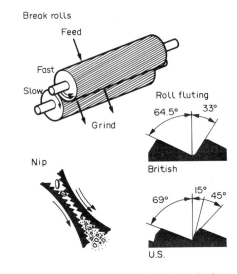

FIG 6.2 Pair of break rolls. In the enlarged view of the 'nip' (*left*) rolls are disposed in the dull-to-dull configuration. Details of typical British and U.S. roll flutings are shown *right*.

resembles an italic V, with one side shorter than the other (Fig. 6.2).

As the flutes are asymmetric the rolls can be run with either the steep (sharp) or shallow (dull) profile disposed towards the nip. The relationship between rolls may thus be described as 'dull-to-dull, sharp-to-sharp, dull-to-sharp or sharp-to-dull' (Fig. 6.2). It is conventional to give the fast roll disposition first in such descriptions.

2. Flaking — flaking rolls are smooth-surfaced and are generally heavier than grinding rolls and they are operated at zero differential. The purpose of this is to increase the surface area of the feedstock, either to facilitate subsequent separation of components (eg. germ from endosperm) or to impart desired product characteristics, as in porridge oats.

Impacting

Grains or milled stocks are thrown against a hard and possibly abrasive surface. This is usually achieved by feeding stocks into the centre of a very high-speed rotor. The process is very versatile; it may be used as a means of dehulling (as of oats), abrading, size-reduction, (as in pin milling), or disinfestation, destroying all stages in the life-cycle of insect pests found in grain and flour.

Fractionating processes

Size

This is an important criterion by which particles are separated. In most milling systems which involve grinding, sieving features at some stage, to separate stocks as final products, or for further appropriate treatment.

Shape

In processes such as oat milling and rice milling, grains are graded on a shape (and size) basis, before treatment, as machine clearances are grain-size dependent; and during the milling process, as small grains which escape treatment need to be re-fed. Shape-sensitive fractionating machines include disc separators and trieur cylinders.

Specific gravity

Particles differing in density may be separated on a fluid medium such as air or water. In the case of water the separation usually depends on one or more component being denser than water and others being less dense. When air is used, the force of an air current supports particles of lesser density to a greater degree than the denser ones, allowing them to be carried upwards and later deposited when the force of the current is reduced. The process is described as aspiration. The lighter particles frequently also have a flat shape, which enhances their buoyancy. Aspiration features in purifiers used in wheat (particularly durum) milling to remove bran from semolina, and in rice milling, to remove pearlings from decorticated grains.

Multiple factors

The paddy separator (see p. 120) is an example of a machine in which several grain characteristics are exploited in effecting their separation. Specific gravity, surface roughness and shape all combine to direct grains into appropriate streams on a tilted vibrating table with a cunningly sculptured surface.

Changes in temperature and/or moisture content

Water can be added with or without substantial change in temperature, it may be added specifically to achieve a required combination of the two physical conditions, as in stabilization of oats, or to change the mechanical properties of the grain components, as in wheat and rice milling, when the temperature is of less importance.

Stocks produced from grains or intermediates milled at very high moisture need to be dried to permit proper processing or safe storage. Drying is performed by heating, and stocks that have been heated may subsequently require cooling.

Fine grinding and air classification

The contents of cells comprising the bulk of storage tissues of many legume cotyledons and cereal endosperms consist essentially of starch granules embedded in a protein matrix. In oats and some legumes an appreciable amount of oil is also present. The spaces among closely packed spherical or near spherical starch granules are wedge shaped, and where protein occupies these spaces it is compressed into the same wedge shape. It has thus been called *wedge protein*. Clearly the size of starch granules determines the sizes of the interstitial wedges. In the case of the Triticeae cereals the wedges among the larger population of starch granules generally have granules of the smaller population (see p. 57) embedded in them.

When wheat endosperm is fragmented by grinding it is usually reduced to a mixture of particles, differing in size and composition (Greer *et al.*, 1951). These may be classified into three main fractions:

1. Whole endosperm cells (singly or in clumps), segments of endosperm cells, and clusters of starch granules and protein (upwards of 35 μm in diameter). This fraction has a protein content similar to that of the parent flour.
2. Large and medium sized starch granules, some with protein attached (15–35 μm in diameter). This fraction has a protein content one half to two thirds that of the parent flour.
3. Small chips (wedges) of protein, and detached small starch granules (less than 15 μm in diameter). This fraction has a protein content approximately twice that of the parent flour (Fig. 6.3).

The proportion of medium-sized and small particles (below 35 μm) in flour milled conventionally from soft wheat is about 50% by weight, but in hard wheat flours it is only 10%. The proportion of smaller particles can be increased at the expense of larger ones by further grinding on, for example, a pinned disc grinder, which consists of two steel discs mounted on a vertical axis, each disc being studded on the inward-facing surface with projecting steel pins arranged in concentric rings that intermesh. One disc, the stator, remains stationary while the other rotates at high speed. Feedstock enters the chamber between the discs at the centre, and it is propelled centifugally by the air current created. The particles impact against the pins and against each other, as a result of which, they are fragmented.

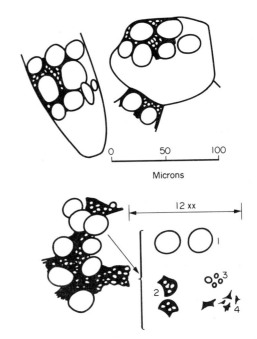

FIG 6.3 *Above*: the two main types of endosperm cell – prismatic (*left*), polyhedral (*right*) – showing large and small starch granules (*white*) embedded in protein matrix (*black*). *Below*: exposed endosperm cell contents (*left*) and products of further breakdown (*right*): 1. detached large starch granules (about 25 μm diameter); 2. 'clusters' of small starch granules and protein matrix (about 20 μm diameter); 3. detached small starch granules (about 7 μm diameter); 4. fragments of free wedge protein (less than 20 μm diameter). 12xx is the representation to scale of the mesh aperture width of a typical flour bolting cloth. (Redrawn from C.R. Jones *et al.*, *J. Biochem Microbiol. Technol. Engng.* 1959, 1:77 and reproduced by courtesy of Interscience Publishers.)

The reduction in particle size due to *fine grinding* further separates the components, as previously described, allowing increased proportions of starch and protein to be concentrated into different fractions.

Particles below about 80 μm are considered to be in the sub-sieve range, and for making separations at 15 μm and 35 μm, the flour as ground, or after fine-grinding, is fractionated by air-classification. This process involves air elutriation, a process in which particles are subjected to the opposing effects of centrifugal force and air drag. Smaller particles are influenced more by the air drag than by centrifugal force, while the reverse is true of the larger particles. The size at which a separation is made is controlled by

TABLE 6.1
Yield and Protein Content of Air-Classified Fractions of Flours With or Without Pinned-Disc Grinding*

Flour	Parent flour protein content (%)	Fine (0–17 μm) Yield (%)	Fine (0–17 μm) Protein content (%)†	Medium (17–35 μm) Yield (%)	Medium (17–35 μm) Protein content (%)†	Coarse (> 35 μm) Yield (%)	Coarse (> 35 μm) Protein content (%)†
Hard wheat							
Unground	13.6	1	17.1	9	9.9	90	13.8
Ground	13.4	12	18.9	41	10.0	47	14.7
Soft wheat							
Unground	7.6	7	14.5	45	5.3	48	8.9
Ground	7.7	20	15.7	71	5.0	9	9.5

* Source: Kent (1965).
† 14% m.c. basis. N × 5.7.

varying the amount of air admitted, or by adjusting the pitch of baffles which divide or 'cut' the airborne stream of particles.

When practised commercially, air classification is generally carried out in the mill. It is customary to effect separations into a protein-rich fraction of less than 15 μm, a starch-rich fraction of 15–35 μm, and a fraction over 35 μm consisting of cells or parts of cells that have resisted breaking into discrete components. Table 6.1 shows typical yields and characteristics of fractions derived from fine-ground and unground flours of hard and soft wheats.

The term *protein shift* has been coined to define the degree of protein concentration achieved with a given feed. Protein shift is the amount of protein shifted into the high-protein fraction plus that shifted out of the lower fractions, expressed as a percentage of total protein in the material fractionated.

Applications for which commercial classified fractions might be used are:

Fine fraction: increasing the protein content of bread flours, particularly those milled from low or medium protein wheats, and in the manufacture of gluten-enriched bread and starch reduced products.

Medium fraction: use in sponge cakes and pre-mix flours.

Coarse fraction: biscuit manufacture where the

uniform particle size and granular nature are advantageous.

Classification can be continued into further fractions by cuts corresponding to larger sizes. As Fig. 6.4 shows, this does not lead, as might be expected, to many fractions varying in size only, but to fractions whose composition also varies.

The highest-protein fraction above 15 μm is that between 44 and 55 μm, in which are concentrated the cells from the outermost layer of starchy endosperm. This *subaleurone* layer contains only few small starch granules embedded in a solid core of protein (Kent, 1966).

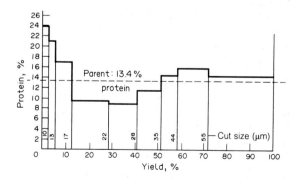

FIG 6.4 Results of air-classification of flour into nine fractions of varying particle size. The flour, milled from CWRS wheat, was pinmilled one pass, and classified at 10, 13, 17, 22, 28, 35, 44, and 55 μm nominal cut sizes. Protein content of the parent flour was 13.4%.

Air classification has been applied to other cereal and legume flours with some success. A rye flour of 8.5% protein was separated into a fine fraction with 14.4% protein and a coarse fraction with 7.3%. Sorghum flour fractions between 3.5% and 16.6% protein were prepared from parent flours of 5.7–7.0% protein. Starting with grits (which are derived from the higher protein, horny parts of the grain) of 9.2–11.9% protein content, fractions between 6.8 and 18.9% protein were obtained (Stringfellow and Peplinski, 1966).

Protein shifting by air-classification has not met with success when attempted in rice. Three factors are held responsible for this: the smallness of the starch granules, the intimate dispersions of the protein bodies and the extreme vitreousness of the endosperm (Deobold, 1972).

Sorghum and millets

Sorghums and millets are considered together as they are both tropical cereals, the majority of the processing of which remains in the hands of subsistence farmers. The grinding of flour is performed by traditional manual methods which occupy much of the day of the women. In the case of sorghum particularly, several industrial milling methods are in use. Because industrial scale operations are relatively new, these may be regarded as somewhat experimental, no single method having yet been established as the standard. Several methods are described in the following text. It is noticeable that, even in domestic processing several principles are involved, and these are similar to those used in the industrial systems.

Domestic processing

The techniques used have been in continuous use for hundreds of years. Although simple, the processes are instructive, both in their own right, and because they reflect the methods from which modern cereal processing has evolved. They are hand operations in which wooden pestle and mortar are used, the abrasive action of pounding on the washed grain freeing the outer pericarp from the remainder of the grain (cf. Ch. 13).

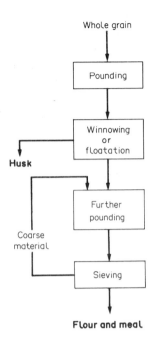

FIG 6.5 Schematic diagram of domestic processing of sorghum.

Water may be used during the continuing grinding, as appropriate to the type of grain used. The decorticated grain is separated from the bran by winnowing, after drying in the sun; or water may be used to separate the components. Further pounding follows to reduce the size of the particles of decorticated grains. Sieving is used to separate material that has been reduced sufficiently, from that needing further treatment.

The simplicity of the process is illustrated schematically in Fig. 6.5.

The principles illustrated are:

1. Use of attrition to break open the grain.
2. Separation of endosperm from the surrounding tissues.
3. Use of water to aid the separation.
4. Sieving to select stocks for appropriate treatment.

Another principle, common to all milling processes, and illustrated by hand milling, is the dependence of the method adopted on the nature of the varieties processed. Variations include the hardness of the endosperm and the thickness of the pericarp. Soft grains break into pieces during decortication and cannot be readily separated from the pericarp. The difficulties are compounded when pericarps are thick (Rooney et al., 1986).

Sufficient flour is produced daily for the needs of the family; three or four hours may be required to produce 1.5–1.6 kg of flour from sorghum at an extraction rate of 60–70% of the initial grain weight. Bran accounts for about 12%, and the same amount is lost. Even longer periods (6 h) may be required to produce a family's daily requirements from millet (Varriano-Marston and Hoseney, 1983) The problem of storage of flour does not arise, and this is just as well as products may have a high moisture content. Further, the continued pounding expresses oil from the embryo and incorporates it into the flour, leading to rancidity on oxidation.

Industrial milling of sorghum

As urban drift accelerates in Tropical countries, strain is increasingly imposed on the domestic production system and industrialization of flour production becomes more attractive.

In Africa the relationship between domestic processing and mechanized milling is a delicate balance that is affected by a number of changes, occurring on that continent. Processing is justified only if it prolongs the storage period, increases convenience and preserves the nutritional quality of the product (Chinsman, 1984).

Processing methods may include adapted wheat flour milling methods and specifically designed abrasive methods. Most begin with a decortication stage using mills with abrasive discs or carborundum stones (Reichert et al., 1982). Wholemeals are produced by use of stone, hammer, pin or roller mills.

In some cases traditional methods of decortication are combined with 'service' milling. Such combinations can improve on the traditional methods alone by as much as 20% extraction rate.

An experimental milling operation using a laboratory Bühler mill gave best results with sorghum conditioned to 20% m/c. Even broomyard sorghum (which has pales attached to the grain, and which cannot be milled by other methods) was successfully processed by this method

(Cecil, 1987). In a study for F.A.O. (Perten, 1977) a decortication rate of 20% was recommended for good consumer acceptance. Pearling at natural moisture content was favoured as tempering reduced throughput, increased breakage of grains and increased ash yield and fat content of the pearled grains. In India, and indeed elsewhere, the grain is conditioned with about 2% of water before pearling (Desikachar, 1977).

As the amount of pearlings increases so the composition changes, reflecting the concentration of fibre in the outer layers, and of protein and fibre in the aleurone layer and the peripheral (subaleurone) starchy endosperm (cf. p. 37). Oil and protein contents of the pearlings are at a maximum when about 12% of the grain has been abraded.

Sorghum endosperm used as a brewing adjunct leaves the mill as coarse grits: they may be produced by impaction following decortication. Embryos are removed during the process (Rooney and Serna-Saldivar, 1991). Removal of embryos after milling is difficult as they are the same size as some of the grits. They can be separated, however, by virtue of their different densities; floatation on water or use of a gravity table are suitable methods.

Reduction to flour particle size may be achieved by roller milling, impaction or pin milling and may or may not include degerming. Maintenance and correct setting of roller mills can be a problem and the use of easily maintained and adjusted special mills is advocated by some. The United Milling System (now Conagra) two stage process is one example. The decortication stage employs a vertical rotor which hurls grains against each other and against a cylindrical screen through which the fragments produced by the impacts pass for collection in a cyclone. This stage was designed to resemble the pounding typical of domestic processing. The disc mill that follows has sawblade elements that are cheap and simple to replace. A two tonnes installation in the Sudan produces flour or grits of 80% extraction rate with ash yield of 0.7–1.1%.

If it is required to remove the embryo before milling, this may be done using machines designed for degerming maize. Alternatively, special machines have been produced for sorghum itself.

Such a device consists of a wire brush rotating within a perforated cylinder.

Industrial milling of millets

Industrial processing of millets is even less common and less developed than that of sorghum. While there are reports of experimental attempts to adapt technology appropriate to other cereals, the concept of industrial scale millet milling is not well established. Industrial production of flour inevitably imposes a need for distribution facilities, and for storage. The inclusion of embryo parts, or even oil expressed from the embryo in flour, reduces storage life, as in sorghum. Any successful process should therefore include a degerming stage. The limited amount of small-scale industrial processing that is carried out consists of abrasive decortication followed by reduction of endosperm with hammer mills or similar devices dependent on attrition.

Dry milling of maize

Dry milling is a relatively minor industry compared with wet milling, by which the majority

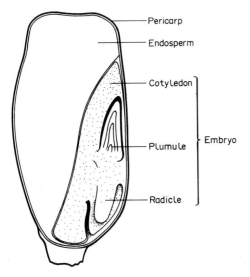

FIG 6.6 Diagram of the maize grain showing the relative sizes of the main anatomical components.

- Pericarp
- Endosperm
- Cotyledon ⎤
- Plumule ⎬ Embryo
- Radicle ⎦

(75% in the U.S.A.) of maize is processed to produce starch. The product with the highest value, coming from dry milling is 'grits'. Two important characteristics of the maize grain influence the production of grits, viz. the large embryo and the presence of horny and mealy endosperm in the same grain (see Fig. 6.6).

The significance of the large embryo lies not only in its failure to contribute to the grits yield but also in its high oil content. Inclusion of this oil in the product, either as a component of embryo chunks or through its expression on to the surface of grits, reduces the shelf-life through its oxidation and consequent rancidity. Variation in endosperm texture is important because grits are essentially derived from the horny parts of the endosperm; softer parts too readily breaking down to flour. In the industrialized world, grits are used mainly in production of, or consumption as, breakfast cereals. They are also used for making fermented beverages. In Africa a fine grit meal ('mealy meal') is an important staple.

Maize dry-milling exploits most of the principles used in grain milling (cf. p. 134), but not all are involved in the same process. Several combinations are described in the text below.

Historical

Dry milling techniques were in use by North and South American Indians in ancient times. Hand-held stones were used initially, but a later development was one hand-held stone ground against a concave bedstone. A further development of this was the hominy block, fashioned from two trees, the stump of one being hollowed out as a mortar, and the springy limb of another nearby, serving as a pestle. The name 'hominy' is derived from a North American Indian word and it describes a coarse ground maize meal mixed with milk or water. It persists today, applied to some of the products of modern maize dry milling, e.g. 'hominy feed' and 'hominy grits'. Later developments in dry processing of maize included the quern, a device common to the processing of many types of grain in Roman times. Querns consisted of two stones; the upper 'capstone' being rotated over the stationary

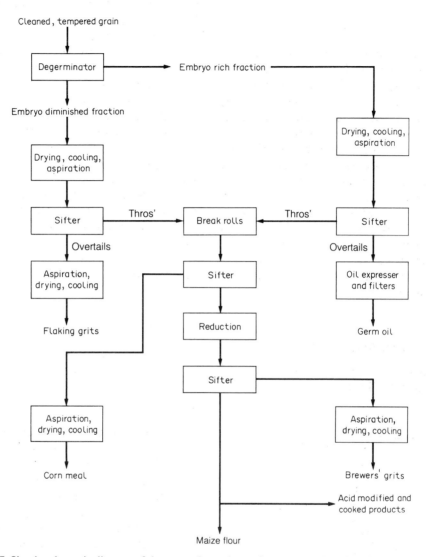

FIG 6.7 Simple schematic diagram of the tempering – degerming process. (Based on Johnson, 1991.)

'netherstone'. Grain was introduced through a hole in the centre of the capstone and it was ground by the abrasive action of the two stones, as it was worked towards the outside, to be collected as a coarse meal.

Today, dry milling is carried out in several ways, the simplest being the production of a coarse 'wholemeal' of 85–95% extraction, in small grist mills. Because it includes the embryo the meal has a limited shelf life.

The tempering–degerming (TD) system

A schematic summary of the tempering–degerming process is shown in Fig. 6.7.

The Beall degerminator

Possibly the most important innovation in dry maize milling was the introduction of degerming stages. The Beall degerminator is unfortunately named, as it neither reverses the process of germination, nor totally or exclusively removes the germ. It was introduced in 1906 and it is used in the majority of U.S. dry milling plants today, as an essential stage in the 'tempering–degerming' (TD) system. Its virtue lies in its potential to produce a high yield of large particle size grits with low fat content and low fibre content (about 0.5%), suitable for manufacture of corn flakes.

The favoured feed stock to the TD system in the U.S.A. is No. 2 yellow dent corn. In Africa white maize is used. After cleaning and tempering to 20% moisture content, it passes to the 'Beall'. This machine consists of a cast iron cone, rotating at about 750 rev/min on a horizontal axis, within a conical, stationary housing, partly fitted with screens and partly with protrusions on the inner surface. The rotor also has protrusions on its outer surface. The maize is fed in at the small end and it works along to the large end, between the two elements. The protrusions on the rotor and the housing rub off the hull and embryo by abrasive action, also breaking the endosperm into particles of various sizes and degrees of purity. The Beall discharges two types of stock: the tail stocks which are too large to pass through the screens, consisting mainly of fragmented endosperm, and the through-stock consisting largely of bran and embryo. The proportions of different sized particles can be controlled by the setting of the Beall (Brekke and Kwolek, 1969).

Drying, cooling and grading

Tail-stock from the Beall degerminator is dried to 15–15.5% moisture content in rotary steam tubes at a temperature of 60°–70°C and cooled to 32°–38°C by aspiration with cold air. The dried stock is sifted to produce a number of particle size fractions. The coarsest fraction, between 3.4 and 5.8 mm, consists of the flaking or hominy grits, originating from the vitreous parts of the endosperm. They may pass through a further aspirator and drier–cooler, before emerging as a finished product. Finer stocks are combined with coarser fractions of the through-stocks from the Beall, for treatment in the milling system.

Milling

The feed to the milling system, *viz.* large, medium and fine hominy, germ roll stock, and meal are mixtures of endosperm, bran and embryo. They are separated by a series of roller milling, sifting and aspiration stages before being dried and emerging as a diverse range of final products. They are fed to the mill, each entering at an appropriate point: the large and medium hominy at the first break, the fine hominy and germ roll stock at the second roll.

The milling is carried out on roller mills, using fluted rolls, a traditional flow containing up to sixteen distinct stages. The grindings with fluted rolls (15–23 cuts per cm rotating at a differential of 1.25:1 or 1.5:1) flatten the embryo fragments, allowing them to be removed by sieving. The products are sifted on plansifters and are aspirated. The mill is divided into a break section, a series of germ rolls and a series of reduction and quality rolls. The break system releases the rest of the embryo as intact particles, and cracks the larger grits to produce grits of medium size. The whole milling system for maize bears some resemblance to the earlier part of the wheat milling system (which is described more fully below — p. 141) as far as B2 reduction roll (2nd quality roll, in the U.S.A.) *viz.* the break, coarse reduction and the scratch systems, but is extended and modified in comparison with this part of the wheat milling to make a more thorough separation of the large quantity of germ present. Modern practice is to use a much shortened system.

The action of the rolls should be less severe at the head end of the mill than at the tail end in order to minimize damage to the germ while simultaneously obtaining maximum yields of oil and oil-free grits.

The finished coarse, medium and fine grits, meal and flour products are dried to 12–14% moisture content on rotary steam tube driers.

The germ concentrate consists largely of heavily damaged embryos with an oil content of 15%. The fat content of grits and flour is 1.5–2.0%.

Finer through-stocks pass through a drier followed by an aspirator and purifier, to remove bran and germ. Germ may be extracted to produce oil, the remainder being compressed into germ cake. Finer fractions of the through-stock consist of endosperm heavily contaminated with bran and germ, the mixture being known as 'standard meal'. The various fractions are combined to give 'hominy feed'

Impact grinding

An alternative to the Beall is the European-developed system in which impact machines such as entoleters (see Fig. 5.7) or turbocrushers are used to detach the embryos. Treatment is performed at natural moisture content 13–15% (cf. 20% for the Beall). It is relatively economical as subsequent drying is not necessary. Separation of germ from endosperm on gravity tables is efficient, but separation of endosperm from bran by aspiration, and of vitreous endosperm from mealy endosperm, is less effective than with the Beall degerminator. The following products are obtained from the entoleter process:

Maize germs, 1–4.5 mm, with 18–25% fat.

Maize grits, 1–4.5 mm, with 1.5% fat, 0.8–1.2% crude fibre content: about 60% of the original maize.

Semolina and flour, which may be made by size reduction of the grits. The mealy endosperm, of higher fat content, reduces to flour particle size more readily than does the vitreous endosperm; thus, flour has a higher fat content — about 3%, and semolina a lower fat content — 0.8–1.3%, than the grits.

Oil extraction from germ

Solvent extraction is generally used in the dry milling industry, although mechanical pressing, e.g. with a screw press, is sometimes used. The germ from the mill is first dried to about 3% m.c. and then extracted while at a temperature of about 121°C. The oil content of the germ is reduced by

extraction from 16–25% to about 6% in the germ cake. The extracted oil is purified by filtering through cloth, using a pressure of 552–690 kN/m² (80–100 lb/in²). The oil, which is rich in essential fatty acids, has a sp. gr. of 0.922–0.925, and finds use as a salad oil. Its high smoke point also makes it suitable for use as a cooking oil.

Composition of products

Chemical composition and nutritional value vary among the tissues of the maize grain, hence their separation leads to a concentration of components into different products (see Table 14.12).

Uses for dry-milled maize products

The products of maize dry milling, their particle size ranges, and average yields are shown in Table 6.2.

TABLE 6.2
Yield and Particle Size Range of Dry-milled Maize Products

Product	Particle size range		Yield (% by weight)
	Mesh*	mm	
Flaking grits	3.5–6	5.8–3.4	12 ⎫
Coarse grits	9–12	2.0–1.4	15 ⎪ 60
Medium grits	12–16	1.4–1.0 ⎱	23 ⎬
Fine grits	16–26	1.0–0.65 ⎰	⎪
Coarse meal	26–48	0.65–0.3	10 ⎭
Fine meal (coarse cones)	48–80	0.3–0.17	10
Maize flour	>80	below 0.17	5
Germ		6.7–0.5	14
Hominy feed		—	11

* Tyler Standard Screen Scale sizes.
Sources: Stiver, Jr (1955); Easter (1969).

Flaking grits are used for the manufacture of the ready-to-eat breakfast cereal 'corn flakes' (cf. Ch. 11). Grits from yellow maize are preferred.

Other uses for dry-milled maize products are discussed elsewhere: porridge (polenta) and ready-toasted cereals (Ch. 11), for bread and other baked foods (Ch. 8), and for industrial purposes (Ch. 15).

Dry milled maize flour is not to be confused with 'corn flour', the term used in the U.K. for maize starch obtained as a product of wet milling.

Milling of common (bread) wheat

Historical

In prehistoric times, the barley and husked wheat (emmer: *Triticum dicoccum*) used for human food were dehusked by pounding the grain in mortars. The invention of rotary grain mills, for grinding ordinary bread wheats (*T. aestivum*) is attributed to the Romans in the second century B.C. Thereafter, until the development of the rollermill in the mid-nineteenth century (see Table 6.3), wheat was ground by stone-milling. Even in Industrial countries today some stone milling is carried out to meet specialist demand.

TABLE 6.3
Significant Dates in the History of Flour-Milling

1753	Wilkinson patented iron rollermills for grinding flour
1839	Roller milling system developed in Budapest
1860	First rollermill system worked in Budapest
1870	Porcelain rollermills found serviceable
1877	Lord Radford of Liverpool and a group of Englishmen visited Budapest and Vienna to see rollermills
1878	First complete rollermill in England at Barlow's, Bilston
1878	Roller milling introduced in America by Washburn Crosby Co.

Stone milling

A stone mill consists of two discs of hard, abrasive stone, some 1.2 m in diameter, arranged on a vertical axis. Types of stone used include French burr from La Ferté-sous-Jouarre, Seine-et-Marne, millstone grit from Derbyshire, German lava, Baltic flint from Denmark, and an artificial stone containing emery obtained from the island of Paxos, Greece. The opposing surfaces of the two stones, which are in close contact, are patterned with series of grooves leading from the centre to the periphery. In operation, one stone is stationary while the other rotates. Either the upper stone ('upper runner') or the lower stone ('under runner') may rotate, but it is usually the former.

Grain fed into the centre ('eye') of the upper stone is fragmented between the two stones, and the ground products issue at the periphery ('skirt'). In western Europe the local soft wheat was ground by a 'low grinding' process, in which the upper stone was lowered as far as possible towards the lower stone, thereby producing a heavy grind from which a single type of flour was made. Attempts have been made, at least as far back as classical Roman times, to make a white flour for breadmaking. At the time of the Norman Conquest (A.D. 1066) in England, stone ground flour was being sieved into fractions — a fine flour called *smedma* and a coarse flour called *gryth* (Storck and Teague, 1952).

Where contrasting conditions existed side by side, as in eighteenth-century France and nineteenth-century Austria, the use of flours of more than one quality became possible. For this purpose a 'high-grinding' system was used, with the upper stone slightly raised, producing a gritty intermediate material from which, by further treatment, flours of diverse quality could be made.

Hard wheat, from the Danube basin, was ideally suited to the high-grinding system, and,

TABLE 6.4
The Ideal Relationship Between Grain Components and Mill Fractions Produced in Milling White Flour from Wheat

Grain component	Mill fraction
Grain (caryopsis)	
(1) Pericarp (fruit coat)	
(a) Outer epidermis (epicarp)	⎫ Beeswing
Hypodermis	⎬
Thin walled cells — remnants over most of grain, but cell walls remain in crease and attachment region; includes vascular tissue in crease	⎭
(b) Intermediate cells	
Cross cells	Bran
Tube cells (Inner epidermis)	
(2) Seed	
(a) Seed coat (testa) and pigment strand	
(b) Nucellar epidermis (hyaline layer)	
(c) Endosperm	
Aleurone layer	
Starchy endosperm	White flour*
(d) Embryo	
Embryonic axis	⎫ Germ
Scutellum	⎭

* In the case of durum milling, 'semolina' should be substituted for "white flour" in the table.

when steam power became available, Hungary became the centre of the milling industry.

Milling of white flour

The majority of flour sold in the U.K. is white. That is to say, it consists almost entirely of starchy endosperm. The ideal dispersal of other grain components is shown in Table 6.4.

Wheat flour has been defined as the product prepared from grain of common wheat (*Triticum aestivum* L.) or club wheat (*Triticum aestivum ssp compactum*), by grinding or milling processes in which the bran and embryo are partly removed and the remainder is comminuted to a suitable degree of fineness. 'Flour fineness' is an arbitrary particle size, but in practice, most of the material described as white flour would pass through a flour sieve having rectangular apertures of 140 μm length of side. Some flour is produced in the milling of durum wheat (*Triticum durum*) but it is not the main product.

The objectives in milling white flour are:

1. To separate the endosperm, which is required for the flour, from the bran and embryo, which are rejected, so that the flour shall be free from bran specks, and of good colour, and so that the palatability and digestibility of the product shall be improved and its storage life lengthened.
2. To reduce the maximum amount of endosperm to flour fineness, thereby obtaining the maximum extraction of white flour from the wheat.

Extraction rate

The number of parts of flour by weight produced per 100 parts of wheat milled is known as the flour yield, or percentage extraction rate. Flour yield is generally synonymous with extraction rate in the U.K.; it is calculated as a percentage of the products of milling of clean wheat, although it may also be expressed as the percentage of flour as a proportion of the dirty wheat received. It is prudent to check which system is in use, wherever possible. In the U.S.A. the term 'yield' is used to express a different

TABLE 6.5
Flour Extraction Rates in Various Countries

Country	Rate (%)	Country	Rate (%)
Austria (1990)†	76	Belgium (1990)†	74
Canada (1988/89)*	76	Denmark (1990/91)†	72
France (1990)†	79	Germany (1990)†	79
Hungary (1988)*	74	Italy (1990)†	74
Malta (1988)*	75	Morocco (1988)*	79
Netherlands (1990)†	81	New Zealand	78
Norway (1990)†	81	S. Africa (1990)*	76–79
Spain (1990)†	72	Sweden (1990/91)†	80
Switzerland (1986)*	76	Tunisia (1989)*	75
U.K. (1990)*	80	Zambia (1988/89)*	75

Sources: * International Wheat Council, † NABIM.

quantity: the weight of clean wheat required to produce 100 lb (1 cwt U.S.; 45 kg) of flour.

The wheat grain contains about 82% of white starchy endosperm which is required for white flour, but it is never possible to separate it exactly from the 18% of bran, aleurone and embryo, and thereby obtain a *white* flour of 82% extraction rate (products basis). In spite of the mechanical limitations of the milling process, extraction rates well in excess of 75% are now achieved. Flour extraction rates prevailing in various countries are shown in Table 6.5. They are calculated from values for wheat milled and flour produced, provided by official bodies in the respective countries.

Mill capacities in the U.K. were formerly expressed in terms of 'sacks/h', a sack being 280 lb . Since metrication in 1976, use of the 'sack' as a unit of weight has been discontinued and mill capacity is now expressed in tonnes (of wheat) per 24 h. The accounting unit in the flour-milling industry is 100 kg. Bran and fine wheat-feed are quoted in metric tonnes. In the U.S.A. capacity is expressed as cwt of flour per 24 h.

Principles

Were it not for the ventral crease in the wheat grain it is likely that an abrasive removal of the pericarp, embryo and aleurone tissue would be a stage in the conventional milling of wheat, as it is in rice milling. The inaccessibility of the outer tissues in the crease region, however, has

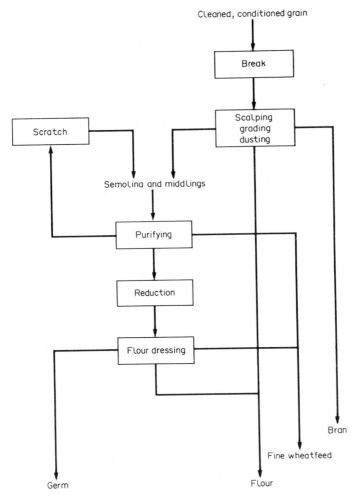

FIG 6.8 A simple schematic diagram of the common-wheat flour-milling process.

generally precluded this approach, leading to the development of the modern roller milling process which solves the problem of the crease by a combination of shearing, scraping and crushing to exploit the differences in mechanical properties between the starchy endosperm, bran and embryo. It is interesting that attention is now being given to the possibility of applying advanced pearling technology as an alternative to this. It is essential, in effecting the required separation, to minimize the production of fine particles of bran, and this basic requirement is responsible for the conditioning process described in Ch. 5, the

complex arrangement of modern flour milling systems, and for the particular design of the specialized machinery used.

The main stages in the milling of wheats, whether durum or common wheats, are grinding on roller mills, sieving and purifying.

Details of different rollermilling possibilities are given earlier in this chapter. In bread wheat milling a break, scratch and reduction system are included. The breaking system serves to open the grain and progressively scrape endosperm from the bran, the scratch system gently scrapes bran from larger endosperm particles, and the

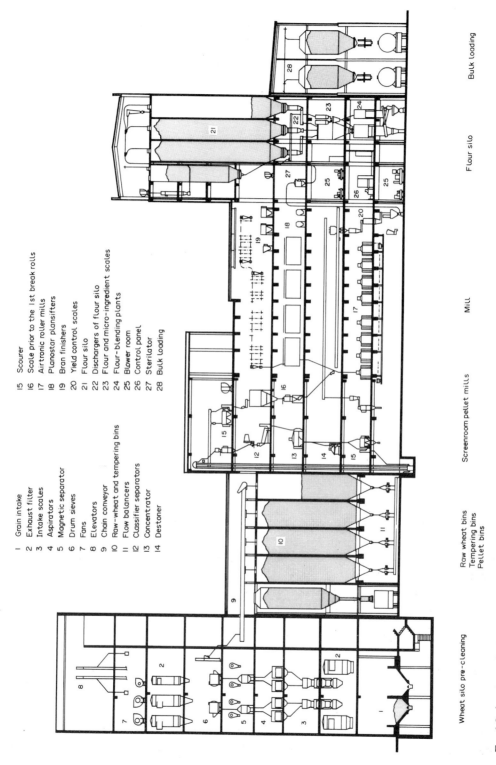

1 Grain intake
2 Exhaust filter
3 Intake scales
4 Aspirators
5 Magnetic separator
6 Drum sieves
7 Fans
8 Elevators
9 Chain conveyor
10 Raw-wheat and tempering bins
11 Flow balancers
12 Classifier separators
13 Concentrator
14 Destoner
15 Scourer
16 Scale prior to the 1st break rolls
17 Airtronic roller mills
18 Planostar plansifters
19 Bran finishers
20 Yield control scales
21 Flour silo
22 Dischargers of flour silo
23 Flour and micro-ingredient scales
24 Flour-blending plants
25 Blower room
26 Control panel
27 Sterilator
28 Bulk loading

Wheat silo pre-cleaning

Raw wheat bins
Tempering bins
Pellet bins

Screenroom pellet mills

Mill

Flour silo

Bulk loading

FIG 6.9 An example of a flour mill configuration, including cleaning, tempering (conditioning), milling, flour storage and loading facilities. Reproduced by courtesy of Bühler Bros., Ltd. Uzwil, Switzerland.

reduction system continues to purify and reduce the size of endosperm particles.

A concept of the flow of the system can be gained from Fig. 6.8 and an example of a mill layout is shown in Fig. 6.9.

The process includes several sieving stages, each performing a different function; names of particular sieving processes include:

Scalping — sieving to separate the break stock from the remainder of the break grind;

Dusting, bolting, dressing — sieving flour from the coarser particles;

Grading — classifying mixtures of semolina and middlings into fractions of restricted particle size range.

Purifying is the separating of mixtures of bran and endosperm particles, according to their terminal velocity in air currents. The process is particularly characteristic of durum milling (see p. 154).

Stocks and materials

The blend of wheat types entering the milling system is known as the *grist*. Its composition is usually expressed in percentages of each wheat type present. It is also described by reference to the product for which it is intended, e.g. a bread grist. Other stocks are described as follows:

Feed — material fed to, or entering, a machine.

Grind — the whole of the ground material delivered by a rollermill.

Break stock — the portion of a break grind overtailing the scalper cover (sieve), and forming the feed to a subsequent grinding stage. The corresponding fraction from the last break grind is the *bran*.

Break release — the throughs of the scalper cover, consisting of semolina (farina), middlings, dunst, and flour.

Tails, overtails — particles that pass over a sieve.

Throughs — particles that pass through a sieve.

Aspirations — light particles lifted by air currents.

Semolina, farina, sizings — coarse particles of starchy endosperm (pure or contaminated with bran and germ). In the U.K. where *Triticum*

durum wheats are not regularly milled the term 'semolina' is used to describe a coarse intermediate stock produced from the break system, in the milling of flour from *T. aestivum* wheats. In the U.S.A., where durum, common and club wheats are milled, the term 'semolina' is reserved for the durum product; and the coarse milling intermediate from the other wheats, equivalent to the U.K. 'semolina' is called 'farina'. Semolina for domestic use (puddings etc.) in the U.K. is obtained from durum wheat.

Middlings, break middlings — endosperm intermediate between semolina and flour in particle size and purity, derived from the break system.

Dunst — this term is used to describe two different stocks:

1. *Break dunst* — starchy endosperm finer than middlings, but coarser than flour, derived from the break system. This stock is too fine for purification but needs further grinding to reduce it to flour fineness.
2. *Reduction dunst* — (*reduction middlings* in the U.S.A.) — endosperm similar in particle size to break dunst, but with less admixture of bran and germ, derived from semolina by rollermilling.

Flour — starchy endosperm in the form of particles small enough to pass through a flour sieve (140 μm aperture). This is the definition of *white flour*, the term flour is also extended to include *brown* and *wholemeal* flours, whose particle size ranges are less well defined. When used to describe the fine product of other cereals the name of the parent cereal usually appears as a prefix, e.g. rye flour.

The modern rollermilling process for making flour is described as a 'gradual reduction process' because the grain and its parts are broken down in a succession of relatively gentle grinding stages. Between grinding stages on rollermills, stocks are sieved and the various fractions conveyed for further grinding if necessary. Although stocks are reground they are never returned to the machine from which they came or any machine preceding it. Another important principle is that flour or wheatfeed made at any point in the process is separated out as soon as possible (this principle is defied in some millstands, in which the products of one roller passage pass directly to another set of rolls mounted immediately below the first; the design is said to lead to savings in energy costs (Anon, 1990)).

Those fractions from other stages that can yield no useful flour are removed from the milling system to contribute to the milling by-products (offals) as *bran* or *wheatfeed* (*millfeed* in the U.S.A., *pollard* in Australia). The flour is also removed from the system as part of the finished product.

Rollermilling operations

The rolls used in wheat flour milling are usually 250 mm diameter and either 800 mm, 1000 mm or 1500 mm long. The feed is distributed evenly over the length of the rolls by a pair of feed rolls which also control the loading.

The succession of grinding stages is grouped into three systems: the *Break* system removes the endosperm from the bran in fairly large pieces, producing as little bran powder as possible. The *Scratch* system removes any small pieces of bran and embryo sticking to the endosperm. A *Sizing* system may be used instead of the scratch system. The *Reduction* system grinds endosperm into flour, at the same time flattening the remaining bran and embryo particles, enabling them to be separated.

The rolls involved in the break and scratch systems are fluted while those used in reduction are usually smooth (fluted rolls are used in head reductions in the U.S.A.).

Break grinding

The Break system consists of four or five 'breaks' or grinding stages, each followed by a sieving stage. Particles differing widely in size (in the grind from any one stage of rollermilling) also differ in composition: particles of starchy endosperm, which tend to be friable, are generally smaller than particles of bran, which tend to be tough and leathery, due to tempering. The process of sieving thus, to some extent, separates particles differing in composition one from another.

A thin curtain of particles is fed into the nip between the rolls; in the first break the particles are whole grains. The flutes shear open the grain, often along the crease, and unroll the bran coats so that each consists of an irregular, relatively thick layer of endosperm closely pressed to a thin sheet of bran (see Fig. 6.10).

A small amount of endosperm is detached from the bran coats, mostly in the form of chunks of up to about 1 mm³ in size (semolina); small fragments of bran are also broken off, but little flour is made.

The first break grind thus consists of a mixture of particles. The largest are the bran coats (break stock), still thickly coated with endosperm; the intermediate-sized particles are either semolina, middlings and dunst, or bran snips (some are free bran, some are loaded with endosperm); the smallest are flour.

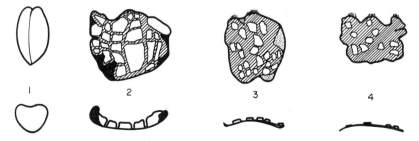

Fig 6.10 Stages in opening out of the wheat grain and the scraping of endosperm from bran by fluted rolls of the Break system. 1. Whole wheat grain; 2. I Bk tails; 3. II Bk tails; 4. III Bk. tails. *Upper row*: plan view (looking down on the inside of the bran in 2–4), *Lower row*: side view. In 2–4, endosperm particles adherent to bran are uncoloured; inner surface of bran, free of endosperm, is hatched; outer side of bran curling over is shown in solid black; Beeswing, from which bran has broken away is shown dotted.

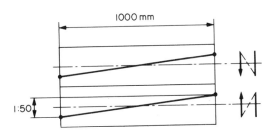

FIG 6.11 Break roll spirals.

The types of particle are separated one from the other according to size by scalping, dusting and grading. The overtails of the wire scalping sieve (bran coats) become the feedstock to the second break. The finer stocks are graded and conveyed to be purified in preparation for appropriate further treatment on reduction mills, except for the flour, which is a finished product and is retained for blending with other machine flours. The feed to the subsequent breaks consists of the break stock scalped from the grind of the previous break.

The scraping action of the break rolls is achieved by a combination of their fluted surface and the speed differential of 2.5:1 at which they run. The flutes run in a spiral: 1:50 (Fig. 6.11) is typical for break rolls running at 550/220 in the first and second breaks and 1:24 for subsequent breaks.

Flutes vary in size from mill to mill, but on a 250 mm diameter roll they might be:

1st break 3.2–4.1 per 10 mm;

2nd break 5.1–5.7 per 10 mm;

3rd break 6.4–7.0 per 10 mm;

4th break 8.6–9.6 per 10 mm;

5th break 10.2–10.8 per 10 mm.

Wear on flutes reduces their sharpness, leading eventually to the need for refluting, the interval at which this occurs varying from about six months upwards.

Many rolls run sharp-to-sharp, but with new rolls, while the extreme sharpness is wearing off, the fast roll may be run in the dull mode. Some millers consider that, if first break rolls are run dull-to-dull, large flakes of bran are produced, while it is claimed that sharp-to-sharp cuts the bran into small flakes which are difficult to clean. Dull-to-dull fluting tends to produce more break flour than sharp-to-sharp at similar releases. In any of these relationships the slow roll serves to 'hold' the stock while it is scraped by the fast roll surface. As the bran passes down the break system the gap between the rolls declines appropriately to the thinner bran flakes.

In each grinding stage the bran becomes progressively cleaner until, following the last break grind, it can yield no more endosperm through further grinding. It may become the finished by-product 'bran' at this stage or it may be subjected to treatment in *bran finishers*. A bran finisher consists essentially of a hollow cylinder, two thirds of which is perforated. Scalpings are fed into the horizontally disposed cylinder and finger beaters, attached to a central shaft, rotated at high speed and the bran skins are propelled against the perforated cover. Some of the remaining endosperm is rubbed off and passes through the apertures. Clean bran overtails the machine. The throughs contain small fragments of bran and require further sieving but, because they are 'greasy' and would blind plansifters, they are dressed on special sifters such as vibratory sifters.

Break release

The break release (throughs of the scalper cover) varies among mills. The criteria determining selected proportions are:

1. the number of break stages;
2. the grist; larger releases on first break are typical for soft wheats, followed by lower releases on succeeding breaks;
3. specifications of the finished products, for example, higher early break releases may be advisable when milling a large proportion of high grade flours.

It is generally considered desirable to release on breaks about 10% more stock than is required in the straight-run (total) flour, subsequently rejecting the 10% from the scratch and reduction systems as fine wheatfeed. Typical break releases, expressed as percentages of the individual break feeds, and of the feed to 1st break; and the usual

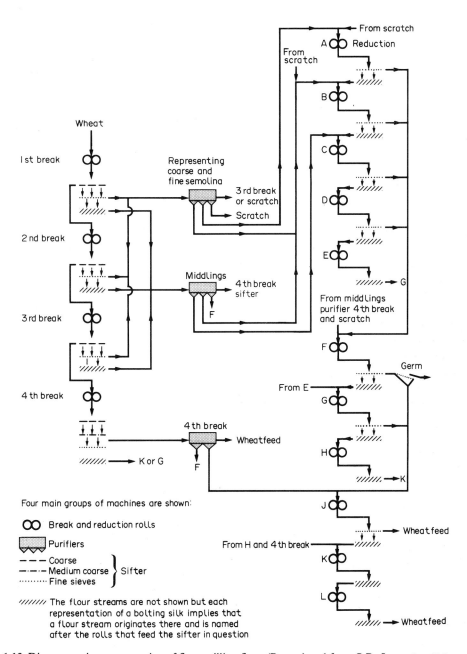

FIG 6.12 Diagrammatic representation of flour milling flow. (Reproduced from C.R. Jones, *Proc. Nutr. Soc.*, 1958, **17**: 9, by courtesy of the Cambridge University Press.)

TABLE 6.6
Typical Break Releases and Scalping Covers

Break	Break release			Scalping cover	
	Percentage of feed to break	Percentage of 1st break feed	Total release	Wire number	Aperture length (mm)
1st	30	30	30	20w	1.00
2nd	52	36	66	20w	1.00
3rd	35	12	78	28w	0.71
4th	14	3	81	36w	0.53

scalping covers (sieves) employed in white flour milling, as quoted by Lockwood (1960) are shown in Table 6.6.

Scratch system

The scratch system separates particles of bran and endosperm which are still stuck together after passing through the break system. These particles are too small to have overtailed to the next break but are not clean enough to go to the reduction system. When operated in British mills it consists of two to (rarely) four grinding stages. The feed to the scratch system contains large particles of semolina and pieces of bran with attached endosperm, but particles are much smaller and more cubical than break feeds. The objectives are to scrape endosperm from the bran without undue production of fine stock, and to reduce oversize semolina. Scratch rolls are more finely fluted than break rolls, e.g. 20 cuts per cm. The disposition of the rolls is sharp-to-sharp as in the later breaks, to ensure maximum cleaning of bran (see Fig. 6.2). Typically the speed differential is 1.8:1.

The grind from the scratch rolls is scalped, dressed, graded and purified in much the same way as that from the breaks, and fractions dispatched respectively to subsequent scratch grinding stages, to offals, to 'flour' (as end product) or to the reduction system.

A variation of the scratch system is the extended use of fluted rolls into the head reductions. This is called a *sizing* system. Such systems are found particularly in purifierless milling (they also feature in durum mills dedicated to producing semolina, but here purifiers also make an extremely important contibution to the mill flow).

Yet another variation is for the coarse semolina to pass to smooth coarse-reduction rolls followed by a *flake disrupter* or *drum detacher* before sieving. The detacher operates on a similar principle to a bran finisher, with finger beaters rotating on a horizontal axis, within a drum. Flakes of endosperm are broken up, enabling them to be separated from branny flakes. This avoids excessive cutting of bran and embryo particles.

Reduction system

The reduction system consists of 8–16 grinding stages, interspersed with siftings for removal of the (reduction) flour made by each preceding grind, and a coarse (tailings) fraction from some grinds.

Grinding in the reduction system is carried out on rollermills differing, in the U.K., in two important respects from the break rollermills:

1. The roll surfaces are smooth, or more often slightly matt (exceptions are those rolls used in the sizing system described above).
2. The speed differential between the rolls is lower, usually 1.25:1, although the fast roll still runs at 500–550 rpm. The grinding effect in reduction mills is one of crushing and shearing (the balance of the two components depending on the smoothness of the roll surface).

The feeds to A and B rolls, despite careful purifying, inevitably contain a few particles of bran; the feeds also contain particles of embryo. By carefully controlling the roll pressure, these can be flattened without fragmentation, and subsequently sifted off as coarse 'cuts' from the grinds. These, and corresponding fractions from subsequent reductions, are fed to certain reduction stages, collectively known as the 'coarse reduction system' (U.K.) or the 'tailing system' (U.S.A.).

A simplified mill flow, illustrating the distribution of stocks is shown in Fig. 6.12, and part of a roller-floor is shown in Fig. 6.13.

The reduction stage in a large mill starts with A and B reductions on the coarse and fine semolina stocks and goes down as far as M reduction. In a U.K. long system, coarse and fine semolinas from the first and second breaks are ground on the A and B rolls to provide dunst for C rolls. This releases a certain amount of flour in the process. Rolls from C downwards (i.e. D, E, G, H, L, M) make up the reduction system and deal with flour stocks from the middle and tail of the mill.

Coarse stocks which overtail from A and B flour dressers are dealt with on the coarse reduction system of B2, F and J rolls.

In the *short system*, the A rolls deal with the largest sized stock, as in the long system, but they produce a feed for B rolls which in turn produce a feed for F rolls. Apart from finishing at H or J rolls, a short system is similar to the long system in that B2 and F are the coarse rolls that deal with branny stocks and germ tailing over from the coarse head reductions.

Factors that have enabled short systems to be developed include:

1. The introduction of the pneumatic mill. Its cooler stocks allowed more work to be done on reduction rolls without overheating the stock.
2. Improvements in the rollermill, including the replacement of brass bearings with cooler trouble-free roller bearings, allowing rolls to run at higher speeds and pressures.
3. The provision of an efficient cooling system for grinding rolls.
4. Improvements in the capacity and dressing efficiency of the plansifter.
5. Availability of flake disrupters.

Short systems have the advantage that less plant is required for a given capacity. Smaller buildings are less expensive to build and maintain, clean and supervize.

Damaged starch

An important function of rollermilling, in addition to those described, is the production of controlled levels of starch damage. 'Damage' in this context has a very specific meaning: it refers to the change resulting to individual starch granules through application of high shear and pressure (cf. p. 62). One effect of starch damage is to increase water absorbing capacity, an important property in making doughs or pastes. For some purposes (e.g. breadmaking) a relatively high level is desirable as this can prolong the storage life of bread. For biscuits (cookies) a high level is undesirable as it increases the tendency of the die-cut dough piece to spread, produces a harder bite and requires more energy for the water to be evaporated off during baking.

Damage can be inflicted in all grinding stages but the greater part occurs during reduction. In recent years the demand has been for higher levels of starch damage as breadmaking methods with short fermentation time require flours with higher water absorption. This has led to heavier reduction grinding; a tendency exaggerated by the adoption of shorter milling systems (i.e. fewer reduction passages) and gristing for bread flours with softer European wheats as an alternative to hard wheats from outside the E.C. As a result of heavier grinding, reduction rolls become very hot, and water-cooled rolls are not unusual. Another consequence of the higher roll pressure is the increase in production of flakes on reduction rolls and the resulting need for flake disrupters in the mill flow. Levels of damaged starch encountered in bread flours may be as high as 40 Farrand units. Damage levels in rye flours tend to be lower than in wheat flours, apparently because of protection afforded by the more pentosans present in rye.

Flake disrupters

The tendency for greater roll speeds and pressures in the reduction grinds has led to an increase in the tendency for particles of endosperm to be compressed into flakes. Without treatment this would lead to the unnecessary retreatment of what is essentially flour. The problem is solved by the introduction of *flake disrupters* into the mill flow. The treatment is given to the grind of head reductions. Flake disrupters can be installed in the pneumatic conveying system. Each consists of a high speed rotor mounted on a vertical axis within a housing.

FIG 6.13 Part of the roller floor of a modern flour mill. (Reproduced by courtesy of Bühler Bros. Ltd, Uzwil, Switzerland).

The rotors consist of two discs held apart at their outer edge by a gallery of pins. The feed enters at the centre of the plates and is thrust outward by centrifugal force. Flakes are broken by impact against the pins, the housing and each other. *Drum detachers* (see p. 148) may be used as an alternative or applied to the grinds of tail reductions.

Sieves

Two principles have been employed in the design of sifters for flour mills. Both depend on the ability of some stocks to pass through apertures of a chosen size while others are prevented by their size from doing so. The earlier system — *centrifugals*, operates on a principle similar to

that described for a *bran finisher*. Beaters revolving within a cylindrical or multi-sided 'barrel' provide turbulence to direct particles against the screen for selection. Centrifugals developed from *reels* in which the clothed cylinder rotated with a bed of stock within it, relying on gravity for the passage of undersize particles.

All modern plants use *plansifters*. Plansifters consist of a series of flat rectangular framed sieves held within chests which, in turn, are suspended on flexible canes (either natural or man-made) in such a way that they can gyrate. Eccentric motion is assured by placement of weights (Fig. 6.14).

Plansifters have the following advantages over centrifugals:

1. they occupy less space;
2. they use less power;

FIG 6.14 Plansifters. (Reproduced by courtesy of Bühler Bros. Ltd. Uzwil, Switzerland.)

3. less sieving surface is required as it is used more efficiently;
4. their capital cost is less;
5. stocks are better dressed;
6. they are compatible with the pneumatic conveying system, which improves their efficiency.

Nylon, polyester and wire covers

Most sieve frames of plansifters, purifiers and small dressing machines were once clothed with woven silk. This has now been replaced by Plated wire, nylon (in plansifters), or polyester (in purifiers). Nylon is less expensive, less subject to change with atmospheric moisture changes and stronger than silk, allowing thinner yarn to be used and hence increasing the open area of sieves.

Wire covers are light, medium or heavy, depending on the type of wire gauge used. Light plated wire is used in flour mills.

Silk covers are described by a combination of a number indicating the number of threads per inch, and an indication of thread gauge. Nylon and polyester screens are described by the aperture sizes. Sieves are prevented from becoming blinded by *sieve cleaners*: small pieces of belting or similar, supported on a grid beneath each cover and able to move freely in a horizontal plane.

Grinding and sieving surface

The loading of rollermills can be defined in terms of the length of roll surface devoted to grinding the stocks. It is now conventional to refer to the length applied to 100 kg of wheat ground per 24 h. Typical lengths in the break system are:

1st break, 0.7 mm; 2nd break, 0.7 mm; 3rd break, 1.07 mm; 4th break, 1.07 mm; 5th break, 0.7 mm. Total break surface in this case is 4.24 mm.

Although the Break surface ranges from 4–6 mm per 100kg of wheat ground in 24 h, the tendency is to increase roll speeds and feeds drastically, and surface allowances as low as 2.1 mm are experienced. Much depends on the type of wheat being milled and the milling system in use.

The allocated surface on reduction rolls depends upon the length of the system. A long system may require up to 40 mm per 100 kg wheat ground while 8–10 mm may be sufficient in a short system.

Sifter allocation depends on the mill size and type of grist. Typical allocations of a 100 tonnes per 24 h mill might be:

1st Bk	1 section	22 sieves
2nd Bk	1 section	22 sieves
3rd Bk +X	1 section	22 sieves
4th Bk	1 section	22 sieves
5th Bk	$\frac{1}{2}$ section	12 sieves
Break Midds	1 section	24 sieves
A reduction	1 section	24 sieves
B reduction	1 section	24 sieves
C reduction	1 section	24 sieves
B_2 reduction	$\frac{1}{2}$ section	12 sieves
D reduction	$\frac{1}{2}$ section	12 sieves
E reduction	$\frac{1}{2}$ section	12 sieves
F reduction	$\frac{1}{2}$ section	12 sieves
G reduction	$\frac{1}{2}$ section	12 sieves
H reduction	$\frac{1}{2}$ section	12 sieves
J reduction	$\frac{1}{2}$ section	12 sieves

This allocation would require two 6 section plansifters (Anon., 1988).

On-line monitoring by NIR

In addition to the tests performed on finished flours in the mill quality control laboratory, some monitoring of important flour characteristics is carried out during production. The main principle involved in such measurements is near-infrared reflectance (NIR) spectroscopy. The method depends on the selective absorption of infrared energy at wavelengths characteristic of the different flour components. Measurements of different components can be made simultaneously on dry flour compressed against a window. This may form the front of a cell presented to a laboratory instrument or part of a continuous sampling loop into which part of the main flow in a pneumatic pipe is diverted. Protein content and moisture content are most successfully measured but claims are made for other parameters including particle size and damaged starch measurements. Measurements by NIR on ground materials have become routine and successful application of the technique even to whole grain is now well established.

Automation

Modern flour mills are highly automated, with microprocessor control of many, if not all, systems being common. Entire mills are capable of running completely unattended for several days.

Energy used in flour production

It has been calculated that the primary energy requirement for milling 1 tonne of flour plus that required to transport the flour in bulk to the bakery is 1.43 GJ, or, if transported in 32 kg bags 1.88 GJ. The primary energy requirement for the fuel used to transport wheat to the U.K. flour mill is calculated as 1.46 GJ/tonne for CWRS, or 0.08 GJ/tonne for home-grown wheat. The total primary energy requirement for growing the wheat (30% CWRS, 40% French, 30% home-grown grist), transportation to the U.K. flour mill, milling the wheat to flour, and transporting this to the bakery is estimated at 6.41 GJ/tonne (Beech and Crafts-Lightly, 1980).

Milling by-products

The principal products, in addition to flour, of the milling process are bran and wheatfeed (both are included in the term millfeed in the U.S.A.). Bran is the coarse residue from the final break grind. Wheatfeed (shorts in the U.S.A., pollard in Australia) is the accumulated residues from the purifiers and reduction grinding. In the U.K. screenings removed during the cleaning of the wheat are generally ground and, unless contaminated with ergot, added to the wheatfeed. The major use of bran and wheatfeed is as a constituent of compound animal feeds.

Types of bran

Bran can be classified on the basis of particle size: *Broad bran* overtails a 2670 µm screen, *ordinary bran* overtails a 1540 µm screen, *fine bran* overtails a 1177 µm screen.

Broad bran generally has the highest value. It is produced by passing ordinary bran through a bran roll. This roll is usually 356 mm diameter; the greater diameter than grinding rolls is demanded by the heavy pressure required.

For reducing the size of bran flakes, cutting rolls, grinders or bran cutters may be used. Ground bran may be added to wheatfeed, allowing both types to be sold as a single product.

Germ

Germ is the milling by-product derived from the embryo (mainly embryonic axis) of the grain. It is not always separated and may be included in wheatfeed. The advantages of separating germ are that: it commands a higher price than wheatfeed when of good quality and its presence in reduction stocks can lead to its oils being expressed into flour during grinding, adversely affecting the keeping and baking properties of the flour.

For effective separation of germ, particles passing the scalping covers but overtailing covers of approximately 1000 µm (depending on wheat type) are sent straight to the semolina purifiers. Here the germ enriched stocks are directed to the A or scratch rolls. The germ becomes flattened while the endosperm elements fragment, providing a means of concentrating the germ as a coarser fraction. Small pieces of embryo find their way to B2, F or J rolls, where they may be retrieved by a similar system to that employed on A rolls, or included in wheatfeed.

Uses of germ include incorporation in speciality breads such as Hovis and Vitbe (flours for this purpose are white, with not less than 10% processed germ added), and preparation of vitamin concentrates due to its high content of (mainly) thiamin, riboflavin, niacin and E vitamins.

Unprocessed germ becomes rancid in two weeks. The storage life can be extended to 12 weeks by defatting, to 20 weeks by drum-drying and to 26 weeks by toasting or steaming. A cocoa substitute which, it is claimed, can replace 50% of the cocoa or chocolate in cakes, biscuits and other dietary products has been made from toasted, defatted wheat germ with added maize sweeteners (U.K. Patent Appl. GB 2031705). The germ is defatted to 1–2% germ oil content and treated with steam at 107°–155°C in the presence of a reducing sugar. The steam treatment is said to develop colour, aroma and taste characteristics resembling those of cocoa.

Milling of brown flour and wholemeal

Wholemeal

In the U.K. wholemeal, by definition, must contain all the products of the milling of cleaned wheat, i.e. it is 100% extraction rate. The crude fibre content usually lies between 1.8 and 2.5% (average 2.2%) but this is not specified in U.K. regulations. In the U.S.A. the 100% product is known as *wholewheat* or *Graham flour*. In India a meal known as *atta* is produced. It is approximately 100% extraction flour, used for making chapattis (see Ch. 13).

Wholemeal may be produced using:

— Millstones — either single or multiple passes with intermediate dressing.
— Roller mills — either a shortened system or by introducing stock diversions in a conventional system.
— Disc grinders — either metal or stone discs.

Wholemeal may also be produced on an impact mill, but this is rarely used for preparation of meal for human consumption.

Brown flours

According to the Bread and Flour Regulations 1984, a brown flour must have a minimum crude fibre content of 0.6% d.b. Brown flours can be of extraction rates between 85% and 98%.

For milling brown flour the white flour process may be modified in various ways, as follows:

1. Releases are increased throughout the break system, by narrowing the roll gap, and by using scalping covers of slightly more open mesh.
2. Operation of the purifiers is altered by adjusting the air valves, altering the sieve clothing and reflowing the cut-offs so that less stock is rejected to wheatfeed, and more goes to the scratch and reduction systems.
3. The scratch system is extended by the use of additional grinding stages.
4. The extraction of flour from the reduction system is increased by more selective grinding, by changing some of the smooth grinding rolls to fluted rolls, and by employing additional grinding stages to regrind offally stock that would be rejected as unremunerative in white flour milling.
5. Additional flour may be obtained by changing the flour sieves to slightly more open mesh sizes.
6. Still additional flour extraction can be obtained, at the sacrifice of good colour, by bringing the wheat on to the 1st. break rolls at 1–1.5% lower moisture content than the optimum for white flour-milling.

Grists

Changes in the composition of the bread grist in the U.K. since 1973, are discussed in Ch. 8. (Fig. 8.1). The proportion of home-grown wheat has increased from about 25% to nearly 80%.

As the home-grown element of the grist has increased, the importance of its quality has also increased. As a result the selective purchasing of favoured varieties has become normal practice.

Steam treatment of wheat

When wheat is treated with live steam, *alpha*-amylase and other enzymes are inactivated, and the protein is denatured so that gluten cannot be recovered (See Ch. 3). Inactivation of amylase is rapid, but not instantaneous; to ensure complete inactivation, the wheat is held at or very near 100°C for 2–4 min in a steamer, which may take the form of a vertical tube well supplied with steam jets. The wheat fills the tube, and the rate of throughput is adjusted to allow the grain an adequate time of treatment. To achieve the same objective, flour may be similarly treated, but with indirect heat and limited amounts of steam, to avoid gelatinization and lumping.

Structure of the U.K. milling industry

Most millers in the U.K. are members of the trade organization NABIM (National Association of British and Irish Millers), which represents 41 companies operating 84 mills. Of these, the three largest members contribute approximately 75% of production.

The proportions of flour types produced have remained fairly consistent, at least over the last five years, figures for which are reproduced in Table 6.7.

Milling of semolina from durum wheat

Durum wheats are milled to produce coarsely ground endosperm particles known as *semolina*, somewhat similar to grits produced by dry milling maize and sorghum. Semolina is used in the manufacture of pasta, it is also cooked unprocessed as the North African staple 'couscous' (See Ch. 10).

The desirable characteristics in a durum semolina include a clear, bright golden appearance, freedom from dark specks originating from impurities, and a high enough content of protein to provide an elastic gluten, thus ensuring the best eating quality of the pasta produced. The maximum ash limit for semolina (U.S.A.) is 0.92%, and for farina 0.6%.

The milling process

Because of the extreme hardness of durum wheat the usual practice is to temper to the high moisture content of 16–16.5% before grinding, further details of tempering are given in Ch. 5.

Grinding

Milling itself consists of three systems, all dependent upon rollermills. The first is the break

TABLE 6.7
U.K. Flour Production by Type

Type of flour	1987/8	1988/9	1989/90	1990/1	1991/2*
Breadmaking					
White	53.4	53.5	54.2	54.6	53.9
Brown	3.4	3.5	3.5	3.4	4.1
Wholemeal	6.8	6.3	6.1	6.1	6.1
Biscuit	13.7	14.6	15.2	14.3	13.8
Cake	1.9	1.9	1.9	1.8	1.6
Pre-packed Household	4.1	3.4	3.1	3.5	3.3
Self-raising	2.7	2.3	2.3	2.2	2.8
For starch manufacture	3.2	3.1	2.8	2.8	3.7
Others	10.8	11.4	10.9	11.3	10.8

*Estimates.
Source: National Association of British and Irish Millers, 1991 and 1992.

FIG 6.15 Purifiers in a modern Durum semolina mill (reproduced by courtesy of Bühler Bros. Ltd. Uzwil, Switzerland).

system and its function is to open the grains and progressively scrape endosperm from the bran, as in the milling of common wheat. Durum milling is characterized by a long break system, with five or six relatively gentle passages. The gentle grinds of the breaks, with rolls set sharp-to-sharp, are designed to release large chunks of endosperm with a minimum of fines. The same number of *detaching* passages as break passages are included. The detacher system is similar to the scratch system used in bread wheat milling, its purpose being not so much to grind but lightly to scrape semolina particles free of any small flakes of bran remaining on them. Detacher rolls have a finer fluting than the corresponding break rolls.

The third rollermilling system featured in durum milling is concerned with *sizing and reduction*. The sizing rolls reduce large endosperm particles to a uniformly smaller size. The rolls are finely fluted. Few smooth reduction rolls are needed in the traditional durum mill as little flour is required. Particles below the size specified for semolina have to be ground to flour size however by reduction rollermills which are characterized by smooth rolls.

Purifying

Stocks leaving the roller mills pass to plansifters to be graded and streamed to machines appropriate to their further treatments. Before further grinding sieved stocks are purified.

Abundant purifiers are characteristic of a mill dedicated to durum wheat milling. A modern purifier is usually a double assembly, in which twin machines work side by side in the same frame (Fig. 6.15). Each half can be adjusted separately and may treat different stocks. A purifier consists of a long oscillating sieve, at a slight downward slope from head to tail, which is divided into four or more sections. The cover becomes progressively coarser from the head to tail. Individually controlled air currents rise through the cover, aspirating stocks as they move over the oscillating sieve cover. When particles of similar density but different particle-size are shaken together, the heavier ones tend to sink. Grading is effected on the basis of size, density and air resistance. Generally, the more bran and less endosperm there is in a particle of a given size, the lower its density and the higher its air resistance will be.

Stratification thus results in the following order:

Light bran and beeswing;
Heavy bran;
Large composites;
Large pure endosperm particles;
Small pure endosperm particles.

Successful purifying depends upon maintaining a continuous layer of pure endosperm along the entire length of the sieve. Thus, only the finest pure endosperm must be allowed to pass through at the beginning, and progressively larger endosperm particles are allowed through as the stocks progress. The lightest impurities are removed by aspiration while overtails are directed to appropriate further treatment. The capacity of a purifier (half the machine) is 0.5–1.4 t/h. It depends on the type of stock being treated, the highest capacities being associated with coarser stocks (Bizzarri and Morelli, 1988).

Within the durum mill, stocks are conveyed through pneumatic lines. The air-flow also serves to cool the mill, particularly the roller mills, which produce heat as a result of grinding. The performance of the mill is continuously monitored by strategically sited weighers, indicating any changes in the output of a particular stage of the process, and providing a warning of faulty tempering or machine condition.

Semolina particle size

The particle-size of semolina and farina in the U.S.A. is such that all the product passes through a No. 20 sieve (840 µm aperture size) but not more than 30% passes a No. 100 sieve (149 µm aperture size).

The degree of grinding and the consequent particle size range considered desirable has changed. With increasing emphasis on continuous processing, a uniform rate of hydration is required, so a uniform particle size is particularly important.

In some mills in the U.S.A. the growing demand for durum flour, which can be processed more quickly than semolina, has been met by installation of additional reduction capacity.

The advantages of rapid hydration must be balanced against the greater starch damage associated with smaller particles, as losses during cooking increase as a function of damaged starch level. A desirable range of semolina particle size is 150–350 µm, but 100–500 µm is a more realistic expectation.

Milling extraction rate

Extraction rate for durum milling may be expressed as *semolina extraction*, or *total extraction* (semolina plus flour). Dick and Youngs (1988) give commercial averages of 60–65% for the granular product with about 3.0% flour, for production in the U.S.A. Figures from the U.S. Department of Commerce suggest higher total extraction rates of 74–78% (mean 75.3%) between 1982 and 1987 (Mattern, 1991). The discrepancy may result from different modes of expression, i.e. on the basis of feed weight or product weight.

Rye milling

The milling methods for rye processing have developed in parallel with those of wheat, but since the end of the last century there have been significant differences between them. The departure began when wheat millers changed from stone to roller grinding. The reasons for the departure are dependent on both the nature of the grain and the requirements of the consumer. Compared with wheat, rye grains have a softer endosperm which breaks down to flour particle size more readily, and greater difficulty is experienced in separating endosperm from bran. Possibly as a consequence of the latter factor, rye consumers have developed a taste for products made from a coarser meal with a higher bran content than in the case of wheat flour. The yield of reasonably pure flour from rye is 64–65%. At increasing extraction rates the flour becomes progressively darker and the characteristic rye flavour more pronounced. In the U.S.A. a good average yield of rye flour would represent a 83% extraction rate (Shaw, 1970).

Rye meal may be of any extraction rate, but rye wholemeal is 100% extraction.

Rye milling varies considerably according to the extraction rate required (wholemeal is produced by two break passages applied to dry grain) and to the geographical region where it is processed; three major regions have been identified as characteristic: North America, Western Europe, and Eastern Europe, including the states of the former U.S.S.R. (Bushuk, 1976).

North American practice is to grind higher protein, smaller sized grains into fine flour with no particular quality requirements.

Western European practice is to grind lower protein, larger sized grains into fine flour with strict quality requirements.

Eastern European rye mills grind all types of rye into a coarse flour according to local quality specifications.

Rye is milled at a moisture content similar to that of soft wheat (14.5–15.5%). Short conditioning (tempering) periods suffice as water penetrates into the grain more quickly than into wheat; between 6 and 15 h are generally used in North America.

An important difference between wheat and rye mills lies in the surface of the rolls: break rolls for rye have shallower and duller grooves to release more break flour. Instead of being smooth or frosted, reduction rolls in a rye mill are finely fluted. This is to cope with the sticky nature of rye endosperm due to the high pentosan content. On smooth rolls excessive flaking would occur. A rye mill in the U.S.A. described by Shaw (1970) has five break passages, a bran duster, and ten reduction passages (Sizings, Tailings and eight Middlings reductions). Purifiers do not feature in rye mills. The specific surface in a modern rye mill would be 18–24 mm per 100 kg of rye milled per 24 h. Rye flour produced in North America may be treated with chlorine (0.19–0.31 g/kg) to improve the colour of white flour.

In Europe, particularly in Germany, the use of combination mills, grinding both wheat and rye, has proved successful (Bushuk, 1976).

Rye milling in the former U.S.S.R. was said to be highly developed (Kupric, 1954). The characteristics of production are: coarser flour granulation, darker flour colour, and high throughput of the plants. To achieve these features coarser flour screens and coarser and deeper roll corrugations were used.

Triticale milling

Milling procedures for triticale resemble those utilized for wheat and rye. Even with the best hexaploid triticales, however, the yield of flour is considerably lower than that of wheat flour. Macri *et al.* (1986) compared milling performance with Canadian Western Red Spring wheat on a Bühler experimental mill. Triticales produced between 58 and 68% flour of 0.44–0.56% ash, while the CWRS gave 71.75% at 0.45% ash. Triticales were tempered to 14.5% mc. Little if any triticale is milled commercially as a separate mill feed.

Rice milling

The main product of rice milling is the dehulled, decorticated endosperm known as *milled rice*, or, after further processing, *polished rice*. Unlike the milled products of most cereals the endosperm remains intact as far as this is possible. *Rice flour* is produced, but this is of minor importance.

As with the processing of most cereals throughout the world the scale of industrial plants is increasing, but, in the case of rice the rate of change is slow and industrial processing accounts for only about half the world crop, the remainder being stored and processed at a village level. At whatever level, the objective of the milling is to remove hull, bran, and embryo as completely as possible without breaking the remaining cores of endosperm. There are two stages, the first being *dehulling* and the second '*whitening*' or *milling*. The moisture at which untreated rice is processed is relatively low, around 11–12% (Sharp 1991) but parboiled rice (see Ch. 5) undergoes heat treatment at 32–38% mc. before drying back to 14%. The method of drying is very important as it influences the degree of breakage occurring during milling. Drying in the sun, or with heated air can lead to cracking and much breakage during milling (Bhattacharya, 1985).

Parboiling is applied to about one fifth of the world's rough rice (cf. Ch. 10). Its effects on milling performance result from changes brought about in both endosperm and hull. In endosperm, starch gelatinization occurs, leading to swelling of the grain which disrupts the 'locking' of the lemma and palea (cf. Ch. 2). Disadvantages of parboiling lie in the additional costs involved, the increased difficulty in removing bran during the whitening process and the measure of discoloration of the endosperm.

Village processing

The *Engleberg huller* is the best known and most widely used small-scale rice mill. It combines the two stages, dehulling and milling, in a single machine. The work is done by a ribbed cylinder rotating on a horizontal axis within a cylindrical chamber, the lower half of which is formed of a screen through which fine material may pass. Rough rice is introduced at one end and passes down the gap between the rotating cylinder and the chamber wall. An adjustable steel blade determines the size of the narrowest gap through which the grain passes and thus determines the level of friction experienced. The rate of flow also serves as a means of controlling the severity of the treatment. The directions of the ribs on the central cylinder vary along its length; the major part of the length has ribs parallel to the cylinder axis but in the early stages dedicated to dehulling, they run in a diagonal curve, helping to feed the stock in as well as to abrade it. The hulls removed in the initial phase help to abrade the grain surface in the later phase. They are ultimately discharged through the slotted screens with fragments of bran, embryo and broken endosperm while the largest endosperm pieces are discharged as overtails of the screen. The proportion of whole endosperm is as variable as the skills of the operators and the condition of the machine. The fine materials are used for feeding domestic stock and, as such, form a valuable by-product.

Small-scale alternatives to the Engelberg huller include two-stage processors in which dehulling is performed by rubber rollers, and whitening is achieved by friction methods. Both processes may be enclosed within a single housing.

Commercial processing

On an industrial scale, processing follows the same principles as those employed at the village level, but it is usual for the processes to be carried out in separate machines. Capacities vary from 1–3 t/h up to 100 t/h but the scale of machinery

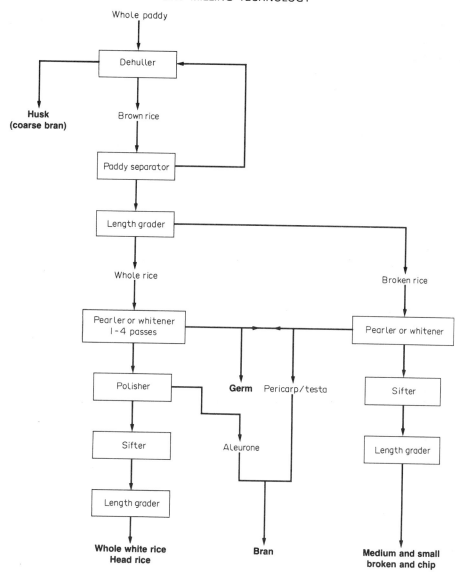

FIG 6.16 Simple schematic diagram of the rice milling process. Products are shown in bold type.

does not vary. The larger mills merely have more of the same machines running in parallel. The process is summarized in Fig. 6.16.

Dehulling

Rubber roll hullers are widely used in modern rice mills. A pair of rolls of equal diameter, run in a similar fashion to the steel rolls of a flour mill, in that both rolls of the pair turn on a horizontal axis, towards the nip, into which the rough rice is fed. Dehulling by an older method, the disc sheller, involves abrasion by discs with surfaces partly covered with coarse emery or carborundum. Two discs are mounted on a vertical axis, the upper one remains stationary while the lower one rotates. Corresponding rings on the outer region of both discs clear each other by a small adjustable gap, through which the rough rice passes, having been fed in from a central hopper.

Optimal hulling results from a clearance of slightly more than half the grain length. Approximately 25% of the grains escape hulling, owing to their small size, and have to be 'returned' for retreatment, not in the same machine but in one with a smaller gap. Such dehullers lead to more breakage of grains than the rubber roller type and they are unsuitable for the automation associated with current practice.

Products of dehullers need to be separated one from another and the usual method for this is *aspiration*. As with other rice machinery, designs abound but the Japanese designed *closed circuit hull aspirator* probably has the most advantages. The air produced by the fan does not leave the machine but is recirculated. A further advantage is that the abrasive hulls do not pass through the fan, thus reducing wear. A more traditional machine performing the function is the *paddy separator*, described in Ch. 5.

Whitening

Having been dehulled, the brown rice is treated to remove bran and embryo. Two principles are involved in design of whitening machines, the most common is the *abrasive cone*. It is used in single or multipass systems, sometimes with the other type of machine, the *friction type*. The cone rotates on a vertical axis, its surface is covered with abrasive material and it is surrounded by a perforated screen. Rubber brakes are installed at intervals around the cone to provide a smaller clearance than that between the rotor and the screen. Raising the cone increases the gap between it and the screen and the brakes can be adjusted independently, to determine the severity of the pearling. The size of cone is inversely related to the number of passes involved in the system.

The innovative vertical whitener (Fig 6.17) employs abrasive rollers and, it is claimed, produces smoother rice with fewer grains breaking during processing.

Friction machines always occur in multipass systems in which abrasive cones also feature. A friction whitener comprises a rotating cylinder within a chamber of hexagonal section, and walls of screening with slotted perforations. The horizontal cylinder has air passages through its centre that pass to the surface as a series of holes in parallel alignment. Rotation of the cylinder causes friction within the grain mass surrounding it, and air from the cylinder cools the stock and supports the discharge of bran. Bran and germ pass out through the perforated screens. A recent innovation introduced in Japanese rice processing is '*humidified friction whitening*'. This concept includes the injection of a small amount of atomized water in the pressurized air stream through the cylinder airways. The process serves a similar purpose to tempering of maize or wheat; it toughens the bran, facilitating its removal. The ultimate result is the reduction in endosperm breakage, and the increased recovery of head rice. Humidification may also be combined with cooling before whitening. The cooling hardens the grain leading to less breakage. In the more modern, larger mills a high degree of automation exists in the whitening process, including microcomputer control.

Polishing

The rice emerging from the whitening process has had the outer layers of bran removed but the inner layers remain. The remaining fragments are removed by the process known as *polishing* or *refining* ('*polishing*' is used by the Japanese to describe the *milling* process — terminology can be very confusing!). In conventional rice mills polishers are like whitening cones, but instead of abrasive coverings, the cone is covered with many leather strips. No rubber brakes are applied.

Polishing extends the storage life of the product as the aleurone layer is removed, thus reducing the tendency for oxidative rancidity to occur in that high-oil tissue.

For a description of coated rice see Ch. 14.

Grading of milled rice products

Milled rice produced by the milling section of commercial rice mills is a mixture of entire and

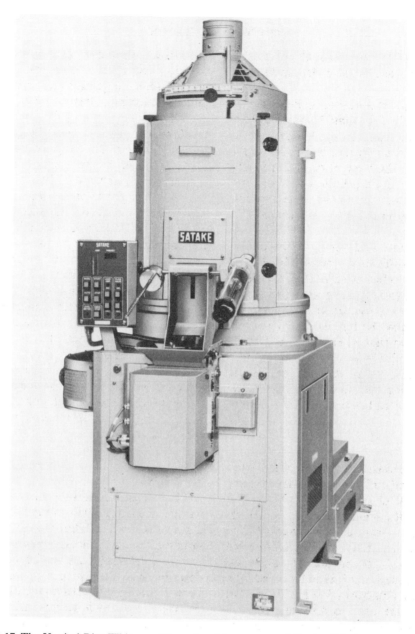

FIG 6.17 The Vertical Rice Whitener. (Reproduced by courtesy of Satake, UK, Ltd.)

broken endosperms of various sizes. The number of grades separated depend on the sophistication of the system and of the market. In developing countries only *'brewers rice'* may be separated from the main product. Brewers rice passes a sieve with 1.4 mm round perforations (F.A.O. definition). Other markets may justify grading into *'head rice'* (grains entire except for the tip at the embryo end), *'large brokens'* and *'small brokens'* as well. Separation is by trieurs or disc separators (see Ch. 5).

Rice flour

Rice flour, bakers' cones and ground rice are the products of grinding milled rice; the head rice is rarely used as a feedstock because of its high value, but the 'second head' product, comprising endosperm particles of about half grain size, is used.

Approximate proportions of products (as percentages of paddy) are husk, 20–30%; milled rice, 63–72%; bran, 7–8%.

Barley milling

Barley is milled to make *blocked barley*, *pearl barley* (Graupen in German), *barley groats* (Greutz), *barley flakes* and *barley flour* for human consumption. Removal of the hull or husk, which is largely indigestible, is an important part of the milling process.

The hardness of barley grains is a characteristic dependent upon type and variety; types with blue coloured aleurone layers tend to be the harder types. For milling purposes, the harder types are preferred, as the objective is generally not to produce flour but to remove the hull and bran by superficial abrasion, yielding particles which retain the shape of the whole grain. With this type of processing, softer grains would tend to fragment, leading to a reduction in the yield of first quality products. Barley for milling should have as low a hull content as possible.

The presence of damaged grains lowers the quality of milling barley. Such grains frequently reveal areas of exposed endosperm where fungal attack may occur, leading to discoloration. Such grains would contribute dark particles to the finished product. Thin grains also lower the milling quality because of the increased proportion of hull.

Processes in barley milling

Following cleaning and conditioning and possibly bleaching e.g. in Germany (not practised in the U.K.) barley is treated in the following sequence to produce the various products:

Blocking.
Aspiration — to remove husk.
Sifting.
Cutting — on groat cutter (for barley groats).

Pearling of blocked barley or large barley groats (to produce pearl barley).
Aspiration, grading, sifting.
Polishing.

For making barley flakes, barley groats or pearl barley are subjected to:

Damping.
Steam cooking.
Flaking — on flaking rolls.
Drying — on a hot air drier.

For making barley flour, blocked or pearl barley is milled on roller mills.

Blocking and pearling

Both blocking (shelling) and pearling (rounding) of barley are abrasive scouring processes, differing from each other merely in degree of removal of the superficial layers of the grain. Blocking removes part of the husk. This process must be accomplished with the minimum of injury to the grains. Pearling, carried out in two stages, removes the remainder of the husk and part of the endosperm. The products of these processes are blocked barley, seconds, and pearled barley, respectively. The three processes remove about 5%, 15%, and 11% respectively, to yield a final product representing about 67% of the grain. Three types of blocking and pearling machine are in general use. The first two are used in Britain, all three are in use in continental Europe:

1. Batch machine, consisting of a large circular stone, faced with emery-cement composition, rotating on a horizontal axis within a perforated metal cage.
2. Continuous-working machine which is of Swedish make, consisting of a rotor faced with abrasive material, rotating on a horizontal axis within a semicircular stator, lined with the same material. The distance between rotor and stator is adjustable.
3. Continuous-working machine comprising a pile of small circular stones rotating on a vertical axis within a metal sleeve, the annular space between the stones and the sleeve, occupied by the barley, being strongly aspirated.

TABLE 6.8
Chemical Composition of Milled Barley Products (Dry Matter Basis)*

Material	Protein (%)	Fat (%)	Ash (%)	Crude fibre (%)	Carbohydrate (%)	Energy kJ/100g
Pearl barley	9.5	1.1	1.3	0.9	85.9	1676
Barley flour	11.3	1.9	1.3	0.8	85.4	1693
Barley husk	1.6	0.3	6.2	37.9	53.9	560
Barley bran	16.6	4.0	5.6	9.6	64.3	—
Barley dust	13.6	2.5	3.7	5.3	74.9	—

* Values calculated from Kent, 1983.

The barley falls between the rotor and the stationary part of each machine; in bouncing from one surface to the other, the husk is split or rubbed off. The degree of treatment, resulting in either blocking or pearling, is governed by the abrasiveness of the stone facing, the distance between rotor and stator, and by the residence period in the machine.

Aspiration of the blocked or pearled grain to remove the abraded portions, and cutting of the blocked barley into grits, are similar to the corresponding processes in oatmeal milling. Cutting of blocked barley is not commonly carried out in the U.K. where it is the practice to pass the whole blocked barley grains into the pearling machine. In Germany, however, the blocked barley is first cut into grits, the grits graded by size, and then rounded in the pearling machine. Pearl barley is polished on machines similar to those used for pearling, but equipped with stones made of hard white sandstone instead of emery composition. The practice of adding talc (magnesium silicate) in polishing, once customary in Germany, has been discontinued. It was never used in Britain.

Bleaching

The bleaching of barley is not permitted by law in Britain, but is generally practised in Germany. Imported barley is preferred to domestic grain for milling in Germany because of its greater hardness and yielding capacity, and it is the foreign barley, the aleurone which has a bluish colour, which is said to require bleaching.

Barley flakes

Pearl barley is converted to flakes by steaming and flaking on large diameter rolls, as described for oats. The flakes have been used as a flavouring ingredient in speciality breads in the U.S.A.

In the U.K. grades of pearl barley are limited to *Pot barley*, *First Pearl* and *Second Pearl*. In Germany an elaborate scheme of grading and sizing exists which divides the grain into as many as twelve different sizes. Much of the pearl barley used in the U.K. is imported (unbleached) from Germany.

Barley flour

Barley flour is milled from pearled, blocked or hull-less barley. Optimum tempering conditions are 13% mc. for 48 h for pearl barley, 14% mc. for 48 h for unpearled, hull-less grain. The milling system uses roller mills with fluted and smooth rolls, and plantsifters, in much the same way as in wheat flour milling. Barley flour is also a by-product of the pearling and polishing processes.

Average extraction rate of 82% of barley flour is obtained from pearl barley representing 67% of the grain, i.e. an overall extraction rate of 55% based on the original whole grain (including the hull). By using blocked grain an overall extraction rate of 59% of the whole grain could be obtained, but the product would be considerably less pure than that produced from pearl barley.

It is possible to mill a mixture of wheat and hull-less barley in ratios between 90:10 and 80:20.

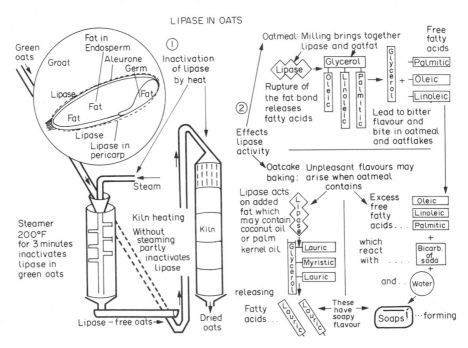

FIG 6.18 Main points in the relationship between fat and the enzyme lipase in oatmeal milling and oatcake baking.

This reduces the tendency for by-products of barley milling to overload the system.

Other milled barley products

Malted barley flour is made by grinding or milling barley malt. The chemical composition of milled barley products is shown in Table 6.8.

Oat milling

The processing of oats in the mill has two complications, one of which arises from the structure of the oat grain, the other from its chemical composition.

The oat grain consists of the groat, or caryopsis, and a surrounding hull or husk. Only the groat is required for the milled products; the hull is indigestible and must therefore be separated and removed in a dehulling (or shelling) stage. In this respect, oats resemble barley and rice.

The groat has a lipid content which is between 2 and 5 times as high as that of wheat, and the groat also contains an active lipase (lipid-splitting enzyme). The lipase is located almost entirely in the pericarp of the groat (Hutchinson *et al.*, 1951), whereas the lipid occurs in the starchy endosperm, aleurone and embryo, the last two being particularly rich sources. There is little, if any, lipid in the pericarp.

Thus, in the intact groat of raw oats the lipase and the lipid do not come into contact with each other and, hence, little or no hydrolysis of the lipid occurs. However, when the grain is milled, the enzyme and substrate come together, and hydrolysis ensues. The substrate consists essentially of acyl glycerols, and hydrolysis leads to the production of glycerol and free fatty acids, mainly oleic, linoleic and palmitic acids. The glycerol is neutral and stable, but the free fatty acids progressively give the product a bitter flavour as they are slowly released. Thus, the inactivation of the enzyme, by a process called stabilization, is an essential step in the milling of oat, a process that is not required in the milling of any other cereal.

If the lipase is not inactivated, then its effects become apparent in oatcakes baked from oatmeal, fat and water. For making oatcakes, the raising agent frequently used is sodium hydrogen carbonate $NaHCO_3$. If free fatty acids are present in the oatmeal, they react with the raising agent to form the sodium salts of fatty acids, which are soaps. The resulting oatcakes thus have a soapy flavour. Furthermore, the fat added to the oatmeal in oatcake baking may be animal fat, e.g. beef dripping, or vegetable fat. All types of fat consist of glycerol plus fatty acids, the particular fatty acids concerned varying from fat to fat. The use of certain vegetable fats, e.g. palm kernel and coconut oils, in oatcake baking is particularly undesirable, if the oatmeal contains an active lipase, because the fatty acids in these fats are chiefly lauric and myristic; the former, when released by lipase, has an unpleasant soapy taste.

A summary of the main points on the relationship between lipase and fat in oatmeal milling and oatcake baking is shown in Fig. 6.18.

Processes in oat milling

Two systems of oat milling have been established. They differ principally as regards the moisture content of the oats at the shelling stage, and in the method of shelling employed, viz. by abrasion on stones, or by impact.

The traditional, or dry-shelling, system was the system formerly used in most oat mills in the U.K., and is still currently in use in a few small mills in Scotland.

The modern green-shelling system has now replaced the traditional dry-shelling system in most U.K. mills. In the U.S.A. one process in use involves whole oats being *pan roasted* before shelling; this imparts a caramelized or toasted flavour to the product.

The sequence of operations in both systems is as follows:

Modern (green shelling)
1. width grading,
2. shelling (by impact),
3. stabilization,
4. kiln-drying,
5. length grading,

Traditional (dry-shelling)
1. width grading,
2. stabilization,
3. kiln-drying,
4. length grading,
5. shelling (on stones),

6. cutting,
7. grinding (for oatmeal, oatflour and oat bran),
8. steaming and flaking (for rolled oats).

Grading

The cleaning operations directed to the removal of foreign matter from oats are similar to those employed for other cereals; they are described in Ch. 5. Before milling, however, oats are subjected to a further treatment because of a problem peculiar to them. The treatment removes oats of narrow width, known as '*light grains*', which consist of a husk enclosing only a rudimentary groat. The length of these grains is not necessarily less than that of normal grains, so they may defy the selective processes of the cleaning treatments. The grading apparatus comprises a perforated cylinder slowly rotating on a horizontal axis. The screen may be slotted or a specially designed mesh. If not removed thin grains cause many problems in the later stages of milling.

Stabilization

For the inactivation of enzymes the grains pass through a continuous cooker into which live steam is injected. The best results are obtained by treating the raw or 'green' oats before kiln-drying, at a moisture content of 14–20% (below 14% inactivation is incomplete). The grain is quickly raised to a temperature of 96°C by injection of steam at atmospheric pressure, and thereafter maintained at that temperature for 2–3 min by controlling the rate of throughput of the steamer.

When the green-shelling process was first introduced, stabilization was carried out on the whole *oats*, as in the dry-shelling process. But in the modern application of the green-shelling process, stabilization is carried out on the *groats* after shelling. Adoption of the current practices is justified mainly on economic grounds, as energy is not expended on heating the husk, from which lipase is absent and which is removed from the high-value product anyway. It might also be argued that inactivation of lipase is more effective on the unprotected *groat*, because of easier access of the inactivation agent to the site of the enzyme.

A check on the completeness of lipase inactivation may be made by applying the tetrazolium test (cf. p. 113). If, after stabilization, a red colour develops during the test at the embryo end of the groat, it follows that the heat treatment was insufficient to inactivate dehydrogenase enzymes, and that other enzymes, notably lipase, have probably survived also. Tests for other enzymes, e.g. tyrosinase, may also be employed. Excessive steaming, beyond that required for enzyme inactivation, is to be avoided, because oxidation of the fat, resulting in oxidative rancidity, may be encouraged by heat treatment.

Kiln-drying

In the traditional dry-shelling system, the *oats*, after stabilization, are dried to 4–8% m.c. Continuous drying is frequently carried out in a Walworth Kiln, in which the hot air is drawn through an annular layer of slowly descending oats. In the modern green-shelling process, the oats are shelled before kiln-drying, and it is therefore the *groats* that are kiln-dried, often by passing them over a series of radiators at a rate which is controlled to give a final moisture content of approximately 12%. The final section of the kiln is a cooling section, in which fresh air is drawn through the groats to ensure that condensation is not present, as this would could cause deterioration during storage. Cooling is also assisted by a further aspiration to separate slivers of husk loosened during kilning. A modern alternative to kilning is micronizing (see Ch. 9).

The purposes of kiln-drying in the traditional dry-shelling system are:

1. To reduce the moisture content to a satisfactory level: 15% for storage, 13% for international trade, about 6% for immediate milling.

2. To facilitate the subsequent shelling of the oats by increasing the brittleness of the husk.
3. To develop in the oats a characteristic flavour, often described as 'nutty'.

In the modern green-shelling system, the function of kilning is limited to the reduction of the moisture content, shelling already having taken place. Herein is a complication of the green-shelling process, because the high temperatures in the later part of the process that have been found necessary for the development of flavour during the kilning of *oats* are quite unsuitable for the kiln-drying of *groats*; the groats no longer have the protection afforded by the husk, as in the case of whole *oats*, and would become burnt and discoloured at such high temperatures. Moreover, high-temperature treatment of the *groats* at low moisture content may lead to the onset of oxidative rancidity. As a result of these limitations, *groats* are kiln-dried at somewhat lower temperatures than those used for whole *oats*, and the products are devoid of the typical oaten flavour. The flavour of oaten products made by the green-shelling process, often described as 'bland', seems insipid to palates familiar with products made by the traditional dry-shelling process.

The high-temperature kilning of *oats* facilitates the subsequent inactivation of any residual lipase by steam in the cooking process to which pinhead meal is subjected before flaking, whereas more severe steam treatment is necessary to inactivate residual lipase in the case of products made by the green-shelling process, in which lower kilning temperatures are used.

Shelling

In the dry-shelling process, the kiln-dried oats are passed between a pair of large circular stones, one stationary the other revolving. The two stones are separated by a distance slightly less than the length of the oat grain: as the grains roll over and up-end the husk is split off in thin slivers. Careful adjustment of the shelling stones, and length grading of the oats before shelling ensure minimum breakage of the groats.

In the green-shelling process, which has almost universally replaced stone-shelling, oats are shelled at natural moisture content, viz 14–18%, on impact machines, such as the Murmac huller. The whole oats are fed into the centre of a high-speed rotor fitted with either blades or fins. The grains are thrown outwards and strike, point-first, a hardened ring, possibly of carborundum, possibly of hardened rubber, attached to the casing of the machine. The combination of high velocity and impact against the ring detach the lemma and palea from the groat. The speed of the rotor is adjustable (typically 1400–2000 rpm) allowing the process to be optimized for a particular feedstock. Both type of oat and moisture content will dictate the conditions required.

A variation of the green-shelling process is the wet-shelling process, invented by Hamring (1950). In this process the whole oats are first damped to 22% m.c. or higher, before shelling by impact. It is claimed that the preliminary moistening decreases the breakage of groats and increases the efficiency of shelling, in comparison with the green-shelling process.

After shelling, the husk slivers are separated from the kernels and unshelled oats by aspiration. The few unshelled oats are then removed on a 'paddy' or inclined table separator (cf. p. 119), on which groats and unshelled oats move in opposite directions. The machine is designed to exploit differences in specific gravity and resiliency (or 'bounce') between oats and groats. The unshelled oats are returned to the shelling stage.

After separation of the husk slivers by aspiration, the groats pass to a 'clipper' or scourer, the abrasive action of which removes surface hairs ('dannack') and any remaining fragments of husk. Further aspiration separates the cleaned groats from the 'oat dust', as the by-product is named.

The groats are graded so that particular size fractions can be directed to further processing appropriate to their sizes, thus large groats are used for flaking into 'jumbo' oatflakes, medium groats are cut to produce quick-cooking flakes, or even for oat bran production and small groats are ground into flour.

Cut groats

Cut groats are the highest volume oat product for human consumption. The machine by which they are produced is the granulator, in which hollow perforated drums rotate on a horizontal axis, typically at 37–40 rpm. On the inner surface, the circular holes that pierce the walls are counter-sunk, so that the elongated groats, having been fed into the drums, naturally fall lengthwise into the holes. As the drum rotates they continue to progress through the holes and project beyond the outer profile of the drum. Outside the drum, and parallel to its axis a series of stationary knives is placed in such a position to execute a series of equidistant cuts on groats passing progressively through the holes, thus producing uniformly sized 'pinhead' granules. The number of groat sections produced depends upon drum speed, but on average, groats are cut into three parts. The holes are kept clear by pins projecting from smaller cylinders mounted parallel to the drums, so that the pins register with the holes at a point on the circumference where the drum is running empty.

Not all groats are perfectly divided and, as a consequence, excessively long fragments and whole groats need to be removed, but first the total stock is aspirated to remove husk fragments and floury material. The amount of fines produced depends on the sharpness of the knives. If the knives are sharp, 1.2–1.3% fines may result but if dull up to 10% is possible (Deane and Commers, 1986). The size selection is made with disk separators or trieur cylinders (p. 116); the good product becomes the liftings and it is now ready for flaking. Oversize portions are returned to the granulator for retreatment.

The cutting process produces a small amount of flour known as 'flow meal'; it is separated by sieving and is used for making dog biscuits.

Flaking

This is carried out by passing steam-cooked groats or parts of groats between a pair of flaking rolls. In the steamer the temperature is raised to 99°–104°C and moisture content rises from 8–10% to 10–12%. Steaming performs two functions at this stage: it completes the inactivation of lipase and prepares the groats for flaking: they enter the roll nip while they are still hot, moist and plastic.

Flaking rolls are much heavier and larger than standard flour milling rolls, typically they are 305–711 mm in diameter and 762–1321 mm long, depending on capacity requirements. Rolls run at zero differential, and at 250–450 rpm, roll pressure and roll gap are maintained hydraulically. Flakes are dried, and cooled and passed over a final sifter and metal detector before packaging at a moisture content of about 10%. The size of the groat or granule that is flaked determines the amount of domestic cooking that the product requires. The smaller the feed to the steam cooker, the greater the proportion of water absorbed and the greater the proportion of starch granules that become gelatinized. The more gelatinization, the less domestic cooking is required. Cooking requirements can also be controlled by varying roll pressures and steam cooking as thinner flakes are more rapidly cooked. Instant or quick-cooking flakes are about 0.25–0.38 mm thick while traditional rolled oats may be 0.5–0.76 mm thick. The product represents 50–60% w/w, of the whole oats processed. The composition of oat flakes is much the same as that of the whole groat as no separation is made of the endosperm, pericarp and embryo during processing.

Oat bran and oat flours

In recent years, particularly in North America, a market for oat bran has developed on account of its high soluble fibre content (see Ch. 14). The response to the demand has come in several different forms, some millers using grinders, others using stones to remove as much as possible of the endosperm from the pericarp wherein the fibre is concentrated. The by-product from the process is 'stripped' flour; it is used in production of extruded products, which are popular as ready-to-eat breakfast cereals.

A more traditional oat flour is produced by grinding stabilized groats, usually in a hammer mill with a screen chosen to suit the specification, and sieving the product to ensure no large pieces. Oat flour produced in this way or by grinding oat flakes, is also used as a feed to the extrusion process.

The method of flour production has an important influence on the viscosity of flour/water slurries produced.

White groats

This product, used for black puddings and haggis, is made by damping groats and subjecting them to a vigorous scouring action in a barley polisher, pearler or blocker before the moisture has penetrated deeply into the grains. Alternatively, the groats may be scoured without any preliminary damping. A proportion of the pericarp is removed by this process. There is also a certain amount of breakage, with the release of oat flour; the latter becomes pasted over the groats (no attempt being made to remove it), which thus acquire a whitened appearance.

References

ANON. (1988) *Mill Processes I*. Module 8 in Workbook Series. Incorporated National Association of British and Irish Millers. London.

ANON. (1990) Special Report. *Milling*. July: 30.

BEECH, G. A. and CRAFTS-LIGHTLY, A. F. (1980) Energy use in flour production. *J. Sci. Fd. Agric.* **31**: 830.

BHATTACHARYA, K. R. (1985) Parboiling of rice. In: *Rice: Chemistry and Technology*. 2nd edn, JULIANO, B. O. (Ed.) Amer. Assoc. of Cereal Chemists Inc. St. Paul MN. U.S.A.

BIZZARRI, O. and MORELLI, A. (1988) Milling of durum wheats. In: *Durum: Chemistry and Technology*, pp. 161–188, FABRIANI, G. and LINTAS, C. (Eds.) Amer. Assoc. of Cereal Chemists Inc., St. Paul, MN. U.S.A.

BREKKE, O. L. and KWOLEK, W. F. (1969) Corn dry-milling: cold tempering and degermination of corn of various initial moisture contents. *Cereal Chem.* **46**: 545–559.

BUSHUK, W. (1976) History, world distribution, production and marketing. In: *Rye, Production, Chemistry and Technology*, pp. 1–7, BUSHUK, W. (Ed.) Amer. Assoc. of Cereal Chemists Inc., St. Paul, MN. U.S.A.

CECIL, J. (1987) White flour from red sorghums. *Milling*. Nov., 17–18.

CHINSMAN, B. (1984) Choice of technique in sorghum and millet milling in Africa. In: *The Processing of Sorghum and Millets: Criteria for Quality of Grains and Products for Human Food*, pp. 83–92, DENDY, D. A. V. (Ed.) ICC Symposium, Vienna 1984.

DEANE, D. and COMMERS, E. (1986) Oat cleaning and processing. In: *Oats: Chemistry and Technology*, pp. 371–412, WEBSTER, F. H. (Ed.) Amer. Assoc. of Cereal Chemists Inc. St Paul, MN. U.S.A.

DEOBOLD, H. J. (1972) Rice flours. In: *Rice: Chemistry and Technology*, pp. 264–271, HOUSTON, D. F. (Ed.) Amer. Assoc. of Cereal Chemists Inc. St. Paul. MN. U.S.A.

DESIKACHAR, H. S. R. (1977) Processing of sorghum and millets for versatile food uses in India. *Proc. Symp. Sorghum and Millets for human food*. ICC. Vienna 1976.

DICK, J. W. and YOUNGS, V. L. (1988) Evaluation of durum wheat, semolina, and pasta in the United States. In: *Durum: Chemistry and Technology*, pp. 237–248, FABRIANI, G. and LINTAS, C. (Ed.) Amer. Assoc. of Cereal Chemists Inc. St. Paul, MN. U.S.A.

EASTER, W. E. (1969) The dry corn milling industry. *Bull. Ass. oper. Millers*, p. 3112.

GREER, E. N., HINTON, J. J. C., JONES, C. R. and KENT, N. L. (1951) The occurrence of endosperm cells in wheat flour. *Cereal Chem.* **28**: 58–67.

HAMRING, E. (1950) Wet shelling process for oats. *Getreide Mehl Brot.* **4**: 177.

HUTCHINSON, J. B., MARTIN, H. F. and MORAN, T. (1951), Location and destruction of lipase in oats. *Nature, Lond.* **167**: 458.

JOHNSON, L. A. (1991) Corn: production, processing, and utilization. In: *Handbook of Cereal Science and Technology*, pp. 55–131, LORENZ, K. J. and KULP, K. (Eds.) Marcel Decker, Inc. NY. U.S.A.

KENT, N. L. (1965) Effect of moisture content of wheat and flour on endosperm breakdown and protein displacement. *Cereal Chem.* **42**: 125.

KENT, N. L. (1966) Subaleurone endosperm cells of high protein content. *Cereal Chem.* **43**: 585–601.

KENT, N. L. (1983) *Technology of Cereals*. 3rd Edn. Pergamon Press, Oxford.

KUPRIC, J. N. (1954) Malomipari Technologia. Cited by ROZSA, T. A. Rye milling. In: *Rye, Production, Chemistry and Technology*, pp. 111–125, BUSHUK, W. (Ed.) Amer. Assoc. of Cereal Chemists. St. Paul, MN. U.S.A.

LOCKWOOD, J. F. (1960) *Flour Milling*. 4th edn. Northern Publn. Co. Ltd, Liverpool.

MACRI, L. J., BALLANCE, G. M. and LARTER, E. N. (1986) Factors affecting the breadmaking potential of four secondary hexaploid triticales. *Cereal Chem.* **63**: 263–267.

MATTERN, P. J. (1991) Wheat. In: *Handbook of Cereal Science and Technology*, pp. 1–54, LORENZ, K. J. and KULP, K. (Eds.) Marcel Dekker Inc, NY. U.S.A.

NATIONAL ASSOCIATION OF BRITISH AND IRISH MILLERS (1991) *Facts and Figures 1991*. NABIM, London.

NATIONAL ASSOCIATION OF BRITISH AND IRISH MILLERS (1992) *Facts and Figures 1992*. NABIM, London.

PERTEN, H. (1977) UNDP/FAO sorghum processing project in the Sudan. Proc. Symp. *Sorghum and Millets for Human Food*, pp. 53–55, ICC, Vienna, 1976.

REICHERT, R. D., YOUNGS, C. G. and OOMAH, B. D. (1982) Measurement of grain hardness and dehulling quality with a multi-sample tangential abrasive dehulling device (TADD). In: *Proceedings of the International Symposium on Sorghum Grain Quality, 1981*, ROONEY, L. W. and MURTY, D. S. (Eds.) International Crops Research Institute for Semi-Arid Tropics. Patancheru, India.

ROONEY, L. W., KIRLEIS, A. W. and MURTY, D. S. (1986) Traditional foods from sorghum: their production, evaluation and nutritional value. *Adv. in Cereal Sci. Technol.* **8**, 317–353.

ROONEY, L. W. and SERNA-SALDIVAR, S. (1991) Sorghum. In: *Handbook of Cereal Science and Technology*, pp. 233–270, LORENZ, K. J. and KULP, K. (Eds.) Marcel Dekker Inc. NY. U.S.A.

SHARP, R. N. (1991) Rice. In: *Handbook of Cereal Science and Technology*, pp. 301–330, LORENZ, K. L. and KULP, K. (Eds.) Marcel Dekker Inc. NY. U.S.A.

SHAW, M. (1970) Rye milling in U.S.A. *Ass. Operative Millers. Bull.* 3203–3207.

SPENCER, B. (1983) Wheat production, milling and baking. *Chemy. Ind.* 1 Aug. 1983.

STIVER, T. E., Jr (1955) American corn-milling systems for degermed products. *Bull. Ass. oper. Millers*. 2168.

STORCK, J. and TEAGUE, W. D. (1952) *Flour for Man's Bread*. Univ. of Minnesota Press. Mineapolis, U.S.A.

STRINGFELLOW, A. C. and PEPLINSKI, A. J. (1966) Air classification of sorghum flours from varieties representing different hardnesses. *Cereal Sci. Today* **11**: 438–440.

VAN RUITEN, H. T. L. (1985) Rice milling: an overview. In: *Rice: Chemistry and Technology*, 2nd edn, JULIANO, B. O. (Ed.) Amer. Assoc. of Cereal Chemists. St. Paul, MN. U.S.A.

VARRIANO-MARSTON, E. and HOSENEY, R. C. (1983) Barriers to increased utilization of pearl millet in developing countries. *Cereal Foods World* **28**: 392.

Further Reading

ANON. (1987–90) Modules of Workbook Series: *4. Instrumentation and Process Control; 8. Mill Processes I & II; 9. Mill Machinery; 10. Product handling, Storage & Distribution; 11. Wheat Intake/Mill Performance/Quality Control; 15. Air, Conveying & Power; 16, U.K. Flour Milling Industry*. National Association of British and Irish Millers, London.

FLORES, R. A., POSNER, E. S., MILLIKEN, G. A. and DEYOE, C. W. (1991) Modelling the milling of hard red winter wheat: estimation of cumulative ash and protein recovery. *Trans. ASAE*, Sept.–Oct., 2117–2122.

HOSENEY, R. C., VARRIANO-MARSTON, E. and DENDY, D. A. V. (1981) Sorghum and millets. *Advances in Cereal Science and Technology* **4**: 71–144.

SHELLENBERGER, J. A. (1980) Advances in cereal technology *Advances in Cereal Science and Technology* **3**: 227–270.

WINGFIELD, J. (1989) *Dictionary of Milling Terms and Equipment*. Association of Operative Millers. U.S.A.

YAMAZAKI, W. T. and GREENWOOD, C. T. (Eds.) (1981) *Soft Wheat*. Amer. Assoc. of Cereal Chemists Inc. St. Paul MN. U.S.A.

YOUNGS, V. L., PETERSON, D. M. and BROWN, C. M. (1982) Oats. *Advances in Cereal Science and Technology* **5**: 49–106.

7

Flour Quality

Introduction

In the milling of cereals by the gradual reduction system (see Ch. 6), flour is produced by every machine in the break, scratch and reduction systems of the normal mill-flow. The stock fed to each grinding stage is distinctive in composition — in terms of proportions of endosperm, embryo and bran contained in it, and the region of the grain from which the endosperm is derived — and each machine flour is correspondingly distinctive in respect of baking quality, colour and granularity, contents of fibre and nutrients, and the amount of ash it yields upon incineration.

By far the most abundant flour consumed in the industrialized world is derived from wheat; because of this, and the unique versatility of wheaten flour, the majority of this chapter is devoted to it. Flours from other cereals are however given some consideration.

In the U.K. today there are no recognized standards for flour grades: each miller makes his grades according to customer's requirements, and exercises his skill in maintaining regularity of quality for any particular grade.

Flour grades

If the flour streams from all the machines in the break, scratch and reduction systems are blended together in their rational proportions, the resulting flour is known as 'straight-run grade'. Other grades are produced by selecting and blending particular flour streams, frequently on the basis of their ash yield or grade colour (measures of their non-endosperm tissue content).

Table 7.1 shows typical proportions of flour streams (expressed as percentages of the wheat) from a well-equipped and well-adjusted mill in the U.K. making flours of fairly low ash yield. The ash yield of the individual flour streams is also shown.

Flour streams with the lowest ash yield (e.g. group 1 in Table 7.1) may be described as 'patent' flour. Those from the end of the milling process with high ash yield are called 'low-grade' in the

TABLE 7.1
Typical Proportions and Ash Yields of Flour Streams

Flour streams	Proportions (% of feed to I Bk)	Ash yield (%, d.m.)
Group 1: High Grade		
A	12.0–21.0	0.35–0.38
B	14.0–17.0	0.35–0.38
C	7.0–10	0.38–0.47
Total group 1	35–40	0.35–0.40
Group 2: Middle Grade		
D	2.5–7.5	0.39–0.70
E	1.7–2.1	0.45–0.89
G	1.3–3.0	0.75–1.47
I Bk	1.5–2.5	0.50–0.72
II Bk	1.5–3.0	0.53–0.69
III Bk	0.0–1.5	0.70–1.00
III Bk bran finisher flour	0.0–2.5	0.70–1.00
X (Scratch)	0.0–0.7	0.70–0.90
I Bk Coarse Midds	3.0–6.0	0.50–0.82
II Bk Coarse Midds	1.5–3.5	0.70–0.84
Total Group 2	25–30	0.70–0.80
Group 3: Low grade		
B2	1.2–2.5	0.40–0.45
F	0.7–1.2	0.58–1.35
H	0.6–1.2	0.60–1.53
J	0.5–0.7	0.88–2.25
IV Bk	2.0–4.0	1.00–2.00
IV Bk finisher flour	0.0–1.0	1.50–2.00
V Bk	0.0–1.0	1.00–2.50
Total Group 3	8–10	1.80–2.3

U.K. or 'clear' flour in the U.S.A. Clear flour is used industrially in the U.S.A. for the manufacture of alcohol, gluten, starch and adhesives (see Ch. 15).

Treatments of wheat flour

Bleaching

Flour contains a yellowish pigment, of which about 95% consists of xanthophyll or its esters, and has no nutritional significance. Bleaching of the natural pigment of wheat endosperm by oxidation occurs rapidly when flour is exposed to the atmosphere, more slowly when flour is stored in bulk, and can be accelerated by chemical treatment. The principal agents used, or formerly used, for bleaching flour are nitrogen peroxide, chlorine, chlorine dioxide, nitrogen trichloride, benzoyl peroxide and acetone peroxide.

Nitrogen peroxide (NO_2)

NO_2 produced by a chemical reaction or by the electric arc process was widely used as a bleaching agent in the early twentieth century. Its use has been discontinued except in the U.S.A and Australia, where it is still legally permitted.

Chlorine

The use of chlorine gas (Cl_2) for treatment of *cake* flour (except wholemeal) is permitted in the U.K. to a maximum of 2500 mg/kg. The chlorine modifies the properties of the starch for high-ratio cake flour (cf p. 178). For cake flours the usual level of treatment is 1000–1800 mg/kg. The Bread and Flour Regulations 1984 do not permit its use in bread flour in the U.K. The use of chlorine is not permitted in most European countries, but it is allowed in flour for all purposes in the U.S.A., Canada, Australia, New Zealand (to 1500 mg/kg) and South Africa (to 2500 mg/kg).

Nitrogen trichloride

This gas (NCl_3), known as 'Agene', was patented as a flour bleach by J. C. Baker in 1921, and replaced chlorine in 1922 as an improving and bleaching agent for breadmaking flour because it was much more effective. Its use was discontinued in the U.S.A. in 1949 and in the U.K. from the end of 1955, after it had been shown by Mellanby (1946) that flour treated with Agene in large doses might cause canine hysteria (although Agene-treated flour has never been shown to be harmful to human health). Nitrogen trichloride reacts with the amino acid methionine, present in wheat protein, to form a toxic derivative, methionine sulphoximine (Bentley *et al.*,1950).

Chlorine dioxide

Chlorine dioxide (ClO_2), known as 'Dyox', is now the most widely used improving and bleaching agent in the U.K., the U.S.A., Australia and Canada. It was first used for these purposes in 1949 in the U.S.A. and in the U.K. in 1955. The gas is produced by passing chlorine gas through an aqueous solution of sodium chlorite. Dyox gas contains a maximum of 4% ClO_2. The chlorine dioxide gas is released by passing air through the solution, and is applied to breadmaking flour at a rate of 12–24 mg/kg (it is permitted in the U.K. up to 30 mg/kg). Chlorine dioxide treatment of flour destroys the tocopherols (cf. Ch. 14). The use of chlorine dioxide is also permitted in Japan.

Benzoyl peroxide

$(C_6H_5CO)_2O_2$ or BzO_2 is a solid bleaching agent which was first used in 1921. It is supplied as a mixture with inert, inorganic fillers such as $CaHPO_4$, $Ca_3(PO_4)_2$, sodium aluminium sulphate or chalk. Novadelox, a proprietary mixture, contains up to 32% of benzoyl peroxide but 16% is the usual proportion. The dosage rate, normally 45–50 mg/kg, is restricted to 50 mg/kg in the U.K. by the Bread and Flour Regulations 1984. The bleaching action occurs within about 48 h. This bleacher has the advantage over gaseous agents that only a simple feeder is required, and storage of chemicals presents no hazard; the fact that it has no improving action is advantageous in the bleaching of patent flours. The treated flour contains traces of benzoic acid, but objection has not been raised. BzO2 is also used in New South Wales, Queensland, the U.S.A, Canada, the Netherlands, New Zealand (up to 40 mg/kg, for pastry flour only) and Japan (up to 300 mg/kg).

Acetone peroxide

Acetone peroxide is a dry powder bleaching and improving agent, marketed as 'Keetox', a blend of acetone peroxides with a diluent such as dicalcium phosphate or starch. The concentration in terms of H_2O_2 equivalent per 100 g of additive plus carrier is 3–10 for maturing and bleaching, or 0.75 for use in doughmaking. Its use has been permitted in the U.S.A. since 1961, and also in Canada, but it is not, as yet, permitted in the U.K. It is used either alone or in combination with benzoyl peroxide. The usual dosage rate is 446 mg/kg on flour basis. Significant dates in the history of flour bleaching are summarized in Table 7.2.

Flour blending for bleaching treatment

Because the various flour streams differ in their characteristics, the optimum level of bleaching treatment varies correspondingly, the lower grade

TABLE 7.2
Significant Dates in the History of Flour Bleaching

1901	Andrews patents flour treatment with NO_2 (chemical process)
1903	Alsop patents flour treatment with NO_2 (electrical treatment)
1909	NO_2 in use
1911	Keswick Convention — unmarked flour to be unbleached
1921	Benzoyl peroxide first used
1921	J. C. Baker patents NCl_3 as flour bleacher
1922	NCl_3 replaces Cl as bleacher for breadmaking flour
1923	Committee appointed to inquire into use of preservatives and colouring matter in food
1924	Committee's activities extended to chemical substances for flour treatment
1927	Committee reported that bleaching and improving agents were in use, and that Cl, NCl_3 and BzO_2 were not among those least open to objection
1949	NCl_3 use discontinued in the U.S.A.
1949	ClO_2 first used in the U.S.A.
1955	NCl_3 use discontinued in the U.K.
1955	ClO_2 first used in the U.K.
1961	Acetone peroxide permitted in the U.S.A. (not in the U.K.)

flours (those nearer the tail end of the break and reduction systems) in general requiring more treatment than the patent flours (cf. p. 170). It is therefore customary to group the machine flours according to quality into three or four streams for treatment. A possible grouping is indicated in Table 7.1 Each group would be given appropriate bleacher treatment: e.g. the lowest 20% of flour might receive treatment at ten times the rate for the best quality 50%. The final grades are then made up by blending two or more of the groups in desirable proportions.

Flours for various purposes

Wheat flour is used for making food products of widely varying moisture content (see Table 7.3).

TABLE 7.3
Flour-based Products and their Moisture Contents

Type of product	Moisture content		Moisture level
	Range (%)	Mean (%)	
Soup	78–80	85	High
Puddings	13–67	45	Medium
Bread	35–40	38	Medium
Cakes	5–30	17	Medium
Pastry		7	Low
Biscuits (cookies, crackers)	1–6	5	Low

Data extracted from McCance and Widdowson (1967).

The proportions in which the various ingredients of baked products are present in the recipe, relative to flour (100 parts), are shown in Table 7.4. Biscuit dough is stiff to permit rolling and flattening; bread dough is a plastic mass that can be moulded and shaped; wafer batter is a liquid suspension that will flow through a pipe.

For comparison with products listed in Table 7.4, a typical wholemeal wheat extruded snack formulation would contain the following amounts of ingredients, in relation to 100 g white flour: 7 g soya protein, 14 g wheat bran, 1.4 g oil, 0.4 g emulsifier, 23 g water, 7 g sugar, 2 g salt, 2 g dicalcium phosphate, 3.6 g milk powder (Guy, 1993).

The flour content of various flour-containing products, as purchased or as consumed, is shown in Table 7.5.

TABLE 7.4
Proportions of Constituents in Recipes for Baked Products§ (Relative to Flour: 100 Parts)

Type of product	Constituents						
	Water	Fat	Salt	Whole egg	Raising agent	Milk powder	Sugar
Yeasted products					(yeast)		
Bread, CBP*	61	0.7	1.8		1.8		
Bread, LFP†	57	0.7	1.8		1.1		
Cream crackers	32	12.5	1.0		0.1–2.0		
Pastry							
Short	25	50	2.0				
Pie	31	43	2.0				
Steak and kidney pudding	30–36	50	0.7				
Puff	40–50	50–70	0.7				
Choux	125	50		150			
Biscuits						(whey)	
Hard sweet	20	17	0.7		1.1‡	2.6	22
Soft	10	32	0.1		0.5‡	2.0	30
Cake							
Plain	50	40		35	3.5‡		40
High ratio	70	65	2.0	60	5.0‡	8	120
Sponge			1.0	170			100
Wafer batter	150	3	0.2		0.3‡		

* Chorleywood Bread Process.
† Long Fermentation Processes.
‡ Mixtures of sodium and ammonium carbonate or bicarbonate.
§ Source: FMBRA.

TABLE 7.5
Flour Content of Flour-Based Foods, as Purchased or Consumed*

Food product	Flour content Range (%)	Mean (%)	Parts of product per 100 pt flour
Crispbread	–	112	90
Biscuit			
semi-sweet	67–82	74	135
ginger nut	43–57	49	205
Bread	53–72	70	145
Short pastry	60–80	65	155
Buns, scones, teacakes	36–57	45	220
Cakes, pastries, choc. wafer	23–40	33	300
Biscuits (chocolate)	8–46	30	330
Puddings	6–40	25	400

* Source: FMBRA. Flour wt at natural m.c. Product wt at m.c. of final product.

The Codex Alimentarius Commission of the United Nations Food and Agriculture Organization issued standard 152 on flour for human consumption in 1985. It defines acceptable sources as *Triticum aestivum* L. bread wheat, and *T. compactum* club wheat, the required protein and moisture contents (at least 7% and not more than 15.5%, respectively), fat acidity, particle size (98% through a 212 μm sieve) and protocol for ash determination. Optional ingredients and approved additives are listed.

In the U.K., flour for human consumption should conform with the nutritional requirements set out in the Bread and Flour Regulations 1984 (cf. p. 293).

For each purpose, flour with particular properties is required: these are secured, in the first place, by choice of an appropriate wheat grist in terms of strong and weak wheats. The average composition of wheat grists used for milling flour for various purposes in the U.K. is shown in Table 4.8. Table 4.8 also shows that, of the total flour milled in the U.K. in 1990/91, 63% was used for bread, 15% for biscuits, 6% for household use, 2% for cakes, 2% for starch manufacture and 12% for 'other products'.

'Other' food products made with wheat flour include pastry, meat pies, sausages, sausage rolls, rusks, pet foods, baby foods, invalid foods, chapatties, buns, scones, teacakes, pizzas, soups

(Ch. 13), premixes, liquorice, batter (for fish frying), chocolate and sugar confectionery, cereal convenience foods, snack foods, breakfast cereals, puddings, gravy powder, blancmange and brewing adjunct. Specific requirements for flours for various purposes are outlined below.

Bread flour

The predominance of wheat flour for making aerated bread is due to the properties of its protein which, when the flour is mixed with water, forms an elastic substance called gluten (cf. Chs 3 and 8). This property is found to a slight extent in rye but not in other cereals.

The property of producing a loaf of relatively large volume, with regular, finely vesiculated crumb structure, is possessed by flours milled from wheats described as 'strong' (cf. Chs 4 and 8). Protein strength is an inherent characteristic, but the amount of protein present can be influenced by the conditions under which wheats are grown. Protein content is also an important determinant of bread quality, there being a positive correlation between loaf specific volume (ml/g) and the percentage of protein present.

Typical characteristics of Chorleywood Bread Process (CBP) flour, Bakers' flour (as used in the bulk fermentation process), and rollermilled wholemeal in the U.K. are shown in Table 7.6.

Maturing and improving agents

The breadmaking quality of freshly milled flour tends to improve during storage for a period of 1–2 months. The improvement occurs more

rapidly if the flour is exposed to the action of the air. During such aerated storage, fat acidity increases at first, owing to lipolytic activity, and later decreases, by lipoxidase action; products of the oxidation of fatty acids appear; the proportion of linoleic and linolenic acids in the lipids falls; and disulphide bonds (-S–S-) decrease in number.

The change in baking quality, known as maturation, or 'ageing', can be accelerated by chemical 'improvers', which modify the physical properties of gluten during fermentation in a way that results in bread of better quality being obtained. Matured flour differs from freshly milled flour in that it has better handling properties, increased tolerance in the dough to varied conditions of fermentation and in producing loaves of larger volume and more finely textured crumb.

Improving agents permitted in the U.K. Bread and Flour Regulations 1984 (SI1984, No.1304), as amended by the Potassium Bromate (Prohibition as a Flour Improver) Regulations 1990 (SI 1990, No. 399) are chlorine, (for cake flour only; not wholemeal), cysteine hydrochloride (920) (all flour except wholemeal), chlorine dioxide (all flour except wholemeal), L-ascorbic acid (vitamin C) (all flour except wholemeal; all bread), and azodicarbonamide (all flour except wholemeal). Besides their improving effect, these substances give a whitened appearance to the loaf because of their beneficial effect on the texture of the crumb. Improving agents do not increase the carbon dioxide production in a fermented dough, but they improve gas retention (because the dough is made more elastic) and this results in increased loaf volume (cf. Ch. 8).

Redox improvers

The action of improvers is believed to be an oxidation of the cysteine sulphydryl or thiol (-SH) groups present in wheat gluten. As a result, these thiol groups are no longer available for participation in exchange reactions with disulphide (-S–S-) bonds — a reaction which is considered to release the stresses in dough — and consequently the dough tightens, i.e. the extensibility is reduced. Alternatively, it has been suggested that the oxidation of -SH groups may lead to the formation

TABLE 7.6
Typical U.K. Bread Flour Analysis 1992

	CBP	Bakers'	Wholemeal
Moisture	14.6%	14.5%	14.6%
Protein	11.0%	12.1%	14.7%
Grade colour	2.1	2.2	—
Falling number	329	334	330
alpha-Amylase*	15FU	22FU	21FU
Starch damage	30FU	34FU	—
Water absorption	60.2%	62.0%	70.2%

* Farrand units, (includes fungal enzyme).
Source: FMBRA.

of new -S–S- bonds which would have the effect of increasing dough rigidity (cf. Ch. 3).

Potassium bromate ($KBrO_3$) has never been allowed in many European countries; it was specifically excluded from the list of permitted additives in the U.K. by the Potassium Bromate (Prohibition as a Flour Improver) Regulations 1990 giving rise to considerable initial difficulties in the baking industry. The changes in the use of oxidizing improvers, consequent upon the deletion of potassium bromate, are considered in Ch. 8 (p. 201). Its use has also been voluntarily discontinued in Japan and it is now little used in New Zealand. Hazards associated with potassium bromate include the fact that, as a strong oxidizing agent, it can cause fire or explosions. It is also toxic and there is strong evidence for its carcinogenicity. At normal levels of addition however, it is not considered to persist at a significant level, into the baked product, when used at permitted levels.

Potassium bromate remains in use in the U.S.A. although an agreement exists between Government and users to reduce usage to a minimum. Although permitted in Canada, its use has declined in recent years in that country (Ranum, 1992). It has been used commercially as a bread improver since 1923. The rate of treatment is 10–45 mg/kg on flour weight. The substance acts as an oxidizing agent after the flour has been made into a dough; it increases the elasticity and reduces the extensibility of the gluten. Treatment with bromate has a similar action to that of ageing or maturing the flour, and enables large bakeries to use a constant fermentation period.

Potassium bromate is added to flour after being suitably diluted with an inert filler such as calcium carbonate or calcium sulphate. Proprietary brands of improver contain 6, 10, 25 or 90% of potassium bromate. The 6% brand is added at the rate of 0.022%. Higher levels of potassium bromate are used in chemical dough development processes (cf. Ch. 8).

Since untreated flour contains 1–8 mg/kg of bromine (Br), the bread made with untreated flour contains 0.7–5.6 mg/kg of natural Br. Flour treatment with 45 mg/kg of bromate leaves a residue of 15 mg/kg of Br in the loaf, increasing the total Br content of the br⟨...⟩ mg/kg.

Use of potassium bromate is permitted ⟨...⟩ mg/kg in the U.S.A.; to 50 mg/kg in Canada, Sweden; to 40 mg/kg in the Soviet Union; and in Eire to about 18 mg/kg. $KBrO_3$ is not allowed in the Netherlands or Australia. The greatest need for bromate occurs in continuous-mix baking, no-time doughs, frozen doughs, and overnight sponges, as used in Cuba and other Latin/American countries. The typical level of addition in these types of baking approaches the 75 mg/kg maximum. (Ranum, 1992).

L-Ascorbic acid (vitamin C), E300, was first used as a bread improver by Jørgensen in 1935. It is now used for this purpose in the U.K., most European countries and elsewhere, particularly in mechanical development processes of breadmaking, such as the Chorleywood Breadmaking Process. The volume increase resulting from use of ascorbic acid is generally less than that obtained with equivalent weight of potassium bromate, and it is more costly. The improving effect of ascorbic acid is mediated by enzymes present in the flour. The functional form is the oxidized form dehydroascorbic acid (DHA), which is highly effective but cannot be used directly as it is unstable. Ascorbic acid is oxidized to DHA through catalytic action of ascorbic acid oxidase. Injection of oxygen during mixing hastens the oxidation, making ascorbic acid more effective (Chamberlain and Collins, 1977). The oxidation to DHA is improved if the head space of the mixing machine contains an oxygen-enhanced atmosphere, e.g. a 50/50 mixture of oxygen and air, equivalent to a mixture of 60% oxygen plus 40% nitrogen (Ch. 8). Under these circumstances, ascorbic acid alone is as effective an oxidizing agent as is a combination of ascorbic acid and potassium bromate used when the dough is mixed under partial vacuum. An enzyme 'DHA reductase' is required for oxidation of sulphydryl (-SH) compounds by DHA.

Ascorbic acid strengthens the gluten; gas retention is thus improved and loaf volume augmented. Ascorbic acid does not hasten proving. The maximum permitted levels (1989) are 50 mg/kg in Belgium and Luxembourg, 100 mg/kg in the etherlands, 200 mg/kg in Canada, Denmark, Italy, Spain, the U.S.A. and the U.K., 300 mg/kg in France. No maximum level is specified in Australia, Greece, Portugal, Germany since ascorbic acid is reckoned to be quite safe, although it is under scrutiny by the COT. Use of ascorbic acid is also permitted in Japan, New Zealand and Sweden.

Azodicarbonamide (1,1′azobisformamide; $NH_2CONNCONH_2$; 'ADA') is a flour maturing agent, marketed as 'Maturox', or 'Genitron', or as 'ADA 20%' ('AK20'). 'Maturox' contains either 10% or 20% of ADA; 'Genitron' contains 20% or 50% of ADA, dispersed in an excipient, generally calcium sulphate and magnesium carbonate. The particle size of ADA is generally 3–5 μm. It was first used in the U.S.A. in 1962 (maximum permitted level 45 mg/kg on flour weight). When mixed into doughs it oxidizes the sulphydryl (-SH) groups and exerts an improving action. Oxidation is rapid and almost complete in doughs mixed for 2.5 min. Short mixing times are thus appropriate. The residue left in the flour is biurea. Flour treated with ADA is said to produce drier, more cohesive dough than that treated with chlorine dioxide, to show superiority in mixing properties, and to tolerate higher water absorption. An average treatment rate is 5 mg/kg (on flour weight) in bulk fermentation and low-speed mixing methods of baking, and 20–25 mg/kg (on flour weight) in the high-speed mixing Chorleywood Bread Process (CBP). The agent does not bleach, but the bread made from treated flour appears whiter because of its finer cell structure. The use of ADA has been permitted, to a maximum level of 45 mg/kg, in the U.K. since 1972. Its usage is also permitted in Canada, New Zealand and the U.S.A., but not in Australia or in EC countries other than the U.K. (1989).

L-Cysteine is a naturally occurring amino acid, is used in the Activated Dough Development process (ADD) (cf. Ch. 8), in which it functions as a reducing agent. The addition of L-cysteine (in the form of L-cysteine hydrochloride or L-cysteine hydrochloride monohydrate) to bread doughs, for this purpose, is permitted by the Bread and Flour Regulations 1984 in the U.K. to a maximum level of 75 mg/kg in bread flours

other than wholemeal; higher levels (up to 300 mg/kg) are permitted in certain biscuit flours. Use of L-cysteine is permitted in Denmark (up to 25 mg/kg), Germany (up to 30 mg/kg), Belgium (up to 50 mg/kg), Australia, New Zealand and the Netherlands (up to 75 mg/kg), Canada (up to 90 mg/kg) and Sweden (up to 100 mg/kg). L-Cysteine is not mentioned as a permitted additive in France, Greece, Italy, Luxembourg, Portugal, Spain or the U.S.A.

Cysteine accelerates reactions within and between molecules in the dough which lead to an improvement in its viscoelastic and gas-holding properties. These reactions normally take place slowly during bulk fermentation but with the addition of cysteine the bulk fermentation period can be eliminated. Cysteine, which is a rapid-acting reducing agent, is used in the ADD in conjunction with slow-acting oxidizing agents, such as ascorbic acid and potassium bromate (where permitted) or azodicarbonamide, which complete the 'activation' commenced by the cysteine. The dough-softening action of cysteine reduces the work input required for the production of fully developed dough.

Blending for improver treatment

The principles applied to bleaching flours of different grades also apply to improver treatment (cf. Table 7.1).

Emulsifiers and stabilizers

'Emulsifiers' and 'stabilizers' are any substances capable of aiding formation of (emulsifiers) or maintaining (stabilizers) the uniform dispersion of two or more immiscible substances. Flours, sold as such, are not allowed to contain emulsifiers but the following are permitted by the Bread and Flour Regulations (1984) to be included in bread: E 322 lecithins; E460 α-cellulose (permitted only in bread for which a slimming claim is made); E466 carboxymethyl cellulose, sodium salt (permitted only in bread for which a slimming claim is made); E471 mono- and di-glycerides of fatty acids; E472(b) lactic acid esters of mono- and di-glycerides of fatty acids; E472(c) citric acid esters of mono- and di-glycerides of fatty acids; E472(e) mono- and di-acetyl tartaric acid esters of mono- and di-glycerides of fatty acids; E481 sodium stearoyl-2–lactylate (cf. Ch. 8); E482 calcium stearoyl-2–lactylate; E483 stearyl tartrate; E481 and E482 are subject to a maximum level of 5000 mg/kg.

Biscuit (cookie, cracker) flour

Biscuit flours for short and semi-sweet biscuits are typically produced from grists containing mainly soft wheats, with some hard wheats included to increase the rate of production in the mill. However, hard wheat flours produce thinner biscuits than those of soft wheats so it is important to use a narrow range of levels of hard wheat in the flour. The level specified will depend on the manufacturer's preference since the biscuit plant will have to produce biscuits of particular sizes and weights to suit the packing plant.

There is no developed gluten network in short biscuit doughs, hence neither the level nor quality of protein is significant in production. However, consistency of quality is critically important in modern, high-speed production plants. Flours would normally be specified to have, say, a range of 1% protein within the typical range for flours (8–10%).

Semi-sweet biscuits have a developed gluten network which is modified during processing, and for these biscuits low protein flours (typically 8.5–9.5%) with weak, extensible glutens are used. At present, sulphur dioxide (SO_2, usually obtained from sodium metabisulphite added at the mixer) is used to increase the extensibility and decrease the elasticity of the doughs. This aids control of the dough sheet and hence biscuit thickness. EC proposals include permission of SO_2 in fine bakery wares, up to 50 mg/kg in the final product. Nevertheless wheat breeders are seeking to develop varieties which perform well without its use.

For wafers, low protein flour milled from weak wheat is suitable. Particle size is an important characteristic; ideally about 55% should be below 40 µm, 35% between 40 and 90 µm, and not more than 10% coarser than 90 µm. Too fine a flour produces light, tender, fragile wafers, while too coarse a flour produces incomplete sheets of unsatisfactory wafers.

Gluten development in wafer batters must be avoided, so flours which have a low tendency to give an aggregated gluten under low shear rates in aqueous flour batters are required. Hence low protein flours with weak extensible glutens are normally specified (cf. p. 186).

Cracker doughs have fully developed gluten networks and protein quality is important in dough processing. Cracker flours with medium protein contents (9.5–10.5%) made mainly from hard wheats are commonly used. Matzos are water biscuits made from unbleached, untreated flour and water only.

Emulsifiers

Emulsifiers are used in biscuits, either as processing aids or as partial replacements for fat. Very low levels (e.g. 0.1% on fat basis) of sodium stearoyl-2-lactylate (E481) in sheeted biscuit doughs produce a smooth, non-sticky surface which aids dough-piece cutting. Lecithin (E322) is commonly used in wafer batters to aid release of the baked wafer from the wafer baking machinery. Emulsifiers which can replace a substantial proportion of fat in biscuits without serious deterioration in product quality are sodium stearoyl-2-lactylate and the diacetyltartaric acid esters of monoglycerides of fatty acids E472e. Lecithin can be used to replace a low level of fat in biscuits.

Flours for confectionery products

Cake flours

Flour in cakes should allow an aerated structure to be retained after the cake has been built up. The stability of the final cake depends largely upon the presence of uniformly swollen starch granules; hence, the granules should be undamaged during milling, free from adherent protein, and unattacked by amylolytic enzymes. These characteristics are found in flour milled from a soft, low-protein wheat of low alpha-amylase activity.

Typical parameter values for cake flour milled in the U.K. would be as follows: Untreated cake flour: 8.5–9.5% protein and a minimum of particles exceeding 90 µm in size. Fine particle size is more important than low protein content for cake quality, giving finer, more even crumb than that given by a coarser flour.

Strong cake flour (for fruit cakes): 12% protein, 20–25 FU starch damage, 0.18% chlorine treatment.

High-ratio flour

In the late 1920s it was discovered, in the U.S.A., that cake flour which had been bleached with chlorine gas to improve its colour permitted production of cakes from formulae containing levels of sugar and liquid each of which is in excess of flour weight. Such flour for use in high sugar/flour ratio and high liquid/flour ratio formulae is known as 'high-ratio flour'. It should also have fine, uniform granularity and low protein content. The chlorination treatment, generally 0.1–0.15% by weight, besides allowing addition of larger proportions of liquid and sugar, reduces elasticity of the gluten and lowers the pH to 4.6–5.1.

Heat treatment of the grain or of the semolina from which the flour is milled has been found to be an effective substitute for the chlorine treatment of high-ratio cake flour (BP nos 1444173 (FMBRA) and 1499 986 (J. Lyons)) (cf. p.171) and cake flours may be similarly treated.

Typical characteristics of high-ratio flour milled in the U.K. would be 7.6–8.4% protein, 20–25 FU starch damage, granularity such that 70% of the particles were below 32 µm in size and a minimum of particles exceeding 90 µm in size. High-ratio flour is particularly suitable for sponge-type goods.

Emulsifiers in cakes

In cake making, emulsifiers such as glycerol monostearate (GMS) and mono- and di-glycerides of fatty acids (E471) are used in soft fats at levels of up to 10% to produce high-ratio shortenings. Certain emulsifiers such as GMS, polyglycerol esters and lactic acid esters of mono-glycerides (E472b) possess remarkable foam-promoting properties so that when added to sponge batters

at a level of about 0.5–1.0% of batter weight, whisking times can be greatly reduced, all-in mixing methods can be used and liquid egg can be replaced with dried egg.

Foam-promoting emulsifiers such as GMS, poly-glycerol esters, propylene-glycol esters (E477) or blends of these, used at about 1% of batter weight, allow a reduction in the fat content of a cake or even substitution of the fat by a smaller quantity of vegetable oil.

Although anti-staling effects of emulsifiers in cakes are not as clearly defined as in bread, sucrose esters (E473), sodium stearoyl lactylate (E481) and poly-glycerol esters (E475) offer some possibilities as a means of minimizing the effects of staling.

Flour for cake premixes

Some cake premixes sold in Britain contain, in powder form, all the ingredients required for a cake, *viz.* flour, fat, sugar, baking powder, milk powder, eggs, flavouring and colour, and need only the addition of water before baking. However, some cake premixes, particularly those sold in the U.S.A., omit the eggs and/or the milk, because lighter cakes of larger volume can be made by the use of fresh eggs instead of dried ingredients.

The type of flour must be suitable for the particular product, flours of high-ratio type generally being used. The fat must have the correct plasticity and adequate stability to resist oxidation. The addition of certain antioxidants to fat, to improve stability, is allowed in Britain, the U.S.A., and elsewhere. Those allowed in Britain under the Antioxidants in Food Regulations 1978 (S.I 1978 No. 105, as amended) for addition to anhydrous oils and fats, and certain dairy products other than butter, for use as ingredients are propyl, octyl or dodecyl gallates up to 100 mg/kg, or butylated hydroxyanisole (BHA) and/or butylated hydroxytoluene (BHT) up to 200 mg/kg (calculated on the fat). Those allowed in the U.S.A. (with permitted levels based on fat or oil content) are: resin guaiac (0.1%), tocopherols (0.03%), lecithin (0.01%), citric acid (0.01%), pyrogallate (0.01%), propylgallate (0.02%) and BHA and/or BHT (0.02%).

In preparing the premixes, the dry ingredients are measured out by automatic measures and conveyed, often pneumatically, to a mixing bin, mixed, and then entoleted (cf. p. 111) to ensure freedom from insect infestation. The fat is then added, and the mixture packaged. If fruit is included in the formula, it is generally contained in a separate cellophane-wrapped package enclosed in the carton.

Flour for fermented goods

For buns, etc. a breadmaking flour is required. Fermentation time is short; the fat and the sugar in the formula bring about shortening of the gluten.

Flour for pastry

A weak, medium strength flour is needed for the production of sweet and savoury short pastes. Flour strength for puff pastry will vary according to the processing methods, with rapid processing methods requiring weaker flours than those used in traditional methods of production. In general, flours for puff pastry should have low resistance to deformation (e.g. low Brabender resistance values) but reasonable extensibility.

Flour from steamed wheat

Flour milled from steamed wheat ('stabilized' flour, in which enzymes have been inactivated) is produced for use in manufacture of soups (Ch. 7), gravies, crumpets, liquorice and as a thickening agent. For these purposes, the flour should form a thick paste when it is heated with water, and the paste should retain its consistency for some time when heated at 90°–95°C. The *alpha*-amylase activity of normal (non-steam-treated) flour is usually high enough to degrade swollen starch granules during the cooking process, resulting in loss of water-binding capacity and formation of thin pastes of low apparent viscosity. The greater water-absorbing capacity of the flour from steamed wheat could also make it a suitable ingredient for canned pet-foods.

As the gluten in the flour from steam-treated wheat has been denatured it is not suitable for breadmaking. The flour may, for many purposes, be regarded as impure starch, and is often used to replace starch in certain types of adhesives, and as a filler for meat products.

The bacteriological status of flour for soups is important and requires not >125 total thermophilic spores per 10 g, not >50 flat sour spores per 10 g, not >5 sulphide spoilage organisms per 10 g, thermophilic anaerobic spores in not >3 tubes out of 6.

Quellmehl

Quellmehl or heat-treated starch, is defined as maize flour or wheat flour of which the starch has undergone hydrothermic (*viz.* steam) treatment resulting in pregelatinization of the starch thereby increasing its swelling capacity by at least 50%.

Flour for sausage rusk

A low protein flour milled from weak wheat, such as U.K.-grown Riband, or a low protein air-classified fraction is required. Desirable characteristics are low maltose figure (not >1.4 by Blish and Sandstedt method), low *alpha*-amylase activity (high Falling Number) and high absorbency.

Batter flour

A low protein flour milled from a grist comprising 90% weak British wheat plus 10% strong wheat is suitable. *Alpha*-amylase activity should be low. Too high a viscosity in the batter caused by excessive starch damage is to be avoided, and therefore the proportion of hard wheat in the grist should be restricted.

Household flour

Household flour is used for making puddings, cakes, pastry, etc. In the U.K. it is milled from a grist consisting predominantly of weak wheats of low protein, such as British or Western European, with admixture of up to 20% of strong wheat to promote flowability and good mixing. Exclusion of sprouted wheat from the grist is important (cf. Ch. 4), as high *alpha*-amylase activity leads to the production of dextrins and gummy substances during cooking, and to sticky and unattractive baked goods.

Self-raising flour

This is a household flour to which raising agents have been added. Choice of sound wheat is important because evolution of gas during baking is rapid and the dough must be sufficiently distensible, and yet strong enough to retain the gas. The moisture content of the flour should not exceed 13.5% in order to avoid premature reaction of the aerating chemicals and consequent loss of aerating power.

Distension of the dough is caused by carbon dioxide which is evolved by the reaction between the raising agents (U.S.: leavening agents), one alkaline and one acidic, in the presence of water. The usual agents used for domestic self-raising flour in the U.K. are 500, sodium hydrogen carbonate ('bicarbonate') ($NaHCO_3$) and acid calcium phosphate, (ACP, E341 calcium tetrahydrogen diorthophosphate) ($CaH_4(PO_4)_2$), and their use was described by J. C. Walker in BP No 2973 in 1865. The usual rate of usage is 1.16% bicarbonate plus 1.61% of 80% grade ACP on flour weight. A slight excess of the acidic component is desirable, as excess of bicarbonate gives rise to an unpleasant odour and a brownish yellow discolouration. The Bread and Flour Regulations 1984 permit the following raising agents in the U.K.: 500, sodium hydrogen carbonate, E341 calcium tetrahydrogen diorthophosphate (monocalcium phosphate (ACP)), E450 disodium dihydrogen diphosphate (sodium acid pyrophosphate (SAPP)), 541 acidic sodium aluminium phosphate (SAP), 570 D-glucono-1,5-lactone and E336 mono potassium-L-(+)-tartrate (cream of tartar).

ACP used at a rate of 1.61% on flour weight adds about 250 mg Ca per 100 g flour; hence, self-raising flour does not require chalk to be added to it (cf. Ch. 14). Phosphate–starch mixtures are known as cream powders, a commonly used commercial example of which consists of sodium acid pyrophosphate (SAPP) diluted with starch and used in the ratio of 2:1 with sodium hydrogen carbonate at a rate of 4.7% on flour weight.

Instantized or agglomerated flour

This a form of free-running flour, readily dispersible in water, made by 'clustering' flour particles in an 'Instantizer'. Uses include the making of sauces and gravies and for thickening and general culinary purposes. In the instantizing process, the flour as normally milled is damped with steam, tumbled in a warm air stream to cause the particles to agglomerate, then dried, sieved, cooled and packed. The U.S. standard for agglomerated flour requires all the flour to pass through a sieve of 840 μm aperture width and not more than 20% to pass through a sieve of 70 μm aperture width. Free-flowing flour is also produced by air classification (cf. Ch. 6).

Flour for export

Besides the specific requirements according to the purpose for which the flour is to be used, flour for export must have low moisture content to prevent development of mould, taint or infestation during its transportation. As a safeguard, the flour should be entoleted. In addition, export flour must conform to any special requirements of the importing country, e.g. regarding the presence or absence of nutrients and improvers, for which the regulations of most other countries differ from those of the U.K. (cf. Ch. 14).

In January 1988 the (U.S.) Food and Drug Administration (FDA) announced guidelines for contamination levels at which flour is seizable. These levels are 75 insect fragments or more per 50 g flour and an average of one rodent hair or more per 50 g flour.

Flours from cereals other than wheat

Besides wheat, all other cereals yield flour when subjected to milling processes as outlined in Ch. 6. The uses for these flours, both commercially and domestically, are many and varied, as the following summary indicates.

Rye flour

Rye flour of various extraction rates is used extensively in Eastern Europe for making a range of breads — both soft breads and crispbreads — using conventional straight dough or sour dough processes.

Rye flour is also used as a filler for sauces, soups and custard powder, and in pancake flour in the U.S.A., and for making gingerbread in France. A mixture of 10% of rye flour with 90% of wheat flour is used for making biscuits and crackers in the U.S.A. The rye flour is said to improve the quality of the products and is less expensive than wheat flour. Rye flour can be fractionated by air-classification; a flour of 8.5% protein content yielded high and low protein fractions of 14.4% and 7.3% protein contents respectively. Rye flour is also used in the glue, match and plastics industries.

Rye flour can also be used for making gun-puffed and shredded ready-to-eat breakfast cereals.

Triticale flour

The use of triticale flour in breadmaking is mentioned in Ch. 8, and its use in making chapattis is referred to in Ch. 13. Other bakery products made with triticale flour include pancakes and waffles.

Barley flour

Barley flour is used in the manufacture of flat bread, for infant foods and for food specialities. It is also a component of composite flours used for making yeast-raised bread (cf. Ch. 8).

Pregelatinized barley flour, which has high absorbent properties, provides a good binder and thickener. Barley breading is made by combining pregelatinized barley flour with barley crunch.

Malted barley flour is made from barley malt (cf. Ch. 9). Malt flour is used as a high diastatic supplement for bread flours which are low in natural diastatic activity, as a flavour supplement in malt loaves, and for various other food products.

Malted barley flour can be air-classified (cf. Ch. 6) to yield protein-rich and protein-poor fractions.

The former finds uses in the food industry, while the latter is reported to make a unique beer. The major food uses for malt products and cereal syrups are in bread, biscuits, crackers, crispbread, breakfast cereals, infant and invalid foods, malted food drinks, pickles and sauces, sugar confectionery and vinegar.

Oat flour

This is made by grinding oatmeal on stones and sifting out the fine material (cf. Ch. 6). Oat flour is also obtained as a by-product of groat cutting. Uses for oat flour include infant foods and for ready-to-eat breakfast cereals, e.g. shredded products made by a continuous extrusion cooking process (Ch. 11) and an extruded gun-puffed product (Ch. 11).

A process for the separation of a protein concentrate by the air-classification (Ch. 6) of oat flour has been described (Cluskey et al., 1973). An ultra-fine fraction with 85–88% protein content ($N \times 6.25$, d.b.) was obtained which comprised 2–5% of the flour by weight. The compound granules of oat starch (Ch. 2) tend to disintegrate upon fine grinding, releasing the individual granules, which measure 2–10 µm. Separation of an almost pure protein fraction would therefore require the use of an extremely fine cut size — less than 2 µm.

Rice flour

This is used in refrigerated biscuit manufacture to prevent sticking; in baby foods, as a thickener; in waffle and pancake mixes, as a water absorbent.

The use of rice flour, in blends with wheat flour, to make bread of acceptable quality, is mentioned in Ch. 8, and also its use, as a component of composite flours, for breadmaking. Rice flour can also be used for making pasta products.

Maize flour

This is used to make bread, muffins, doughnuts, pancake mixes, infant foods, biscuits, wafers, breakfast cereals, breadings, and as a filler, binder and carrier in meat products.

Dry milled maize flour is not to be confused with 'corn flour', the term used in the U.K. for maize starch obtained as a product of wet-milling.

The inclusion of maize flour in composite flour used for breadmaking, and its use, alone, to make bread of a sort in Latin America, are mentioned in Ch. 8. The use of maize flour, in blends with wheat semolina, to make pasta products is mentioned in Ch. 10 and for making extrusion-cooked ready-to-eat breakfast cereals in Ch. 11.

Industrial uses for maize flour are noted in Ch. 15.

Sorghum and millet flours

Sorghum flour is used as a component of composite flour for making bread in those countries in which sorghum is an indigenous crop (cf. Ch. 8).

The use of flour or wholemeal from sorghum and some of the millets to make porridge, roti, chapatti, tortilla and other products is described in Ch. 13.

Sorghum flour finds industrial uses, e.g. as core-binder, resins, adhesives, and in oil-well drilling (cf. Ch. 15).

Composite flours

The Composite Flour Programme was established by the Food and Agriculture Organization in 1964 to find new ways of using flours other than wheat — particularly maize, millet and sorghum — in bakery and pasta products, with the objective of stimulating local agricultural production, and saving foreign exchange, in those countries heavily dependent on wheat imports (Kent, 1985).

Quality control and flour testing

Testing protocols, and acceptable degrees of reproducibility (agreement between laboratories) and repeatability (agreement between replicate determinations by the same operator using the same equipment), have been established, usually through collaborative testing, by various standardizing organizations. Most countries have national Standards Organizations (e.g. British Standards Institute, BSI), and international standards are also produced by e.g. the International Association of Cereal Science and Technology (ICC), International Standards Organization (ISO) and American Association of Cereal Chemists (AACC).

Tests may be applicable to whole grains or derived products, in the case of whole grains it may be necessary to grind them to achieve an appropriate particle size distribution.

For valid comparisons to be made it is necessary to observe proper sampling procedures (ICC 130, AACC 64) and to normalize results of most tests to a constant moisture basis. Either a dry matter basis or a 14% moisture basis is usually adopted. As test procedures become more stringent, protocols increasingly demand that test samples contain a consistent dry weight of sample, requiring an adjustment of the actual mass taken to compensate for moisture variation, rather than subsequent correction. Moisture content must thus be determined by an acceptable method, such as determination of weight-loss when the ground product is heated at 100°C for 5 h *in vacuo* or at 130°C for 1 h (flour) or 1.5 h (ground wheat) at atmospheric pressure. (ICC 101/1). These methods are suitable for moisture contents up to 17%. Oven methods are primary methods as they determine directly the required parameter; secondary methods may also be used, they measure a property which varies as a function of the required parameter, and thus require calibration against a standard. Secondary methods are frequently more rapid but less accurate than oven methods; they include electrical conductivity and Near Infrared Reflectance Spectroscopy (NIRS) methods (ISO 202, covers moisture and protein determinations). NIRS determinations are based on absorption of NIR energy at specific wavelengths by OH bonds in water molecules. The same is true of protein determinations where the peptide bonds between amino acids define the critical wavelengths. Considerable mathematical processing of signals and measurements at reference wavelengths are necessary to ensure accurate indications of the required parameters.

Parameters dependent on the nature of the grains milled

Protein content

Protein content of whole grains, meals and flours may be calculated from nitrogen contents determined by the Kjeldahl method (AACC 46) in which organic matter is digested with hot concentrated sulphuric acid in the presence of a catalyst. Ammonia, liberated by addition of an excess of alkali to the reaction product, is separated by distillation and estimated by titration. A convenient apparatus is the Tecator Kjeltec 1030 Auto System (ISO 1871). In white wheat flour, protein content is estimated by multiplying N_2 content by 5.7, and in many references this factor is used for other wheat products and other cereals. However FAO/WHO (1973) recommend factors appropriate to individual foodstuffs. Those relating to cereals are given in Table 7.7.

TABLE 7.7
Factors Used for Converting Kjeldahl Nitrogen to Protein Values

Wheat fraction	Factor	Cereal	Factor
Wholemeal flour	5.83	Maize	6.25
Flours, except wholemeal	5.70	Rice	5.95
Pasta	5.70	Barley, oats, rye	5.83
Bran	6.31		

For routine estimations of protein (and moisture) content NIRS methods are now very reliable for whole grains as well as their derivatives. Determinations are carried out at intake and on-line during processing, in many cereals plants throughout the world.

For determining the amount of wheat protein contributing to gluten, the Glutomatic instrument (Falling Number Co) and method may be used. A dough is prepared from a sample of flour or ground wheat, and a solution of sodium chloride. Wet gluten is isolated by washing this dough with a solution of buffered sodium chloride and, after removal of excess water, weighed, or dried and weighed according to whether wet or dry gluten content is required (ICC 137, AACC 56–81B).

Sedimentation tests

These provide a useful indication of the suitability of a flour (it is more usually performed on a ground wheat) for breadmaking. The Zeleny test (Pinkney *et al.*, 1957) (AACC 56–60, ICC 116) has been adopted in a number of countries for protein evaluation. It depends on the superior swelling and flocculating properties, in a dilute lactic acid solution, of the insoluble proteins of wheats with good breadmaking characteristics (Frazier, 1971). In the U.K. a better guide to breadmaking properties has consistently been obtained using the SDS sedimentation test (Axford *et al.*, 1979) rather than the Zeleny test. The SDS test has been standardized as BS 4317, part 19, and it has been adopted for evaluation of *T. durum* quality as ICC 56–70 and AACC 151. It consists of an initial suspension and shaking of ground material in water, to which sodium dodecyl sulphate is later added. Following a series of carefully timed inversions of the cylinder containing the suspension, it is allowed to stand for 20 min after which the height of sediment is read. The test is performed under controlled temperature conditions.

Enzyme tests

One of the most important enzymes influencing flour quality is *alpha*-amylase. Its activity may be determined directly, using the method of Farrand (1964) or McCleary and Sheehan (1987), or indirectly, as a result of its solubilizing effect on starch, leading to a reduction in paste viscosity. The most widely adopted method uses the Falling Number apparatus to detect starch liquefaction in a heated aqueous suspension of flour or (more usually) ground grain (ICC Standard Method 107, AACC 56–81B). As enzyme activity can vary dramatically among individual grains it is essential to grind a large sample of grain (at least 300 g) in preparing a representative meal, and to mix it thoroughly before taking the test sample (7 g at 15% m/c) from it. Wholemeals also require regrinding and thorough mixing, and flours have to be free of lumps. Following the preparation of a suspension in a special tube, this is introduced into the apparatus and the test proceeds automatically. The suspension is heated and stirred at a programmed rate for 60 sec, after which a plunger is allowed to fall through it. The Falling Number is the number of seconds from the start of heating to the coming to rest of the plunger.

A similar principle underlies stirring tests with the Rapid Visco-Analyser.

Used on grain or grain products to measure *alpha*-amylase, both instruments probably also respond to hydrolysis of other viscous components such as proteins and cell-wall components, but the effects of these are usually comparatively small.

Heat damage test for gluten

The effect of overheating on gluten is measured directly by a method introduced by Hay and Every (1990). Described as the glutenin turbidity test, the procedure measures the loss in solubility of the fraction of glutenins normally soluble in acetic acid. Dilute acetic acid extracts are precipitated by addition of alkaline ethanol and the precipitate is quantified by spectrophotometric measurement of turbidity allowing the degree of damage to be assessed by comparison with standards. Good correlations have been found between loss of turbidity and reductions in baking quality.

Pigmentation

The yellow colour of durum semolinas is highly valued. Under ICC 152 the carotenoid pigments are extracted at room temperature with water-saturated butanol for photometric evaluation of optical density of the clear filtrate against a β-carotene standard.

End-use tests

In spite of much research no single test has yet been devized which can reliably predict the breadmaking properties of an unknown wheat or flour. Hence for this application, and for many other applications of cereals, the most reliable means of evaluating a sample is to subject it to the intended end-use itself, or to a scaled down version of the same. Several bread baking tests appear under AACC 10 and ICC 131, rye flour is tested by AACC 10–70. Biscuit (cookie) and foam-type cake tests also appear under the 10-heading. Pasta semolinas are also subjected to approved small scale tests (AACC 66-41 and -42).

Machinability test

In adopting a test for bread wheats eligible for Intervention price support, the EC has not standardized a breadmaking test but instead has defined flour of breadmaking quality as flour which produces dough which does not 'stick' to the blades or the bowl of the mixer in which the dough is mixed, nor to the moulding apparatus.

Extraneous matter test ('filth test')

The rodent hair and insect fragment count of flour is determined by digesting the flour with acid and adding the cooled digest to petrol in a separating funnel. The hair and insect fragments are trapped at the petrol/water interface, and can be collected and identified microscopically.

Tests for characteristics dependent mainly on processing conditions

Ash test BS 4317 Part 10

The ash test (incineration of the material in a furnace at a specified temperature and under prescribed conditions, and the weighing of the resultant ash) is widely used as a measure of milling refinement because pure starchy endosperm yields relatively little ash, whereas bran, aleurone and germ yield much more. The ash test can be carried out very precisely, but, as the endosperms of different wheats vary in mineral content, a given ash value can correspond to different levels of bran content. The test is not suitable for indicating the content of non-starchy-endosperm components in flours to which chalk has been added.

Grade Colour

The Grade Colour test, performed, for example, with the Kent–Jones and Martin Colour Grader, can be used to estimate the degree of contamination of white flour with bran particles. In the test, the intensity of light in the 530 nm region, reflected from a standardized flour/water paste, in a glass cell, is compared with that reflected from a paste of a reference flour. The grade colour was said to be unaffected by variation in content of flour pigment (xanthophyll) (cf. p. 171). It has now been demonstrated however, that this is not so (Barnes, 1978), thus diminishing the value of the test as a means of quantifying bran content.

Tristimulus methods

Use of an instrument designed to simulate the visual response of the human eye has found favour in some applications, as an alternative to the Grade Colour system. The instrument depends upon complex mathematical transforms to produce values in three arbitrary spectral ranges, X, Y and Z, which cannot be produced by any real lights (Hunter and Harold, 1987). Users can derive a Whiteness index to suit their needs, by selecting from these values. Like the Grade Colour system it responds to factors other than bran content — probably particle-size — and it thus has a limited usefulness (Evers, 1993).

Damaged starch

The amount of starch that is mechanically damaged influences a flour's ability to absorb water. The content of damaged starch in flour is estimated by methods which measure either the digestibility or the extractability of the starch. Digestibility-based methods measure the amount of hydrolysis effected by added amylase enzymes; extractability-based methods measure the amount of amylose present in an aqueous extract by its reaction with iodine in potassium iodide. The iodine/amylose complex may be assayed colorimetrically (e.g. McDermott, 1980), amperometrically or potentiometrically (e.g. Chopin SD 4 method). Only the damaged granules are susceptible to amylase at temperatures below gelatinization temperature and appreciable leaching of amylose occurs only from damaged granules under the test conditions.

Methods such as those of Farrand (1964),

Donelson and Yamazaki (1962) and Barnes (1978) rely on assaying reducing sugars (mainly maltose) produced by the action of *alpha*-amylase derived from malt flour. Another method (Gibson *et al.*, 1992) employs amyloglucosidase to further hydrolyze oligosaccharides to glucose, which is then determined by the effect of a derivative on a chromogen.

Damaged starch granules may be recognized microscopically by a red coloration with Congo Red (colour index: 22120) stain (cf. Ch. 6), and have been described, from their microscopical appearance, as 'ghosts' by Jones (1940). Undamaged granules do not stain with Congo Red. A method of damaged starch determination, in which damaged and undamaged granules are measured separately taking advantage of their different staining reactions with a fluorescent dye, has been developed, using image analysis for making the measurements.

Particle size analysis

Test sieving by hand or sieve shakers tends to be unreproducible on flour, but the Alpine air-jet sieve, in which negative pressure below the sieve assists particles through the mesh, and clears the mesh with reversed air-jets, gives more reproducible analyses. Sedimentation methods depend upon the faster settling rates of larger particles, in a non-aqueous solvent; the Andreassen pipette is an example of a simple device using this principle (ICC 127), another is the Simon Sedimentation Funnel. Sedimentation methods are not much used today in flour quality control, having been superseded by more reproducible methods. The Coulter counter is a device which rapidly measures the volume of thousands of individual particles as they pass through an orifice. Each particle is measured by the change in electrical resistance that it causes by displacing its own volume of the electrolytic solution (non-aqueous in the case of flour) in which the particles are suspended. Reproducibility is very good with this method (Evers, 1982) which is suitable for flours and starches but not wholemeals. Laser diffraction-based instruments provide a means of making rapid comparisons among flours. There are many instruments of this type available, some capable of operating on dry samples. The distributions that they indicate may not agree well among different instruments or with those of the other methods described.

Physical tests on doughs and slurries

The physical characteristics of doughs and slurries are important in relation to the uses of flours. Pseudo-rheological characteristics are investigated mainly with the following:

The Brabender Farinograph (D'Appolonia and Kunerth, 1984) measures and records the resistance of a dough to mixing as it is formed from flour and water, developed and broken down. This resistance is called consistency. The maximum consistency of the dough is adjusted to a fixed value by altering the quantity of water added. This quantity, the water absorption (cf Ch. 8), may be used to determine a complete mixing curve, the various features of which are a guide to the strength of the flour (AACC 54–21, ICC 115).

The Brabender Extensograph (Rasper and Preston, 1991) (ICC 114, AACC 54-10) records the resistance of dough to stretching and the distance the dough stretches before breaking. A flour–salt–water dough is prepared under standard conditions in the Brabender Farinograph and moulded on the Extensograph into a standard shape. After a fixed period the dough is stretched and a curve drawn, recording the extensibility of the dough and its resistance to stretching (see Fig. 7.1). The dough is removed and subjected to a further two stretches. The Extensograph has replaced the Extensometer in the Brabender instrument range but the older instrument is still widely used for testing biscuit flours.

The Chopin Alveograph (AACC 54–30) uses air pressure to inflate a bubble of dough until it bursts; the instrument continuously records the air pressure and the time that elapses before the dough breaks.

The Brabender Amylograph (ICC 126 for wheat and rye flours) continuously measures the resistance to stirring of a 10% suspension of flour in water while the temperature of the suspension is raised at a constant rate of 1.5°C/min from 20°–95°C and then maintained at 95°C (Shuey and Tipples, 1980). It is of use in testing flour for soups, etc., for which purpose the viscosity of the product after gelatinization is an important characteristic (cf. Ch. 6), and for adjusting the malt addition to flours for breadmaking (cf. p. 62).

The Rapid Visco Analyser (RVA), produced by Newport Scientific, in Australia, may be regarded as a derivative of the Amylograph.

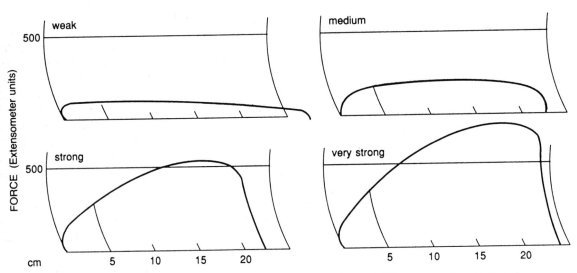

FIG 7.1 Extensometer curves of unyeasted doughs made from flours of different strengths as indicated.

Measurements of viscosity are made using small samples, containing 3–4 g of starch, in periods which may be as short as 2 min. Use of disposable containers and mixer paddles eliminates the need for careful washing of the parts between tests. As with the Amylograph, the characteristics of starch pastes and the effects of enzymes on them can be recorded on charts. They can also be transferred in digital form, direct to a data-handling computer.

The capacity for measuring the liquefying effects of enzymes on viscous pastes enables the RVA to be used for detecting the products of sprouted grains in cereal meals.

True rheological instruments. In recent years frustration with instrument-dependent units obtained with some of the above methods, together with the poor reproducibility from one instrument to another of the same type, has led cereal chemists to turn to true rheological measurements. Suitable instruments for use with doughs, slurries and gels, derived from flours, include the Bohlin VOR (viscometric, oscillation and relaxation), the Carri Med CSL Rheometer and the Rheometrics RDA2. In addition to providing excellent reproducibility, these instruments, which are also used on many non-food, and many non-cereal food materials, allow comparisons to be made across a wide range of substances. Although they are expensive, and may never feature prominently in routine testing, they will undoubtedly enable the development of tests that can be performed on simpler, dedicated instruments (Faridi and Faubion, 1990).

Storage of flour

It has been recommended that, for long periods of conservation, flour should be stored in a closed atmosphere (Bellenger and Godon, 1972). In these conditions, flour acidity increases owing to accumulation of linoleic and linolenic acids, which are slowly oxidized; reduction of disulphide groups (-S–S-) is slow, and there is little increase in sulphydryl groups (-SH); solubility of gluten protein decreases; as a result, changes in baking strength are only minor.

Flour is stored commercially in bags or in bulk bins. Bags of flour in the U.K. contain 50 or 32 kg when packed; the multiwall kraft paper bags are stacked, often several tiers high, on palleting. The harshness of treatment to be expected during filling and handling influences the number of plys in the walls of the chosen bags. Using single-spout packers approximately 300–350 bags/h can be filled. Using multi-spout packers, with up to 8 spouts, 600–800 bags/h are possible.

The hazards to flour in storage include those to wheat in storage, viz. mould and bacterial attack, and insect infestation (cf. Ch. 6) and also oxidative rancidity (cf. Ch. 3) and eventual deterioration of baking quality.

Freedom from insect infestation during storage can be ensured only if the flour is free from insect life when put into store, and if the store itself is free from infestation. Good housekeeping in the mill and the milling of clean grain should ensure that the milled flour contains no live insects, larvae or eggs, but as a precautionary measure flour is often passed through an entoleter before being bagged off, or going to bulk bins. The entoleter (BP No 965267) is a machine consisting of a rotor rapidly rotating within a fixed housing. The flour is fed in centrally and is flung with considerable impact against the casing. At normal speeds of operation (2900 rev/min for flour) the machine effectively destroys all forms of insect life and of mites, including eggs (cf. Ch. 5). The insect fragments, however, are not removed from the flour by the entoleter.

The optimum moisture content for the storage of flour must be interpreted in relation to the length of storage envisaged, and to the prevailing ambient temperature and r.h., remembering that flour will gain or lose moisture to the surrounding atmosphere unless packed in hermetically sealed containers. For use within a few weeks, flour can be packed at 14% m.c., but at moisture contents higher than 13% mustiness, due to mould growth, may develop, even if the flour does not become visibly mouldy. At moisture contents lower than 12%, the risk of fat oxidation and development of rancidity increases. The reactions involved in oxidative rancidity are catalyzed by heavy metal ions, such as Cu^{2+}.

The expected shelf life of plain (i.e. non-self-raising) white flour packed in paper bags and stored in cool, dry conditions and protected from infestation is 2–3 years. The rate of increase in acidity increases with temperature rise and with fall in flour grade (i.e. as the ash residue increases). Hence, the shelf-life of brown and wholemeal flours is shorter than that of white flour.

Stored at 17°C (62°F), the shelf-life of brown flour of 85% extraction rate and of wholemeal (100% extraction rate) is closely related to the moisture content and temperature. Brown flour, for instance, should keep 9 months at 14% m.c., 4–6 months at 14.5% m.c., 2–3 months at 15.5% m.c. For wholemeal stored under the most favourable conditions, a shelf life of 3 months may be expected, or of 12 months if the product has been entoleted.

Flour blending

Blending of finished flours is widely practised on the continent of Europe and it is becoming increasingly popular in the U.K. It can ensure greater uniformity in a product and it can provide flexibility in response to requirements for flours of unusual specification, through the blending of separately stored flours of various types. Blending can be performed as a batch process or on a volumetric basis. The simpler volumetric method depends upon flours being discharged from two or more bins into a common conveyor, their discharge rates being controlled to provide the respective proportions required. The more accurate batch method involves the use of weighers to deposit required weights of products from each of the selected bins into a batch mixer. An additional advantage of the batch system is that improvers can be added at the same time as the flours are mixed. The mixers used may be of the ribbon type where the blend is continually tumbled, or the air mixer type in which the blend is agitated by air injected into a holding bin.

Bulk storage and delivery of flour

The storage of flour in bulk bins, and delivery in bulk containers, has advantages over storage and delivery in bags. Although constructional costs of bulk storage facilities are high, the running costs are low because manhandling is much reduced, and warehouse space is better utilized.

The capacity of bins for storing flour in bulk is 70–100 t. Packing pressure inside the bin increases with bin area, not with bin height; a bin area of 5.6 m^2 is satisfactory. Normally bins are constructed of concrete or metal. Wooden bins are liable to become infested. The choice of construction material is a personal one but steel is currently most popular as metal bins are cheaper (unless capacity is over 20,000 tonnes), they do not crack, they are easily installed and relocated, and they are immediately usable on completion of construction (Anon., 1989). The inner surfaces must be smooth to allow stock to slide down the walls. Steel bin walls are usually coated with shellac varnish and lower parts may be painted with a low-friction polyurethane paint. Concrete surfaces are ground and coated with several coats of sodium silicate wash to provide a seal. The shape of bins is again a matter of choice. Circular bins are cheaper as lighter gauge steel may be used, however, there is more space wasted between cylindrical bins than between rectangular bins. A problem that can arise, when flour is discharged from the base of the bin, is bridging of stock: this can be avoided by good hopper design and efficient dischargers. Bins may be filled and emptied pneumatically; fluidizing dischargers use 0.8–1.1 m^3/min of low-pressure air (20–70 kN/m^2) to fluidize the flour, causing it to behave as a liquid and to flow down a reduced gradient to the outlet. Mechanical (worm or screw type), and vibratory dischargers may also be used to assist discharge of flour from bins. When flour and air are present in appropriate proportions there is a risk of dust explosions if a source of ignition is also present. In all flour conveying and handling situations, it is essential to avoid sources of ignition arising. Additional precautions include the incorporation of explosion relief panels into bin tops. Similar panels are recommended in the areas of buildings surrounding the bins.

Flour was first delivered in bulk in the 1950s and by 1987, 65% of flour delivered in the U.K. was in bulk (Anon., 1989).

Bulk wagons for transport can be filled at the

mill by gravity feed or by blowline, or, most efficiently, by fluidized delivery from an outload bin directly above the vehicle. By this method flow rates of 250–300 t/h can be achieved. Discharge of the vehicle is assisted by air pressure; some larger tankers have a fluidizing pad in the base of each hopper. Compressors mounted either on the vehicle or at the customer's premises blow the delivery direct to the storage bins at the bakery.

Mini-bulk containers, holding up to about 2 t, may be used in some mills for transport and delivery of products — mainly bran and germ, but in some cases flour also.

References

AMERICAN ASSOCIATION OF CEREAL CHEMISTS INC. (1962) *Cereal Laboratory Methods*, 7th edn, Amer. Assoc. of Cereal Chemists Inc., St. Paul MN., U.S.A.

ANON. (1989) Flour milling correspondence course. *Product Handling, Storage and Distribution. (Module 10).* Incorporated National Association of British and Irish Millers, London.

AXFORD, D. W., McDERMOTT, E. E., and REDMAN, D. G. (1979) A note on the sodium dodecyl sulphate test for breadmaking quality: comparison with Pelshenke and Zeleny tests. *Cereal Chem.* **56**: 582–584.

BARNES, W. C. (1978) The rapid enzymic determination of starch damaged wheat. *Staerke* **30**: 115–119.

BELLENGER, P. and GODON, B. (1972) Influence de l'aeration sur l'evolution de diverse caracteristiques biochimiques et physicochimiques. *Ann. Technol. Agric.* **21**: 145.

BENTLEY, H. R., McDERMOTT, E. E., MORAN, T., PACE, J. and WHITEHEAD, J. K. (1950) Toxic factor from 'Agenized' protein. *Nature, Lond.* **165**: 150.

CHAMBERLAIN, N. and COLLINS, T. H. (1977) The Chorleywood bread process: The importance of air as a dough ingredient. *FMBRA Bull.* **4** (Aug): 122.

CLUSKEY, J. E., WU, Y. V., WALL, J. S. and INGLETT, G. E. (1973) Oat protein concentrates from a wet-milling process: preparation. *Cereal Chem.* **50**: 475.

D'APPOLONIA, B. L. and KUNERTH, W. H. (1984) *The Farinograph Handbook*, 3rd edn, American Assoc of Cereal Chemists Inc. St Paul, MN. U.S.A.

DONELSON, J. R. and YAMAZAKI, W. T. (1962) Note on a rapid method for the estimation of damaged starch in soft wheat flours. *Cereal Chem.* **39**: 460–462.

EVERS, A. D. (1982) Methods for particle-size analysis of flour: a collaborative test. *Lab. Practice* **31**: 215–219.

EVERS, A. D. (1993) On-line quantification of bran particles in white flour. *Food Sci Technol. Today* **7**, 23–26.

FAO/WHO (1973) *Energy and Protein Requirements*. Report of a joint FAO/WHO *Ad Hoc* Expert Committee. FAO Nutrition Meetings Report Series, No. 52, WHO Technical Report Series, No. 522.

FARIDI, H, and FAUBION, J. M. (1990) *Dough Rheology and Baked Product Texture*, Van Norstrand Rheinhold. NY.

FARRAND, E. A. (1964) Modern bread processes in the United Kingdom with special reference to α-amylase and starch damage. *Cereal Chem.* **41**: 98–111.

FRAZIER, P. (1971) *A Physico-chemical Investigation into the Mechanism of the Zeleny Test*. PhD thesis.

GIBSON, T. S., AL QUALLA, H, and McCLEARY, B. V. (1992) An improved enzymic method for the measurement of starch damage in wheat flour. *J. Cereal Sci.* **15**: 15–27.

GUY, R. (1993) Ingredients. In: *The Technology of Extrusion Cooking.* N. FRAME. (Ed.) Blackie, Glasgow.

HAY, R. L. and EVERY, D. (1990) A simple glutenin turbidity test for the determination of heat damage in gluten. *J. Sci. Food Agric.* **53**: 261–270.

HOLLAND, B, WELCH, A. A., UNWIN, I. D., BUSS, D. H., PAUL, A. A. and SOUTHGATE, D. A. T. (1991) *McCance and Widdowson's The Composition of Foods*, 5th edn, The Roy. Soc of Chem. Cambridge.

HUNTER, R. S. and HAROLD, R. W. (1987) *The Measurement of Appearance*. 2nd Edn. John Wiley & Sons, NY. U.S.A.

JONES, C. R. (1940) The production of mechanically damaged starch in milling as a governing factor in the diastatic activity of flour. *Cereal Chem.* **15**: 133–169.

KENT, N. L. (Technical Ed.) (1985) *Technical compendium on composite flours*. Economic Commission for Africa, Addis Ababa.

McCANCE, R. A. and WIDDOWSON, E. M. (1967) *The Composition of Foods*. Med. Res Coun., Spec. Rpt Ser. 297. 2nd Imp., H.M.S.O., London.

McCLEARY, B. V. and SHEEHAN, H. (1987) Measurement of cereal alpha-amylase: a new assay procedure. *J. Cereal Sci.* **6**: 237–251.

McDERMOTT, E. E. (1980) The rapid, non-enzymic determination of damaged starch in flour. *J. Sci Food Agric.* **31**: 405–413.

MELLANBY, E. (1946) Diet and canine hysteria. *Brit. Med. J.* **ii**: 885.

MINISTRY OF AGRICULTURE, FISHERIES AND FOOD (1984) *The Bread and Flour Regulations 1984 Statutary Instruments 1984, No. 1304, as amended by the Potassium Bromate (Prohibition as a Flour Improver) Regulations 1990 (SI 1990, No. 399)* H.M.S.O., London.

PINKNEY, A. J., GREENAWAY, W. T. and ZELENY, L. (1957) Further developments in the sedimentation test for wheat quality. *Cereal Chem.* **34**: 16.

RANUM, P. (1992) Potassium bromate in bread baking. *Cereals Foods World* **37**: 253–258.

RASPER, V. F. and PRESTON K. R. (1991) *The Extensograph Handbook*. Amer. Assoc of Cereal Chemists Inc. St. Paul, MN. U.S.A.

SHUEY, W. C. and TIPPLES, K. H. (1980) *The Amylograph Handbook*. Amer. Assoc. of Cereal Chemists Inc. St Paul, MN. U.S.A.

Further Reading

ANON. (1980) *Glossary of Terms for Cereals and Cereal Products*. British Standards Institution, London.

ANON. (1989) *Glossary of Baking Terms*, American Institute of Baking. Manhattan, KS. U.S.A.

ANON. (1990) *Flour Treatments & Flour Products. Module 12 in Workbook Series*, National Association of British and Irish Millers, London.

DENGATE, H. N. (1984) Swelling, pasting and gelling of

wheat starch. *Advances in Cereal Science and Technology.* **6**: 49–82.

FARIDI, H. and RASPER, V. F. (1987) *The Alveograph Handbook.* Amer. Assoc. of Cereal Chemists Inc. St. Paul MN. U.S.A.

FARRAND, E. A. (1972) Controlled levels of starch damage in a commercial U.K. bread flour and effects on absorption, sedimentation value and loaf quality. *Cereal Chem.* **49**: 479.

GRAVELAND, A., BOSVELD, P., LITCHENDONK, W. J. and MOONEN, J. H. E. (1984) Structure of glutenins and their breakdown during dough mixing by a complex oxidation-reduction system. pp. 59–68 In: *Gluten Proteins.* A. GRAVELAND and J. H. E. MOONEN (Eds), Inst. Cereals, Flours and Bread, TNO Wageningen, The Netherlands.

GREER, E. N. and STEWART, B. A. (1959) The water absorption of wheat flour; relative effects of protein and starch. *J. Sci. Food Agric.* **10**: 248–252.

MACRITCHIE, F. (1980) Physicochemical aspects of some problems in wheat research. *Advances in Cereal Science and Technology,* **3**: 271–326.

MARTIN, D. J. and STEWART, B. G. (1987) Dough stickiness in rye-derived wheats. *Cereal Foods World.* **32**: 672–673.

MISKELLY, D. M. and MOSS, H. J. (1985) Flour quality requirements for Chinese noodle manufacture. *J. Cereal Sci.* **3**: 379–387.

OSBORNE, B. G., FEARN, T. and HINDLE, P. H. (1993) *Practical Near Infrared Spectoscopy* 2nd Edition. Longmans, Harlow, Essex.

PAYNE, P. I., NIGHTINGALE, M. A. KRATTIGER, A. F. and HOLT, L. M. (1987) The relationship between HMW glutenin subunit compostion and the breadmaking quality of British-grown wheat varieties. *J. Sci. Food Agric.* **40**: 51–65.

POMERANZ, Y., BOLLING, H. and ZWINGELBERG, H. (1984) Wheat hardness and baking properties of wheat flours. *J. Cereal Sci.* **2**: 137–143.

POMERANZ, Y. (1983) Single, universal, bread baking test — why not? pp. 685–690 In: *Progress in Cereal Chemistry and Technology,* J. HOLAS and J. KRATOCHVIL (Eds.) Elsevier Science Publishers, N.Y.

SCHNEEWEISS, R. (1982) *Dictionary of Cereal Processing and Cereal Chemistry.* Elsevier Scientific Publishing Co. Amsterdam.

8

Bread-baking Technology

Principles of baking

Primitive man, a nomadic hunter and gatherer of fruits and nuts, started to settle down and abandon his nomadic life when, in Neolithic times, he discovered how to sow the seeds of grasses and, in due time, reap a crop of 'cereal grains'. With this change in his way of life came the beginnings of civilization which, in western Europe, is based on a diet relying on wheat, wheaten flour, and the baked products made from flour, the principal product being bread.

The function of baking is to present cereal flours in an attractive, palatable and digestible form.

While wheat is the principal cereal used for breadmaking, other cereals, particularly rye, are also used to some extent. The first part of this chapter will consider breadmaking processes and bread in which wheat flour or meal is the sole cereal. The use of other cereals will be discussed later (p. 211).

Use of milled wheat products for bread

Bread is made by baking a dough which has for its main ingredients wheaten flour, water, yeast and salt. Other ingredients which may be added include flours of other cereals, fat, malt flour, soya flour, yeast foods, emulsifiers, milk and milk products, fruit, gluten.

When these ingredients are mixed in correct proportions, three processes commence:

— the protein in the flour begins to hydrate, i.e. to combine with some of the water, to form gluten (cf. pp. 70 and 174). Flour consists of discrete and separate particles, but the gluten is cohesive, forming a continuous three-dimensional structure which binds the flour particles together in a 'dough'. The gluten has peculiar extensible properties: it can be stretched like elastic, and possesses a degree of recoil or spring;

— air bubbles are folded into the dough. During the subsequent handling of the dough these bubbles divide or coalesce. Eventually the dough comes to resemble a foam, with the bubbles trapped in the gluten network;

— enzymes in the yeast start to ferment the sugars present in the flour and, later, the sugars released by diastatic action of the amylases on damaged starch in the flour, breaking them down to alcohol and carbon dioxide. The carbon dioxide gas mixes with the air in the bubbles and brings about expansion of the dough. "Bread is fundamentally foamed gluten" (Atkins, 1971).

Three requirements in making bread from wheat flour are formation of a gluten network and the creation of air bubbles within it; the incorporation of carbon dioxide to turn the gluten network into a foam; and the development of the rheological properties of the gluten so that it retains the carbon dioxide while allowing expansion of the dough; and, finally, the coagulation of the material by heating it in the oven so that the structure of the material is stabilized. The advantage of having an aerated, finely vesiculated crumb in the baked product is that it is easily masticated.

Corresponding with these requirements, there

are three stages in the manufacture of bread: mixing and dough development, dough aeration, and oven baking. The method of dough development and aeration that has been customary since the time of the Pharaohs is panary fermentation by means of yeast.

Ingredients

Flour

Good breadmaking flour is characterized as having:

— protein which is adequate in quantity and which, when hydrated, yields gluten which is satisfactory in respect of elasticity, strength and stability;
— satisfactory gassing properties: the levels of amylase activity and of damaged starch (cf. pp. 183, 185) should be adequate to yield sufficient sugars, through diastatic action, to support the activity of the yeast enzymes during fermentation and proof;
— satisfactory moisture content — not higher

than about 14% to permit safe storage, and satisfactory colour, and should meet specifications regarding bleach and treatment (cf. pp. 171–172).

These requirements are met by the type of wheat called 'strong' (cf. pp. 81, 92, 174), viz. wheat having a reasonably high protein content. Wherever possible, home-grown wheat is used for breadmaking, and this is the situation, for example, in Canada and in the U.S.A., where such strong wheats, e.g. CWRS, HRS, are readily available.

In the U.K., however, the home-grown wheat is, or until recently was, characteristically weak, viz. of low protein content, and would not, by itself, yield flour from which bread, of the kind to which U.K. consumers are accustomed, could be made. It was therefore customary for flour millers in the U.K. to mill breadmaking flour from a mixed grist of strong and weak wheats, the strong wheat component being imported, generally from Canada, and the weak component being home-grown U.K. wheat. Until the early 1960s, the average breadmaking grist in the U.K.

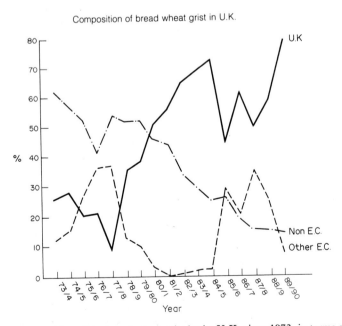

FIG 8.1 Average composition of the bread wheat grist in the U.K. since 1973, in terms of U.K. wheat, other EC wheat, and non-EC wheat (data from MAFF, H-GCA, and NABIM).

would consist of 60–70% of imported strong wheat plus 20–30% of weak home-grown wheat (with a small proportion of 'filler' wheat of medium strength, cf. p. 87) — see Fig. 8.1, yielding a white flour of about 12% protein content.

The imported Canadian wheat is more expensive than the home-grown U.K. wheat and, in consequence, there was a strong urge to decrease the ratio of strong to weak wheat. This change was made possible in a number of ways, one of which was the advent of the CBP (cf. p. 203) because, among other advantages, the CBP permitted the use of a flour of about 1% lower protein content to produce bread of quality equivalent to that produced by the BFP (cf. p. 201).

Additional impetus to reduce still further the proportion of imported strong wheat in the bread grist followed the entry of the U.K. into the EC, and the imposition of a heavy import levy, which has run as high as £120–130 per tonne, on the cost of wheat imported from third (i.e. non-EC) countries. Various measures have been adopted whereby the proportion of home-grown (or EC-grown) wheat in the breadmaking grist could be further increased, while maintaining loaf quality. They include:

— breeding stronger wheats with higher yielding potential for growing in the U.K. and other EC countries. Examples of such promising new varieties are Avalon and Mercia. Moreover, the considerable increase in the size of the U.K. wheat harvest in recent years has provided the flourmiller with the possibility of obtaining adequate supplies of these newer varieties of good breadmaking quality;
— awarding of remunerative premiums to growers for higher protein home-grown wheats which are poorer yielders than low protein wheats;
— use of vital gluten as a bread ingredient (cf. p. 195);
— supplementation of flours from lower-protein home-grown wheats with air-classified high protein fractions of flour (cf. p. 132);
— use of high levels of fungal *alpha*-amylase (cf. p. 196).

Figure 8.1 shows that the proportion of imported non-EC wheat (mostly Canadian CWRS wheat)

in the U.K. bread-wheat grist has fallen from about 70% in 1960 to about 15% in 1990 (with a corresponding increase in the home-grown wheat proportion), with a considerable saving in the cost of the raw material. By 1992, some millers were supplying breadmaking flour milled entirely from home-grown U.K. and EC wheats, with no non-EC component, but with the addition of 2% or perhaps 2.5% of vital gluten.

A similar reduction in the imported non-EC (strong) wheat content of the breadmaking grist has also occurred in other countries.

One possible complication associated with the lowering of the strong/weak wheat ratio in the bread grist is the reduced proportion of damaged starch in the flour because of the frequent association of strength with hardness (as in the imported Canadian wheat) and, conversely, of weakness with softness (as in the EC-grown wheats). It is desirable that the content of damaged starch should be maintained at a reasonably high level, and this requirement can be met by adjustments to the milling process (cf. p. 149). However, it is a fortunate coincidence that the two varieties of wheat classified by breadmaking quality and widely grown in the U.K. at the present time, Avalon and Mercia, both have a hard textured endosperm, and thus go some way towards avoiding this complication.

Leavening

Leavened baked goods are preferred in all countries where wheat is available as a staple food. Leavening can be achieved in several ways, including the following:

1. Whisking egg into a foam with flour and other ingredients. This method is used in production of sponge and other cakes.
2. Water vapour production as in Scandinavian flat breads and puff pastry.
3. Yeast.
4. Baking powder.

Yeast and baking powder are the most important. Each is appropriate for its own range of products, and in some cases, such as doughnuts, coffee cake, and pizza-dough, either may be used alone or in combination.

Baking powders

Baking powders depend upon sodium bicarbonate as a source of CO_2 gas, which may be liberated by the action of sodium acid phosphate, monocalcium phosphate, sodium aluminium phosphate or glucono-δ-lactone. One hundred grams of baking powder generates 15 mg (or 340 mM, or 8.2 l) of CO_2. Some is released at dough temperature and the remainder during baking.

Yeast

The quantity of yeast used is related inversely to the duration of fermentation, longer fermentation systems generally employing somewhat lower levels of yeast and also lower dough temperatures. Thus, 1% of yeast on flour wt would be used for a 3 h straight dough system with the dough at 27°C, whereas 2–3% of yeast on flour wt would be required for a no-time dough at 27°–30°C. Yeast activity increases rapidly with temperature, and its level of use is therefore reduced if the temperature is increased within a fixed time process. In addition to providing CO_2 as a leavening agent, yeast also affects rheological properties of dough through the lowering of pH by CO_2 production, evolution of alcohol, and the mechanical effects of bubble expansion. Further, yeast contributes significantly to the flavour and aroma of baked products.

Yeast is used in several different forms: compressed, cream (liquid), dried into pellets, and instant active powders.

In recent years, attitudes to yeast production have become more enterprizing. Specialized strains have been selected and bred to meet newly identified criteria. This has resulted partly from changing technologies within the baking industries and partly from new means of genetic manipulation.

Examples of these innovations are the replacement of conventional spore fusion by protoplast fusion, and genetic engineering through the use of recombinant DNA (rDNA) for introduction of an advantageous segment of the genetic materials of one strain to the genome of another. Strains that have an excellent performance in sugar-rich doughs normally show a poor performance in lean doughs, but the subject of a European Patent (EP 0 306 107 A2) is a yeast that performs well both in high sucrose conditions and also in 'lean' conditions, where maltose is the available substrate. The technique involved was the introduction of genes coding for increased activity of the two enzymes maltose permease and maltase (alpha-glucosidase), allowing best use to be made of the limited quantities of maltose available in a lean dough.

Ability to ferment sugars anaerobically remains the major criterion of selection, but meeting this under different conditions has led to the introduction of specialized strains. The conditions that provide the challenge include the requirements: (a) to be supplied and stored in a dry form with a longer life than the traditional compressed form; (b) to retain high activity in high sugar formulations and (c) to retain activity in yeast-leavened frozen doughs.

Dried yeasts

Until the early 1970s, two strains of *Saccharomyces cerivisiae* were used widely. The yeast was grown to a nitrogen level of 8.2–8.8% (on a dry basis), and an A.D.Y. (active dry yeast), which was grown to a nitrogen content of 7.0%. Thus, in the pelleted product, it had only 75–80% of the gassing activity of the compressed yeasts (when compared on the same m.c. basis). New products available since that time have allowed the gap to be narrowed, although it does still exist.

Three forms of dried yeast are now available: A.D.Y., and the powdered products Instant A.D.Y. (I.A.D.Y.) and protected A.D.Y. (P.A.D.Y.).

A.D.Y. must be rehydrated in warm water (35°–40°C) before it is added to dough, while I.A.D.Y. and P.A.D.Y. can be added to dry ingredients before mixing. In fact, this results in more productive gassing. During storage, dried yeasts are subject to loss of activity in oxygen. The improved strains are supplied in vacuum packs or in packs with inert gas in the headspace

(I.A.D.Y.), or in the presence of an antioxidant (P.A.D.Y.). P.A.D.Y. features in complete mixes containing flour and other ingredients but the flour present must be at a very low m.c. to avoid moisture transfer and reduction in the level of production against oxidation.

High sugar yeast

Products such as Danish pastries, doughnuts and sweet buns have a high sugar content. The high osmotic pressures involved are not tolerated by standard yeast strains, but good strains are available as I.A.D.Y. products. Japanese compressed yeasts can also withstand high osmotic conditions.

Frozen dough yeasts

The production of breads from frozen doughs, at the point of sale, has increased dramatically and has created a requirement for cryoresistant yeasts. Most yeasts withstand freezing, but deteriorate rapidly during frozen storage. The best cryoresistant strains perform well in sweet goods but less well in lean doughs. The requirement has not been fully satisfied (Reed and Nagodawithana, 1991).

Salt

Salt is added to develop flavour. It also toughens the gluten and gives a less sticky dough. Salt slows down the rate of fermentation, and its addition is sometimes delayed until the dough has been partly fermented. The quantity used is usually 1.8–2.1% on flour wt, giving a concentration of 1.1–1.4% of salt in the bread. Salt is added either as an aqueous solution (brine) or as the dry solid.

Fat

Fat is an essential ingredient for no-time doughs, such as the CBP. Added at the rate of about 1% on flour wt, fat improves loaf volume, reduces crust toughness and gives thinner crumb cell walls, resulting in a softer-textured loaf with improved slicing characteristics. Fat also keeps the bread soft and palatable for a longer period, which is equivalent to an anti-staling effect (Hoseney, 1986).

During storage of flour, free fatty acids accumulate owing to the breakdown of the natural fats, and the gluten formed from the protein becomes less soluble and shorter in character. When flour that has been stored for a long time, e.g. a year, at ambient temperature is used for the CBP, the fat level should be increased to about 1.5% on flour wt.

Sugar

Sugar is generally added to bread made in the U.S.A., giving an acceptable sweet flavour, but it is not usually added to bread in the U.K. However, sugar may be included in prover mixes.

Vital gluten

Vital wheat gluten, viz. gluten prepared in such a way that it retains its ability to absorb water and form a cohesive mass (cf. pp. 70, 174), is now widely used in the U.K. and in other EC countries as an ingredient of bread:

— at levels of 0.5–3.0% on flour wt to improve the texture and raise the protein content of bread, crispbread, and speciality breads such as Vienna bread and hamburger rolls;
— to fortify weak flours, and to permit the use by millers of a wheat grist of lower strong/weak wheat ratio (particularly in the EC countries) by raising the protein content of the flour (cf. p. 193);
— in starch-reduced high protein breads (cf. p. 209), in which the gluten acts both as a source of protein and as a texturing agent;
— in high-fibre breads (cf. p. 209) now being made in the U.S.A., to maintain the texture and volume.

In the U.S.A., about 70% of all vital gluten is used for bread, rolls, buns and yeast-raised goods (Magnuson, 1985). Vital gluten is also used as a binder to raise the protein level in meat products, e.g. sausages, and in breakfast cereals (e.g. Kelloggs Special K), breadings, batter mixes, pasta foods, pet foods, dietary foods and textured vegetables products (t.v.p.).

The origin of the gluten is of little importance when used to raise the flour protein content by only 1–2%: thus, U.K.-grown wheat can be used to provide vital gluten, thereby further reducing the dependence on imported strong wheat. The vital gluten is generally added to the flour at the mill, particularly in the case of wholemeal. (McDermott, 1985).

Gluten flour

This is a blend of vital wheat gluten with wheat flour, standardized to 40% protein content in the U.S.A.

Fungal amylase

Besides the use of low levels (e.g. 7–10 Farrand Units) of fungal amylase to correct a deficiency in natural cereal *alpha*-amylase and improve gassing (cf. p. 198), fungal amylase, sold under such trade names as MYL-X and Amylozyme, has a marked effect in increasing loaf volume when used at much higher levels as a bread ingredient in rapid breadmaking systems. Use of high levels is possible because the fungal amylase has a relatively low thermal inactivation temperature. The fungal amylase starts to act during the mixing stage, when it causes a softening of the dough, which must be corrected by reducing the amount of doughing water, so as to maintain the correct dough consistency. Use of high levels of fungal amylase in the BFP would not be desirable, as the dough softening effect would be too severe. Hence, addition of fungal amylase at these high levels is made by the baker, and not at the mill.

The fungal amylase continues to act during the early part of the baking process, attacking gelatinized starch granules, improving gas retention, and helping the dough to maintain a fluid condition, thus prolonging the dough expansion time and increasing loaf volume. The increase in loaf volume is directly related to the level of fungal amylase addition up to about 200 Farrand Units.

The effect of the addition of about 120 Farrand Units of fungal amylase is so powerful that it may permit the use of flour of up to 2% lower protein content with no loss in loaf quality.

A similar increase in loaf volume could be produced by addition of a variety of commercial carbohydrase enzyme preparations (Cauvain and Chamberlain, 1988).

Soya flour

Enzyme-active soya flour is widely used as a bread additive, at a level of about 0.7% on flour wt. Advantages claimed for its use include: beneficial oxidizing effect on the flour, bleaching effect on flour pigments (β-carotene) due to the presence of lipoxygenase, increase in loaf volume, improvement in crumb firmness and crust appearance, and extension of shelf life (Anon., 1988a) (cf. p. 215)

The improving action and bleaching properties of enzyme-active soya flour are due to peroxy radicals that are released by a type-2 lipoxygenase, which has an optimum activity at pH 6.5. Enzyme-active soya flour has two effects in a flour dough: it increases mixing tolerance, and it improves dough rheology, viz. by decreasing extensibility and increasing resistance to extension. The action of the lipoxygenase is to oxidize the linoleic acid in the lipid fraction of the wheat flour, but the action only occurs in the presence of oxygen (Grosch, 1986).

Improving agents

The use and effects of improving agents — potassium bromate, ascorbic acid, azodicarbonamide, L-cysteine — have been discussed in Ch. 7.

Physical treatments

The breadmaking quality of flour can be improved also by physical means, e.g. by controlled heat treatment (cf. p. 113) or by an aeration process, in which flour is whipped with water at high speed for a few minutes and the batter then mixed with dry flour. Improvement is brought about by oxidation with oxygen in the air, probably assisted by the lipoxidase enzymes (cf.

p. 68) present in the flour. A similar improving effect can be obtained by overmixing normal dough (without the batter stage): cf. the Chorley-wood Bread Process (p. 203).

Doughmaking

Water absorption

The amount of water to be mixed with flour to make a dough of standard consistency is usually 55–61 pt per 100 pt of flour, increasing in proportion to the contents of protein and damaged starch (cf. pp. 62, 174) in the flour.

Flour contains protein, undamaged starch granules and damaged starch granules, all of which absorb water, but to differing degrees. Farrand (1964) showed that the uptake of water, per gram of component, was 2.0 g for protein, 0–0.3 g for undamaged starch, and 1.0 g for damaged starch. Thus, flours from strong wheat (with higher protein content) and from hard wheat (with a higher damaged starch content) require more water than is needed by flours from weak (lower protein) or soft (less damaged starch) wheats to make a dough of standard consistency.

Besides the protein and starch, the soluble part of the hemicellulose (pentosan) forming the walls of the endosperm cells also absorbs water.

The water used in dough-making should have the correct temperature so that, taking account of the flour temperature and allowing for any temperature rise during mixing, the dough is made to the correct final temperature. When using a process such as the CBP (cf. p. 203) in which the temperature rise during mixing may be as much as 14°C, it may be necessary to cool the doughing water.

It is important, particularly in plant bakeries, to maintain constant dough consistency. This may be done by adjusting the level of water addition automatically or semi-automatically. Determination of water absorption of the flour by means of the Brabender Farinograph is described on p. 186.

A flour with high water absorption capacity is generally preferred for breadmaking. Apart from increasing the proportion of strong wheat (high

protein) in the grist — which is uneconomic, the most convenient way of increasing water absorption is to increase the degree of starch damage. The miller can bring this about by modifying the milling conditions (cf. p. 149).

Fermentation

The enzymes principally concerned in panary fermentation are those that act upon carbohydrates: alpha-amylase and beta-amylase in flour, and maltase, invertase and the zymase complex in yeast. Zymase is the name that was formerly used for about fourteen enzymes.

The starch of the flour is broken down to the disaccharide maltose by the amylase enzymes; the maltose is split to glucose (dextrose) by maltase; glucose and fructose are fermented to carbon dioxide and alcohol by the zymase complex.

Some of the starch granules in flour become mechanically damaged during milling (cf. pp. 62, 149, 185), and only these damaged granules can be attacked by the flour amylases. It is therefore essential that the flour should contain adequate damaged starch to supply sugar during fermentation and proof. When the amylase enzymes break down the damaged starch, water bound by the starch is released and causes softening of the dough. This situation must be borne in mind when calculating the amount of doughing water required, the amount of water released being dependent not only on the level of damaged starch, but also on the alpha-amylase activity, length of fermentation time, and dough temperature. Excessive levels of starch damage, however, have an adverse effect on the quality of the bread (cf. p. 150): loaf volume is decreased, and the bread is less attractive in appearance.

There are small quantities of sugar naturally present in flour (cf. p. 55) but these are soon used up by the yeast, which then depends on the sugar produced by diastatic action from the starch.

During fermentation about 0.8 kg of alcohol is produced per 100 kg of flour, but much of it is driven off during the baking process. New bread is said to contain about 0.3% of alcohol. Secondary products, e.g. acids, carbonyls and esters, may affect the gluten or impart flavour to the bread.

Amylase

Both *alpha*- and *beta*-amylases catalyze the hydrolysis of starch, but in different ways (cf. p. 67).

Normal flour from sound wheat contains ample *beta*-amylase but generally only a small amount of *alpha*-amylase. The amount of *alpha*-amylase, however, increases considerably when wheat germinates. Indeed, flour from wheat containing many sprouted grains may have too high an *alpha*-amylase activity, with the result that, during baking, some of the starch is changed into dextrin-like substances. Water-holding capacity is reduced, the crumb is weakened, and the dextrins make the crumb sticky (cf. p. 67). However, flour with too high a natural *alpha*-amylase activity could be used for making satisfactory bread by microwave or radio-frequency baking methods (cf. p. 206). Another possibility would be to make use of an *alpha*-amylase inhibitor, e.g. one prepared from barley, as described in Canadian Patent No. 1206157 of 1987 (Zawistowska *et al.*, 1988).

The functions of starch in the baking of bread are to dilute the gluten to a desirable consistency, to provide sugar through diastasis, to provide a strong union with gluten, and by gelatinization to become flexible and to take water from the gluten, a process which helps the gluten film to set and become rigid.

Gas production and gas retention

The creation of bubble structure in the dough is a fundamental requirement in breadmaking. The carbon dioxide generated by yeast activity does not create bubbles: it can only inflate gas cells already formed by the incorporation of air during mixing.

Adequate gas must be produced during fermentation, otherwise the loaf will not be inflated sufficiently. Gas production depends on the quantity of soluble sugars in the flour, and on its diastatic power. Inadequate gassing (maltose value less than 1.5) may be due to an insufficiency of damaged starch or to a lack of *alpha*-amylase; the latter can be corrected by adding sprouted wheat to the grist, or malt flour, or fungal amylase, e.g. from *Aspergillus oryzae* or *A. awamori*, to the flour (cf. p. 196). Fungal amylase is preferred to malt flour because the thermal inactivation temperature of fungal amylase is lower (75°C) than that of cereal *alpha*-amylase (87°C), and its use avoids the formation of gummy dextrins during baking and the consequent difficulties in slicing bread with a sticky crumb.

Gas retention is a property of the flour protein: the gluten, while being sufficiently extensible to allow the loaf to rise, must yet be strong enough to prevent gas escaping too readily, as this would lead to collapse of the loaf. The interaction of added fat with flour components also has a powerful effect on gas retention.

Dough development

Protein

The process of dough development, which occurs during dough ripening, concerns the hydrated protein component of the flour. It involves an uncoiling of the protein molecules and their joining together, by cross-linking, to form a vast network of protein which is collectively called gluten. The coils of the protein molecules are held together by various types of bonds, including disulphide (–SS–) bonds, and it is the severing of these bonds — allowing the molecule to uncoil — and their rejoining in different positions — linking separate protein molecules together — that constitutes a major part of dough development.

Sulphydryl (–SH) groups (cf. pp. 66, 174) are also present in the protein molecules as side groups of the amino acid cysteine. Reaction between the –SH groups and the –SS– bonds permits new inter- and intra-protein/polypeptide relationships to be formed via –SS– bonding, one effect of this interchange being the relaxation of dough by the relief of stress induced by the mixing process.

While gluten is important in creating an extensible framework, soluble proteins in the dough liquor may also contribute to gas retention by forming an impervious lining layer within cells, effectively blocking pin-holes in cell walls (Gan *et al.*, 1990).

Dough ripening

A dough undergoing fermentation, with intermittent mechanical manipulation, is said to be ripening. The dough when mixed is sticky, but as ripening proceeds, it becomes less sticky and more rubbery when moulded, and is more easily handled on the plant. The bread baked from it becomes progressively better, until an optimum condition of ripeness has been reached. If ripening is allowed to proceed beyond this point a deterioration sets in, the moulded dough gets shorter and possibly sticky again, and bread quality becomes poorer. A ripe dough has maximum elasticity after moulding and gives maximum spring in the oven; a green or underripe dough can be stretched but has insufficient elasticity and spring; an overripe dough tends to break when stretched.

If the optimum condition of ripeness persists over a reasonable period of time the flour is said to have good fermentation tolerance. Weak flours quickly reach a relatively poor optimum, and have poor tolerance, whereas strong flours give a higher optimum, take longer to reach it, and have good tolerance. Addition of improvers or oxidizing agents to the flour can speed up the rate at which dough ripens and hence shorten the time taken to achieve optimum development.

Dough stickiness

Certain agronomic advantages and improved disease resistance in wheat have been achieved by incorporating genes from rye. The short arm of the rye chromosome 1R has been substituted for the short arm of the homologous group 1 chromosome of wheat. However, the doughs made from the flour of many of the substitution lines have a major defect in that they are intensely sticky. This stickiness is not due to overmixing, excess water or excess amylolytic activity: the factor responsible for the stickiness, introduced with the rye chromosome, has not yet been identified (Martin and Stewart, 1991).

Proteolytic enzymes

Besides the enzymes that act on carbohydrates, there are many other enzymes in flour and yeast, of which those that affect proteins, the proteolytic enzymes, may be of importance in baking. Yeast contains such enzymes, but they remain within the yeast cells and hence do not influence the gluten.

The proteolytic enzymes of flour are proteases. They have both disaggregating and protein solubilizing effects, although the two phenomena may be due to distinct enzymes.

The undesirable effect on bread quality of flour milled from wheat attacked by bug (cf. p. 9) is generally considered to be due to excessive proteolytic activity. Inactivation temperature is lower for proteolytic enzymes than for diastatic enzymes, and heat treatment has been recommended as a remedy for excessive proteolytic activity in buggy wheat flour. However, it is difficult to inactivate enzymes by heat treatment without damaging the gluten proteins simultaneously.

Surfactants

These substances act as dough strengtheners, to help withstand mechanical abuse during processing, and they also reduce the degree of retrogradation of starch (cf. pp. 62 and 209). They include calcium and sodium stearoyl lactylates (CSL, SSL) and mono- and di-acetyl tartaric esters of mono- and di-glycerides of fatty acids (DATEM), and are used at levels of about 0.5% on flour wt (Hoseney, 1986). The Bread and Flour Regulations 1984 permit the use of SSL, up to a maximum of 5 g/kg of bread, in all bread, and of DATEM esters, with no limit specified, in all bread.

Stearoyl-2-lactylates

Calcium stearoyl-2–lactylate (CSL) and sodium stearoyl-2–lactylate (SSL) are the salts of the reaction product between lactic and stearic acids. CSL ('Verv') and SSL ('Emplex') are dough improving and anti-staling agents; they increase gas retention, shorten proving time and increase loaf volume. They increase the tolerance of dough to mixing, and widen the range over which good-quality bread can be produced. The use of CSL or SSL permits the use of a considerable proportion of non-wheat flours in 'composite flours' to make bread of good quality by ordinary procedures (cf. p. 214). A typical composite breadmaking flour would contain (in parts) wheat flour 70, maize or cassava starch 25, soya flour 5, CSL 0.5–1.0, plus yeast, sugar, salt and water. The nutritive value of such bread has been shown to be superior to that of bread containing only wheat flour, salt, yeast and water. Use of CSL and SSL has been permitted in the U.S.A. since 1961.

Colour of bread crust and crumb

The brown colour of the crust of bread is probably due to melanoidins formed by a non-enzymic 'browning reaction' (Maillard type) between amino acids, dextrins and reducing carbohydrates. Addition of amino acids to flours giving pale crust colour results in improvement of colour. The glaze on the crust of bread is due, in part, to starch gelatinization which occurs when the oven humidity is high. An under-ripe dough which still contains a fairly high sugar content will give a loaf of high crust colour: conversely, an over-ripe dough gives a loaf of pale crust colour.

The perceived colour of bread crumb is influenced by the colour, degree of bleach, and extraction rate of the flour; the use of fat, milk powder, soya flour or malt flour in the recipe; the degree of fermentation; the extent to which the mixing process disperses bubbles within the dough and the method of panning — cross-panning and twisting to increase light-reflectance.

Bread aroma and flavour

The aroma of bread results from the interaction of reducing sugars and amino compounds, accompanied by the formation of aldehydes. Aroma is also affected by the products of alcoholic and, in some cases, lactic acid fermentation — organic acids, alcohols, esters. The flavour of bread resides chiefly in the crust.

Commercial processes for making white bread

A white pan loaf of good quality is characterized by having sufficient volume, an attractive appearance as regards shape and colour, and a crumb that is finely and evenly vesiculated and soft enough for easy mastication, yet firm enough to permit thin slicing. A more open crumb structure is characteristic of other varieties, e.g. Vienna bread and French bread. The attainment of good quality in bread depends partly on the inherent characteristics of the ingredients — particularly the flour — and partly on the baking process.

In the U.K., white bread comprised about 52% of the total bread eaten in the home in 1989. Methods used for commercial production of white bread differ principally according to the way in which the dough is developed. This may be:

— biologically, by yeast fermentation. Examples: bulk (long) fermentation processes (Straight dough system; Sponge and dough system);
— mechanically, by intense mixing and use of oxidizing agents. Examples: J. C. Baker's 'Do-Maker' process and AMFLOW processes (continuous); Chorleywood Bread Process; Spiral Mixing Method;
— chemically, by use of reducing and oxidizing agents. Example: Activated Dough Development (ADD) process.

In the bulk fermentation process, some of the starch, after breakdown to sugars, is converted to alcohol and carbon dioxide, both of which are volatile and are lost from the dough (cf. p. 197). The bulk fermentation process is thus a wasteful method, and processes which utilize mechanical or chemical development of the dough offer considerable economic advantages, as there is less breakdown of the starch, as well as being much more rapid.

Other rapid methods include the Continental No-time process (or Spiral Mixing Method), the Emergency No-time process and the Aeration or Gas-injection process.

Bulk (long) fermentation process

The bread is made by mixing a dough from flour, water, yeast, fat and salt, allowing the dough to rest at a temperature of 26°–27°C while fermentation and gluten ripening take place, and then baking in the oven.

Straight dough system

In a representative procedure, the ingredients for a 100-kg 3-h dough would be 100 kg of flour, with probably 1 kg of yeast, 2 kg of salt, 1 kg of fat and 55–57 kg of water at a temperature that will bring the mixture to about 27°C after mixing. Until the prohibition of its use in the U.K. in April, 1990 (MAFF, 1990), potassium bromate would generally have been added to the flour by the miller at a rate of 15 mg/kg. The yeast is dispersed in some of the water, the salt dissolved in another portion. All these ingredients, together with the rest of the water, are then blended and mixed in a low speed mixer during 10–20 min, during which there may be a temperature rise of 2°C. The resulting dough is set aside while fermentation proceeds.

As an alternative to treatment of the flour with potassium bromate, the miller can use ascorbic acid at a rate of 15 mg/kg, or azodicarbonamide at a level of 5 mg/kg. If the bulk fermentation time is 1 h or longer, no further treatment of the flour with oxidizing agents by the baker is required. For use in a bulk fermentation process of 30 min or less, the baker would probably add a further 50–100 mg/kg of ascorbic acid or 20–30 mg/kg of azodicarbonamide, as one of the functions of oxidizing agents is to shorten the fermentation time.

The Bread and Flour Regulations 1984 permit the use, in the U.K., of up to 200 mg/kg of ascorbic acid in all bread, and of up to 45 mg/kg of azodicarbonamide in all bread except wholemeal. To avoid the accidental breaching of these Regulations, a Code of Practice has been agreed, by which millers will add not more than 50 mg/kg of ascorbic acid and/or 10 mg/kg of azodicarbonamide to flour at the mill, while improver manufacturers will ensure that their products, when added to flour at the recommended levels, will add not more than 150 mg/kg of ascorbic acid and/or 35 mg/kg of azodicarbonamide.

After about 2 h in a 3 h fermentation process, the dough is 'knocked back', i.e. manipulated to push out the gas that has been evolved in order to even out the temperature and give more thorough mixing.

After another hour's rising, the dough is divided into loaf-sized portions and these are roughly shaped. The dough pieces rest at about 27°C for 10–15 min. ('1st proof') and are then moulded into the final shape, during which the dough is mechanically worked to tighten it so that the gas is better distributed and retained, and placed in tins. The final mould is very important in giving good texture in bulk-fermented bread. It is during the rounding and moulding processes that the bubble structure, resulting in a satisfactory crumb structure, is developed: bubbles that have been inflated during the fermentation are subdivided to produce a greater number of smaller bubbles.

The dough rests again in the tins for the final proof of 45–60 min at 43°C and 80–85% r.h., and it is the carbon dioxide evolved during the final proof that inflates the dough irreversibly. The dough is then baked in the oven at a temperature of 235°C for 20–40 min, depending on loaf size, with steam injected into the oven to produce a glaze on the crust.

A number of changes take place as the temperature of the dough rises at the beginning of baking:

— the rate of gas production increases;
— at about 45°C the undamaged starch granules begin to gelatinize and are attacked by alpha-amylase, yielding fermentable sugars;
— between 50° and 60°C the yeast is killed;
— at about 65°C the beta-amylase is thermally inactivated;
— at about 75°C the fungal amylase is inactivated;
— at about 87°C the cereal alpha-amylase is inactivated;
— finally, the gluten is denatured and coagulates, stabilizing the shape and size of the loaf.

Sponge and dough system

When the bulk fermentation process is used in England, the straight dough system is generally employed. But in the U.S.A., to some extent in Scotland, and occasionally in England, a sponge and dough system is used.

The sponge and dough system differs from the straight dough system in that only part of the flour is mixed at first with some or all of the yeast, some or all of the salt, and sufficient water to make a dough, which is allowed to ferment for some hours at 21°C. The sponge (as this first dough is called) is then broken down by remixing, and the remainder of the flour, water and salt, and all the fat added to make a dough of the required consistency. Addition of oxidizing agents — ascorbic acid alone or ascorbic acid plus azodicarbonamide — would usually be made at the dough stage. The dough is given a short fermentation at 27°C before proving and baking. Further details of the procedure are to be found in *The Master Bakers' Book of Breadmaking* (Brown, 1982). The sponge and dough system is said to produce bread that has a fuller flavour than that made by the straight dough system. A Bakers' Grade flour of, say, 12% protein content is suitable for the dough stage, but the sponge stage requires a stronger flour of, say, 13% protein content. About 65% of all bread in the U.S.A. is made by the sponge and dough system.

Plant baking

In the small bakeries, most of the processes of dividing, moulding, placing in the proving cabinets and the oven and withdrawing therefrom are carried out by hand. However, disadvantages of hand-processing are lack of uniformity in the products and the excessive amount of labour involved. In large bakeries, machines carry out all these processes. Mixers are of the closed bowl high-speed type, consuming a total quantity of energy (including no-load power) of 7.2–14.4 kJ/kg (1–4 Wh/kg) and taking 15–20 min. Dough dividers divide the dough by volume: automatic provers have built-in controls giving correct temperature and relative humidity. In the smaller bakeries, peel, reel or rack ovens have now largely replaced drawplate ovens. In larger bakeries, so-called 'travelling' ovens are used. In these, the doughs are placed on endless bands which travel through the oven. The oven is tunnel-shaped and possibly 18.3 m (60 ft) long.

The Bulk Fermentation Process is now used to make probably not more than 10% of all commercially-made bread in the U.K.: the principal users would be plant bakeries in Scotland and small bakeries, some of which prefer the sponge and dough system.

Mechanical development processes

Continuous doughmaking

A further stage in the mechanization of breadmaking is represented by the continuous breadmaking process, exemplified by the Wallace and Tiernan Do-Maker Process, based on the work of J. C. Baker, which was formerly used in the U.K. and the U.S.A. About 35% of U.S. bread was made by the Baker process in 1969. In the process, first used in the U.K. in 1956 (see Anon., 1957), the flour, spouted from a hopper, is continuously mixed with a liquid pre-ferment or 'brew' in electronically regulated quantities. The pre-ferment is a mixture consisting of a sugar solution with yeast, salt, melted fat and oxidizing agents which is fermented for 2–4 h. The dough is allowed no fermentation time, but instead is subjected to intense mechanical mixing whereby the correct degree of ripeness for proving and baking is obtained. In the absence of fermentation, it is essential to incorporate an appropriate quantity of oxidizing agent into the dough. The dough is extruded through a pipe, cut off into loaf-sized portions, proved and baked. The Do-Maker Process gives bread with a characteristic and very even crumb texture. Considerable time is saved, in comparison with bulk fermentation processes.

The AMFLOW process has an overall similarity to the Do-Maker process, but features a multistage pre-ferment containing flour, and a horizontal, instead of a vertical, development chamber.

In the U.S.A., the Do-Maker and AMFLOW

processes are being replaced by the sponge and dough processes or by the CBP. The Do-Maker and AMFLOW processes are no longer used in the U.K.

Chorleywood Bread Process (CBP)

This is a batch or continuous process in which dough development is achieved during mixing by intense mechanical working of the dough in a short time, and bulk fermentation is eliminated. The process was devized in 1961 by cereal scientists and bakers at the British Baking Industries Research Association, Chorleywood, Herts, England (Chamberlain et al., 1962; Axford et al., 1963). It is necessary to use a special high-speed mixer for mixing the dough. The process is characterized by:

— the expenditure of a considerable, but carefully controlled, amount of work (11 Wh/kg; 40 J/g) on the dough during a period of 2–4 min;

— chemical oxidation with ascorbic acid (vitamin C) alone or with potassium bromate (if allowed) at a relatively high total level, viz. 100 mg/kg, or with azodicarbonamide at a level of 20–30 mg/kg or with both ascorbic acid and azodicarbonamide;

— addition of fat (about 0.7% on flour wt) — this is essential — of which 5% (0.035% on flour wt) should be high melting point fat which will still be solid at 38°C;

— use of extra water (3.5% more than normal, based on flour wt);

— absence of any pre-ferment or liquid ferment;

— a higher level of yeast, 2% on flour wt, than is used in the BFP;

— a first proof of 2–10 min, after dividing and rounding, followed by conventional moulding and final proof.

The level of work input is critical, but is dependent on the genetic make-up or strength of the wheat. Thus, while a figure of 11 Wh/kg is applicable to dough made from an average grist, there are certain U.K. wheat varieties, e.g. Fresco, the flour from which, if used alone, would require 17–20 Wh/kg. If such varieties are included in the bread grist, it will be desirable for the miller to formulate the grist in such a way that the work input requirement is maintained at, or near, 11 Wh/kg. For an average grist, the quality of the bread, in respect of loaf volume and fineness of crumb structure, improves at work input levels from 7 to 11 Wh/kg, but at 13 Wh/kg or more the structure of crumb deteriorates. Work input level is monitored by a watt–hour meter and a counter unit attached to the mixer motor. The total work input required — dough wt in kg × 11 — is set on the counter unit, and the mixer motor is automatically switched off when the determined amount of work has been performed.

One reason for the intense and rapid mixing is that it brings the molecules rapidly into contact with the oxidizing agents. During the intense mechanical development, a gas bubble structure is created in the dough which, provided the dough is properly handled, is expanded in proof and becomes the loaf crumb structure.

Final dough temperature after mixing should be 28°–30°C, but as the mixing process causes a temperature rise of 14°–15°C, the doughing water may have to be cooled, and a water-cooling unit is generally a part of the plant.

The use of potassium bromate is no longer permitted in the U.K.; if potassium bromate is not used as an oxidizing agent, the use of ascorbic acid instead, at a level of 100 mg/kg, may not be adequate to maintain loaf volume if the dough is mixed under partial vacuum. Under these conditions, the use of azodicarbonamide at a level of 20–30 mg/kg, in addition to the ascorbic acid, would be beneficial (cf. p. 176).

To achieve the full oxidation potential of the ascorbic acid, an adequate concentration of oxygen in the mixing machine bowl is essential: this requirement may be partially met by eliminating the vacuum, so that the dough is mixed in air, or, more effectively, by filling the head space of the mixer with an oxygen-enriched atmosphere, e.g. 60% oxygen/40% nitrogen, equivalent to a 50/50 mixture of oxygen and air.

Further improvement in loaf volume, if potassium bromate is not being used, could be achieved by changing to a flour of slightly higher protein content or by adding vital gluten, by increasing

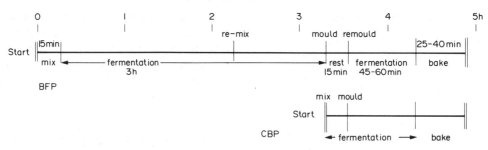

FIG 8.2 Time required to make bread by the traditional 3-h bulk fermentation process (BFP) and by the Chorleywood Bread Process (CBP).

the yeast level, by adding fungal amylase or by adding an emulsifier.

Equally satisfactory results can be obtained by replacing some or all of the fat with emulsifiers such as diacetyl tartaric acid esters (DATEM) or sodium stearoyl-2–lactylate at a level of 0.1–0.3% on flour wt.

The additional water required by the CBP, as compared with the BFP, is the consequence of the absence of bulk fermentation in the CBP. During fermentation, the breakdown of starch to sugars releases water, previously held by the damaged starch. The dough is softened by this released water and allowance is made for this effect, in the BFP, when calculating the amount of doughing water required. In the CBP, water is not released to the same extent because there is little or no breakdown of damaged starch, and without the released water the dough would tend to be too tight unless the amount of doughing water were increased. The extra water added in the CBP thus largely corresponds with the additional solids in the CBP that have not been lost (as in the BFP) during bulk fermentation. It leads to an increase of less that 1% in the moisture content of the bread.

Additional yeast is required in the CBP to compensate for the lower rate at which fermentation proceeds at the beginning of the final proof, because, in the absence of a bulk fermentation, the yeast has not been activated as it is in the BFP. Another reason for use of extra yeast in the CBP is that the dough is relatively denser at the start of proof than it is in the BFP.

It is claimed that bread made by the CBP is indistinguishable in flavour or crumb structure from bread made by bulk fermentation, and that it stales less rapidly (Axford *et al.*, 1968)

Advantages claimed for the CBP, besides the avoidance of bulk fermentation, are:

— an additional yield of about 7 pt of dough per 100 pt of flour, leading to an increase in yield of bread of 4%, and a net saving on raw material costs;
— a saving of 60% in processing time (see Fig. 8.2)
— a saving of 70% of space, previously occupied by fermenting dough;
— a reduction of about 70% in the amount of dough being processed at any one time and consequently a considerably reduced loss in case of plant breakdown;
— a lower staling rate;
— a greater amenability to control.

Moreover, by use of the CBP, flour with a protein content lower by about 1% than that required for the bulk fermentation process can be baked into bread with no loss of bread quality (cf. p. 193).

About 80% of the bread baked commercially in the U.K. in 1986 was being made by the CBP. The process is particularly favoured for the plant baking, for hot bread shops, and for in-store bakeries, and it is adaptable to the making of all kinds of bread.

A mechanical dough development process, resembling the CBP, is used in New Zealand. About 90% of bread was being made in that country by this process in 1983 (Mitchell, 1983). The CBP is also used in about 30 other countries, in some of which, e.g. South Africa, it is probably the most widely used process.

Chemical development process

The optimum work input for mechanical dough development is lowered if a proportion of the disulphide bonds are broken chemically by the introduction of a reducing agent.

The *Activated Dough Development* process (ADD) achieves dough development without either bulk fermentation or mechanical development. A relatively rapid-acting reducing agent, L-cysteine, and a relatively slow-acting oxidizing agent, potassium bromate, or a mixture of potassium bromate and ascorbic acid, are added at the dough mixing stage, using conventional, low-speed, mixing equipment. All the ingredients, which include 2% of yeast, 0.7–1.0% of fat, and extra water, as for the CBP, are mixed together for 10–20 min, and the dough temperature after mixing should be 28°–30°C. The reducing agent accelerates the uncoiling and reorientation of the protein molecules and the oxidizing agent follows up by stimulating the formation of cross links to stabilize the desired elastic three-dimensional gluten network.

During mixing, air is entrained in the dough, starting the process of cell formation which is continued throughout the subsequent stages of dough dividing, rounding, first proof of 6–10 min, and final moulding. During the final proof of 45–55 min (as for the BFP and the CBP) sufficient gas to inflate the dough is produced by activity of the yeast.

The ADD requires the use of a Bakers' Grade flour of about 12% protein content as used in the BFP; the lower protein content flours used in the CBP are not suitable for the ADD. Apart from this restriction on flour quality, the ADD offers most of the advantages over the BFP that are claimed by the CBP and, in addition, the ADD does not require the use of a special high-speed mixer.

The ADD was used by hot-bread shops, in-store bakeries and family bakers, and accounted for 5–10% of all bread made commercially in the U.K.

The usual levels of addition of reducing and oxidizing agents are 35 mg/kg on flour wt of L-cysteine hydrochloride (corresponding to about 27 mg/kg of L-cysteine) with 25 mg/kg of potassium bromate plus 50 mg/kg of ascorbic acid. These levels are based on the use of flour of 12% protein content that has already been treated at the mill with up to 20 mg/kg of potassium bromate.

The ADD process was introduced to the baking industry by the British Baking Industries Research Association in 1966, but the process could not be used commercially in the U.K. until 1972, when the use of L-cysteine hydrochloride was permitted by the Bread and Flour (Amendment) Regulations 1972 (MAFF, 1972). Potassium bromate and ascorbic acid were listed in the Bread and Flour Regulations 1984 (MAFF, 1984) as permitted improvers, but potassium bromate was removed from the list of improvers permitted in the U.K. in April, 1990 (MAFF, 1990), thereby necessitating the reformulation of additives used in the ADD. Replacement of potassium bromate by additional ascorbic acid or azodicarbonamide (within the maximum permitted limits) is not an ideal solution, as the balance between the oxidizing and reducing agents is upset, and hence, ADD is no longer a viable method in the U.K.

EC policy regarding additives

Additives will only be included in a permitted list if a reasonable technological need is demonstrated, and if this need cannot be achieved by other means that are economically and technologically practicable. Furthermore, the additives must present no hazard to health at the levels of use proposed, and they must not mislead the customer. 'Need' is understood to mean preservation of nutritional quality; the meeting of special dietary requirements; enhancement of keeping quality, stability and organoleptic properties; or providing aid in manufacture, processing, preparation, treatment, packaging, transport or storage. Specified additives are to be allowed only in specified foods, and at levels not exceeding those required to achieve the desired effect (Spencer, 1989).

Other rapid methods

No-time continental process

This process, also known as the Spiral Mixing Method, avoids a long bulk-fermentation, the use of a high-speed mixer, and the addition of L-cysteine and potassium bromate. In this process, all the ingredients are mixed together for 8–11 min in a special open-bowl mixer equipped with a spiral-shaped beater; the action of the mixer is faster and more vigorous than that of the low-speed mixers used in the BFP, but not so intense as that which is necessary for the CBP. The mixing action incorporates more air, and hence oxygen, in the dough, thereby improving cell creation and increasing the effectiveness of ascorbic acid.

The ingredients used would include a higher level of yeast (3% on flour wt) than used in other processes, 2% of salt, about 60 pt of water per 100 pt of flour, fat, ascorbic acid, emulsifier, fungal amylase and sugar. The dough temperature aimed at is 26°–28°C. A short period of bulk fermentation (15–30 min) follows mixing. This process is widely used on the Continent of Europe and is being used increasingly in the U.K., particularly in small bakeries.

Emergency no-time dough

This is a short system, somewhat resembling the No-time Continental process, that is used occasionally in the U.S.A. and the U.K., particularly for emergency production. The dough is made, using a larger amount of yeast, e.g. 2.5% on flour wt, and a higher temperature, e.g. 30°–32°C, than are usual for normal fermentation systems, and is immediately scaled off. Final moulding follows after about 15 min, and the dough pieces are proved for 1 h at 43°C before baking. The bread has a coarse, thick-walled crumb structure, and it stales rapidly.

Aeration (gas-injection) process

In 1860 Dauglish described a rapid bread-making method in which a dough was made by mixing soda water (water charged with carbon dioxide gas) and flour under pressure. When the pressure was released the dough expanded and was immediately divided and baked. The whole process, including baking, took 90 min. A modern equivalent is the Oakes Special Bread Process, a continuous system in which carbon dioxide gas is injected into the developing dough. Neither process is in commercial use for making standard bread.

Microwave and radio frequency baking

The use of microwave (MW) energy for baking bread was investigated at the Flour Milling and Baking Research Association (FMBRA), Chorleywood, England, (Chamberlain, 1973). Microwave energy, generated by a magnetron, and transmitted by radiation of frequency from about 900 MHz upwards, penetrates the dough very rapidly and cooks the loaf uniformly throughout.

Another source of energy that has been investigated is radio frequency (RF) energy, of about 27 MHz, which similarly heats the loaf rapidly throughout.

Commercial application of MW baking has not so far (1992) been possible because of the unavailability of a thermostable material for the pans which has the mechanical properties of metal but is freely permeable to microwave radiation. It is reported that the RF method can be operated with conveyors, baking pans and foil containers made of metal, and RF ovens are now commercially available for making bread, biscuits and other cereal-based products.

In both MW and RF baking, the dough is held for only a short period within the temperature range at which the activity of alpha-amylase is unwelcome, thereby permitting the use of a wheat grist of higher alpha-amylase activity than would be acceptable for conventional baking. Moreover, in conventional baking, the gases evolved are rapidly lost unless the protein content of the flour is high enough — say 10.5% or more — to confer adequate strength to the walls of the crumb cells. In both MW and RF baking, however, the rate of gas production exceeds the rate of gas loss (because the dough is heated rapidly throughout): hence, high protein content in the flour is not obligatory. In fact, flour of 7.5% protein content was used experimentally at the FMBRA to produce bread by MW baking that compared favourably with bread conventionally baked from flour of normal protein content. Thus, a further impetus towards the commercial application of RF heating would be a substantial price differential in favour of west-European wheat (of lower protein content, and often of high alpha-amylase activity) as against imported, non-EC, strong wheat.

Baking by MW or RF alone produces crustless bread. A crust can be developed by applying thermal radiation, in the form of hot air, simultaneously during a total baking time of less than 10 min for a standard 800 g loaf. The Air Radio Frequency Assisted (ARFA) oven therefore combines air radio frequency and convected hot air in a technique developed by the Electricity Council Research Centre, Capenhurst, England (Anon., 1987, 1988b). However, there is to date (1992) no known commercial application of ARFA for breadmaking.

Frozen dough

The use of frozen dough, which can conveniently be stored, has recently increased in popularity, e.g. for in-store bakeries. The best results are obtained if ascorbic acid is used as the oxidant, and if the doughs are frozen before fermentation and then stored at a constant temperature to avoid problems associated with the melting of ice crystals.

Bread cooling

The cooling of bread is a problem in mechanical production, particularly when the bread is to be wrapped and/or sliced before sale. Bread leaves the oven with the centre of the crumb at a temperature of about 96°C and cools rapidly. During cooling, moisture moves from the interior outwards towards the crust and thence to the atmosphere. If the moisture content of the crust rises considerably during cooling, the texture of the crust becomes leathery and tough, and the

attractive crispness of freshly baked bread is lost. Extensive drying during cooling results in weight loss (and possible contravention of the Weights and Measures Act in the U.K.), and in poor crumb characteristics. The aim in cooling is therefore to lower the temperature without much change in moisture content. This may be achieved by subjecting the loaves to a counter-current of air conditioned to about 21°C and 80% r.h. The time taken for cooling 800 g loaves by this method is 2–3 h.

Automation

Recent developments that increase the efficiency of the plant baking process include the use of load cells for weighing the ingredients and controlling ingredient proportions in the mixer, and the use of the microelectronics for temperature corrective feedback and consistency corrective feedback (Baker, 1988). Another development is the introduction of computer-programmed mixers and plants.

Bread moisture content

There is no legal standard for the moisture content of bread in the U.K. The moisture content of American and of Dutch bread must not exceed 38%. In Australia the maximum permitted moisture content in any portion weighing 5 g or more is 45% for white bread, 48% for brown and wholemeal. In New Zealand, 45% is the maximum moisture content similarly permitted in any bread.

Bread weights

In the U.K., standard bread weights were 1 lb and 2 lb until 6 May 1946, when weights were reduced to 14 oz and 28 oz. From 1 May 1978 loaves sold in the U.K. weighing more than 300 g (10.6 oz) were required to weigh 400 g or a multiple of 400 g.

In Belgium, loaves weighing more than 300 g must weigh 400 g or multiples of 400 g. In Germany, prescribed weights for unsliced bread were 500 g and then multiples of 250 g up to

2000 g, and above 2000 g by multiples of 500 g. Prescribed weights for sliced bread were 125 g, 250 g, and then by multiples of 250 g to 1500 g, and then by multiples of 500 g to 3000 g.

From 1980, enforcement of bread weight regulation in the U.K. has taken place at the point of manufacture rather than, as formerly, at the point of sale, and is based on the average weight of a batch rather than on the weight of an individual loaf.

Yield of bread

Using the CBP, 100 kg of white flour at 14% m.c. produce an average 180 loaves of nominal weight 800 g (average 807 g) containing an average of 39% of moisture (total bread yield from 100 kg of flour: 145 kg of bread). Thus, a nominal 800 g loaf is made from an average of 556 g of flour at natural m.c. (478 g on dry basis) and contains on average 492 g of dry matter (i.e. 14 g of dry matter are contributed by non-flour constituents).

Energy consumption in making bread

The CBP uses 40 J/g of dough, or 35.7 kJ per 800 g loaf, for the mixing process. The bulk fermentation process uses about one-fifth of this amount for mixing, but some additional energy is used in heating the water for doughmaking.

In the baking process, the heat required comprises the heat needed to raise the temperature of the dough piece from that of the prover (about 40°C) to that at the oven exit (about 96°C); the latent heat of evaporation of the water changed to steam, and the heat required to raise the temperature of that steam to that of the oven; and the heat required to raise the temperature of the pan to the oven exit temperature.

Thus, for baking a nominal 800 g loaf (with flour at 14% m.c., water absorption 60.7% on flour wt, and oven loss 65 g), about 400 kJ (379 Btu) theoretical are required. This figure is made up of about 300 kJ (284 Btu) for the loaf itself plus about 100 kJ (95 Btu) for the pan. Oven efficiency depends on oven type and quantity of steam used for conditioning the oven atmosphere, and

averages 40%, thus giving a practical requirement of 1 MJ (948 Btu) per nominal 800 g loaf plus pan: 750 kJ (711 Btu) for the 800 g loaf alone, equivalent to 937 kJ (889 Btu) per kg of bread (without pan).

Additional energy is used in conveyors, final proof, cooling, slicing, wrapping of bread, but these amounts are small in relation to the energy used for baking (Cornford, S. J., private communication, 1979).

A breakdown of the total energy requirements for making a white loaf, including energy used in growing the wheat, milling the wheat, baking the bread, and selling the bread, is shown in Table 8.1.

TABLE 8.1
Energy Requirements for Making a White Loaf

	Percentage of total energy required	
Growing wheat		
Tractors	5.3	
Fertilizers	11.1	
Drying, sprays	3.0	19.4
Milling the wheat		
Direct fuel and power	7.4	
Other	2.1	
Packaging	1.3	
Transporting	2.0	12.8
Baking		
Direct fuel and power	30.2	
Other items	17.3	
Packaging	9.0	
Transporting	7.8	64.3
Shops	3.4	3.4
		99.9

Data from Leach (1975).

Other kinds of bread

Brown and wholemeal breads

When using the bulk fermentation process, the level of fat used in brown and wholemeal breads is generally raised to about 1.5% on flour wt (as compared with 1% for white bread) because the fat requirement for brown flour and wholemeal is more variable than that for white flour. With the CBP, it is essential to raise the fat level in this way for brown and wholemeal breads.

A short fermentation system is generally used for wholemeal bread. For example, the dough might be allowed to ferment for 1 h before knocking back, plus 30 min to scaling and moulding, at an appropriate yeast level and temperature.

Wheat germ bread

This is made from white flour with the addition of not less than 10% of processed germ (cf. p. 153) which has been heat-treated to stabilize the lipid content and to destroy glutathione, a component which has an adverse effect on bread quality. A fermentation process to inactivate glutathione, as an alternative to heat treatment, was described in Australia in 1940.

Gluten bread: High-protein bread

These breads are made by supplementing flour with a protein source, such as wheat vital gluten, whey extract, casein, yeast, soya flour. Procea and Slimcea are proprietary breads in which the additional protein is provided by wheat gluten. However, most bread now made in the U.K. and in numerous other countries contains a small amount of added vital wheat gluten (cf. p. 195) depending on the protein content of the flour.

High-fibre bread

This bread has both higher fibre content and fewer calories per unit than normal bread. The high fibre content is achieved by addition of various supplements, such as cracked or kibbled wheat, wheat bran, or powdered cellulose. A type of cellulose used in the U.S.A., called Solka-Floc, is delignified alpha-cellulose obtained from wood; usage levels are 5–10%. The use in the U.K. of alpha-cellulose or the sodium salt of carboxymethyl cellulose in bread, for which a slimming claim is made, is permitted by the Bread and Flour Regulations 1984 (MAFF, 1984).

Granary bread

This is a proprietary bread made from a mixture of wheat and rye which has been allowed to sprout, kiln dried and rolled. To this is added barley malt.

Speciality breads

Other 'special' breads include pain d'épice, fruit breads, malt loaves, mixed grain bread, bran bread, etc.

Bread staling and preservation

Bread staling

Staling of bread crumb is not a drying-out process: loss of moisture is not involved in true crumb staling. The basic cause of staling is a slow change in the starch, called retrogradation, at temperatures below 55°C from an amorphous to a crystalline form, the latter binding considerably less water than the former. This change leads to a rapid hardening, a toughening of the crust and firming of the crumb, loss of flavour, increase in opaqueness of the crumb, migration of water from crumb to crust, and to shrinkage of the starch granules away from the gluten skeleton with which they are associated, with consequent development of crumbliness (Hoseney, 1986).

The rate at which staling proceeds is dependent on the temperature of storage: the rate is at a maximum at 4°C, close to the temperature inside a domestic refrigerator, decreasing at temperatures below and above 4°C. Staling can be prevented if bread is stored at temperatures above 55°C (although this leads to loss of crispness and the probability of rope development) or at −20°C, e.g. in a deep freezer.

As the amylose (straight-chain) portion of the starch is insolubilized during baking or during the first day of storage, it is considered that staling is due to heat-reversible aggregation of the amylopectin (branched-chain) portion of the starch.

However, starch crystallization cannot account for all the crumb firming that occurs at temperatures above 21°C, and it has been suggested by Willhoft (1973) that moisture migration from protein to starch occurs, leading to rigidification of the gluten network, and contributing to crumb firming. See reviews of the subject by Radley (1968) and Elton (1969); see also Pomeranz (1971, 1980), and Pomeranz and Shellenberger (1971).

Emulsifiers like monoglycerides retard the rate of starch retrogradation. Monoglycerides, when added at the mixing stage, first react with free, soluble amylose to form an amylose–lipid complex. When the rate of addition exceeds 1%, all the free amylose is complexed and the monoglycerides begin to interact with the amylopectin, thereby retarding retrogradation (Krog *et al.*, 1989).

Rope

Freshly-milled flour contains bacteria and mould spores, but these normally cause no trouble in bread under ordinary conditions of baking and storage. Mould spores and the vegetative forms of bacteria are killed at oven temperature, but spores of some of the bacteria survive, and may proliferate in the loaf if conditions are favourable, causing a disease of the bread known as 'rope'. Ropy bread is characterized by the presence in the crumb of yellow-brown spots and an objectionable odour. The organisms responsible are members of the *B. subtilis* var. *mesentericus* group and *B. licheniformis*. Proliferation of the bacteria is discouraged by acidic conditions in the dough, e.g. by addition of 5.4–7.1 g of ACP or 9 ml of 12% acetic acid per kg of flour.

Bread preservation

The expected shelf-life of bread made in the U.K. is about 5 days for white, 3–4 days for wholemeal and brown, and 2–3 days for crusty. Thereafter, bread becomes unacceptable because of staling, drying out, loss of crispness of crust, or mould development.

Mould development can be delayed or prevented by addition of propionic acid or its sodium, calcium or potassium salts, all of which are permitted by the Bread and Flour Regulations 1984 at levels not exceeding 3 g/kg flour (calculated as propionic acid), or by addition of sorbic acid (not permitted in the U.K. or the U.S.A.). Use of propionic acid to delay mould development in

bread is also permitted legally in Denmark, Germany, Italy, the Netherlands and Spain.

Other means of preservation include the use of sorbic acid-impregnated wrappers, γ-irradiation with 5×10^5 rad, or infra-red irradiation.

Gas-packaging, with an atmosphere of carbon dioxide, nitrogen or sulphur dioxide replacing air, has been used in an attempt to extend the mould-free shelf life of baked goods. However, even in an 'anaerobic environment' of 60% CO_2 plus 40% N_2 the fungi *Aspergillus niger* and *Penicillium* spp. may appear after 16 days unless the concentration of oxygen can be kept at 0.05% or lower. The use of impermeable packaging to prevent entry of oxygen is very expensive; a less expensive alternative is to include an oxygen absorbent, such as active iron oxide, in the packaging. By this means, the mould-free shelf life of crusty rolls has been increased to 60 days (Smith *et al.*, 1987).

A so-called '90-day loaf' is packaged in nylon-polypropylene laminate and the interior air partly replaced by carbon dioxide. The packaged loaf is then sterilized by infra-red radiation. None of these methods prevents the onset of true staling.

Freezing

Freezing of bread at −20°C is the most favourable method of preserving the freshness of bread. Suitably packed, the bread remains usable almost indefinitely.

Part-baking of soft rolls and French bread is a technique now widely used in hot bread shops, in-store bakeries, bake-off units in non-bakery shops, catering establishments, and domestically. In this process, doughs for soft rolls would be proved at about 43°C and 80% r.h., those for French bread at about 32°C and 70% r.h. The proved doughs are then baked just sufficiently to kill the yeast, inactivate the enzymes and set the structure, but producing little crust colour or moisture loss. Temperature for part-baking would be about 180°C. The part-baked product can then be deep-frozen at, say, −18°C, or stored at ambient temperature until required. The final bake-off at the point of use, at a temperature of about 280°C, defrosts the frozen product, increases the crust colour to normal, and reverses all the staling that may have taken place.

Use of cereals other than wheat in bread

The statement of Atkins (1971) that "bread is fundamentally foamed gluten" can be correctly applied only to bread in which wheat flour or meal is the sole or dominant cereal, because the protein in other cereals, on hydration, does not form gluten which is comparable in rheological properties with wheat gluten.

The flour of cereals other than wheat is used for breadmaking in two ways: either blended with wheat flour, in a form sometimes known as 'composite flour' (cf. p. 214), or as the sole cereal component. Bread made from composite flour employs conventional baking processes, but when a non-wheat flour is used alone it is usual to make use of a gluten substitute (cf. p. 215). Rye flour, however, is exceptional in that bread made from it as the sole cereal component does not require the addition of a gluten substitute.

Rye

Rye flour and meal are used for the production of numerous types of bread, both soft bread and crispbread, and rye is regarded as a bread grain in Germany and in most Eastern European countries (Drews and Seibel, 1976).

In Europe, 'rye bread' is made from all rye flour; 'rye/wheat bread' contains not less than 50% of rye flour, with wheat flour making up the remainder; 'wheat/rye bread' implies a blend of not less than 50% of wheat flour plus not less than 10% of rye flour.

Factors that influence the baking potential of rye flour include variety, environmental conditions of growth and fertilizer use, activity of amylase, protease and pentosanase enzymes, and functions of carbohydrates and proteins.

Soft bread

Under certain conditions rye grains germinate in the harvest field and then exhibit increased enzymic activity which may be undesirable for breadmaking purposes. Rye flour with high maltose figure (e.g. 3.5) and low amylograph value (350 or less) is of poor baking quality. Rye flours with Falling number (cf. p. 183) below 80 produce loaves with sticky crumb, but rye flour with FN 90–110 can be processed into acceptable bread with the aid of additives, acidifiers and emulsifiers to compensate for the effects of sprout damage. Such additives would include an acidifier to adjust the pH to 4.0–4.2, 2% of salt (on flour wt), 0.2–0.5% of emulsifier, 1–3% of gelatinized flour (Gebhardt and Lehrack, 1988).

Deterioration of baking quality of rye during storage is better indicated by glutamic acid decarboxylase activity than by Falling Number (Kookman and Linko, 1966).

The possibility that rye may be infected, in the field, with ergot (*Claviceps purpurea*) has been discussed elsewhere (cf. p. 15).

The protein in rye flour is less important than the protein in wheat flour. The rye protein, when hydrated, does not form gluten because the proportion of the protein that is soluble is much larger in rye than in wheat (up to 80% soluble in rye sour dough as compared with 10% soluble protein in wheat dough), and because the high content of pentosans inhibits the formation of gluten (Drews and Seibel, 1976). Conversely, the pentosans and starch in rye are much more important than in wheat (Telloke, 1980). The pentosans, which comprise 4–7% of rye flour, and the starch have an important water-binding function in forming the crumb structure of rye bread. The pentosans, in particular, play a role in raising the viscosity of rye dough. Rye flour can be fractionated according to particle size to yield fractions which vary in starch and pentosan contents. These fractions can then be blended to give an optimal ratio of pentosan to starch. A pentosan:starch ratio of 1:16 to 1:18 is considered ideal.

The starch of rye gelatinizes at a relatively low temperature, 55°–70°C, at which the activity of alpha-amylase is at a maximum. In order to avoid excessive amylolytic breakdown of the starch, a normal salt level is used for making rye soft bread, and the pH of the dough is lowered by acid modification in a 'sour dough' process, preferably by lactic acid fermentation with species of *Lactobacillus*.

Straight dough process

Yeast is used for leavening, and the dough is acidified by adding lactic acid or acidic citrates. The dough is mixed slowly to prevent too much toughening that could be caused by the high viscosity of the pentosans.

Sour-dough process

Sour doughs, containing lactic acid bacteria, were probably in use for making bread as long ago as 1800 B.C. in Eastern Mediterranean regions, the process spreading to Germany between the 1st and 6th centuries A.D., where it was used mainly by monks and guilds (Seibel and Brümmer, 1991).

The sour dough is a sponge-and-dough process. A starter dough is prepared by allowing a rye dough to stand at 24°–27°C for several hours to induce a natural lactic acid fermentation caused by grain micro-organisms. Alternatively, rye dough is inoculated with sour milk and rested for a few hours, after which a pure culture of organic acids (acetic, lactic, tartaric, citric, fumaric) is added to simulate the flavour of a normally soured dough. Part of the mature sour dough is retained as a subsequent starter, while the remainder is mixed with yeast and rye flour or rye wholemeal, or a blend of rye and wheat flours, and acts as a leavening agent in the making of rye sour bread (Drews and Seibel, 1976).

The flavour of San Francisco sour dough bread is due largely to lactic and acetic acids which are produced from D-glucose by *Lactobacillus*, a bacterium active in sour dough starter. The starter is fermented for 2 h at 24°–26°C and then held in a retarder for 10 h at 3°–6°C. The dough, which incorporates 2–10% of starter (on flour wt basis) besides vital gluten (1–2%), shortening, yeast (2.5–4.0%), salt, sugar, yeast food and water, is fermented for 30–60 min, scaled, rested for 12–15 min, proved for 5–8 h, and baked (1 lb loaves) for 30–35 min at 204°–218°C. However, this long drawn-out process can be avoided by using commercially available free-flowing sour dough bases (Ziemke and Sanders, 1988; Seibel and Brümmer, 1991).

All doughs containing rye flour have to be acidified because sour conditions improve the swelling power of the pentosans and also partly inactivate the amylase, which would otherwise have a detrimental effect on the baking process and impair normal crumb formation (Seibel and Brümmer, 1991).

Conventional dough improvers are not widely used in making rye or rye/wheat bread. Instead, pregelatinized potato flour or maize starch or rice starch may be added at a level of 3% (on flour wt). These materials have high water-binding capacity and increase the water absorption of the dough. Other substances used with similar effect include hydrocolloids and polysaccharide gums such as locust bean and guar gums.

Staling of rye bread is less serious than that of wheat bread, and shelf-life may be extended in various ways: by the addition of malt flour or pregelatinized potato flour or starch; by the use of sour dough; or by wrapping while still warm.

Pumpernickel

Pumpernickel is a type of soft rye bread made from very coarse rye meal by a sour dough process. A very long baking time (18–36 h) is used, with a starting temperature of about 150°C being gradually lowered to about 110°C. Pumpernickel has a long shelf-life.

Crispbread (Knackerbrot)

Rye crispbread is generally made from rye wholemeal or flaked rye, using water or milk to mix the dough, and may be fermented with yeast (brown crispbread) or unfermented (white crispbread). The traditional method used in Sweden is to mix rye meal with snow or powdered ice; expansion of the small air bubbles in the ice-cold foam raises the dough when it is placed in the oven. It is desirable to use rye flour or meal of low alpha-amylase activity.

In one process, a dough made from rye wholemeal, yeast, salt and water is fermented for 2–3 h at 24°–27°C. After fermentation, the dough is mixed for 5–6 min, proved for 30–60 min, sheeted, dusted with rye flour, and cut to make pieces about 7.6 × 7.6 cm (3 × 3 in) in size, which are baked for 10–12 min at 216°–249°C. The baked pieces are stacked on edge and dried in a drying tunnel for 2–3 h at 93°–104°C to reduce the moisture content to below 1%.

Ryvita is a crisp bread made from lightly salted rye wholemeal.

Flat breads

A new type of crispbread product appeared in the 1990s under the trade name 'Cracotte', manufactured from wheatflour by a continuous extrusion cooking process. In this process the flour of about 16% m.c. is heated and sheared to form a fluid melt at 130°–160°C in which starch forms the continuous phase. After extrusion, it forms a continuous strip of expanded foam (specific volume 7–10 ml/g). Individual biscuits are cut from the strip and packed like crispbreads. This type of product, which has gained popularity in many countries, may be manufactured from any cereal type, and variations have appeared which included rye, rice and maize, either as the minor or major component in blends with wheat.

Triticale

The breadmaking characteristics of flour made from early strains of triticale were discouraging, although bread quality could be improved by addition of dough conditioners. However, bread of good quality has been made from recent triticale selections. Bread baked commercially with 65% of wheat flour blended with 35% of triticale stoneground wholemeal was first marketed (as 'tritibread') in the U.S.A. in 1974.

Triticale flour has been tested extensively in Poland for breadmaking. The best results, using a blend of 90% triticale flour plus 10% of rye four, were obtained with a multi-phase (pre-ferment, sour-dough) process in which the pre-ferment was made with the rye flour (10% of the total flour) with water to a pre-ferment yield of 400%, and a fermentation time of 24 h at 28°–29°C. The sour used triticale flour (50% of the total flour) with 1–2% of yeast (on total flour basis) and water to give a sour yield of 200%. This was

fermented for 3 h at 32°C. The rest of the triticale flour was then added, with salt, 1.5% on flour wt, and water, to give a dough yield of 160–165%, and then fermented for 30 min at 32°C. The loaves were baked at 235°–245°C (Haber and Lewczuk, 1988). Bread made from all-triticale flour stales more rapidly than all-wheat bread.

Bread made from 50:50 or 75:25 blends of triticale flour and wheat flour had higher specific volumes (4.8; 4.9 ml/g) than the bread baked from all wheat flour (4.4 ml/g); no deleterious effect on crumb characteristics, viz. grain and texture, resulted from the admixture of triticale flour (Bakhshi et al., 1989).

Barley and oats

During World War II, when supplies of imported wheat were restricted in Britain, the Government authorized the addition of variable quantities (up to 10% of the total grist in 1943) of barley, or of barley and oats, to the grist for making bread flour (cf. p. 93). For this purpose, the barley was generally blocked (cf. p. 162) to remove the husk, and the oats were used as dehusked groats (cf. p. 166).

Good quality bread has been made in Norway from a blend of 50% of wheat flour of 78% extraction rate, 20% of barley flour of 60% extraction rate, and 30% of wheat wholemeal, using additional shortening (Magnus et al., 1987).

For use in bakery foods in the U.S.A., cleaned oat grain is steam-heated to about 100°C and then held in silos to be 'ovenized' by its own heat for about 12 h. This process preserves the mineral and vitamin contents, and conditions the oats. The grain is then impact dehulled (cf. p. 167) without previous kilning (McKechnie, 1983).

Rice

Bread of acceptable quality has been made from a blend of 75 parts of wheat flour (12.1% protein content, 14% m.c. basis) with 25 parts of rice flour which had been partly gelatinized by extrusion. The rice flour was milled from rice grits (7.32% protein content, 14% m.c. basis) which were pregelatinized to 76.8% by extrusion, using a Creusot Loire BC-45 twin screw extruder. The beneficial effects of extrusion treatment appeared to be due to thermal modification of the starch in the rice flour (Sharma *et al.*, 1988). The volume of loaves baked from this blend was below that of all-wheat flour loaves, but in other respects the bread was judged acceptable.

The low contents of sodium, protein, fat and fibre and the high content of easily digested carbohydrate favour the use of rice bread as an alternative to wheat bread for persons suffering from inflamed kidneys, hypertension and coeliac disease (cf. p. 297).

The volume of loaves of yeast-leavened bread made from 100% of rice flour is improved by the addition of hydroxypropylmethyl cellulose, a gum which creates a film with the flour and water that retains the leavening gases and allows expansion (Bean, 1986).

Stabilized and extracted rice bran can provide nutritional fortification, when used at levels up to 15%, for bakery products such as yeast-raised goods, muffins, pancake mixes and biscuits. The rice bran contributes flavour, increases water absorption without loss of volume, adds significant amounts of essential amino acids, vitamins and minerals, but does not affect mixing tolerance or fermentation. The blood cholesterol-lowering capabilities of rice bran are a further inducement for its use (Hargrove, 1990).

Maize

Maize flour is used to make bread of a sort in Latin America, and also for pancake mixes, infant foods, biscuits and wafers. Pregelatinized maize starch may be used as an ingredient in rye bread (cf. p. 212).

Bread made with composite flour

Flour milled from local crops can be added to wheat flour to extend the use of an imported wheat supply and thereby save the cost of foreign currency. This arrangement is particularly appropriate for developing countries which do not grow wheat.

Satisfactory bread can be made from such composite flour, viz. a blend of wheat flour with flour of other cereals such as maize, sorghum, millet or rice, or with flour from root crops such as cassava.

The flour of the non-wheat component acts as a diluent, impairing the quality of the bread to an extent depending on the degree of substitution of the wheat flour. A higher level of substitution is possible with a strong wheat flour than with a weak one.

Possible levels of substitution, as percentage by weight of the composite flour, are 15–20% for sorghum flour and millet flour, 20–25% for maize flour. Somewhat higher levels of substitution may be possible by the use of bread improvers or by modifying the bread-making process.

A blend of 70% of wheat flour, 27% of rice flour and 3% of soya flour made acceptable bread, provided surfactant-type dough improvers were used. A more economical blend, producing acceptable bread, is 50% of wheat flour, 10% of rice flour and 40% of cassava flour. Rice starch can also be used, e.g. a blend of 25% of rice starch with 75% of wheat flour yielded acceptable bread (Bean and Nishita, 1985).

Bread of acceptable quality is being made in Senegal and Sudan from a blend of 70% of imported wheat flour of 72% extraction rate and 30% of flour milled locally from white sorghum to an extraction rate of 72–75%.

The water absorption of a blend of 15% of millet flour with 85% of wheat flour is about 3% higher than that of the wheat flour alone, and extra water must therefore be added. Acceptable bread can be made at an even higher rate of substitution, viz. 30%, by modifying the bread-making process in various ways, e.g. by delaying the addition of the millet flour until near the end of the mixing process; by the use of improvers, such as calcium stearoyl lactylates or tartaric esters of acetylated mono- and di-glycerides of stearic acid; or by increasing the addition of sugar and fat to 4% (each) on composite flour wt.

When using a blend of maize flour and wheat flour to make bread it is desirable to increase the addition of water by about 2% for each 10% substitution of the wheat flour by maize flour, and to increase the amount of yeast to about 1.5

times that suitable for wheat flour alone. The use of hardened fat or margarine (2% on flour wt) is recommended to achieve good bread quality. Addition of emulsifiers, such as lecithin, stearates or stearoyl lactylates, is also recommended.

Dried Distillers' Grains (DDG)

This is a by-product in the production of distilled alcohol. During the fermentation of grains such as maize, rye, barley, the starch is converted to alcohol, carbon dioxide and other products, while the nutrients, e.g. protein, fibre, fat, vitamins, minerals, remain in the residue and are concentrated three-fold in the DDG. DDG has potential for use in bread (at levels up to 15%) and other baked products. Addition of 10% of DDG to wheat flour raised the protein content from 15 to 17%, and made a valuable contribution of amino acids such as threonine, serine, glutamic acid, alanine, methionine, leucine, histidine and lysine (Reddy et al., 1986).

Soya bread

A bread with good flavour, good storage properties (up to two weeks) and a fine to medium crumb structure has been made from a blend of wheat flour with 30–40% of soya flour. With a protein content of 19% and a reduced carbohydrate content, such bread is particularly suitable for diabetics (Anon., 1988a).

Bread made with gluten substitutes

It is no longer true that wheat gluten is necessary to make white bread of good quality. Acceptable bread has been made from sorghum flour — without any wheat flour and therefore without gluten — by the use of a gluten substitute, xanthan gum, a water-soluble polymer of high viscosity, which functions in the same way as gluten. Xanthan gum is made by fermenting carbohydrates with a bacterium, *Xanthomonas campestris*. The resulting viscous broth is pasteurized, precipitated with isopropyl alcohol, dried and ground. Xanthan gum is a form of bacterial cellulose, viz. a 1,4-linked beta-D-glucose polymer. 1% solutions are thixotropic, appearing gel-like at rest, but mixing, pouring, pumping easily (Anderson and Audon, 1988). Xanthan gum is already finding a use as a thickening agent in fast foods. To achieve the best results with sorghum flour, the xanthan gum should be soaked in water before incorporation in the dough to give the bread a more open structure, and salt should be added to improve the flavour of the bread. Sorghum/xanthan bread retains its freshness for at least six days — longer than wheat bread (Satin, 1988).

Yeasted rice flour breads, using 100% rice flour or 80% of rice flour plus 20% of potato starch, but without any wheat flour, could be made if the binding function of the wheat gluten was replaced by a gum — carboxy methyl cellulose (CMC) at a level of 1.6% on flour + starch basis, or hydroxy propyl methyl cellulose (HPMC) at a level of about 3% (on flour + starch basis). Bread made in this way met reference standards for wheat (white) bread for sp. vol., crumb and crust colour, Instron firmness and moisture content. The CMC or HPMC has the viscosity and film-forming characteristics to retain gas during proofing that are usually provided by gluten (Bean and Nishita, 1985; Ylimaki et al., 1988).

A mixture of galactomannans (hydrocolloids) from carob, guar and tara seed has been used as a gluten substitute. A blend of three parts of carob seed flour (locust bean: *Ceratonia siliqua*) and one part of tara seed flour, 75–100 μm particle size, was favoured, giving good volume yield, crumb structure and flavour (Jud and Brümmer, 1990). The use of tara gum would not be permitted in the U.K., while carob and guar gums would be permitted only for coeliac sufferers.

Another kind of gluten substitute, particularly for use in developing countries, can be made by boiling a 10% suspension of flour milled from tropical plants, e.g. cereal or root, in water until the starch gels, and then cooling. This material is then added to a blend of non-wheat flour, sugar, salt, yeast, vegetable oil, then mixed for 5 min, allowed to rise, and baked. The starch gel functions like gluten, trapping the gas evolved by yeast action (Anon., 1989).

References

ANDERSON, D. M. W. and AUDON, S. A. (1988) Water-soluble food gums and their role in product development. *Cereal Foods Wld*, **33** (10): 844, 846, 848–850.

ANON. (1957) Breadmaking processes now available. *Northwest Miller*, **257** (13): 13.

ANON. (1987) Are you turned on to radio frequency? *Br. Baker*, **184** (38): 36, 39.

ANON. (1988a) Soya, the wonder bean. *Milling*, **181** (July): 13–15.

ANON. (1988b) Radio frequency cuts baking times by half. *Br. Baker*, Nov.: 46, 48.

ANON. (1989) F.A.O. sees wheatless bread as boost to developing nations. *Milling and Baking News*, 21 March, **68** (3): 18–19.

ATKINS, J. H. C. (1971) Mixing requirements of baked products. *Food Manuf.* Feb.: 47.

AXFORD, D. W. E., CHAMBERLAIN, N., COLLINS, T. H. and ELTON, G. A. H. (1963) The Chorleywood process. *Cereal Science Today*, **8**: 265.

AXFORD, D. W. E., COLWELL, K. H., CORNFORD, S. J. and ELTON, G. A. H. (1968) Effect of loaf specific volume on the rate of staling bread. *J. Sci. Food Agric.*, **19**: 95.

BAKER, A. P. V. (1988) Baking progress. *Food Process. U.K.*, **57** (April): 37.

BAKHSHI, A. K., SEHGAL, K. L., PAL SINGH, R. and GILL, K. S. (1989). Effect of bread wheat, durum wheat and triticale blends on chapati, bread and biscuit. *J. Food Sci. Technol.*, **26** (4): 191–193.

BEAN, M. M. (1986) Rice flour — its functional variations. *Cereal Foods Wld*, **31** (7): 477–481.

BEAN, M. M. and NISHITA, K. D. (1985) Rice flours for baking. In: *Rice: chemistry and technology*. JULIANO, B. O. (Ed.). Amer. Assoc. Cereal Chem., St Paul, Minn., U.S.A.

BROWN, J. (Tech. Ed.) (1982) *The Master Bakers' Book of Breadmaking*. Nat. Assoc. of Master Bakers, Confectioners and Caterers, London.

CAUVAIN, S. P. and CHAMBERLAIN, N. (1988) The bread improving effect of fungal α-amylase. *J. Cereal Sci.* **8**: 239–248.

CHAMBERLAIN, N. (1973) Microwave energy in the baking of bread. *Food Trade Rev.*, Sept.: 8; *Brit. Baker*, **167** (July 13): 20.

CHAMBERLAIN, N. (1988) Forging ahead. *Food Process. U.K.*, **57** (April): 14, 16, 19, 20.

CHAMBERLAIN, N., COLLINS, T. H. and ELTON, G. A. H. (1962) The Chorleywood Bread Process. *Bakers' Dig.*, **36** (5): 52.

DREWS, E. and SEIBEL, W. (1976) Bread-making and other uses around the world. In: *Rye: production, chemistry and technology*, BUSHUK, W. (Ed.) Ch. 6. Amer. Assoc. Cereal Chem., St Paul, Minn., U.S.A.

ELTON, G. A. H. (1969) Some quantitative aspects of bread staling. *Bakers' Dig.*, **43**(3): 24.

FARRAND, E. A. (1964) Flour properties in relation to the modern bread processes in the United Kingdom, with special reference to *alpha*-amylase and starch. *Cereal Chem.*, **41**: 98–111.

GAN, Z., ANGOLD, R. E., WILLIAMS, M. R., ELLIS, P. R., VAUGHAN, J. G. and GALLIARD, T. (1990) The microstructure and gas retention of bread dough. *J. Cereal Sci.*, **12**: 15–24.

GEBHARDT, E. and LEHRACK, U. (1988) The processing quality of rye in the German Democratic Republic. *Bäker u. Konditor*, **36**(2): 55–57.

GROSCH, W. (1986) Redox systems in dough. In: *Chemistry and physics of baking*. Ch. 12, pp. 155–169, BLANSHARD, J. M. V., FRAZIER, P. J. and GALLAIRD, T. (Eds.). Royal Soc. Chem., London.

HABER, T. and LEWCZUK, J. (1988) Use of triticale in the baking industry. *Acta aliment. Polon.*, **14** (3–4): 123–129.

HARGROVE, K. (1990) Rice bran in bakery foods. *Amer. Inst. Baking Res. Dept Tech. Bull.*, **12** (2), Feb.: 1–6.

HOSENEY, R. C. (1986) *Principles of cereal science and technology*. Amer. Assoc. Cereal Chem., St Paul, Minn., U.S.A.

JUD, B. and BRÜMMER, J.-M. (1990) Production of gluten-free breads using special galactomannans. *Getreide Mehl Bröt*, **44** (June): 178–183.

KOOKMAN, M. and LINKO, P. (1966) Activity of different enzymes in relation to the baking quality of rye. *Cereal Sci. Today*, **11**: 444.

KROG, N., OLESEN, S. K., TOERNAES, H. and JOENSSON, T. (1989) Retrogradation of the starch fraction in wheat bread. *Cereals Foods Wld*, **34**: 281.

LEACH, G. (1975) *Energy and Food Production*. IPC Science and Technology Press, Guildford, Surrey.

MAGNUS, E. M., FJELL, K. M. and STEINSHOLT, K. (1987) Barley flour in Norwegian bread. In: *Cereals in a European Context*, pp. 377–384, MORTON, I. D. (Ed.) Chichester: Ellis Horwood.

MAGNUSON, K. M. (1985) Uses and functionality of vital wheat gluten. *Cereals Foods Wld*, **30** (2): 179–181.

MARTIN, D. J. and STEWART, B. G. (1991) Contrasting dough surface properties of selected wheats. *Cereal Foods Wld*, **36** (6): 502–504.

McDERMOTT, E. E. (1985) The properties of commercial glutens, *Cereal Foods Wld*, **30** (2): 169–171.

McKECHNIE, R. (1983) Oat products in bakery foods. *Cereal Foods Wld*, **28** (10): 635–637.

MINISTRY OF AGRICULTURE, FISHERIES and FOOD (1962) *Preservatives in Food Regulations* 1962. Statutory Instruments 1962, No. 1532, H.M.S.O., London.

MINISTRY OF AGRICULTURE, FISHERIES and FOOD (1972) *The Bread and Flour (Amendment) Regulations 1972*. Statutory Instruments 1972, No. 1391. H.M.S.O., London.

MINISTRY OF AGRICULTURE, FISHERIES and FOOD (1984) *The Bread and Flour Regulations 1984*. Statutory Instruments 1984, No. 1304. H.M.S.O., London.

MINISTRY OF AGRICULTURE, FISHERIES and FOOD (1990) *Potassium bromate (Prohibition as a Flour Improver) Regulations 1990*. Statutory Instruments 1990, No. 399, H.M.S.O., London.

MITCHELL, T. A. (1983) Changes in raw materials and manufacturing. *Baker and Millers' J.*, Feb. 15, 18.

POMERANZ, Y. (1980) Molecular approach to breadmaking — an update and new perspectives. *Bakers' Dig.* April: 12.

POMERANZ, Y. and SHELLENBERGER, J. A. (1971) *Bread Science and Technology*, Avi Publ. Co. Inc., Westport, Conn., U.S.A.

RADLEY, J. A. (1968) *Starch and its derivatives*, 4th edn. Chapman and Hall, London.

REDDY, N. R., PIERSON, M. D. and COOLER, F. W. (1986) Supplementation of wheat muffins with dried distillers' grain flour. *J. Fd Qual.* **9** (4): 243–249.

REED, G. and NAGODAWITHANA, T. W. (1991) *Yeast technology*, 2nd edn. Van Norstrand Rheinhold, N. Y. U.S.A.

SATIN, M. (1988) Bread without wheat. *New Scient.* 28 April: 56–59.

SEIBEL, W. and BRÜMMER, J.-M. (1991) The sourdough process for bread in Germany. *Cereal Foods Wld*, **36** (3): 299–304.

SHARMA, N. R., RASPER, V. F. and VAN de VOORT, F. R. (1988) Pregelatinized flours in composite blends for bread-making. *Can. Inst. F.S.T.J.* **21** (4): 408–414.

SMITH, J. P., CORAIKUL, B. and KOERSEN, W. J. (1987) Novel approach to modified atmosphere packaging of bakery products. In: *Cereals in a European Context*, MORTON, I. D. (Ed). pp. 332–343. Chichester: Ellis Horwood.

SPENCER, B. (1989) Flour improvers in the EEC — harmony or discord? *Cereal Foods Wld* **34** (3) 298–299.

TELLOKE, G. W. (1980) Private communication.

WILLHOFT, E. M. A. (1973) Recent developments on the bread staling problem. *Bakers' Dig.*, **47** (6): 14.

YLIMAKI, G., HAWRYSH, Z. J., HARDIN, R. T. and THOMSON, A. B. R. (1988) Application of response surface methodology to the development of rice flour yeast breads: objective measurements. *J. Fd. Sci.*, **53** (6): 1800–1805.

ZAWISTOWSKA, U., LANGSTAFF, J. and BUSHUK, W. (1988) Improving effect of a natural α-mylase inhibitor on the baking quality of wheat flour containing malted barley flour. *J. Cereal Sci.*, **8**: 207.

ZIEMKE, W. H. and SANDERS, S. (1988) Sourdough bread. *Amer. Inst. Baking, Res. Dept. Tech. Bull.* **10** (Oct.): 1–4.

Further Reading

BAILEY, A. (1975) *The Blessings of Bread*, Paddington Press Ltd, London.

BEECH, G. A. (1980) Energy use in bread baking. *J. Sci. Fd Agric.*, **31**: 289.

FANCE, W. J. and WRAGG, B. H. (1968) *Up-to-date Bread-making*, McLaren & Sons Ltd, London.

FOWLER, A. A. and PRIESTLEY, R. J. (1980) The evolution of panary fementation and dough development — a review. *Fd Chem.*, **5**: 283.

HALTON, P. (1962) The development of dough by mechanical action and oxidation. *Milling*, **138**: 66.

HUTCHINSON, J. B. and FISHER, E. A. (1937) The staling and keeping quality of bread. *Bakers' Nat. Ass. Rev.*, **54**: 563.

KATZ, J. R. (1928) Gelatinization and retrogradation of starch in relation to the problems of bread staling. In: *A Comprensive Survey of Starch Chemistry*, WALTON, R. P. (Ed.) **1**: 100. Chemical Catalog Co. Inc., New York

KENT-JONES, D. W. and MITCHELL, E. F. (1962) *Practice and Science of Breadmaking*, 3rd edition, Northern Publ. Co. Ltd, Liverpool.

LANGRISH, J., GIBBONS, M. and EVANS, W. G. (1972) *Wealth from Knowledge*, Macmillan, London.

MATZ, S. A. (1972) *Baking Technology and Engineering*, 2nd edition, Avi Publ. Co. Inc., Westport, Conn., U.S.A.

MAUNDER, P. (1969) *The Bread Industry in the United Kingdom*, Unversity of Nottingham, Loughborough.

MULDERS, E. J. (1973) The odour of white bread. *Agric. Res. Rep. (Versl. Landbouwk, Onderz)*, p. 798. Wageningen, Netherlands.

NATIONAL ASSOCIATION OF BRITISH AND IRISH MILLERS LTD (1989) *Practice of flour milling*, N.A.B.I.M., London.

POMERANZ, Y. (Ed.) (1971) *Wheat: Chemistry and Technology*, 2nd edition, Amer. Ass. Cereal Chem., St Paul, Minn., U.S.A.

PYLER, E. J. (1973) *Baking Science and Technology*, 2nd edition, Siebel Publ. Co., Chicago.

SCHOCH, T.J. and FRENCH, D. (1945) *Fundamental Studies on Starch Retrogradation*, Office of the Quartermaster General, U.S.A

UNITED NATIONS ECONOMIC COMMISSION FOR AFRICA (1985) *Technical Compendium on Composite Flours*, Tech. Ed. KENT, N. L., ECONOMIC COMMISSION FOR AFRICA, ADDIS ABABA.

WIENER, J. and COLLIER, D. (1975) *Bread*, Robert Hale, London.

WILLIAMS, A. (Ed.) (1975) *Breadmaking, the Modern Revolution*, Hutchinsons Benham, London.

9

Malting, Brewing and Distilling

Introduction

The essential process involved in brewing is the conversion of cereal starch into alcohol to make a palatable, intoxicating beverage. Fermentation is mediated by yeasts appropriate to the cereal or cereals involved. Most yeasts used belong to the species *Saccharomyces cerevisiae*, which now includes the 'bottom yeast' previously classified as *S. carlsbergensis* (Reed and Nagodawithana, 1991).

Two processes are involved: the starch has first to be converted to soluble sugars by amylolytic enzymes, and second, the sugars have to be fermented to alcohol by enzymes present in yeast. In the first process the enzymes may be produced in the grains themselves (endogenously) or exogenously, in other organisms present. Alternatively they may be added as extracts.

The process in which the grain's own enzymes are employed is known as malting. This comprises a controlled germination during which enzymes capable of catalyzing hydrolysis, not only of starch, but also other components of the grain, are produced. The most significant are the proteases and the β-glucanases, as the products resulting from their activities affect the qualities of the beverage.

Other organisms are employed as a source of enzymes in the production of saké — a beer produced from rice. Enzymes are added in solution, particularly when it is required to hydrolyze the starch etc. present in endosperm grits or flours, themselves incapable of enzyme production. Such adjuncts may provide any proportion of the total starch, depending on legislation relevant to the country of origin and the description of the product. Consequently, added enzymes may contribute different proportions of the enzyme complement.

The alcohol content of the liquor produced by fermentation is limited by the tolerance of the yeasts. Probably the most tolerant yeasts are used in saké production. They can survive alcohol contents of about 20% although the product is sold in a diluted form.

Distillation allows the concentration of alcohol into drinks described as spirits, the special character of which depends on flavours imparted by the processing or added to a distillate, the added flavours usually being extracts from other plant sources.

For alcohol production from plant material, sugars must be present, as in fleshy fruits, or other substrates from which fermentable sugars can be produced. Starch is such a substrate, so all cereals can be used for beer production. In the West, the most commonly used cereal is barley but substantial quantities are derived from maize (beer in central America), rye (kvass beer in the former U.S.S.R), rice (saké in Japan and shaoshinchu in China), sorghum (beer in Africa). Triticale may be used as an adjunct in beers.

Malting

During malting, large molecular weight components of the endosperm cell walls, the storage proteins and the starch granules, are hydrolyzed by enzymes, rendering them more soluble in water.

All cereals are capable of undergoing malting but barley is particularly suitable because the adherent pales (lemma and palea — see Ch. 2) provide protection for the developing plumule, or acrospire, against damage during the necessary handling of the germinating grains. Further, the husk (pales) provides an aid to filtration when the malt liquor is being removed from the residue of insoluble grain components. A third advantage of barley lies in the firmness of the grain at high moisture content.

Both two-row and six-row barleys (see Ch. 2) are suitable: the former are generally used in Europe, the latter in North America. Distinct varieties were formerly grown for malting. They were lower yielding than varieties grown for feeding. Modern malting varieties have high yields and are thus suitable for the less demanding alternative uses also.

The characteristics required of a malting barley are:

1. High germination capacity and energy, with adequate enzymic activity.
2. Capacity of grains modified by malting to produce a maximum of extract when mashed prior to fermentation.
3. Low content of husk.
4. High starch and low protein contents.

The above qualities can be affected by husbandry and handling as well as by genetic factors: loss of germination capacity can result from damage to the embryo during threshing, or overheating during drying or storage.

Provided that grains are ripe, free from fungal infestation and intact, the yield of malt extract should be directly related to starch content.

High-nitrogen barley is unsuitable for malting because:

1. Starch content is lower.
2. Longer malting times are required.
3. Modification never proceeds as far as in low-nitrogen barleys.
4. The greater quantities of soluble proteins lead to haze formation and may provide nutrients for bacteria and impair the keeping quality of the beer. The average nitrogen content of malting barley is 1.5%; some 38% of this appears in the beer in the form of soluble nitrogen compounds, the proportion of the total nitrogen entering the beer being somewhat larger from two-row than from six-row types.

Dormancy

Harvest-ripe barley may not be capable of germination immediately. While this is advantageous in the field, protecting the crop against sprouting in the ear, it is clearly a problem in relation to malting, which depends on germination occurring. The mechanism of dormancy is not fully understood and indeed it is unlikely that a single cause is involved in all cases; in many instances it has been shown that germination is inhibited by inability of the embryo to gain access to oxygen. A distinct phenomenon known as 'water sensitivity' can arise during steeping if a film of water is allowed to remain on the surface of grains. The water contains too little dissolved oxygen to satisfy the needs of the developing embryo and it acts as a barrier to the passage of air. Dormancy declines with time and storage is thus not just a means of holding sufficient stocks of grain, it is an essential part of the process of malting. During the storage of freshly harvested barley tests are performed to detect the time at which dormancy has declined sufficiently for malting proper to commence. Both 'dormancy' and 'water sensitivity' are defined in relation to the test performed. In one test 100 grains are germinated on filter papers with 4 and 8 ml of water, the difference between viability and the germination on 4 ml of water is called dormancy while the difference between the levels of germination on the different volumes of water is the water-sensitivity. Factors involved in controlling and breaking dormancy have been reviewed by Briggs (1978).

Barley malting operations

The practical steps in malting are shown schematically in Fig. 9.1.

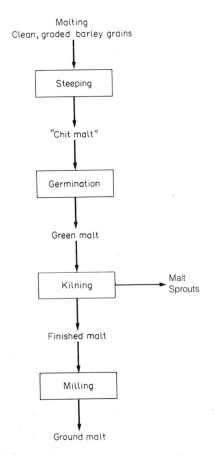

FIG 9.1 Diagrammatic summary of the malting process.

Selected barley is 'steeped', usually by immersion in water, for a period chosen to achieve a particular moisture level. The water is drained from the grain, which germinates. Conditions are regulated to keep the grain cool (generally below 18°C) and to minimize water losses. As the grain germinates the coleoptile (acrospire) grows beneath the husk and pericarp while the 'chit' (coleorhiza, root sheath) appears at the base of the grain, and is split by the emerging rootlets. Fig. 9.2.

At intervals the grain is mixed and turned to provide more uniform growth opportunities, and to prevent the roots from matting together. As the embryo grows it produces hormones including gibberellic acid, stimulating production of hydrolytic enzymes in the scutellum and aleurone layer, leading to 'modification' of the starchy endosperm. The malting process is regulated by the initial choice of barley, the duration of growth, the temperature, the grain moisture content, changes in the steeping schedule, and by use of additives. When modification is sufficient it is stopped by kilning the 'green malt', that is, by drying and cooking it in a current of hot dry air. The dry, brittle culms are then separated and the finished malt is stored. Dry malt is stable on storage and, unlike barley, it is readily crushed. The conditions of kilning are critical in determining the character of the malt: it can cause a slight enhancement of the levels found in green malt or completely destroy it. Malt contains relatively large quantities of soluble sugars and nitrogenous substances and, if it has been kilned at low temperatures, it contains high levels of hydrolytic enzymes. When crushed malt is mixed with warm water the enzymes catalyze hydrolysis of the starch, other polysaccharides, proteins and nucleic acids accessible to them, whether from the malt or from materials mixed with it. Malt also confers colour, aroma and flavour to the product. The solution of the products of hydrolysis extracted from the malt is the 'wort'. It forms the feedstock for fermentation for brewing or distillation (Briggs, 1978).

One of the benefits derived from the application of technology in malting has been the reduction in time required to produce satisfactory malts. The amount of time saved can be inferred from the diagrams in Fig. 9.3. It is clear that most saving has occurred during the present

FIG 9.2 Diagrammatic longitudinal sections through barley grains in the early stages of germination. 1. imbibed grain, 2. rootlets emerged, 3. Rootlets and coleoptile emerged. From Briggs (1978), *Barley* (Fig. 1.11). Reproduced by courtesy of Chapman and Hall Ltd.

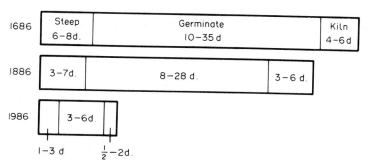

FIG 9.3 Diagram showing the reduction, over three centuries, in periods needed for malting and the three stages involved in the process.

century, and that savings are made in all stages, although the greatest benefits have been achieved in the germination stage.

Steeping

Traditional malting included a ditch steep followed by germination in heaps on the floor. It was a labour-intensive process as the heaps or 'couches' required frequent turning. It was also time consuming.

Current practice varies according to the size of operation and the preference of the maltster but self-emptying steep tanks have replaced the ditch steep. Vessels may be flat-bottomed or conical-bottomed tanks. They have facilities for water filling and emptying, and compressed air blowers provide both aeration and 'rousing' and mixing during the steep. Together, these processes combine to remove carbon dioxide which accumulates as a result of respiration of grains and micro-organisms associated with them. Vigorous aeration immediately after discharging the barley into the steep tank also serves to raise dust, chaff and light grains to the surface for removal. These are accumulated and sold for stock-feed.

Aeration, damping and temperature have to be carefully controlled in order that germination occurs at the required rate and to the required degree. For poorly modified traditional pale lager, European 2-rowed barley needs steeping to 41–43% m/c, while for a pale-ale malt 43–45% is appropriate. For a high nitrogen barley

(say 1.8% N), destined for a vinegar factory, 46–49% m/c may be preferred. Higher moisture levels induce faster modification but greater losses are incurred (Briggs, 1978). Anaerobic conditions are dangerous as they favour fermentation by micro-organisms present. The ethanol produced can harm the grain. On the other hand, excessive aeration leads to chitting under water and, consequentially, an undue rise in moisture content.

It is desirable to replace steeping water with fresh, between 1 and 4 times, as phosphates and organic compounds, including alcohol, accumulate and the microbial population grows. During the first steep, dissolved oxygen is depleted from steep-water at a rate of 1 mg/kg per hour but it rises 10-fold by the third steep. Complete depletion is possible within one hour. Temperature is controlled to a degree by the water temperature to between 10° and 16°C.

There are many variations in steeping practice including steeping in running water ('Bavarian') and the 'flushing' regime in which immersions are frequent but cover the grains for only a few minutes. Modern steep tanks are filled with barley to a depth of 1.2 m. This increases to 1.8 m when the grain is swollen. The uniform depth of flat-bottomed tanks provides for more uniform aeration and CO_2 removal than in conical bottomed vessels.

Following steeping, the grain is transferred to germinating vessels. This may be by 'wet-casting' whereby it is pumped in water suspension or

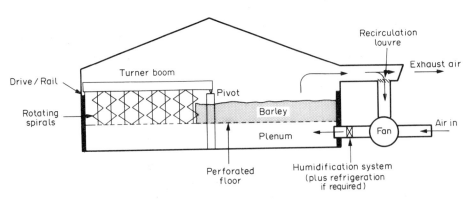

FIG 9.4 Diagrammatic section through a circular germination vessel. Source: Gibson 1989. Reproduced by courtesy of Aberdeen University Press, Ltd.

'dry-casting' after draining. If additives are to be added it is convenient to do so during transfer. These may include gibberellic acid and potassium bromate. The former hastens malting, especially if bruised grains are present, and potassium bromate reduces respiration and hence the rise in temperature that accompanies it. It is also said to inhibit proteolysis and control colour development in the malting grain, as well as reducing malting loss by reducing root growth.

Germination

Early mechanical maltings had rectangular germination vessels, but this was followed in due course by the drum, which provided ideal control but is limited to about 100 tonnes and has a high unit cost. Later the circular germination vessel was developed and capacities rose to 500 tonnes. Features common to all mechanical maltings are an automatic means of turning the germinating grains and a means of aeration. Turning may be performed by rotating spirals that are moved slowly through the grain mass. In the rectangular vessels they move end to end on booms while in the circular type vertical turners rotate on a boom or alternatively they remain stationary while the grain is transported past on a rotating floor. It is usual for aeration to be provided by air passing up (usually) or down, from or into a 'plenum' beneath the floor which is constructed of slotted steel plates. The layout of a circular germination vessel is shown in Fig. 9.4.

Temperature and humidity are controlled by humidifying and refrigerating or warming the air passing through. Air volumes passing are of the order of 0.15–0.2 m^3/sec/tonne of barley. Temperatures of 15°–19°C are common. Microprocessor control of conditions is commonplace (Gibson, 1989).

The danger of microbial contamination is high as air passing with high humidity is ideal for growth of bacteria and fungi and nutrients are plentiful. As well as introducing health hazards microbes gain preferential access to oxygen, thus inhibiting the germination and modification of the barley for which the system is designed. In some plants the germination and kilning are carried out in a single vessel and this has the advantage of the microbes being killed by the heat of kilning. Cleaning the dry residue is easier than complete removal of the wet remains of germinating grain.

Kilning

The objectives of kilning are to arrest botanical growth and internal modification, to reduce moisture for grain storage, and to develop colour and flavour compounds in the malt. Kilning is responsible for 90% of the energy consumption of the entire malting process unless a heat recovery is in use, when the proportion may be reduced to 75–80%.

For kilning, ambient air is heated by the preferred fuel and passed under positive or negative pressure through the bed of grains. A plenum below the floor is provided as in the earlier stages of the malting process. A recent innovation in kiln design is the double-deck version. In this system green malt is loaded on to the upper deck and transferred to the lower deck after partial drying. Warmed air is passed first through the drier, lower bed before passing, unsaturated, to the upper bed where it is capable of removing moisture from the green malt.

The depth of green malt in a modern kiln is 0.85–1.2 m. For maximum efficiency the bed should be level and uniformly compacted, a condition readily achieved by the automatic or semi-automatic loading machinery available today.

Conditions for kilning are determined by the nature of the end product required. Variables include the extract potential, moisture content, colour and enzyme activity. Curing temperatures range from 80° to 100°C. In modern kilns the maltster is assisted in monitoring and controlling conditions by control programmes incorporated into automated systems.

Enzymes survive high curing temperatures best if the malt is relatively dry, but under these conditions colour and flavour development are minimal. They develop mainly as a result of Maillard reactions occurring between the reducing groups of sugars and amino groups, mainly of amino acids. Development of additional flavour compounds is favoured by the combination of high temperatures with wet malts (Briggs, 1978). Small contributions to malt colour may be caramelized sugars and oxidized polyphenols and other contributions come from aldehydes, ketones, alcohols, amines and miscellaneous other substances including sulphur-containing compounds and nitrogenous bases. These are discussed in detail by Briggs (1978) and Palmer (1989). Peated distillery malts, for Scotch whisky manufacture, take up many substances from the peat smoke and contain various alkanes, alkenes, aldehydes, alcohols, esters, fatty acids, aromatic and phenolic substances, including phenol and cresols. The use of peat as a fuel for kilning originated in cottage industry enterprises in the crofts of the Scottish Highlands where peat was the only available fuel for domestic heating. In lowland distilleries peat heating has now been superseded by a succession of fuels, currently oil or gas. Direct heating by these fuels is not permitted as they can lead to introduction of nitrosamines into the product. Peatiness is a valued character of malt whiskies however and direct heating with peat is the rule in their production.

For production of brewing malts a major economic consideration is brewer's extract. This is measured as hot water extract available as the soluble nitrogen required to maintain fermentation and beer quality properties. An analysis of

TABLE 9.1
Analytical Characteristics of Barley and Ale Malt

Constituent	Barley	Ale malt
Moisture content %	15	4
Starch %	65	60
β-D-glucan %	3.5	0.5
Pentosans %	9.0	10.0
Lipid %	3.5	3.1
Total nitrogen %	1.6	1.5
Total soluble nitrogen %	0.3	0.7
α-Amino nitrogen %	0.05	0.17
Sucrose %	1.0	2.0
Minerals %	approx 2	approx 2
Colour °EBC	under 1.5	5.0
Hot water extract (1°/kg)	150	305
Diastatic power (*beta*-amylase °L)	20	65
Dextrinizing unit (*alpha*-amylase)	under 5	30
Endo β-glucanase (IRV units)	under 100	500

Source: Palmer, 1989. Reproduced by courtesy of Aberdeen University Press.

TABLE 9.2
Characteristics of a Selection of Malts

Malt type	Extract (°/kg)	Moisture (%)	Colour (°EBC)	Final kilning temperature (°C)
Ale*	305	4.0	5.0	100
Lager*	300	4.5	2.0	80
Cara pils	265	7.0	25–35	75
Crystal malt	268	4.0	100–300	75
Amber malt	280	2.0	70–80	150
Chocolate malt	268	1.5	900–1200	220
Roasted malt	265	1.5	1250–1500	230
Roasted barley	270	1.5	1000–1550	230

* Of all the malts listed only ale and lager malts contain enzymes.
Source: Palmer, 1989. Reproduced by courtesy of Aberdeen University Press.

ale malt compared with that of barley is shown in Table 9.1. The characteristics of brewers malts of various types are given in Table 9.2.

Ageing

Before use it is necessary to mill kilned malts but it is customary to delay this process to permit moisture equilibration. Kilning effects drying rapidly and in individual grains a gradient exists from a higher inner to a lower outer (husk) moisture content. Differences in a malt of 3–5% m.c. may be 4 or 5% (1–3% outside to 5–8% inside). Unless equilibrated, agglomeration of the damper parts can reduce extraction potential and undue fragmentation of dry husk leads to haze in the extract (Pyler and Thomas, 1986). Storage for up to 3 months may be used. As specifications become increasingly sensitive to moisture content the conditions of storage progressively include humidity control (Palmer, 1989).

Energy consumption and other costs

The Energy Technology Support Unit published a report in 1985 of a survey of United Kingdom maltsters. Specific energy consumed per tonne of malt ranged from 2.48 to 6.81 GJ, with a weighted average of 3.74 GJ. Quoted costs included fuel and electricity. Power costs comprised grain handling and process requirements.

Estimates of proportionate costs, including these values are given by Gibson (1989) as:

fuel, 25–30%;
electricity, 15–20%;
wages, 15–20%;
repairs/maintenance, 10–20%;
miscellaneous, 15–25%.

As with many wet processes of cereals the costs of water treatment before discharge are increasing as standards become more stringent. The Biological Oxygen Demand (BOD) load from a 30,000 tonnes per annum maltings is equivalent to a population of about 9000 people (Gibson, 1989).

Malt production

Palmer (1989) reports that about 17 million tonnes of barley are used annually, world-wide, to produce 12 million tonnes of malt and about 970 million hectolitres of beer. This represents about 10% of world barley production.

By-products of malting

The main by-product of malting is called 'malt sprouts'. They are separated from kilned malt by passing the malt through revolving reels or a wire screen. They account for 3–5% of product and they are incorporated into stock feeds. Typically they contain 25–34% N-compounds, 1.6–2.2% fat, 8.6–11.9% fibre, 6.0–7.1% ash and 35–44% N-free extract (Pomeranz, 1987).

Non-brewing uses of malt

Milled barley malt is used as a high diastatic supplement for bread flours which are low in natural diastatic activity, and as a flavour supplement in malt loaves. Malt extracts and syrups are produced by concentrating worts by evaporation. Malt is also used in the manufacture of malt vinegar.

Adjuncts

Although malt derived from barley is generally considered to be the superior feedstock for brewing and distilling, it is common practice in many countries to supplement malt with alternative sources of soluble sugars or starch capable of conversion to soluble sugars.

The principal adjuncts, as such non-malt additives are described, are rice, maize grits and cereal starches. Adjuncts contribute virtually no enzymes to the wort so hydrolysis of their starch depends upon those enzymes present in the malt to which they are added. Use of adjuncts is common practice in the U.S.A. and this is one reason for the preference there for the higher enzyme-containing six-row barleys.

In the U.S.A. 38% of total materials used in brewing (excluding hops) was reported to be contributed by adjuncts (Pyler and Thomas, 1986). Of this 46.5% was corn grits, 31.4% rice and 0.7% barley. Sugars and syrups accounted for the remaining 21.4%.

The form in which rice is added is the broken grains that do not meet the requirements of milled rice. As the quality of the products of fermentation is little affected by the nature of the adjunct, the choice is usually made purely on economic grounds. This is not related only to the price per tonne of the adjunct because the yield of extract is not the same from each. Tests for extraction carried out in the laboratory generally give higher values than those obtained in the commercial practice. Pyler and Thomas quote 78% for rice and 74% for maize grits in the brewhouse and 87–94% and 85–90% respectively in the laboratory (American Soc of Brewing Chemists procedure). Maize grits also contain higher levels of fat and protein than rice, both of which constituents are considered undesirable.

Other adjuncts used are: refined maize starch, wheat and wheat starch, rye, oats, potatoes, tapioca, triticale, heat treated (torrefied or micronized) cereals, cereal flakes. Micronization involves heating grains to nearly 200°C by infrared radiation while torrefaction achieves similar temperatures by use of hot air (Palmer, 1989). In grains treated by either method the vapourized water produced disrupts the physical structure of the endosperm, denaturing protein and partially gelatinizing starch. Digestibility is thus increased and these products are also used in cattle feeds and whole-grain baked products.

Solubility of proteins is decreased and some flavour may be introduced through their use if adequate treatment temperatures are used in processing them. Heat treated cereals can be added to malt before grinding; their extract yield is increased if they are precooked before use. Investigations with extruded cereals gave poorer extract yields than traditional adjuncts (Briggs et al., 1986).

Sorghum was used more in the U.S.A. when maize was in short supply during World War II, and it is used to a significant extent in Mexico today. It has a lower fat and protein content and a higher extract than maize and it thus has some merit.

Barley and wheat starch have a lower gelatinization temperature than maize and rice starch, hence digestion may occur at mash temperatures. It is usual, however, to premash maize, rice, wheat and barley etc. by cooking with a small amount of malt before adding them to the mash.

Addition of barley provides a means of reducing the nitrogen content of the wort. It is disallowed, however, by the German beer law for the production of bottom-fermented beers in which only barley-malt, hops, yeast and water are allowed. Top-fermented beers follow the same regulations but wheat malt may be included (Narziss, 1984). For special beers, pure beet-cane-invert-sugar is allowed.

Malts from other cereals

In Africa many malts are produced from sorghum and, to a lesser extent, from millets. In the Republic of South Africa commercial production is of the order of 1000 m litres annually and home brewing may be of the same order (Novellie, 1977).

Wheat malt is used in the production of wheat-malt beers. Examples of these and their characteristics are shown in Table 9.3.

Malts made on a pilot scale from U.K.-grown triticales were evaluated by Blanchflower and Briggs (1991). Viscosities of resulting worts were high due to pentosans, particularly arabinose and xylose. Hot water extracts after five days germination were 302–324 litre degree per kg. Filtered

TABLE 9.3
Beers Made From Wheat-Malt and their Characteristics

Type	Character	Origin	Alcohol (% v/v)	Flavour features
Weizenbeer	Lager/Ale	Bavaria	5–6	Full bodied, low hops
Weisse	Lager	Berlin	2.5–3	Light flavoured
Gueuze–Lambic	Acid ale	Brussels	5+	Acidic
Hoegards wit	Ale	East of Brussels	5	Full bodied, bitter

Source: Pomeranz, 1987, citing information from A. A. Leach of the Brewers' Society, U.K.

worts were turbid owing to proteinaceous materials. Malt yields were between 87 and 90%

Brewing

Beer

Wort production

The starting material for brewing may be pure (usually barley) malt, or a mixture of malt and adjunct. If solid adjuncts are to be included they may be milled with the malt. The coarsely ground material, on hydration with brewing liquor, produces a brewers' extract from the solubles, and a filter bed from the husk. The quality of the filter bed depends on the size of the husk particles; they should not be too fine. The process is known as mashing and it is carried out in vessels called mash tuns.

After an initial rest for hydration to be completed the temperature is raised above the gelatinization temperature of the starch. This renders the starch much more susceptible to digestion by amylase enzymes, to produce soluble sugars. The process of conversion, begun during malting, thus continues during this phase.

It is now necessary to separate the liquid wort from the solid remains of the malt and adjuncts, and this is done by a process called 'lautering'. The spent grains act as a filter bed when the mixture is transferred to a lauter tub, which has a perforated bottom. The spent grains accumulate on this and allow the liquid to pass through while retaining the fine solids. The sugary liquid is known as the 'sweet wort'; it is supplemented with syrups at this stage if such adjuncts are to be used to increase the amount of fermentable sugars. Hops are also added at this stage. As well as adding flavour they serve to sterilize the wort and participate in reactions that precipitate proteins responsible for haze when the wort is boiled.

Boiling may continue for 1.5–2 h. During boiling the humulones, or α-acids are isomerized to the bitter iso-α-acids. As the yield of bitter iso-α-acids extracted from the hops by boiling may be as low as 30%, a modern procedure is to replace part of the raw hops by a pre-isomerized hop extract, which is added to the beer after fermentation. It is common for half the hops to be added at the beginning of the period and half at the end. The wort is cooled (to 15.5°–18°C for British ales or to 4°–7°C for pilsners and lagers), filtered and transferred to pitching tanks where it is 'pitched' with yeast (i.e. yeast is added). Air is also passed into the hopped wort to provide a supply of oxygen for the yeast which rapidly becomes active (added oxygen removes the lag phase that would otherwise occur).

Fermentation

Yeasts vary in their behaviour during fermentation, some strains tend to flocculate, as a result they trap CO_2 and rise to the top. Others, which do not flocculate, sink to the bottom. Several styles of lagers are produced by bottom fermentation while many types of ales and stouts are produced using top fermentation. Examples are given in Table 9.4.

An efficient type of fermenter is a deep cylindrical vessel with a conical base into which the yeast eventually sediments. In such vessels the liberation of carbon dioxide provides efficient agitation, allowing cycles of filling, fermentation, emptying and cleaning to be accomplished in 5 days at 12°C or 2.5 days at 18°C.

TABLE 9.4
Classical Beers of the World Classified According to Yeast Types Used in their Brewing

Type	Character	Origin	Alcohol (% v/v)	Flavour features
Bottom-fermented				
Münchener	Lager/ale	Munich	4–4.8	Malty, dry, mod. bitter
Vienna (Märzen)	Lager	Vienna	5.5	Full-bodied, hoppy
Pilsner	Lager	Pilsen	4.5–5	Full-bodied, hoppy
Dortmunder	Lager	Dortmund	5+	Light hops, dry, estery
Bock	Lager	Bav., U.S., Can	6	Full-bodied
Dopplebock	Lager/ale	Bavaria	7–13	Full-bod., estery, winey
Light beers	Lager	U.S.	4.2–5	Light-bodied, light hops
Top-fermented				
Saissons	Ale	Belgium, France	5	Light, hoppy, estery
Trappiste	Ale	Bel. Dutch Abbeys	6–8	Full-bodied, estery
Kölsch	Ale	Cologne	4.4	Light, estery, hoppy
Alt	Ale	Düsseldorf	4	Estery, bitter
Provisie	Ale	Belgium	6	Sweet, ale-like
Ales	Ale	U.K., U.S., Can., Aus	2.5–5	Hoppy, estery, bitter
Strong/old ale	Ale	U.K.	6–8.4	Estery, heavy, hoppy
Barley wine	Ale/wine	U.K.	8–12	Rich, full, estery
Stout (Bitter)	Stout	Ireland	4–7	Dry, bitter
Stout (Mackeson)	Stout	U.K.	3.7–4	Sweet, mild. lact. sour
Porter	Stout	London, U.S., Can	5–7.5	V. malty, rich

Source: Pomeranz, 1987, using information from A. A. Leach of the Brewers' Society, U.K.

Fermentation continues for 7–9 days, producing ethanol and carbon dioxide. The gas may be collected for sale or for adding back when the beer is bottled or casked. The reaction is exothermic and the temperature would rise unduly if not controlled. The yeast increases during fermentation through asexual reproduction and in consequence this constitutes an additional by-product. During fermentation the pH drops from 5.2 to 4.2 as a result of acetic and lactic acids synthesized by bacteria inevitably introduced with the yeast. The green (jargon for young) beer is run off from the aggregated yeast cells and cooled to precipitate further haze-producing proteins and aged before filtering and carbonating. Lagers are stored at 0°C for some weeks ('lagering') before packaging into bottles or kegs. An alternative treatment is for the green beer to be run into casks, primed with sugars to permit secondary fermentation and 'fined' with isinglass for sale as 'naturally conditioned' beer; a product peculiar to the British Isles. (The brewing process is summarized schematically in Fig. 9.5.)

Saké

Saké production was reported to be 15 × 10⁵ kl in 1985.

Two thirds of the world's saké production takes place in Japan.

An essential difference between beer and saké is that for saké the natural enzymes present in the grain are not employed in solubilizing the starch, indeed they are expressly inactivated before the saccharifying phase of saké brewing. Enzymes are of course needed and these are derived from the fungus *Aspergillus oryzae*, they are provided from a culture of that mould known as 'koji'. Koji contains 50 different enzymes including the *alpha-* and *beta-*amylases present in malt and an additional amylolytic enzyme amyloglucosidase (see Fig. 3.12, p. 67), capable of hydrolyzing starch polymers to glucose. The balance of amylase to protease is inflenced by cultural conditions, higher temperatures favouring amylase production. The process is summarized in Fig. 9.6.

The yeast used in saké production is a specialized strain of *Saccharomyces cerevisiae*.

Unlike the husk of barley in the malting process, that of rice is not valued for its filtration properties, nor indeed any other qualities. It is removed by milling (see Ch. 6). The degree of polishing is severe; removing 25–30% of the brown rice weight, in extreme cases 50%.

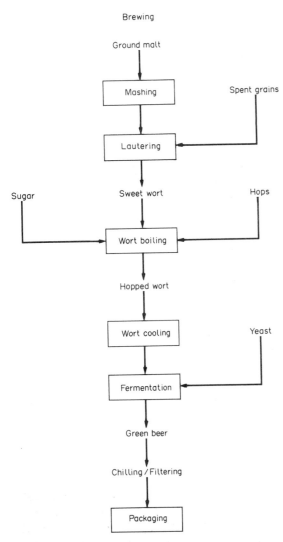

Brewing

Ground malt

Mashing

Spent grains

Lautering

Sweet wort   Hops

Sugar

Wort boiling

Hopped wort

Wort cooling   Yeast

Fermentation

Green beer

Chilling / Filtering

Packaging

Fig 9.5 Schematic summary of the brewing process

Following milling the rice is washed, steeped and steamed for 30–60 min. For 1 tonne of rice approximately 25 kg of water are used (Yoshizawa and Kishi, 1985).

Koji is added to the main mash in a seed mash: steamed rice in which an innoculum of *A. oryzae* has been cultured. Steamed rice and seed mash are added in equal proportions, water and saké yeast are also added. Quantities involved in a mash are usually 2–7 tonnes but over 10 tonnes is possible. The seed mash is acid as a result either of lactic acid bacteria present or of added lactic acid. The acid conditions inhibit wild micro-organisms. After two days at 12°C the yeast population reaches the high concentration originally present in the seed mash (10^8 cells per g) and an addition of an equal amount of steamed rice and water is again made. A third addition is made the following day and fermentation increases in vigour, raising the temperature from about 9°C to 15°–18°C. By the 15–20th day after the final addition the alcohol content rises to 17–19% and fermentation virtually ceases. The high alcohol content is attributable to:

1. the tolerance of the yeast;
2. the low sugar content of the mass at any time;
3. the solid matrix of the mash and
4. the proteolipids in koji.

The alcohol content of the liquid is further increased to 20% before filtration and pasteurization of the saké. Maturation takes 3–8 months after which water is added to give 15–16% alcohol content. Bottling takes place after filtration through activated carbon to improve colour and flavour.

Distilling

Distillation is a process of evaporation and recondensation used for separating liquids into various fractions according to their boiling temperature ranges. In the context of beverages it is used to produce potable spirits with an alcohol content above that of fermented drinks.

Spirits produced from grains are of two major categories: whisky (or whiskey, according to its origin!) and neutral spirits. In whiskies care is taken to retain flavours and colour carefully introduced during production, while in neutral spirits care is taken to avoid introduction of flavours and colour during production, although flavours may be added later, for example, in gin.

Whiskies of several types exist, their names denoting the carbohydrate source from which they are derived and the manner of their production. The essential characteristics of some are cited below.

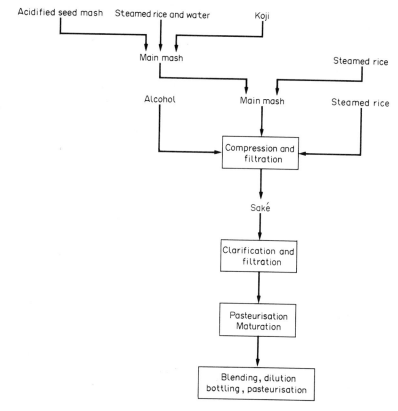

FIG 9.6 Schematic summary of saké production.

Scotch Malt Whisky is produced by traditional methods, using malt as the sole carbohydrate source (by legal definition).

Grain Whisky is the component of blended Scotch whisky that is made from a mash of cooked grain, and saccharified by the action of enzymes from malted barley. The grain may be maize, barley or wheat.

In Bourbon whiskey the grain in the mash consists of at least 51%, and usually 60–70%, maize. Typically some rye is included to impart a spicy, estery flavour. Although included, malt contributes only 10–15% of total carbohydrate.

Rye whiskey contains at least 51% of rye, Irish whiskey is made predominately from malted or unmalted barley, with wheat, rye or oats making up the remaining 20%. Canadian whiskies are mainly blends of neutral grain whiskies (90%) and Bourbon or rye (Nagodawithana, 1986).

Traditional malt whisky

Well modified, peated malt is dried at relatively low temperatures (to retain enzyme activity), milled and mashed at temperatures of 64°–65°C. The first worts are run off before mashing at a higher temperature of 70°C and running off of a second wort. Both worts become the liquor that is fermented. Two further worts are produced from mashes at 80° and 90°C, and are used as mashing liquor for subsequent grists. Batches of 5–10 tonnes of malt are typical.

Fermentation is preceded by cooling to 20°–21°C, and is initiated by pitching with distiller's yeast. Yeasts are of the high attenuation type (tolerating high alcohol levels) grown in molasses and ammonium salts. They are also selected to produce appropriate flavours. After 36–48 h, alcohol content reaches 8% and the temperature reaches 30°C.

The contents of the fermentor are transferred to the first copper pot still (wash still) which produces a distillate of more than 20% alcohol and known as 'low wines'. This is further distilled in a second copper pot still — the spirit still. From this three fractions are collected: the first or 'foreshots' contains volatile components with undesirable flavour characteristics, the third fraction is known as 'feints', it is mixed with the foreshots and added to the low wines from the next batch. It is the middle fraction which is collected as the basis of the marketable product. It contain 68% alcohol but it has to mature in oak casks, over a period of at least three years before sale (Bathgate, 1989).

Scotch grain whisky

Prior to Britain's entry into the European Community the most used raw material was maize imported from North or South America or South Africa. As 'third country' products these currently attract import levies which have discouraged their use in favour of wheat. Barley is a less attractive alternative as beta-glucans released during cooking produce undesirably high viscosity and spirit yields are low. Wheat usage has increased from 8000 tonnes in 1982/83 to 495,000 tonnes in 1988/89, the latter quantity representing over 80% of the distiller's starch requirements (Anon., 1990). The preferred wheat type is soft because of its lower protein content and the better flavoured spirit that can be produced from it. More than half of the wheat used usually comes from Scotland, the remainder coming from elsewhere in the U.K. The mode of use of the grain varies among distilleries, in some the grain is cooked whole prior to mashing while in others it is milled first. Mashing may be carried out in batches or as a continuous process. Individual preferences also influence the nature of the malt used, the variation being in the degree of kilning.

The malt is milled and suspended in cold water before being added to the freshly cooked cereal. The temperature of 65°C of the combined slurries is appropriate for enzymic conversion of the starch. Malt accounts for 8–15% of total solids present before saccharification. Following con-

version, wort is not separated and fermentation occurs in the whole cooled mash.

As with malt whisky the alcohol content is increased by a two stage distillation, but in this case to the higher alcohol content of 94%. Distillation is carried out, not in pot stills but by the more economical continuous distillation method. Many continuous stills are based on the Coffey

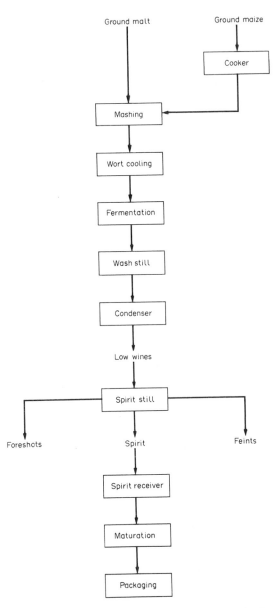

FIG 9.7 Schematic summary of Scotch whisky production.

Still, named after its inventor (Bathgate, 1989). A summary of Scotch grain whisky production is given in Fig. 9.7.

Bourbon whiskey

Maize and other cereals to be mashed are ground in a hammer mill prior to addition of malt and water. Water is added at the rate of 6–7.4 l/kg. and some stillage may be included to adjust the pH to 5. Although a total of 10–15% of malt is finally present only 1–5% is added initially, the remainder ('conversion malt') being added following cooling after the cook. Cooking consists of raising the temperature to 70°C and holding it for 30–60 mins. It is cooled to 63°C for the conversion of starch to sugars to be completed. Further cooling follows before the whole mash is pumped to a fermentor where it is pitched with 2% v/v yeast. The rise in temperature due to fermentation is not allowed to exceed 35°C. Fermentation takes about 72 h, the resulting product is known as 'drop beer' and this is distilled in a continuous column distillation system (Nagodawithana, 1986).

Neutral spirits

The beer for neutral spirits distillation is produced as economically as possible as flavour is an undesirable attribute and expensive means of producing flavours (as in whisky) are unnecessary. Only the chemistry of conversion of starch to sugars and of sugars to alcohol has to be considered. The cheapest source of starch is used. In countries where maize can be grown that cereal is used but elsewhere home-grown cereals may be cheaper.

It is more economical to use enzymes derived from micro-oganisms than those from malt. Fungal and bacterial enzymes are thus used in this process. A further advantage is that amyloglucosidase is available from microbial sources. Adjustment of pH is achieved by use of chemicals, calcium hydroxide being added to the suspension of ground cereal introduced into the cooker to achieve pH 6.3 and dilute hydrochloric acid adjusting the pH to 4.5 before saccharification.

Temperatures during cooking may be 100°C under atmospheric conditions or 160°C if high pressure cooking is used. The pressure cooking saves time (5 min instead of 30–60 min at the highest temperature).

Amylases are added at appropriate temperatures as the mash cools: bacterial at 85°C, fungal at 63°C and amyloglucosidase at 60°C. Fermentation is achieved by yeasts selected for their high rate of alcohol production and their tolerance of high alcohol levels, possibly 8–9%. Continuous distillation of the resulting beer produces an alcohol content of 95%.

By-products of Brewing and Distilling

The largest quantity of by-products consists of the solids remaining after the final sparging, leaving spent grains in the case of brewing and distiller's dark grains. They are valued as cattle-feed. Yeast is saleable and CO_2 may be worth harvesting for sale if produced in sufficient quantity. Fusel oils, including the following components: furfural, ethyl acetate, ethyl lactate, ethyl decanoate, n-propanol, iso-butanol and amyl alcohol, are used in the perfume industry (Walker, 1988).

References

ANON. (1990) H-GCA Weekly Digest 15/1/90.

BATHGATE, G. N. (1989) Cereals in Scotch whisky production. In: Cereal Science and Technology, Ch. 4, PALMER, G. H. (Ed.) Aberdeen University Press.

BLANCHFLOWER, A. J. and BRIGGS, D. E. (1991), Quality characteristics of triticale malts and worts. J. Sci. Food Agric. 56: 129–140.

BRIGGS, D. E. (1978) Barley. Chapman and Hall Ltd, London.

BRIGGS, D. E., WADESON, A., STATHAM, R. and TAYLOR, J. F. (1986) The use of extruded barley, wheat and maize as adjuncts in mashing. J. Inst Brew. 92: 468–474.

GIBSON, G. (1989). Malting plant technology. In: Cereal Science and Technology, Ch. 5, PALMER, G. H. (Ed.) Aberdeen Univ. Press.

MARTIN, D. T. and STEWART, B. G. (1991) Contrasting dough surface properties of selected wheats. Cereal Foods World 36: 502–504.

NAGODAWITHANA, T. W. (1986) Yeasts: their role in modified cereal fermentations. Adv. in Cereal Sci. Tech. 8: 15–104.

NARZISS, L. J. (1984) The German beer law. J. Inst. Brewing 90: 351–358.

NOVELLIE, L. (1977). Beverages from sorghum and millets.

In: *Proc. Symp. Sorghum and Millets for Human Food.* IACC Vienna 1976. Tropical Products Institute. London.

PALMER, G. H. (Ed.) (1989). Cereals in malting and brewing. In: *Cereal Science and Technology*, Ch. 3, Aberdeen Univ. Press.

POMERANZ, Y. (1987) Barley. In: *Modern Cereal Science and Technology*, Ch. 18, POMERANZ, Y. (Ed.). UCH Publishers Inc. NY. U.S.A.

PYLER, R. E. and THOMAS, D. A. (1986) Cereal research in brewing: cereals as brewers' adjuncts. *Cereal Foods World* **31**: 681–683.

REED, G. and NAGODAWITHANA, T. W. (1991) *Yeast Technology*. Van Norstrand Rheinhold. N. Y. U.S.A.

WALKER, E. W. (1988) *By-products of distilling. Ferment*, pp. 45–46, Institute of Brewing Publication (cited by Palmer, 1989).

YOSHIZAWA, K. and KISHI, S (1985) Rice in brewing. In: *Rice: Cemistry and Technology*, JULIANO, B. O. (Ed.) American Assoc of Cereal Chemists Inc. St Paul MN. U.S.A.

Further reading

AISIEN, A. O., PALMER, G. H. and STARK, J. R. (1986) The ultrastructure of germinating sorghum and millet grains, *J. Inst. Brew.* **92**: 162–167.

BAMFORTH, C. W. (1985) Biochemical approaches to beer quality. *J. Inst. Brew.* **91**: 154–160.

BRIGGS, D. E., HOUGH, J. S., STEVENS, R. and YOUNG, T. W. (1981) *Malting and Brewing Science*, 2nd edn. Chapman and Hall, N.Y. U.S.A.

COOK, A. H. (Ed.) (1962) *Barley and Malt.* Academic Press, London.

MATZ, S. A. (1991) *The Chemistry and Technology of Cereals as Food and Feed*, 2nd edn. Avi. Van Norstrand Rheinhold N.Y. U.S.A.

REED, G. (1981) Yeast — a microbe for all seasons. *Adv. Cereal Sci. Technol.* **4**: 1–4.

SHEWRY, P. R. (Ed.) (1992) *Barley: Genetics, Biochemistry, Molecular Biology and Biotechnology.* C. A. B. International, Wallingford, Oxon.

10

Pasta and Whole Grain Foods

Pasta

Pasta is the collective term used to describe products such as macaroni, spaghetti, vermicelli, noodles etc. which are traditionally made from the semolina milled from hard durum wheat (*T. durum*) (cf. p. 154 for milling of semo). The highest quality pasta products are made from durum wheat alone, but other hard wheat, e.g. CWRS or HRS, can be substituted, but only at the expense of quality. The Plate wheat Taganrog used for puffing (cf. p. 87) is not liked for pasta as it tends to yield a brownish-coloured product.

Durum wheat grown in Britain has been successfully used for pasta, although parcels with high alpha-amylase activity are unsuitable. U.K.-grown durum wheat will probably be blended in small quantities into a grist in which imported durum wheat will predominate.

Historically, pasta products are believed to have been introduced into Italy from China in the 13th century, but were first produced in Europe in the 15th century, in Germany (Pomeranz, 1987).

Degree of cooking

Pasta products, as sold, could be described as 'uncooked', 'partially cooked', or 'fully cooked' (pre-cooked), depending on the degree to which the protein is denatured and the starch gelatinized.

Traditionally, pasta products are prepared in an uncooked state, because dry, uncooked pasta can be stored at room temperatures for long periods while maintaining its highly glutinous properties.

Cooking quality of pasta

Particle size of the semolina is an important characteristic: it is recommended that at least 90% of the particles should fall between 150 μm and 340 μm in size. Particles larger than 340 μm impede the activity of enzymes in the dough (Milatovic, 1985). Good quality of the pasta may be defined as the ability of the proteins to form an insoluble network capable of entrapping swollen and gelatinized starch granules (Feillet, 1984), and is thus related to the composition of the protein. A strong correlation (+0.796) has been found between pasta cooking quality and acid-insoluble residue protein (Sgrulletta and De Stefanis, 1989); a highly significant correlation was also found between good cooking quality and the total sulphydryl plus disulphide (SH + SS) content of glutenins extracted by sodium tetradecanoate solution (Kobrehel and Alary, 1989; Alary and Kobrehel, 1987), and also with certain low mol. wt glutenin subunits, particularly the band 45 of the electrophoretic pattern. On the other hand, poor cooking quality was associated with the 42 band, and it is suggested that this association could be a useful indicator for the breeding of durum wheat for pasta-cooking qualities (du Cros and Hare, 1985; Autran *et al.*, 1989).

In uncooked pasta, the protein is largely undenatured, and most of the starch is ungelatinized. The processing conditions of dough moisture content and extrusion temperature were carefully controlled to prevent protein denaturation and starch gelatinization.

Traditional (kneading/sheeting) process

In the traditional manufacturing process in use at the beginning of the 20th century, a dough is first made by mixing semolina and water to give a moisture content of about 30% in the dough. A moisture content of 30% is needed to ensure that the viscosity of the dough is low enough to avoid the use of excessive pressure within the extruder. The dough is then kneaded, brought to a temperature of about 49°C, and is then extruded through a hydraulic press to form a thin sheet which can then be cut into strips and the strips carefully dried. A temperature not above 49°C is necessary to prevent cooking the product and avoid denaturing the protein. A modification of this process, in which the dough was extruded through special dies to make shaped products — rods, strips, tubes, etc. — was first introduced about 50 years ago (Pomeranz, 1987).

Extrusion process

In the extrusion process, the moisture content of the dough would be about 28% and the extrusion temperature, i.e. the temperature of the extrusion barrel, would be around 54°C, low enough to avoid cooking the product. The extrusion barrel would be provided with a jacket in which water could be circulated to maintain the required temperature.

It is claimed that the kneading/sheeting process has the advantage of producing a springy, elastic texture in the product, leading to good eating quality, while advantages claimed for the extrusion process are that it produces a pasta with a firmer texture in a larger variety of shapes, that it is more efficient and less time-consuming, and that the extruded dough has a lower moisture content requiring a shorter drying time to complete the preparation of the pasta. However, pasta made by the extrusion process lacks the laminated structure and desirable eating qualities of pasta made by the kneading/sheeting process, and the colour of the pasta is less appealing (U.S. Pat. No. 4,675,199).

Drying of pasta

Finally, in the traditional process, the extruded product is dried to about 12.5% m.c. In Italy, drying might be done outdoors, but elsewhere special equipment is used in which the temperature and relative humidity of the air can be carefully controlled. The rate of drying must be correct: drying at too low a rate could lead to the development of moulds, discoloration and souring, whilst drying too rapidly could cause cracking ('checking' in the U.S.A.) and curling. Until recently, drying temperatures were about 50°C, and the drying process took 14–24 h.

A three-stage drying process, comprising pre-drying, sweating, and drying, has been described (Antognelli, 1980). In the pre-drying stage, air at 55°–90°C circulates round the product and dries it to 17–18% m.c. in about 1 h. Moisture migrates from the centre to the periphery, where it evaporates, resulting in a marked moisture gradient in the product. In the sweating period that follows, the pasta is rested to allow moisture equilibration between the inner core of the pasta and the surface to occur. In the drying stage, diminishing periods of hot-air circulation alternate with periods of sweating. Temperatures in the drying stage are 45°–70°C, and the total drying time is 6–28 h.

In the late 1970s, the processing of macaroni was improved by drying at temperatures above 60°C, using shorter processing times, and sterilizing the product during drying.

A rapid drying process using microwave energy has been described (Katskee, 1978). A pre-drying stage uses hot air at 71°–82°C to reduce the m.c. of the pasta to 17.5% in 30 min. In the microwave stage that follows, the product is fully dried to the target m.c. in 10–20 min. The microwave system operates at about 30 kW, an air temperature of 82°C and 15–20% r.h. The final, or cooling and equalization stage, operated at 70–80% r.h., brings the total drying time up to 1.5 h. Advantages claimed for the microwave drying process, besides the considerable saving of time, include space-saving, improved product quality — cooking quality, colour enhancement, and microbiological quality — and lower installation and operating costs.

Recent processing changes

Within the last 20 years or so, numerous processing changes have been introduced to the pasta-manufacturing industry which are directed towards achieving various objectives, including the following:

— improvement in pasta quality, e.g. by reducing cooking loss, avoidance of cracking, improving the colour of the product, elimination of dark specks, destruction of micro-organisms, improving the appearance and firmness of the product (Milatovic, 1985; Manser, 1986; Abecassis et al., 1989a; Euro. Pat. Appl. No. 0,267,368; U.S. Pat. Nos 4,539,214 and 4,876,104);
— improvement in nutritional quality, e.g. by better retention of vitamins and other nutrients (Manser, 1986; Euro. Pat. Appl. No. 0,322, 153);
— improvement in shelf-life of the product (U.S. Pat. Nos 4,828,852 and 4,876,104);
— reduction in drying time, with consequent saving in energy expenditure (Abecassis, 1989b);
— faster-cooking products, e.g. a rapidly rehydratable pasta, or a microwave-cookable product (Euro. Pat. Appl. Nos 0,267,368, 0,272,502 and 0,352,876; U.S. Pat. Nos 4, 539,214 and 4,540,592);
— a ready-to-eat pasta, or an instant pre-cooked pasta (U.S. Pat. Nos 4,540,592 and 4,828,852).

Another objective has been the diversification of the starting material, i.e. the replacement of some (or all) of the durum semolina with other cereal materials or even with non-cereal materials (Molina et al., 1975; Mestres et al., 1990).

These objectives have been addressed, and in most cases attained, by modification of the traditional process in various ways, such as:

— modification of the moisture content of the product before drying (Abecassis et al., 1989a,b);

— modification of the temperature of the doughing water, the temperature in the cooking zone, and the extrusion temperature (Milatovic, 1985; Euro. Pat. Appl. Nos 0,267,368 and 0,288,136);
— partially pre-cooking the dough by pre-conditioning (Euro. Pat. Appl. No. 0,267,368);
— soaking partially-cooked pasta in acidified water (U.S. Pat. No. 4,828,852);
— use of higher drying temperatures (Manser, 1986; Abecassis et al., 1989a; Euro. Pat. Appl. Nos 0,309,413, 0,322,053 and 0,352,876);
— back-mixing the dough, combined with high drying temperature (Brit. Pat. Spec. No. 2,151,898; U.S. Pat. No. 4,540,592);
— venting the product between the cooking and the forming zones (Euro. Pat. Appl. No. 0,267,368);
— pre-drying and drying as two separate stages (Milatovic, 1985);
— use of superheated steam for simultaneously cooking and drying (U.S. Pat. No. 4,539,214);
— high-temperature after-treatment to make pre-cooked pasta (Abecassis et al., 1989a,b; Mestres et al., 1990; U.S. Pat. No. 4,830,866);
— pasteurization (U.S. Pat. No. 4,876,104);
— extrusion into N_2/CO_2 or into vacuum to improve storability (Euro. Pat. Appl. No. 0,146,510; U.S. Pat. No. 4,540,590);
— use of extrusion cooking instead of ordinary extrusion (U.S. Pat. No. 4,540,592);
— use of additives, e.g. emulsifiers, in the dough (Euro. Pat. Appl. Nos 0,352,876 and 0,288,136).

Use of non-wheat materials

In some countries, e.g. France and Italy, the material used for making pasta must be durum semolina, but other countries allow the use of soft wheat flour, maize flour, or various other diluents. Countries which are non-wheat producers, e.g. many African countries, are interested in pasta made from non-wheat materials, e.g. maize, sorghum. Preparation of such pasta is difficult because of the lack of gluten which is formed when wheat is the starting material, and which contributes to dough development during mixing and extrusion, and thus prevents disaggregation of the pasta during cooking in boiling water (Feillet, 1984; Abecassis *et al.*, 1989a). It has been suggested that the lack of gluten can be overcome by blending pregelatinized starch or corn flour before adding water and mixing, or by gelatinizing some of the starch during mixing or extruding (Molina *et al.*, 1975). A compromise is to use a blend of semolina and maize flour, from which good quality pasta can be obtained, provided the maize flour has fine granularity (less than 200 μm) and a low lipid content (not higher than 2% on d.b.). By submitting the dried pasta to thermal treatment at 90°C for 90–180 min, a blend of 70 maize flour: 30 durum wheat semolina was found to be satisfactory (Mestres *et al.*, 1990).

L-Ascorbic acid as additive

The use of L-ascorbic acid as an additive for improving the quality of pasta has been suggested (Milatovic, 1985). Ascorbic acid, which is oxidized to dehydroascorbic acid, a strong reducing agent, inhibits the destruction of naturally occurring pigments, leading to improvement in the colour of the product by inhibiting lipoxygenase activity. Addition of 300 mg/kg of L-ascorbic acid to the doughing water improved the colour of the pasta made from all soft wheat flour or from 50 soft wheat flour: 50 semolina, and also reduced the loss of solid matter and protein from the cooked product.

Doughing water temperature

The temperature of the doughing water is important. Water at 36°–45°C is normally used for cold dough making, 45°–65°C for a warm system using high temperature drying, and 75°–85°C for very warm processing. If egg is an additive, the doughing water temperature should not exceed 50°C, since albumen coagulates at 49°C. Similarly, when ascorbic acid is added, the water temperature should not exceed 55°C, otherwise degradation of the additive will be accelerated. In the warm system, the gluten hydrates rapidly, reducing the length of the dough preparation time. Use of the very warm system is restricted to preparation of pasta from semolina or flour having more than 32% wet gluten content. This system produces the most rapid hydration of gluten and gelatinization of the starch, but of course cannot be employed if ascorbic acid is included, because the latter decomposes at these temperatures and has no beneficial effect.

Glyceryl monostearate as additive

The use of glyceryl monostearate as a flour modifier is proposed in order to permit the extrusion of the dough at a lower m.c., viz. 28%, than is customary in making uncooked pasta. The temperature of the dough at extrusion is above 54°C, but not so high as to cause gelatinization of the starch. Other flow modifying agents suggested are whey solids and sulphydryl reducing substances such as L-cysteine, glutathione, sodium bisulphite or calcium sulphite at levels of 0.025–0.1% by wt (Euro. Pat. Appl. No. 0,288,136).

Prevention of starch leaching from uncooked pasta

High temperature (above 74°C) drying of pasta causes denaturation of the protein, thereby entrapping the starch and rendering the pasta stable to starch leaching in the presence of cold water. If dried at the lower, traditional, temperature, the protein is not denatured and, as a result, starch leaches out, making a gummy, mushy product, unless the pasta is immediately contacted with extremely hot water, to set the protein matrix. An uncooked pasta which can withstand exposure to cold water without leaching of starch may be made by the addition of low-temperature coagulatable materials such as albumin, whole egg, whey protein concentrates, but preferably egg white, added at a level of 0.5–3.0% by wt. Sulphydryl reducing agents, e.g. cysteine, glutathione, which reduce disulphide (–SS–) bonds to sulphydryl groups (–SH), thereby facilitating the irreversible denaturation of the gluten, can be added at levels of 0.02–0.04% by wt: their addition is essential at drying temperatures below 74°C. A modern drying process would use two stages: drying at 71°–104°C for 2–4 h, followed by further drying at 32°–71°C for 0–120 min, but with the use of a high velocity air current (150 ft³/min), the drying times could be shortened to 15–30 min for the first step plus 30–120 min for the second. Pasta made in this way can be rehydrated by soaking in cold water, and then cooked in about 2 min by conventional boiling or by microwave energy (Euro. Pat. Appl. No. 0,352,876).

'Cooking value' of pasta

'Cooking value' is a measure of the texture or consistency of the pasta after cooking. The cooked pasta should offer some resistance to chewing but should not stick to the teeth. To retain these qualities and to avoid loss of nutrients, current procedures limit drying temperature to not above 60°C, but such drying takes a long time — 16–24 h. Higher drying temperatures, while still avoiding nutrient loss and adversely affecting cooking value, can be used in a step-wise process, e.g. by isothermal application of heat at temperatures between 40° and 94°C while maintaining the vapour pressure within the pasta (A_w) below 0.86.

In one such drying schedule, the product passes through eight areas in which the temperatures are (successively) 40°, 50°, 60°, 70°, 80°, 84°, 94° and 70°C, taking about 3 h to complete the drying. The product, with 13% m.c., is then cooled to 25°C (Euro. Pat. Appl. No. 0,322,053).

Micro-organism control

Control of micro-organisms in stored pasta products is essential. The use of high temperature (above 60°C) drying to improve pasta product sterility has been described (Milatovic, 1985).

Another process is described in which dough is made from flour (durum wheat semolina, or whole soft wheat flour plus corn flour, rice flour or potato flour) about 75 parts, with about 25 parts of whole egg, but with no water. Additives such as wheat gluten, soya protein isolate, alginates and surfactants may be added. The dough is sheeted through a series of rollers to a thickness of 0.03 in, while keeping the m.c. about 24%

(below 24% m.c. cracking may occur, while above 30% m.c. the dough becomes too soft and elastic). The sheeted dough is cut into pieces which are partially dried in 10–60 sec by, e.g. infrared heating lamps and hot air at 204°C. The product is then steamed and pasteurized to kill micro-organisms, cooled to 0°–10°C and packed in sterile trays with injection of CO_2/N_2 (25:75 to 80:20). Such packaged products should be storable at 4°–10°C for at least 120 days (U.S. Pat. No. 4,876,104).

Quick-cooking pasta

A process described for making a quick-cooking pasta starts by making a dough from flour or semolina of which at least 90% is derived from durum wheat, with the addition of 0.5–5.0% by wt of an edible emulsifier, such as glyceryl or sorbitan monostearate, lecithin or polysorbates. Water, at 79°C, is added to bring the m.c. to 22–32%. The ingredients are mixed at 76°–88°C, extruded through a co-rotating twin-screw extruder at 60°–88°C. The extruder comprises three zones: a cooking zone, a venting zone, and a forming zone. When passing through the venting zone, the material may be subjected to a degree of vacuum (2.5–5 psi) to draw off excess moisture. The product can be cooked by microwave energy (Euro. Pat. Appl. No. 0,272,502).

Another process for making rapidly rehydratable pasta simultaneously cooks and dries the extruded dough (made from semolina/flour and water) by the use of superheated steam at a temperature of 102°–140°C for 7–20 min. The product is in an unexpanded condition and can be packaged without further cutting or shaping (U.S. Pat. No. 4,539,214).

A process to eliminate darkened specks in a quick-cooking pasta pre-cooks a mixture of pasta flour and water in a pre-conditioner before it passes through an extruder with three zones: cooking, venting, extruding. The temperature of the mixture is kept below 101°C in the cooking zone to prevent the formation of darkened specks symptomatic of burning during cooking (Euro. Pat. Appl. No. 0,267,368).

Pre-cooked pasta

A method of preparing instant, pre-cooked pasta is disclosed in U.S. Pat. No. 4,540,592. The pasta dough is completely gelatinized in a co-rotating, twin-screw extrusion cooker, incorporating at least one high-shear back-mixing cooking zone, and using high temperature and pressure so that the final product rehydrates to a cooked pasta instantly in hot water.

Shelf-stable, pre-cooked pasta

Preparation of a pre-cooked product which is shelf-stable for long periods is described (U.S. Pat. No. 4,828,852). The starting material can be durum semolina or flour, soft wheat flour, corn flour, pregelatinized corn flour, rice flour, waxy rice flour, pre-cooked rice flour, potato flour, pre-cooked potato flour, lentil flour, pea flour, soya flour, kidney or pinto beans, Mung bean flour, corn starch, wheat starch, potato starch, pea starch etc. A dough of 20–28% m.c. is extruded to make pasta of 1.0–2.0 mm thickness. The pasta is then boiled in acidified water (using acetic, malic, fumaric, tartaric, phosphoric or adipic, but preferably lactic or citric acid) and then soaked in water acidified to pH 3.8–4.3 to 61–68% m.c. The partially cooked pasta is drained and then coated with an acidified cream at pH 4.1–4.4. The acidification gives a better shelf life, and thickens the cream to improve coating. The product is then flush-packaged with an inert gas, or is vacuum packaged. Finally, the containers are sealed and heated to 90°–100°C for 20–40 min to complete the cooking of the pasta.

U.K. Code of Practice

A Code of Practice for dry pasta products in the U.K. requires all pasta products (other than those containing egg, additional gluten or other additives) to conform to the following standard: protein 11.5% (min.), m.c. 12.5% (max.) when packed, ash 0.60–0.85%. Degree of acidity and colour are also specified.

Couscous

Couscous is a type of pasta product made in Algeria from a durum semolina and water paste which is dried and ground. The product is size-graded, and is similar in size to that of very coarse semolina. Thus prepared, the couscous has excellent keeping qualities.

Pasta composition

The composition of pasta made from durum semolina in the U.K. is shown in Table 10.1.

Pasta consumption

The *per capita* consumption of pasta products in 1989 was (in kg) 21 in Italy, 6 in France, Greece, Portugal, Switzerland, 5 in Germany FR, 4 in Sweden, 3.5 in Austria and the U.K., 3 in Spain, 2 in Finland and Ireland, 1.5 in Belgium, Denmark and The Netherlands, and 0.2 in Norway (European Food Marketing Directory, 1991).

Rice substitutes

Bulgur

Bulgur consists of parboiled whole or crushed partially debranned wheat grains, and is used as a substitute for rice, e.g. in pilaf, an eastern European dish consisting of wheat, meat, oil and herbs cooked together. The ancient method of producing bulgur, which is referred to as *Arisah*

TABLE 10.1
Nutrients in Pasta* (per 100g)

Carbohydrates, g	75	Thiamin, mg	0.09	Calcium, mg	10
Protein, g	12	Riboflavin, mg	0.1	Iron, mg	1.2
Lipoprotein, g	1.8	Niacin, mg	2.0	Phosphorus, mg	144
Calories	380				

* Source: *Home Economics* 1972, Dec.: 24.

in the Old Testament, consists of boiling whole wheat in open vessels until it becomes tender. The cooked wheat is spread in thin layers for drying in the sun. The outer bran layers are removed by sprinkling with water and rubbing by hand. This is followed by cracking the grains by stone or in a crude mill.

A product resembling the wheat portion of the pilaf dish was developed in the U.S.A. in 1945 as an outlet for part of the U.S. wheat surplus. In one method of manufacture of bulgur, described by Schäfer (1962), cleaned white or red soft wheat, preferably decorticated, is cooked by a multistage process in which the moisture content is gradually increased by spraying with water and raising the temperature. Eventually, when the m.c. has reached 40%, the wheat is heated at 94°C and then steamed for 1.5 min at 206.85 kN/m^2 (30 lb/in^2) pressure so that the cooked product is gummy and starchy. The starch is partially gelatinized. The m.c. is then reduced to about 10% by drying with air at 66°C, and the dried cooked wheat is pearled or cracked. One brand of the whole grain product is called Rediwheat in the U.S.A.; the crushed is Cracked-Bulgur (Nouri, 1988).

Other processes for making bulgur are reviewed by Shetty and Amla (1972).

In a continuous system used in the U.S.A. for the production of bulgur, the wheat is soaked in a succession of three tanks in which the m.c. is raised progressively to 25–30% during 3.5–4 h in the first tank, to 35–40% m.c. during 2.5 h in the second and to 45% m.c. during 2–2.5 h in the third, giving a total soaking time of 8 h. The grain is then cooked with steam at a pressure of 1.5–3 bar for 70–90 sec and then dried to 10–11% m.c. The outer layers are removed, the grain lightly milled, and the product sieved to separate large from small bulgur (Certel et al., 1989).

During the soaking and cooking stages, a proportion of the vitamins and other nutrients present in the outer layers of the grain are mobilized, and move to the inner part of the grain, in a similar way to that described for parboiled rice (q.v.). Thus, removal of the outer layers of the grain, after soaking, cooking and drying, does not much reduce the nutritive value of the bulgur.

Bulgur is sent from the U.S.A. to peoples in the Far East, as part of the programme of American aid to famine areas. In 1971, 227,000 t of bulgur were produced in the U.S.A., of which 5% was used domestically, the remainder being used in the Foods for Peace Program. The staple food of people in these areas had always been boiled rice; the process of breadmaking was unknown; wheat and wheat flour were therefore unacceptable foods. Bulgur provided a cheap food that was acceptable because it could be cooked in the same way as rice and superficially resembled it. The level of bulgur exports from the U.S.A. for the food aid programme is now about 250,000 t per year (Certel et al., 1989).

Bulgur can be stored for 6–8 months under a wide range of temperature and humidity conditions, and its hard and brittle nature discourages attack by insects and mites.

Nutritive value

The nutritive value of bulgur is similar to that of the wheat from which it is made. Fat, ash and crude fibre levels are slightly lower, but protein level is unchanged. Retention of thiamin and

TABLE 10.2
Nutrients in Bulgur* (per 100 g)

Carbohydrates,	75.7	Thiamin, mg	0.28	Phosphorus, mg	430
Protein, g	11.2	Riboflavin, mg	0.14	Potassium, mg	229
Fat, g	1.5	Niacin, mg	5.5	Magnesium, mg	160
Energy, kcal	354	Vitamin B$_6$, mg	0.32	Calcium, mg	22.6
		Pantothenic acid, mg	0.83	Iron, mg	7.8
		Folacin, mg	0.038	Zinc, mg	4.4
				Iodine, mg	14

* Source: Protein Grain Products International, Washington, D.C.

niacin is about 98% of the original, that of riboflavin about 73%. Iron and calcium contents are slightly increased, while phosphorus content is decreased by the parboiling process. The nutrient composition of bulgur is shown in Table 10.2.

Bulgur of acceptable quality has also been prepared from triticale (Singh and Dodda, 1979) and from maize (Certel *et al.*, 1989).

The consumption of bulgur in Turkey is estimated at 20–30 kg/head/annum. Turkey is said to export about 100,000 t of bulgur per annum (Certel *et al.*, 1989).

WURLD Wheat

Peeled bulgur wheat is a light-coloured, low-fibre bulgur, made to resemble rice by removal of the bran in a lye-peeling operation. The wheat grain is treated with sodium hydroxide, and the loosened bran removed by vigorous washing with water. The grain is then treated with warm dilute acetic acid to restore the surface whiteness, and dried. This treatment leaves the aleurone layer intact. Peeled bulgur made from a red wheat is known as WURLD wheat, a U.S. Department of Agriculture *W*estern *U*tilization *R*esearch *L*aboratory *D*evelpment (Shepherd *et al.*, 1965).

Ricena

Ricena is a rice substitute, originating in Australia, made from wheat by a relatively inexpensive patented process which gives a yield of about 65% (Brit. Pat. No. 1,199,181). The wheat is washed, cleaned, steamed under pressure and dried. The product sells at the price of low-grade rice, although the protein, iron and vitamin B_1 contents of ricena exceed those of milled rice.

Parboiled rice

The parboiling (viz. *part-boiling*) of rice is an ancient tradition in India and Pakistan, and consists of stepping the rough rice (paddy) in hot water, steaming it, and then drying it down to a suitable moisture content for milling. The original purpose of parboiling was to loosen the hulls, but

in addition the nutritive value of the milled rice is increased by this treatment, because the water dissolves vitamins and minerals present in the hulls and bran coats and carries them into the endosperm. Thus, valuable nutrients which would otherwise be lost with the hulls and bran in the milling of raw rice are retained by the endosperm. It has been shown that rice oil migrates outwards during parboiling: therefore in parboiled rice the oil content of the milled kernel is lower, and bran content higher, than in raw rice.

By gelatinizing the starch in the outer layers of the grain, parboiling seals the aleurone layer and scutellum so that these fractions of the grain are retained in milling to a greater degree in milled parboiled than in milled raw rice (Hinton, 1948). Parboiling toughens the grain and reduces the amount of breakage in milling. Moreover, parboiled rice is less liable to insect damage than is milled raw rice, and has an improved storage life.

Conversion

Conversion of rice is the modern commercial development of parboiling. In the H. R. Conversion Process, wet paddy is held for about 10 min in a large vessel which is evacuated to about 635 mm (25 in) of mercury. The paddy is then steeped for 2–3 h in water at 75–85°C introduced under a pressure of 552–690 kN/m² (80–100 lb/in²). The steeping water is drained off and the paddy is heated under pressure for a short time with live steam in a steam-jacketed vessel. The steam is blown off, and the pressure in the vessel reduced to 711–737 mm (28–29 in) of vacuum. The product is then vacuum dried to about 15% m.c. in the steam-jacketed vessel, or it can be air-dried at temperatures not exceeding 63°C. After cooling, the converted paddy is tempered in bins for 8 h or more to permit equilibration of moisture, and is then milled.

In the Malek process, paddy is soaked in water at 38°C for 4–6 h, steamed at 103.4 kN/m² (15 lb/in²) pressure for 15 min, dried and milled. The product is called Malekized rice.

Redistribution of thiamin (vitamin B_1) in rice parboiled by the Malek process was investigated by Hinton (1948). By microdissection of raw and Malekized rice grains, followed by microanalysis of the fractions, Hinton showed that the thiamin content of the endosperm adjoining the scutellum increased from 0.4 to 5.2 iu/g, while that of the scutellum decreased from 44 to 9 iu/g, and that of the pericarp/aleurone from 10 to 3 iu/g. The thiamin content of the whole endosperm increased from 0.25 to 0.5 iu/g. The redistribution of thiamin was due to the inward passage of water through the thiamin-containing layers rendered permeable by heat to the vitamin.

In a process called *double-steaming*, raw paddy is presteamed for 15–20 min and then steeped in water for 12–24 h. The water is drained off, and the product is steamed again and then dried.

Parboiling may induce discoloration of the grains or development of deteriorative flavour changes. A suggested remedy for reducing flavour changes is steeping in sodium chromate solution (0.05%). It has been shown that the additional chromium present in the milled rice is not absorbed when the rice is consumed, and thus poses no health hazard (Pillaiyar, 1990). A remedy suggested for bleaching is steeping in sodium metabisulphite solution (0.32%), but this treatment lowers the availability of thiamin.

If steeping time exceeds about 8 h, the steep water tends to develop an off odour which is picked up by the rice. The odour is due to the activity of anaerobic bacteria in the steep water. Various methods have been suggested to avoid

this problem. If the steep water is maintained at a temperature of 65°C, the steeping time can be limited to 4 h, and odour development is avoided.

The soaking conditions must be chosen carefully so as to hydrate the grain sufficiently for it to be gelatinized on subsequent heating, but so as to avoid splitting the hull — which leads to excessive hydration and leaching of nutrients. Rate of hydration increases with temperature of the steeping water, but above about 75°C the rate of hydration increases so rapidly that splitting of the hull occurs. Thus, the solution is to limit the moisture level in the grain to 30–32% either by soaking at 70°C or lower, or by starting the soaking at 75°C and allowing the material to cool naturally during soaking. This method gives the fastest hydration without complications (Bhattacharya, 1985).

A process devized by F. H. Schüle GmbH does not involve steaming. Rough rice is soaked for 2–3 h in water at a medium temperature while the hydrostatic pressure is raised to 4–6 kg/cm^2 by admitting compressed air. The pressure is released, and the rice is cooked in water at about 90°C. The rice is then pre-dried in a vibratory drier and subsequently dried by three passes through columnar driers with successively decreasing air temperature (Bhattacharya, 1985).

A high-temperature, short-time process for parboiling rice has been described (Pillaiyar, 1990). Paddy steeped to about 24% m.c. is simultaneously parboiled and dried in a sand-roaster

TABLE 10.3
Nutrient Composition of Rice* (g/100g)

Product	Moisture	Protein	Fat	Crude fibre	Carbohydrate	Ash	Calories
Brown rice	12	7.5	1.9	0.9	76.5	1.2	360
White rice	12	6.7	0.4	0.3	80.1	0.5	363
Parboiled rice	10	7.4	0.3	0.2	81.1	0.7	369

Product	Ca	P	Fe	Na	(mg/100g) K	Thiamin	Riboflavin	Niacin	Tocopherol
Brown rice	32	221	1.6	9	214	0.34	0.05	4.7	29
White rice									
Unenriched	24	94	0.8	5	92	0.07	0.03	1.6	—
Enriched	24	94	2.9	5	92	0.44	—	3.5	—
Parboiled rice									
Enriched	60	200	2.9	9	150	0.44	—	3.5	—

* Source: U.S. Dept. Agric., Agricultural Research Service, *Agricultural Handbook* 8 (1963).

in which the paddy is mixed with hot sand and remains in contact with the sand for about 40 sec, after which the sand and the rice are separated by sieving. During this process the rice loses about 10% of moisture.

For a comprehensive survey of other processes, see Bhattacharya (1985).

Consumption

Parboiled or converted rice is readily consumed in India and Pakistan, while in the U.S.A. it is used as a ready-to-eat cereal, as canned rice, and as a soup ingredient. Elsewhere, however, parboiled rice is not popular.

Nutrient value

Specimen figures for the chemical composition of brown rice, white (polished) rice and parboiled rice are given in Table 10.3; the vitamin contents of milled (unconverted) and parboiled (converted) rice are shown in Table 10.4. The folic acid content of rice is increased from 0.04 to 0.08 µg/g by conversion, whilst that of the polishings is reduced from 0.26–0.40 to 0.12 µg/g (De Caro et al, 1949).

TABLE 10.4
Vitamin Contents of Rice (µg/g)

Material	Thiamin	Riboflavin	Niacin
Paddy rice	3.5–4.0	0.4–0.5	52.3–55.0
Converted parboiled rice	1.9–3.1	0.3–0.4	31.2–47.8
Milled, polished, non-parboiled rice	0.4–0.8	0.15–0.3	14.0–25.0

* Sources: Kik (1943); Kik and van Landingham (1943).

References

ABECASSIS, J., FAURE, J. and FEILLET, P. (1989a) Improvement of cooking quality of maize pasta products by heat treatment. *J. Sci. Food Agric.* 47(4): 475–485.

ABECASSIS, J., CHAURAND, M., METENCIO, F. and FEILLET, P. (1989b) Effect of moisture content of pasta during high temperature drying. *Getreide Mehl Brot* 43(Feb): 58–62.

ALARY, R. and KOBREHEL, K. (1987) The sulphydryl plus disulphide content in the proteins of durum wheat and its relationship with the cooking quality of pasta. *J. Sci. Food Agric.* 39: 123–136.

ANTOGNELLI, C. (1980) The manufacture and applications of pasta as a food and as a food ingredient: a review *J. Food Technol.* 15: 125.

AUTRAN, J. C., AIT-MOUH, O. and FEILLET, P. (1989) Thermal modification of gluten as related to end-use properties. In: *Wheat is Unique*, pp. 563–593, POMERANZ, Y. (Ed.). Amer. Assoc of Cereal Chemists Inc. St. Paul MN. U.S.A.

BHATTACHARYA, K. R. (1985) Parboiling of rice. In: *Rice: Chemistry and Technology*, Ch. 8, pp. 289–348, 2nd edn. JULIANO, B. O. (Ed.). Amer. Assoc. of Cereal Chemists Inc. St. Paul MN. U.S.A.

BRITISH PATENT SPECIFICATION NOS 2,151,898 (back-mixing of dough); 1,199,181 (ricena).

CERTEL, V. M., MAHNKE, S. and GERSTENKORN, P. (1989) Bulgur—nichte eine türkishe Getreide spezialität. *Mühle Mischfuttertechnik*, 126(6 July): 414–416.

DE CARO, L., RINDI, G. and CASELLA, C. (1949) Contents in thiamine, folic acid, and biotin in an Italian converted rice. *Abst. Commun. 1st Int. Cong. Biochem.* 31.

DU CROS, D. L. and HARE, R. A. (1985) Inheritance of gliadin proteins associated with quality in durum wheat. *Crop Sci.* 25: 674–677.

European Food Marketing Directory, 2nd edn (1991) Euromonitor plc., London.

EUROPEAN PATENT APPLICATIONS NOS 0,146,510 (extrusion into N₂/CO₂); 0,267,368 (quick-cooking pasta); 0,272,502 (quick-cooking pasta); 0,288,136 (shaped pasta products); 0,322,053 (drying pasta); 0,352,876 (shelf-stable, microwave-cookable pasta).

FEILLET, P. (1984) Present knowledge on biochemical basis of pasta cooking quality. Consequence for wheat breeders. *Sci. Aliment.* 4: 551–566.

HINTON, J. J. C. (1948) Parboiling treatment of rice. *Nature, Lond.* 162: 913–915.

HOUSTON, D. F. (1972) *Rice: Chemistry and Technology*. JULIANO, B. O. (Ed.) Amer. Assoc. of Cereal Chemists Inc. St. Paul MN. U.S.A.

KATSKEE, A. L. (1978) Microwave macaroni drying. *Macaroni J.* June: 12.

KIK, M. C. (1943) Thiamin in products of commercial rice milling. *Cereal Chem.* 20: 103.

KIK, M. C. and VAN LANDINGHAM, F. B. (1943) Riboflavin in products of commercial rice milling and thiamin, riboflavin and niacin content of rice. *Cereal Chem.* 20: 563–569.

KOBREHEL, K. and ALARY, R. (1989) The role of a low molecular weight glutenin fraction in the cooking quality of durum wheat pasta. *J. Sci Food Agric.* 47: 487–500.

MANSER, J. (1986) Einfluss von Trocknungs-Höchst-Temperaturen auf die Teigwarenqualität. *Getreide Mehl Brot.* 40: 309–315.

MESTRES, C., MATENCIO. F. and FAURE, J. (1990) Optimizing process for making pasta from maize in admixture with durum wheat. *J. Sci. Food Agric.* 51: 355–368.

MILATOVIC, L. (1985) The use of L-ascorbic acid in improving the quality of pasta. *Int. J. Vitam. Nutr. Res. Suppl.* 27: 345–361.

MOLINA, R. MAYORGA, I., LACHANCE, P. and BRESSANTI, R. (1975) Production of high protein quality pasta products using semolina-corn–soya-flour mixture. 1. Influence of thermal processing of corn flour on pasta quality. *Cereal Chem.* 52: 240–247.

NOURI, N. (1988) Bulgur—ein Beitrag zur Vollwert-und vegetarischen Ernährung. *Getreide Mehl Brot.* 42: 317–319.

PILLAIYAR, P. (1990) Rice parboiling research in India. *Cereal Foods Wld* 35: 225–227.

POMERANZ, Y. (1987) *Modern Cereal Science and Technology*, Ch. 20: Extrusion products. VCH Publishers Inc., NY.

SCHÄFER, W. (1962) Bulgur for underdeveloped countries. *Milling* **139**: 688.

SGRULLETTA, D. and DE STEFANIS, E. (1989) Relationship between pasta cooking quality and acetic acid insoluble protein of semolina. *J. Cereal Sci.* **9**: 217–220.

SHEPHERD, A. D., FERREL, R. E., BELLARD, N. and PENCE, J. W. (1965) Nutrient composition of bulgur and lye-peeled bulgur. *Cereal Sci Today* **10**: 590.

SHETTY, M. S. and AMLA, B. L. (1972) Bulgur wheat. *J. Fd. Sci. Technol.* **9**: 163.

SINGH, B and DODDA, L. M. (1979) Studies on the preparation and nutrient composition of bulgur from triticale. *J. Food Sci.* **44**: 449.

U.S. PATENT SPECIFICATION NOS 4,539,214 (rapidly rehydratable pasta); 4,540,590 (extrusion into vacuum); 4,540,592 (pre-cooked pasta); 4,675,199 (extrusion/compression process); 4,828,852 (shelf-stable, pre-cooked pasta); 4,830,866 (pre-cooked pasta); 4,876,104 (long shelf-life pasta).

Further Reading

ANON. (1987) Bulgur wheat production. *Amer. Miller Process.* **92**(10): 17–18.

AUTRAN, J. C. and GALTERIO, G. (1989) Association between electrophoretic composition of proteins, quality characteristics and agronomic attributes of durum wheats. II. Protein-quality associations. *J. Cereal Sci.* **9**: 195–215.

JULIANO, B. O. (Ed.) (1985) *Rice: Chemistry and Technology*, 2nd edn. Amer. Assoc. of Cereal Chemists Inc. St. Paul MN. U.S.A.

LICHODZIEVSKAJA, I. B., BRANDO, E. E., CHIMIZOV, J. I., MUCHIN, M. A., TOLSTOGUZOV, V. B., MUSCHIOLIK, G. and WEBERS, V. (1989) Herstellung von Teigwaren aus Backmehl unter Anwendung von Polysaccharid-Zusätzen. *Nahrung* **33**: 191–201.

MORGAN, A. I. Jr, BARTA, E. J. and GRAHAM, R. P. (1966) WURLD wheat — a product of chemical peeling. *Northwest Miller* **273**(6): 40.

11

Breakfast Cereals and Other Products of Extrusion Cooking

Breakfast Cereals

All cereals contain a large proportion of starch. In its natural form, the starch is insoluble, tasteless, and unsuited for human consumption. To make it digestible and acceptable it must be cooked. Breakfast cereals are products that are consumed after cooking, and they fall into two categories: those made by a process that does not include cooking and which therefore have to be cooked domestically (hot cereals) and those which are cooked during processing and which require no domestic cooking. The first class of products is exemplified by various types of porridge, the second by products which are described as 'ready-to-eat' cereals.

Besides the distinction regarding the need for domestic cooking as against readiness for consumption, breakfast cereals can also be classified according to the form of the product, and according to the particular cereal used as the raw material.

Cooking of cereals

If the cereal is cooked with excess of water and only moderate heat, as in boiling, the starch gelatinizes and becomes susceptible to starch-hydrolyzing enzymes of the digestive system. If cooked with a minimum of water or without water, but at a higher temperature, as in toasting, a non-enzymic browning (Maillard) reaction between protein and reducing carbohydrate may occur, and there may be some dextrinization of

the starch. Cooking by extrusion at low moisture causes the starch granules to lose their crystallinity, but they are unable to swell as in the normal gelatinization process in excess water. However, when they are exposed to moisture during consumption they hydrate and swell to become susceptible to enzymic digestion.

Hot cereals

Porridge from oats

Porridge is generally made from oatmeal or oatflakes (rolled oats or 'porridge oats'), the manufacture of which was described on p. 167. The milling process to make oatmeal includes no cooking (unless the oats are stabilized to inactivate the enzyme lipase: cf. p. 165), and the starch in oatmeal is ungelatinized; moreover, the particles of oatmeal are relatively coarse in size. Consequently, porridge made from coarse oatmeal requires prolonged domestic cooking, by boiling with water, to bring about gelatinization of the starch. Oatmeal of flour fineness cooks quickly, but the cooked product is devoid of the granular structure associated with the best Scotch porridge.

Rolled oats are partially cooked during manufacture; the pinhead oatmeal from which rolled oats is made is softened by treatment with steam and, in this plastic condition, is flattened on flaking rolls. Thus, porridge made from rolled oats requires only a brief domestic cooking time to complete the process of starch gelatinization.

The amount of domestic cooking required by

rolled oats is dependent to a large extent on the processes of cutting, steaming and flaking, which are interrelated. The size of the pinhead oatmeal influences rate of moisture penetration in the steamer; smaller particles will be more thoroughly moistened than large particles by the steaming process, and hence the starch will be gelatinized to a greater degree, and the steamed pinhead meal will be softer. For a given roller pressure at the flaking stage, this increase in softness will result in thinner flakes being obtained from smaller-sized particles of pinhead meal. During the domestic cooking of porridge, the thinner flakes will cook more rapidly than thicker flakes because moisture penetration is more rapid.

Thin flakes would normally be more fragile than thick ones, and more likely to break during transit. However, thin flakes can be strengthened by raising the moisture content of the pinhead meal feeding the steamer, thereby increasing the degree of gelatinization of the starch. Gelatinized starch has an adhesive quality, and quite thin flakes rolled from highly gelatinized small particle-size pinhead meal can be surprisingly strong.

The average thickness of commercial rolled oats is generally 0.30–0.38 mm (0.012–0.015 in); when tested with Congo Red stain (which colours only the gelatinized and damaged starch granules; cf. p. 185), about 30% of the starch granules in rolled oats appear to be gelatinized.

Ready-cooked porridge

In the search for porridge-like products which require even less cooking than rolled oats, a product called 'Porridge without the pot' has been made. Porridge can be made from this material merely by stirring with hot or boiling water in the bowl: it consists of oatflakes of a special type. As compared with ordinary rolled oats, these flakes are thinner, stronger, and contain starch which is more completely gelatinized. They could be manufactured by steaming the pinhead oatmeal at a somewhat higher moisture content than normal, rolling at a greater pressure than normal, and using heated flaking rolls.

Another type of porridge mix, known as Ready-brek, consists of a blend of two types of flakes

in approximately equal proportions: (a) ordinary rolled oats made from small particle-size pinhead oatmeal, and (b) very thin flakes of a roller-dried batter of oatflour and water, similar to products of this nature used for infant feeding. When this porridge mix is stirred with hot water, the thin flakes form a smooth paste while the rolled oats, which do not completely disperse, provide a chewy constituent and give body to the porridge.

The preparation of an instant reconstitutable oatflake is disclosed in U.S. Pat. No. 4,874,624. The product is made by conditioning normal oatflakes with water to 18.5% m.c., extrusion cooking them (cf. p. 246) at high pressure for 10–120 sec to an exit temperature of 95°C, cutting the extrudate into pellets, and flaking the pellets on rolls, and then drying the flakes to 2–12% m.c. The flakes so processed may be blended (70:30) with normal oatflakes which have been steamed to inactivate enzymes.

Specification for oatmeal and oatflakes

Quality tests for milled oat products include determination of moisture, crude fibre and free fatty acid (FFA) contents, and of lipase activity. A recommended specification is a maximum of 5% acidity (cf. p. 164) due to FFA (calculated as oleic acid, and expressed as a percentage of the fat) and a nil response for lipase activity. Other suggested tests are for arsenic (which could be derived from the fuel used in the kiln), lead and copper which catalyze oxidation of the fat (Anon., 1970).

Raw oats normally contain an active lipase enzyme, and, with the fat content of oats being some 2–5 times as high as that of wheat, it is desirable that the lipase should be inactivated during the processing of oats, to prevent it catalyzing the hydrolysis of the fat, which would lead to the production of bitter-tasting free fatty acids. Lipase is inactivated by the stabilization process, as described above in Ch. 6, a most important safeguard in keeping the quality of the oatmeal.

Porridge from other cereals

In Africa, maize grits or hominy grits are used to make porridge, by boiling with water. In Italy, maize porridge, made from fine maize grits or coarse maize meal, and flavoured with cheese, is called 'polenta'.

Barley meal is used for making a type of porridge in many countries in the Far East, the Middle East, and North Africa (cf. p. 13).

Wholemeal flour made from sorghum or millet may be cooked with water to make a porridge-like food in African countries and in India. Porridge made from parched millet grain in the Soviet Union is called *Kasha*.

Ready-to-eat cereals

While porridge-type cereals have been consumed for many years, the development of ready-to-eat cereals is relatively recent. Ready-to-eat cereals owe their origin to the Seventh Day Adventist Church, whose members, preferring an entirely vegetable diet, experimented with the processing of cereals in the mid-nineteenth century.

A granulated product, 'Granula', made by J. C. Jackson in 1863, may have been the first commercially available ready-to-eat breakfast cereal. A similar product, 'Granola', was made by J. H. Kellogg by grinding biscuits made from wheatmeal, oatmeal and maizemeal. Mass acceptance of ready-to-eat cereals was achieved in countries such as the U.S.A. by means of efficient advertizing.

Processing

The stages in the processing of ready-to-eat cereals would include the preparation of the cereal by cleaning, and possibly pearling, cutting or grinding; the addition of adjuncts such as salt, malt, sweeteners and flavouring materials; mixing with sufficient water to give a paste or dough of the required moisture content; cooking the mixture; cooling and partially drying, and shaping the material by, e.g. rolling, puffing, shredding, into the desired form, followed by toasting, which also dries the material to a safe m.c. for packaging.

Batch cooking

Until recently, the cooking was carried out in rotating vessels, 'cookers', into which steam was

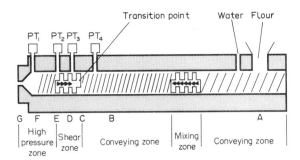

FIG. 11.1 Diagram of an extrusion cooker, showing its components and zones. (Reproduced from Guy, 1989, by courtesy of the A.A.C.C.)

injected, and the system was a batch process. The batch cooking process has now been largely superseded by continuous cooking processes in which cooking and extrusion through a die are both carried out in a single piece of equipment — a cooking extruder or extrusion-cooker (see Fig. 11.1). Extrusion-cooking is a high-temperature, short-time (HTST) process in which the material is plasticized at a relatively high temperature, pressure and shear before extrusion through a die into an atmosphere of ambient temperature and pressure (Linko, 1989a).

Continuous cooking

Continuous cooking methods have many advantages over batch methods: for example, continuous methods require less floor space and less energy in operation; they permit better control of processing conditions, leading to improved quality of the products. Moreover, batch cooking methods were usually restricted to the use of whole grain or to relatively large grain fragments, whereas extrusion cooking can also utilize much finer materials, including flour.

Extrusion cookers

An extrusion cooker is a continuous processing unit based on a sophisticated screw system rotating within the confines of a barrel. Raw materials are transported into a cooking zone where they are compressed and sheared at elevated temperatures and pressures to undergo a melt transition and form a viscous fluid. The extruder develops the fluid by shearing the biopolymers, particularly the starch (Guy and Horne, 1988; Guy, 1991) and shapes the fluid by pumping it through small dies.

The equipment may consist of single- or twin-screws with spirally-arranged flights for conveying, and special kneading and reversing elements for creating high pressure shearing and kneading zones. In order to achieve the high temperatures necessary for the melt transition, the raw materials require large heat inputs. These are achieved by the dissipation of mechanical energy from the screw caused by frictional and viscous effects, by the injection of steam into the cereal mass, and by conduction from the heated sections of the barrel or screw, using heating systems based on electrical elements, steam or hot fluids.

In extrusion cookers with twin screws, the screws may be co-rotating or counter-rotating. Further, there are many variations possible in screw design relating to physical dimensions, pitch, flight angles, etc. and, in the twin-screw, the extent to which the separate screws on each shaft intermesh (Fichtali and van de Voort, 1989). The main difference between single- and twin-screw extruders concerns the conveying characteristics of the screws.

Single-screw extruders were first used to manufacture ready-to-eat breakfast cereals in the 1960s, but they had problems with the transport of slippery or gummy materials because they rely on the drag flow principle for conveying the materials within the barrel. The problems of slippage can be overcome to some degree by using grooves in the barrel walls (Hauck and Huber, 1989). The single-screw extruder has a continuous channel from the die to the feed port, and therefore its output is related to the die pressure and slippage. The screw is usually designed to compress the raw materials by decreasing the flight height, thereby decreasing the volume available in the flights. At relatively high screw speeds the screw mixes and heats the flour mass, and a melt transition is achieved permitting the softened starch granules to be developed by the shearing action of the screw.

This transition usually occupies a fairly broad zone on the screw but the use of barrel heaters and steam injection in preconditioning units can help to induce sharper and earlier melt transitions or to increase the throughput (Harper, 1989). Considerable back-mixing may occur in the channel of the screw, giving a fairly broad residence-time distribution.

All twin-screw extruders have a positive pumping action and can convey all types of viscous materials with efficiency and narrow residence-time distributions. Special zones can be set up along the screw to improve the mixing, compression and shearing action of the screws. Co-rotating twin-screw extruders, which have self-wiping screws and higher operating speeds than counter-rotating machines, are currently the predominant choice of extruders for use in the food industry (Fichtali and van de Voort, 1989). The physical changes to the raw materials occurring within the single- and twin-screw extruders are basically the same and have the same relationships to temperature, shearing forces and time. However, control of the process is simpler in the twin-screw machines because the output is not affected by the physical nature of the melt phase being produced within the screw system, and the back-mixing can be more tightly controlled, giving better overall control and management of the process.

Flaked products from maize

Maize (for 'corn flakes'), wheat or rice are the cereals generally used for flaking.

In the traditional batch process for making cornflakes, a blend of maize grits — chunks of about 0.5–0.33 of a kernel in size – plus flavouring materials, e.g. 6% (on grits wt) of sugar, 2% of malt syrup, 2% of salt, possibly plus heat-stable vitamins and minerals, is pressure-cooked for about 2 h in rotatable batch cookers at a steam pressure of about 18 psi to a moisture content of about 28% after cooking. The cooking is complete when the colour of the grits has changed from chalky-white to light golden brown, the grits have become soft and translucent, and no raw starch remains.

The cooked grits are dried by falling against a

counter-current of air at about 65°C under controlled humidity conditions, to ensure uniform drying, to a moisture content of about 20%, a process taking 2.5–3 h, and are then cooled and rested to allow equilibration of moisture. The resting period was formerly about 24 h, but is considerably less under controlled humidity drying conditions. The dried grits are then flaked on counter-rotating rollers, which have a surface temperature of 43°–46°C, at a pressure of 40 t at the point of contact, and the flakes thus formed are toasted in tunnel or travelling ovens at 300°C for about 50 sec. The desirable blistering of the surface of the flakes is related to the roller surface temperature and to the moisture content of the grits, which should be 10–14% m.c. when rolled. After cooling, the flakes may be sprayed with solutions of vitamins and minerals before packaging.

Extruded flakes

These, made from maize or wheat, are cooked in an extrusion cooker, rather than in a batch pressure cooker, and can be made from fine meal or flour rather than from coarse grits. The dry material is fed continuously into the extrusion cooker, and is joined by a liquid solution of the flavouring materials — sugar, malt, salt, etc. These are mixed together by the rotation of the screw and conveyed through the heated barrel, thereby becoming cooked.

The material is extruded through the die in the form of ribbons which are cut to pellet size by a rotating knife. The pellets are then dried, tempered, flaked and toasted as described for the traditional method (Fast, 1987; Fast and Caldwell, 1990; Hoseney, 1986; Midden, 1989; Rooney and Serna-Saldivar, 1987).

Flaked products from wheat and rice

Wheat flakes

These are traditionally made from whole wheat grain, which is conditioned with water to about 21% m.c. and then 'bumped' by passing through a pair of smooth rollers set so that the roll gap is slightly narrower than the width of the grain. Without fragmenting the grain, bumping disrupts the bran coat, assisting the penetration of water. Flavouring adjuncts — sugar, malt syrup, salt — are then added and the grain is pressure cooked at about 15 psi for 30–35 min. The cooked wheat, at 28–30% m.c., emerges in big lumps which have to be 'delumped', and then dried from about 30% m.c. to 16–18% m.c. After cooling to about 43°C, the grain is binned to temper for a short time, and flaked (as for maize). Just before flaking, the grain is heated to about 88°C to plasticize the kernels and prevent tearing on the flaking rolls. The flakes leave the rolls at about 15–18% m.c. and are then toasted and dried to 3% m.c.

Rice flakes

To make rice flakes using the traditional process, the preferred starting material is head rice (whole de-husked grains) or 2nd heads (large broken kernels). Flavouring adjuncts are similar to those used with maize and wheat. The blend of rice plus adjuncts is pressure-cooked at 15–18 psi for about 60 min. The moisture content of the cooked material should not exceed 28% m.c., otherwise it becomes sticky and difficult to handle. Delumping, drying (to about 17% m.c.: at lower moisture contents the particles shear; at higher moisture contents the flaking rolls become gummed up), cooling, tempering (up to 8 h) and flaking are as for wheat flakes.

In the toasting of the rice flakes, more heat is required than for making wheat flakes. The moisture content of the feed and the heat of the oven are adjusted so that the flakes blister and puff during toasting; accordingly, the discharge end of the oven is hotter than the feed end. The moisture content of the final product is 1–3% m.c.

The process for making rice flakes by extrusion resembles that described for maize and wheat, except that a colouring material is added to offset the dull or grey appearance caused by mechanical working during extrusion. The lack of natural colour is emphasized if the formulation is low in sugar or malt syrup as sources of reducing sugars that could participate in a Maillard reaction.

Puffed products

Cereals may be puffed in either of two ways: by sudden application of heat at atmospheric pressure so that the water in the cereal is vaporized *in situ*, thereby expanding the product (oven puffing); or by the sudden transference of the cereal containing superheated steam from a high pressure to a low pressure, thereby allowing the water suddenly to vaporize and cause expansion (gun puffing). The key to the degree of puffing of the cooked grain is the suddenness of change in temperature or pressure (Hoseney, 1986).

The preferred grains for puffing are rice, wheat, oats or pearl barley, which are prepared by cleaning, conditioning and depericarping (e.g. by a wet scouring process). Flavouring adjuncts (sugar, malt syrup, salt, etc.) are added as for flaked products.

Oven-puffed rice

Oven-puffed rice is made from raw or parboiled milled rice which is cooked, with the adjuncts, for 1 h at 15–18 lb/in^2 in a rotary cooker until uniformly translucent. It is dried to 30% m.c., tempered for 24 h, dried again, this time to 20% m.c., and subjected to radiant heat to plasticize the outside of the grain. The grain is 'bumped' through smooth rolls, just sufficiently to flatten and compress it, and then surface dried to about 15% m.c., and tempered again for 12–15 h at room temp. The bumped rice then passes to the toasting oven, where it remains for 30–90 sec. The temperature in the oven is about 300°C in the latter half of the oven-cycle — as hot as possible short of scorching the grains. Due to the bumping, which has compressed the grains, and the high temperature, the grains immediately puff to 5–6 times their original size. The puffed grains are cooled, fortified with vitamins and minerals, if required, and treated with antioxidants (Hoseney, 1986; Juliano and Sakurai, 1985). Kellog's 'Rice Krispies' is a well-known brand of oven-puffed rice. Kellogg's 'Special K', containing 20% of protein, is made in a similar way to Rice Krispies up to the drying before bumping. The material is then wetted, coated with the enrichment, and bumped more heavily than for Krispies, then oven-puffed and toasted. The high protein enrichment may be vital wheat gluten, defatted wheat germ, non-fat dry milk, or dried yeast, plus vitamins, minerals and antioxidants (Juliano and Sakurai, 1985).

Gun-puffed rice

Long-grain white rice or parboiled medium-grain rice is generally used for gun-puffing, although short-grain, low-amylose ('waxy') rice is used for gun-puffing in the U.S.A., and parboiled waxy rice may be used in the Philippines. Puffed parboiled rice has a darker, less acceptable colour and tends to undergo oxidative rancidity faster than puffed raw rice, but parboiled rice requires less treatment, viz. lower steam pressure and temperature, than raw rice (Juliano and Sakurai, 1985).

A batch of the prepared grain is preheated to 521°–638°C and fed to the puffing gun, a pressure chamber with an internal volume of 0.5–1.0 ft^3, which is heated externally and by injection of superheated steam, so that the internal pressure rapidly builds up to about 200 lb/in^2 (1.379 MN/m^2) at temperatures up to 242°C, and the starch in the material becomes gelatinized. The pressure is suddenly released by opening the chamber of the puffing gun. The material is 'shot' out, expansion of water vapour on release of the pressure blowing up the grains or pellets to several times their original size. The puffed product is dried to 3% m.c. by toasting, then cooled and packaged (Fast, 1987; Fast and Caldwell, 1990; Juliano and Sakurai, 1985).

For satisfactory puffing, the starch should have plastic flow characteristics under pressure, and hence the temperature should be high enough. Moreover, the material at the moment before expansion requires cohesion to prevent shattering and elasticity to permit expansion. The balance between these two characteristics can be varied by adding starch, which has cohesive properties.

Extruded gun-puffed cereals can be made from oat flour or maize meal, with which tapioca or rye flour can be blended. This material is fed into the cooking extruder and a solution of the adjuncts — sugar, malt syrup, salt — is added with more water. The dough is cooked in the cooking extruder and is transferred to a forming extruder in which a non-cooking temperature — below 71°C — is maintained. The extruded collets are dried from 20–24% m.c. to 9–12% m.c. and then gun-puffed at 260°–427°C and 100–200 lb/in^2 pressure as previously described (Fast and Caldwell, 1990; Rooney and Se Saldivar, 1987). A 10–16-fold expansion results.

Puffed wheat

A plate wheat called Tagenrog is the type of wheat preferred for puffing on account of its large grain size which gives high yields of large puffs, but durum or CWRS wheat may also be used.

The wheat is pretreated with about 4% of a saturated brine solution (26% salt content) to toughen the bran during preheating and make it cohesive, so that the subsequent puffing action blows the bran away from the grain, thereby improving its appearance. Alternatively, the bran can be partly removed by pearling on carborundum stones. The puffing process is similar to that described for rice (Fast and Caldwell, 1990).

Continuous puffing

Using a steam-pressurized puffing chamber, the prepared grain is admitted through valves and subsequently released through an exit pore without loss of pressure in the chamber (U.S. Pat. No. 3,971,303).

Shredded products

Wheat is the cereal generally used, a white, starchy type, such as Australian, being preferred. The whole grain is cleaned and then cooked in boiling water with injection of steam for 30–35 min until the centre of the kernel changes from starchy white to translucent grey, and the grain is soft and rubbery. The moisture content is 45–50%, and the starch is fully gelatinized. The cooked grain is cooled to room temperature and rested for up to 24 h to allow moisture equilibra-tion. During this time, the kernels firm up because of retrogradation of the starch: this firming is essential for obtaining shreds of adequate strength. The conditioned grain is fed into shred-ders consisting of a pair of metal rolls — one is smooth, the other has grooves between which the material emerges as long parallel shreds. The shreds are detached from the grooves by the teeth of a comb and fall onto a slowly travelling band, a thick mat being built up by the superimposition of several layers. The mat is cut into tablets by a cutter which has dull cutting edges: the squeezing action of the cutter compresses the shreds and makes them adhere to one another. The tablets are baked at 260°C in a gas-heated revolving oven or a conveyor-belt oven, taking about 20 min. The major heat input is at the feed end; the biscuits increase in height as moisture is lost in the middle section, while colour is developed in the final section. The moisture content of the biscuits is about 45% entering the oven, about 4% leaving the oven. The biscuits may be further dried to 1% m.c., passed through a metal detector, and then packaged (Fast, 1987; Fast and Caldwell, 1990).

Shredded products may also be made from the flour of wheat, maize, rice or oats which would be cooked in batches or by continuous extrusion cooking. Flavouring and nutritional adjuncts may be added. After cooking, cooling and equilibra-tion for 4–24 h, the material is shredded and baked as described above. When using maize or rice to make a shredded product, however, it is desirable to produce a degree of puffing to avoid hardness. This is achieved by using a lower temperature in the first part of the baking, followed by an extremely high temperature in the last part.

Granular products

A yeasted dough is made from a fine wholemeal or long extraction wheaten flour and malted barley flour, with added salt. The dough is fermented for about 6 h and from it large loaves are baked. These are broken up, dried and ground to a standard degree of fineness.

Sugar-coated products

Flaked or puffed cereals, prepared as described, are sometimes coated with sugar or candy. The process described in BP No. 754,771 uses a sucrose syrup containing 1–8% of other sugars (e.g. honey) to provide a hard transparent coating that does not become sticky even under humid conditions. The sugar content of corn flakes was raised from 7 to 43% by the coating process, that of puffed wheat from 2 to 51%. As an alternative to sugar, use of aspartame as a sweetener for breakfast cereals is described in U.S. Pat. No. 4,501,759 and U.S. Pat. No. 4,540,587, while the use of a dipeptide sweetener is disclosed in U.S. Pat. Nos. 4,594,252, 4,608,263 and 4,614,657.

Keeping quality of breakfast cereals

The keeping quality of the prepared product depends to a large extent on the content and keeping quality of the oil contained in it. Thus, products made from cereals having a low oil content (wheat, barley, rice, maize grits: oil content 1.5–2.0%) have an advantage in keeping quality over products made from oats (oil content: 4–11%, average 7%). Whole maize has high oil content (4.4%), but most of the oil is contained in the germ which is removed in making grits (cf. p. 138).

The keeping quality of the oil depends on its degree of unsaturation, the presence or absence of antioxidants and pro-oxidants, the time and temperature of the heat treatment, the moisture content of the material when treated, and the conditions of storage.

Severe heat treatment, as in toasting or puffing, may destroy antioxidants or induce formation of pro-oxidants, stability of the oil being progressively reduced as treatment temperature is raised, treatment time lengthened, or moisture content of the material at the time of treatment lowered. On the other hand, momentary high-temperature treatment, as at the surface of a hot roll in the roller-drying of a batter, may produce new antioxidants by interaction of protein and sugar (non-enzymic browning, or Maillard reaction); such a reaction is known to occur, for example, in the roller-drying of milk, and may be the explanation of the improved antioxidant activity of oat products made from oats after steam treatment for lipase inactivation — stabilization (cf. pp. 165, 244). Similarly, enzyme-inactivated, stabilized wheat bran has a long shelf life: this material can be used for breakfast cereals, snack foods, extruded products which need a high fibre content and extended shelf-life (Cooper, 1988).

The addition of synthetic antioxidants, such as BHA or BHT, to the prepared breakfast cereal, or to the packaging material as an impregnation, as practised in the U.S.A., is not at present permitted in the U.K. (cf. p. 178). In Japan, where addition of antioxidants is not permitted, use is made of oxygen absorbers to restrict the onset of oxidative rancidity (Juliano, 1985).

Another form of deterioration of breakfast cereals after processing and packaging is moisture uptake which causes loss of the distinctive crisp texture. Moisture uptake is prevented by the use of the correct type and quality of moisture vapour-proof packaging materials (Fast, 1987).

Oat bran

A boost has been given to the use of oat bran in breakfast cereals following the discovery that this material has a hypocholesterolaemic effect in the human (de Groot et al., 1963), that is, it causes a lowering of the concentration of plasma cholesterol in the blood. As high blood cholesterol has been associated with the incidence of coronary heart disease — for each 1% fall in plasma cholesterol, coronary heart disease falls by 2% (Nestel, 1990) — a dietary factor that will reduce blood cholesterol is to be welcomed.

The content of soluble fibre is much higher in oat bran (10.5%) than in wheat bran (2.8%); this may be an important factor in the cholesterol-lowering activity of oat bran (not shown by wheat bran), and it has been suggested that a hemicellulose , beta–D-glucan, which is the major constituent of the soluble fibre, is the cholesterol-lowering agent, acting by increasing the faecal excretion of cholesterol (Illman and Topping, 1985; Oakenfull, 1988; Seibert, 1987).

Oat bran is obtained by milling oat flakes that have been made from stabilized oat kernels (groats) as described elsewhere (cf. p. 168). It can be used as an ingredient in both hot and cold breakfast cereals. A method for making a ready-to-eat cereal from cooked oat bran is disclosed in U.S. Pat. No. 4,497,840. It can also be incorporated into bread: addition of 10–15% of oat bran to white wheat flour yielded bread of satisfactory quality (Krishnan et al., 1987).

Rice bran, and in particular the oil in rice bran, has also been shown to have a plasma cholesterol

TABLE 11.1
Chemical Composition of Ready-to-Eat Breakfast Foods (per 100 g as sold)

Food	Water (g)	Energy (kJ)	Carbohydrates Starch (g)	Carbohydrates Sugars (g)	Protein (g)	Fat (g)	Ash (g)	Dietary fibre (g)	Source of data
Grape Nuts	3.5	1475	67.8	12.1	10.5	0.5		6.2[9]	10
Kellogg's									
Corn Flakes	3	1550	75	8	8[3]	0.7	3	1[5]	11
Frosties	3	1600	48	40	5[3]	0.5	2	0.6[5]	11
Rice Krispies	3	1600	75	10	6[3]	0.9	3.5	0.7[5]	11
Ricicles	3	1600	50	39	4[3]	0.6	2	0.4[5]	11
Coco Pops	3	1600	48	39	5[3]	0.9	2.5	1[5]	11
All-Bran	3	1150	28	18	14[3]	3.5	5	24[5]	11
Bran Buds	3	1200	27	23	13[3]	3	4.5	22[5]	11
Bran Flakes	3	1350	45	18	12[3]	2	4	13[5]	11
Sultana Bran	7	1300	37	27	10[3]	2	3.5	11[5]	11
Smacks	3	1600	36	49	7[3]	1.5	1	2.5[5]	11
Fruit 'n Fibre	6	1500	43	22	9[3]	5	2.5	7[5]	11
Toppas	5	1450	52	21	10[3]	1.5	1	7[5]	11
Special K	3	1550	60	15	15[3]	1	3	2.5[5]	11
Country Store	8	1500	47	22	9[3]	4.5	2.5	6[5]	11
Start	3	1550	51	28	8[3]	2	2.5	5[5]	11
Nabisco									
Shredded Wheat	9	1525	75.3	2.7	10.3[1]	2.3	1.5	11.4	12
Quaker Oats									
Puffed Wheat	2	1360	68.8[6]	1.3[7]	13.1[1]	1.25	1.6	7.7[5]	13
Oat Krunchies	2.5	1600	71.1[6]	15.9[7]	10.5[1]	7.3	4.5	n/a	13
Sugar Puffs	2	1554	88[6]	51	6.0	1.2	0.8	3.5	13
Ryvita									
Corn Flakes		1565	71.8	9.1	8.3[3]	1.3	n/a	9.3	14
Morning Bran		1180	48.1		14.0[1]	3.8	n/a	18.5	14
Weetabix									
Weetabix	5.6	1427	60.6	5.2	11.2[3]	2.7	2.2	12.9	15
Bran Fare	4.0	962	31.5	nil	17.1[3]	4.8	5.3	37.5	15
Toasted Farmhouse Bran	3.0	1293	36.7	10.2	12.6[3]	3.0	4.8	20.0	15
Weetaflake	3.7	1490	51.7	18.9	9.7[3]	3.1	n/a	10.5	15

[1] $N \times 5.7$.
[2] $N \times 5.95$.
[3] $N \times 6.25$.
[4] $N \times 6.31$.
[5] Non-starch polysaccharides, soluble plus insoluble.
[6] As available monosaccharides.
[7] Total sugars as sucrose.
[8] Enriched to this level.
[9] Southgate method.
[10] McCance and Widdowson's Composition of Foods (4th edn; 3rd Supp.) 1988, reproduced with the permission of the Royal Society of Chemistry and the Controller of HMSO.
[11] Data courtesy of Kellogg Co. of Great Britain Ltd (1993).
[12] Data courtesy of Nabisco Ltd (1982).
[13] Data courtesy of Quaker Oats Ltd (1990).
[14] Data courtesy of The Ryvita Co Ltd (1990).
[15] Data courtesy of Weetabix Ltd (1990).

lowering effect. Rice bran was not so effective as oat bran in lowering plasma total cholesterol, but rice bran favourably altered the ratio of high density lipoprotein (HDL) to low density lipoprotein (LDL), a sensitive lipid index of future coronary heart disease (Nestel, 1990).

Preliminary work indicates that the beta-glucan in the soluble fibre of a waxy, hull-less barley cultivar also has hypocholesterolaemic effects, and the extracted beta-glucans from barley have possible use as a fibre supplement in baked products (Klopfenstein et al., 1987; Newman et al., 1989).

Nutritive value of breakfast cereals

The nutritive value of breakfast cereals, as compared with that of the raw materials from which they were made, depends very much on the processing treatment involved, remembering that all heat treatment processes cause some modification or loss of nutrients. Thus, while extrusion cooking may cause the loss of essential amino acids, it also inactivates protease inhibitors, thereby increasing the nutritional value of the proteins.

The chemical composition of some ready-to-eat breakfast cereals manufactured in the U.K. is shown in Table 11.1.

Shredded wheat, made from low protein, soft wheat has a protein content considerably lower than that of puffed wheat, which is made from a high-protein hard wheat, such as durum or CWRS wheat.

Proteins and amino acids

All cereal products are deficient in the amino acid lysine, but the deficiency may be greater in ready-to-eat cereals than in bread because of the changes that occur in the protein at the high temperature treatment. The protein efficiencies of wheat-based breakfast cereals (relative to casein = 100), as determined by rat-growth trials, have been reported as: −15.3 for extrusion puffed; 1.8–16.3 for flaked-toasted; 2.8 for extrusion toasted; and 69.9 for extruded, lightly roasted. The differences were partly explained by the loss

of available lysine as Maillard reaction products (McAuley et al., 1987). However, lysine deficiency is of less importance in ready-to-eat cereals than in bread because the former are generally consumed with milk, which is a good source of lysine. Moreover, some ready-to-eat breakfast cereals have a protein supplementation.

Carbohydrates

The principal carbohydrate in cereals is starch, the complete gelatinization of which is desirable in processed foods, such as ready-to-eat cereals. Whereas ordinary cooking at atmospheric pressure requires the starch to have a moisture content of 35–40% to achieve complete gelatinization, the same occurs at feed moisture levels of less than 20% in extrusion cooking at 110°–135°C (Asp and Björck, 1989; Linko 1989a). Extrusion cooking increases the depolymerization of both amylose and amylopectin by random chain splitting. The susceptibility of starch to the action of alpha-amylase increased in the following sequence: steam cooking (least), steam flaking, popping, extrusion cooking, drum drying (most) (Asp and Björck, 1989).

Calorific value

The calorific value of most ready-to-eat cereals as eaten is higher than that of bread (975 kJ/100 g; 233 Cal/100 g), largely on account of the relatively lower moisture content of the former. Compared at equal moisture contents, the difference in calorific value is small. Fat and cholesterol contents may be lower than those of some other cereal foods.

The processes involved in the manufacture of ready-to-eat cereals cause partial hydrolysis of phytic acid (cf. p. 295); the degree of destruction increases at high pressures: about 70% is destroyed in puffing, about 33% in flaking.

Enzymes

Enzymes, which are proteins, are generally inactivated partially or completely during extrusion cooking. Thus, peroxidase was completely inactivated by extrusion cooking at 110°–149°C of material with 20–35% m.c., although there was residual activity if cooked at lower m.c. Under relatively mild conditions of extrusion cooking some activity of alpha-amylase, lipase and protease could be retained, and such residual activity could influence product keeping quality and shelf life (Linko, 1989a,b). Conversely, wheat flour with abnormally high alpha-amylase activity that would be unsuitable for conventional bread-making can be processed by extrusion cooking in which, provided the conditions are correctly chosen, the enzyme is quickly and totally inactivated, permitting the production, from such flour, of flat bread, snacks, biscuits, etc. (Cheftel, 1989).

Minerals and vitamins

The content of some of the minerals and vitamins in ready-to-eat breakfast cereals is shown in Table 11.2. Information about other vitamins is meagre. About 50% of the thiamin (vitamin B_1) is destroyed during the manufacture of shredded wheat and in extrusion cooking, while nearly 100% is destroyed during puffing and flaking. These processes have little effect on riboflavin (vitamin B_2), niacin, pyridoxin and folic acid. Extrusion cooking caused a loss of 11–21% of vitamin E, while in products enriched with wheat germ, extrusion cooking caused losses of 50–66% of vitamin E (Asp and Björck, 1989). Many of the ready-to-eat breakfast cereals manufactured in the U.K. are enriched with vitamins, as shown in Table 11.2; some are enriched with iron, and some with protein (viz. with the high protein fraction of wheat and oat flour, defatted wheat germ, soya flour, non-fat dry milk, casein or vital wheat gluten) and with the vitamins B_6 (pyridoxin), D_3, C and E. Some ready-to-eat breakfast cereals made in the U.S.A. are enriched also with vitamins A and B_{12}.

Incorporation of the vitamin supplements may be accomplished in various ways: at the cooking stage, at extrusion, by surface spraying after processing, or by incorporation in a sugar coating, the method chosen depending on the relative stability of the individual vitamins. Incorporation of the protein supplement may similarly be made

at a number of points which are chosen to avoid subjecting the protein material to excessive heat treatment: by incorporation as a dry supplement at the extrusion stage, or by coating the product with a batter of wheat gluten.

Consumption of breakfast cereals

The consumption of ready-to-eat breakfast cereals in the U.K. has shown a steady growth from quite a small figure before World War II to about 4.2 kg/head/an. in 1972, increasing further to 5.0 kg/head/an. in 1978, and to 6.5 kg/head/an. in 1988. The consumption of oat products for hot cereals in the U.K. was 0.6 kg/head/an. in 1984, but increased to 0.9 kg/head/an. in 1988, possibly in response to the claim that oat bran has a blood-cholesterol lowering effect.

The total tonnage of packeted breakfast cereals marketed in the U.K. in 1988 was 383,758 t, of which 38% was wheat based, 29% maize based, and the remainder based on other cereals or on a mixture of cereals (Business Monitor, 1989).

In the U.S.A. in 1971 about 0.75 million tonnes of breakfast cereal were produced, of which about 35% was puffed, 35% flaked, 10% shredded and about 20% hot cereal. Between 1980/81 and 1988/89 the total quantity of hot cereal (excluding corn grits) sold in the U.S.A. increased from 0.16 to 0.20 million tonnes, most of the increase being accounted for by oat-based products, the proportion of which increased from 71.6% in 1980/81 to 81.2% in 1988/89.

The average consumption of breakfast cereals (ready-to-eat plus hot) in the U.S.A. in 1971 was about 3.4 kg/head/an. Of the cereals used, wheat, bran or farina comprised about 37%, oatmeal or oat flour 30%, maize grits 22% and rice 11%. However, by 1985, the consumption of ready-to-eat cereals alone had increased to 4.1 kg/head/an. (Anon., 1986), slightly lower than in the U.K., with the consumption of maize-based ready-to-eat cereals in the U.S.A. increasing from 2.72 to 3.63 kg/head/an. between 1970 and 1980. Between 1974 and 1983 domestic consumption of ready-to-eat cereals in the U.S.A. grew by about 2% per annum, and growth increased further to 3.3% per annum between 1983 and

TABLE 11.2

Mineral and Vitamin Content of Ready-to-Eat Breakfast Foods (per 100g as sold)

Food	Minerals							Vitamins								Source
	Na (g)	K (mg)	Ca (mg)	Fe (mg)	P (mg)	Zn (mg)	Mg (mg)	Thiamin (mg)	Niacin (mg)	Riboflavin (mg)	Pyridoxin (mg)	Folic acid (μg)	D (μg)	B_{12} (μg)	C (mg)	of data§
Grape Nuts	0.59	310	37	9.5	250	4.2	95	1.3*	17.6*	1.5*	1.8*	350*		5 *		10
Kelloggs																
Corn Flakes	1.1	100	5	7.9*	50	0.3	10	1.2*	15 *	1.3*	1.7*	333*	2.8*	0.8*		11
Frosties	0.8	60	5	7.9*	30	0.2	10	1.2*	15 *	1.3*	1.7*	167*	2.8*	0.8*		11
Rice Krispies	1.2	160	10	7.9*	160	1.1	50	1.2*	15 *	1.3*	1.7*	167*	2.8*	0.8*		11
Ricicles	0.8	100	5	7.9*	100	0.7	30	1.2*	15 *	1.3*	1.7*	167*	2.8*	0.8*		11
Coco Pops	0.9	220	20	7.9*	140	0.9	50	1.2*	15 *	1.3*	1.7*	167*	2.8*	0.8*		11
All-Bran	0.9	1000	60	8.8*	750	6.6	220	0.9*	11.3*	1 *	1.3*	125*	2.1*	0.6*		11
Bran Buds	0.5	950	60	8.8*	700	6.3	220	0.9*	11.3*	1 *	1.3*	125*	2.1*	0.6*		11
Bran Flakes	0.9	600	40	11.7*	400	2.9	140	1.2*	15 *	1.3*	1.7*	333*	2.8*	0.8*		11
Sultana Bran	0.8	700	40	8.8*	350	2.3	110	0.9*	11.3*	1 *	1.3*	125*	2.1*	0.6*		11
Smacks	0.01	180	10	7.9*	150	1	50	1.2*	15 *	1.3*	1.7*	167*	2.8*	0.8*		11
Fruit 'n Fibre	0.7	450	40	11.7*	240	1.8	70	1.2*	15 *	1.3*	1.7*	167*	2.8*	0.8*		11
Toppas	0.01	350	30	6 *	250	2	70	0.9*	11.3*	1 *	1.3*	125*	2.1*	0.6*		11
Special K	0.9	250	60	23.3*	200	1.8	50	2.3*	30 *	2.7*	3.3*	333*	8.3*	1.7*		11
Country Store	0.6	500	60	8 *	300	1.6	80	0.7	8 *	0.9 *						11
Start†	0.6	300	60	17.5*	220	18.8*	60	1.8*	22.5*	2 *	2.5*	250*	6.3*	1.3*		11
Nabisco																
Shredded Wheat	0.01	376	49	3.6	328			0.3	4.5	0.05						12
Quaker Oats																
Puffed Wheat			26‡	4.6‡	350‡	3.0		0.3	5	0.13	0.14					13
Oat Krunchies			50.5	3.6				0.4	4.3	0.1	0.15	57				13
Ryvita																
Corn Flakes	0.65			12*				1.2*	18.0*	1.6*		300*	2.5*	2.0*		14
Morning Bran				12*				1.2*	18.0	1.6*		300*	2.5*	2.0*		14
Weetabix																
Weetabix	0.375	375	35	7	290	2	120	0.7*	10.0*	1.0*	0.2*				5.0	15
Bran Fare	0.16	1100	105	14	1100	7	140	0.7	21.0	0.3	1.2					15
Toasted Farmhouse Bran	0.87	725	70	44	620	4	180	1.6*	19.0*	1.8*						15
Weetaflake	0.375	375	35	7	290	2	120	0.7*	10.0*	1.0*	0.2				5.0	15

* Enriched to these levels.
† Also contains vitamin E, 12.5 mg/100 g.
‡ Paul and Southgate, 1978.*
§ For references see Table 11.1.

1985. Recent estimates (1987) suggest an even larger growth rate of 4–5%, attributed to an advertizing campaign (Fast, 1987).

Other extrusion-cooked products

Extrusion-cooking is a high-temperature, short-time (HTST) process in which material is transported, mixed and plasticized at relatively high temperature, pressure and shear before extrusion to atmospheric temperature and pressure through a die. Because the process is able to bring about gelatinization, solubilization and complex formation of starches, polymerization of proteins, partial or complete inactivation of enzymes (according to the severity of the operating conditions), reduction of microbial load, and production of particular forms of texture, the process can be used to make, from cereals, many food products, besides ready-to-eat breakfast foods, such as pet foods, flat bread, snacks, croutons, soup bases, drink bases, biscuits, confectionery and breadings. Besides these end products, extrusion cooking is also used to make intermediate products for further processing, both for food and also for non-food use (Fichtali and van de Voort, 1989; Linko, 1989a).

In 1987, about 3 million tonnes of products were made by extrusion cooking in the U.S.A. (Hauck and Huber, 1989).

Pet foods

The largest product group manufactured by extrusion cooking is pet foods, both dry expanded and semi-moist, to replace canned food and dog-biscuits (Harper, 1986, 1989).

'Flat bread'

The extrusion cooker takes over the function of the oven, producing expansion of dough pieces, formation of structure, partial drying, and formation of flavour and colour. Thus, the process is suitable for making bread, described as 'flat bread' (Kim, 1987). The flat breads produced by extrusion cooking are imitations of crisp breads (although the word 'bread' is really a misnomer). The typical crumb-pore structure results from expansion of, and evaporation of water from, the plasticized mass on exiting from the die, when the pressure drops instantly from about 150 kg/cm^2 at 160°C, resulting in drying from about 18% m.c. to about 8% m.c. The extrudate emerges as a flat strip which is cut into slices, roasted in an oven, and cooled. The exactness of expansion in terms of height and weight is most important, as the slices must fit exactly into packages of prescribed volume and weight.

Pellets

Pellets, or half-snacks (unexpanded half-products), can be made in various shapes — flat, tube, shell, ring, screw, wheel — and are generally made in a single screw extrusion cooker, using a single die and cooling of the barrel, and are dried to 5–10% m.c. The pellets are later fried, during which they expand 6–8 fold, lose water, pick up 15–25% of fat, and increase in weight (Colonna *et al.*, 1989; Meuser and Wiedmann, 1989).

Modified starch

Extrusion cooking can be used to treat starch at a relatively low moisture content, e.g. about 40% m.c., to bring about derivatization, plasticization, and drying to make thin-cooking, pregelatinized and substituted and chemically modified starches for use in food, and also for non-food uses, e.g. in the paper and textile industries (Meuser and Wiedmann, 1989).

Brewing adjunct

Extrusion cooked cereals have potential use as brewing adjuncts. The starch granule structure is destroyed during extrusion cooking, so that the starch is more easily hydrolyzed by enzymes in the mashing process (Smith, 1989).

High dextrose-equivalent syrups

These can be made by extrusion cooking of starch with the addition of thermostable alpha-amylase (Smith, 1989).

High alpha-amylase activity flour

Flour having an abnormally high alpha-amylase activity, unsuitable for making bread by conventional baking methods, can be processed by extrusion cooking — which causes rapid and complete inactivation of the enzyme — to make flat bread, snacks and biscuits, while indigenous raw materials can be processed by extrusion cooking in developing countries to make stable and nutritionally balanced foods, such as biscuits and pre-cooked flours for preparation of gruels, porridges and infant foods (Cheftel, 1989; Linko, 1989b).

Other uses for extrusion cooking

Other suggested uses for extrusion cooking include the texturizing of vital wheat gluten; the pre-treatment of wheat bran for the extraction of hemicellulose; the treatment of wheat flour, as an alternative to chlorination, for use in high-ratio sponge cake (Kim, 1987).

References

ANON. (1970) Cereal specifications. *Milling*, Oct.: 16.

ANON. (1986) Ready-to-eat cereal industry report. *Investment report* No. 609910. Kidder, Peabody and Co. Inc.

ASP, N.-G. and BJÖRCK, I. (1989) Nutritional properties of extruded foods. In: *Extrusion Cooking*, MERCIER, C., LINKO, P. and HARPER, J. (Eds). Amer. Assoc. Cereal Chemistry, St Paul, MN, U.S.A.

BRIT. PAT. SPEC. NO. 754,771 (sugar coating).

BUSINESS MONITOR, PAS 4239. Miscellaneous Foods, 1989. Business Statistics Office.

CHEFTEL, J. C. (1989) Extrusion cooking and food safety. In: *Extrusion Cooking*, MERCIER, C., LINKO, P. and HARPER, J. (Eds.). Amer. Assoc. Cereal Chem., St. Paul, MN., U.S.A.

COLONNA, P., TAYEB, J. and MERCIER, C. (1989) Extrusion cooking of starch and starchy products. In: *Extrusion Cooking*, MERCIER, C., LINKO, P. and HARPER, J. (Eds.). Amer. Assoc. Cereal Chem., St Paul, MN, U.S.A.

COOPER, H. (1988) Milling moves. *Food Processing*, April: 41–42.

DE GROOT, A. P., LUYKEN, R. and PIKAAR, N. A. (1963) Cholesterol lowering effect of rolled oats. *Lancet* 2: 303.

FAST, R. B. (1987) Breakfast cereals: processed grains for human consumption. *Cereal Fds Wld*, 32 (3): 241.

FAST, R. B. and CALDWELL, E. F. (Eds) (1990) *Breakfast cereals and how they are made*. Amer. Assoc. Cereal Chem., St Paul, MN, U.S.A.

FAST, R. B., LAUHOFF, G. H., TAYLOR, D. D. and GETGOOD, S. J. (1990) Flaking ready-to-eat breakfast cereals. *Cereal Fds Wld*, 35 (3): 295.

FICHTALI, J. and van de VOORT, F. R. (1989) Fundamental and practical aspects of twin screw extrusion. *Cereal Fds Wld*, 34 (11): 921–929.

GUY, R. C. E. (1991) Structure and formation in snack foods. *Extrusion communique*, 4 (Jan.–March): 8–10.

GUY, R. C. E. and HORNE, A. W. (1988) Cereals for extrusion cooking processes: a comparison of raw materials derived from wheat, maize and rice. *35th Technology Conference 1988. Biscuit, Cake, Chocolate and Confectionery Alliance*, 45–49.

GUY, R. C. E. and HORNE, A. W. (1989) The effects of endosperm texture on the performance of wheat flours in extrusion cooking processes. *Milling*, 182 (Feb.): ix–xii.

HARPER, J. M. (1986) Processing characteristics of food extruders. In: *Food Engineering and Process Applications*, Vol. 2, Unit operations, Le MAGUER, M. and JELEN, P. (Eds) Elsevier Appl. Sci. Publ., London.

HARPER, J. M. (1989) Food extruders and their applications. In: *Extrusion Cooking*, MERCIER, C., LINKO, P. and HARPER, J. M. (Eds.). Amer. Assoc. Cereal Chem., St Paul, MN, U.S.A.

HAUCK, B. W. and HUBER, G. R. (1989) Single screw vs twin screw extrusion. *Cereal Fds Wld*, 34 (11); 930–939.

HOLLAND, B., UNWIN, I.D. and BUSS, D. H. (1988) Cereals and cereal products. 3rd Suppl to McCance and Widdowson's *The Composition of Foods*, 4th edn. R. Soc. Chem. and Min. Agric. Fish. Food.

HOSENEY, R. C. (1986) Breakfast cereals. *Principles of Cereal Science and Technology*. Ch. 13. Amer. Assoc. Cereal Chem., St Paul, MN, U.S.A.

ILLMAN, R. J. and TOPPING, D. L. (1985) Effects of dietary oat bran on faecal steroid excretion, plasma volatile fatty acids and lipid synthesis in rats. *Nutr. Res.* 5: 839.

JULIANO, B. O. (Ed.) (1985) *Rice: Chemistry and Technology*, 2nd edn. Amer. Assoc. Cereal Chem., St Paul, MN, U.S.A.

JULIANO, B. O. and SAKURAI, J. (1985) Miscellaneous rice products. In: *Rice: Chemistry and Technology* 2nd edn, JULIANO, B. O. (Ed.) Amer. Assoc. Cereal Chem., St Paul, MN, U.S.A.

KIM, J. C. (1987) The potential of extrusion cooking for the utilisation of cereals. In: *Cereals in a European Context* pp. 323–331, MORTON, I. D. (Ed.) Chichester, Ellis Horwood.

KLOPFENSTEIN, C. F. and HOSENEY, R. C. (1987) Cholesterol-lowering effect of β-glucans enriched bread. *Nutr. Rep. Int.* 36: 1091.

KRISHNAN, P. G., CHANG, K.C. and BROWN, G. (1987) Effect of commercial oat bran on the characteristics and composition of bread. *Cereal Chem.* 64: 55.

LINKO, P. (1989a) The twin-screw extrusion cooker as a versatile tool for wheat processing. In: *Wheat is Unique*, Ch. 22, POMERANZ, Y. (Ed.) Amer. Assoc. Cereal Chem., St Paul, MN, U.S.A.

LINKO, P. (1989b) Extrusion cooking in bioconversions. In: *Extrustion Cooking*, Ch. 8, MERCIER, C., LINKO, P. and HARPER, J. (Eds), Amer. Assoc. Cereal Chem., St Paul, MN, U.S.A.

McAULEY, J. A., HOOVER, J. L. B., KUNKEL, M. E. and ACTON, J. C. (1987) Relative protein efficiency ratios for wheat-based breakfast cereals. *J. Fd Sci.* 52 (July–Aug.): 1111.

MEUSER, F. and WIEDMANN, W. (1989) Extrusion plant design. In: *Extrusion Cooking*, Ch. 5, MERCIER, C., LINKO, P. and HARPER, J. (Eds) Amer. Assoc. Cereal Chem., St Paul, MN, U.S.A.

MIDDEN, T. M. (1989) Twin screw extrusion of corn flakes. *Cereal Fds Wld*, **34** (11): 941.

NESTEL, P. J. (1990) Oat bran, rice bran. *Food Australia*, **42** (7): 342.

NEWMAN, R. K., NEWMAN, C. W. and GRAHAM, H. (1989) The hypocholesterolaemic function of barley beta-glucans. *Cereal Fds Wld*, **34** (10): 883–886.

OAKENFULL, D. (1988) Oat bran. Does oat bran lower plasma cholesterol and, if so, how? *CSIRO Food Res. Q.*, **48**: 37–39.

PAUL, A. A. and SOUTHGATE, D. A. T. (1978) *McCanee and Widdowson's 'The Composition of Foods'*, 4th ed., H.M.S.O., London.

ROONEY, L. W. and SERNA-SALDIVAR, S. O. (1987) Corn-based ready-to-eat breakfast cereals. In: *Corn: Chemistry and Technology*, WATSON, S. A. and RAMSTAD, P. E. (Eds). Amer. Assoc. Cereal Chem., St Paul, MN, U.S.A.

SEIBERT, S. E. (1987) Oat bran as a source of soluble dietary fibre. *Cereal Fds Wld*, **32** (8): 552–553.

SMITH, A. (1989) Extrusion cooking: a review. *Food Sci. Technol. Today* **3** (3): 156–161.

U.S. PAT. SPEC. Nos. 3,971,303 (continuous puffing); 4,497,840 (cooked oat bran cereal); 4,501,759 (aspartame sweetener); 4,540,587 (aspartame sweetener); 4,594,252 (dipeptide sweetener); 4,608,263 (dipeptide sweetener); 4,614,657 (dipeptide sweetener); 4,874,624 (reconstitutable oatflakes).

Further Reading

BROCKINGTON, S. F. and KELLY, V. J. (1972) Rice breakfast cereals and infant foods. In: *Rice: Chemistry and Technology*, 1st edn, pp. 400–418, HOUSTON, D. F. (Ed.), Amer. Assoc. Cereal Chem., St Paul, MN, U.S.A.

FAST, R. B. and CALDWELL, E. F. (Eds.) (1990) *Breakfast Cereals and How They Are Made*. Amer. Assoc. Cereal Chem., St Paul, MN, U.S.A.

GUY, R. C. E. (1986) Extrusion cooking versus conventional baking. In: *Chemistry and Physics of Baking*, pp. 227–235. Spec. Pub. 56, R. Soc. Chem., London.

GUY, R. C. E. (1989) The use of wheat flours in extrusion cooking. In: *Wheat is Unique*, Ch. 21, POMERANZ, Y. (Ed.) Amer. Assoc. Cereal Chem., St Paul, MN, U.S.A.

JOHNSON, I. T. and LUND, E. (1990) Soluble fibre. *Nutr. and Fd Sci.* 123: 7–9.

LUH, B. S. and BHUMIRATANA, A. (1980) Breakfast rice cereals and baby foods. In: *Rice: Production and Utilization*, pp. 622–649, LUH, B. S. (Ed.) Avi Publ. Co. Inc., Westport, CT, U.S.A.

MILLER, R. C. (1988) Continuous cooking of breakfast cereals. *Cereal Fds Wld*, **33** (3): 284–291.

MORTON, I. D. (Ed.) (1987) *Cereals in a European Context*. Chichester, Ellis Horwood.

POMERANZ, Y. (Ed.) (1987) *Modern Cereal Science and Technology*, Ch. 20: *Extrusion Products*. VCH Publishers Inc., New York.

POMERANZ, Y. (Ed.) (1989) *Wheat is Unique*. Amer. Assoc. Cereal Chem., St Paul, MN, U.S.A.

WATSON, S. A. and RAMSTAD, P. E. (Eds.) (1987) *Corn: Chemistry and Technology*. Amer. Assoc. Cereal Chem., St Paul, MN, U.S.A.

12

Wet Milling: Starch and Gluten

Purpose of wet milling

Wet milling of cereal grains differs fundamentally from dry milling in being a maceration process in which physical and chemical changes occur in the nature of the basic constituents — starch, protein and cell wall material — in order to bring about a complete dissociation of the endosperm cell contents with the release of the starch granules from the protein network in which they are enclosed. In dry milling, the endosperm is merely fragmented into cells or cell fragments with no deliberate separation of starch from protein (except in protein displacement milling by air-classification, which is a special extension of dry milling; cf. p. 132).

Although the grains of all cereals contain starch, those most widely processed by wet milling are wheat and maize. Other cereals which are less frequently wet milled are rice, sorghum and millet, while experimental work has been carried out to separate starch and protein from triticale and rye.

Wheat

Means for separating starch from wheat by wet milling processes have been known from classical times. Marcus Porcius Cato (234–149 B.C.) described a process in which cleaned wheat was steeped for 10 days in twice its weight of water, the water poured off, the soaked wheat slurried, enclosed in a cloth, and the starch milk pressed out. The residue of gluten, bran and germ would have been discarded or used as animal feed.

All wet processes for manufacture of starch and gluten comprise the steps of extracting the crude starch and crude protein, purifying, concentrating, and drying the two products. In order to obtain gluten in a relatively pure form (in addition to starch) it is necessary to separate the gluten from the bran and germ. This may be done by first milling the wheat by conventional dry processes and using the white flour as the starting material for the wet process.

Until the development of recent methods, processes starting with flour were mostly variants on three long-established methods:

Martin process: dough is kneaded under water sprays. The gluten agglomerates, and the starch is washed out.

Batter process: a flour–water batter is dispersed in more water so that the gluten breaks down into small curds. The gluten is separated from the starch milk by screening.

Alkali process: flour is suspended in an alkaline solution (e.g. 0.03 N sodium hydroxide) in which the protein disperses. Starch is removed by tabling and centrifuging, and the protein precipitated by acidifying to pH 5.5. The protein product is in a denatured condition, i.e. non-vital.

A number of modern processes start with flour: the Canadian process (1966) resembles the alkali process, but suspends the flour in 0.2 M ammonium hydroxide; the Far-Mar-Co process (U.S. Pat. No. 3,979,375) is similar to the Martin process, but the dough moves through a tube in which it is washed and mixed; the Alfa-Laval/Raisio process (Stärke, 1978, **30**:8) resembles the batter process, but uses centrifuging, decantation and hydrocyclones for separating the starch from the gluten; the Koninklijke Scholten-Honig process

(BP 1,596,742) also resembles the batter process and uses hydrocyclones. A process by Walon (U.S. Pat. No. 4,217,414) uses bacterial alpha-amylase to solubilize the starch; the gluten is not denatured and can be separated as 'vital gluten'.

Some modern processes start with wheat but differ from Cato's method in using a steep-liquor containing 0.03–0.7% of sulphur dioxide to inhibit development of micro-organisms. After draining off the steep liquor, the wet grain is coarsely milled and slurried with water. The bran and germ are separated by screening, and the heavy starch granules separated from the light gluten curd by sedimentation and centrifuging.

In the Pillsbury process (Br. Pat. No. 1,357,669) grain is steeped in an acid medium with application of vacuum or carbon dioxide to remove the air pocket at the base of the crease where micro-organisms might develop. In the Far-Mar-Co process (U.S. Pat. No. 4,201,708) wheat is soaked in water and flaked. The flakes are disintegrated and the resulting bran-germ and endosperm particles are hydrated and form a dough-like mass which is tumbled and manipulated in water to separate and recover vital gluten, starch and bran-germ components.

In all processes, the starch and gluten are dried using, in the case of gluten, methods such as freeze-drying which do not denature the gluten.

'Vital' gluten, or undenatured gluten, is gluten separated from wheat by processes which permit the retention of the characteristics of natural gluten, viz. the ability to absorb water and form an extensible, elastic mass. Commercial glutens are produced in the U.K., Europe, Australia and Canada (McDermott, 1985). In the U.K., some 260,000–270,000 tonnes of wheat were used in 1988/89 for the manufacture of starch and vital gluten, chiefly by the dough (Martin) or batter processes, which start with flour. So far as is known, processes such as the Pillsbury and Far-Mar-Co processes that start with wheat grain are not currently being used in the U.K.

With a yield of about 0.075 t of gluten from each tonne of wheat, the quantity of gluten produced in the U.K. in 1988/89 would have been some 20,000 t. The demand for vital gluten in the U.K. considerably exceeds this figure, and

is met by imports of gluten, chiefly from the European continent. Imports of gluten in 1988/89 amounted to 37,000 t, giving a total availability of about 57,000 t. World production of vital gluten in 1986 is reported as 253,000 t (Godon, 1988), of which 130,000 t were produced in western Europe and 54,000 t in the U.S.A., Canada, Mexico and Argentina.

Uses for vital wheat gluten

Vital gluten is used as a protein supplement:

— at levels of 0.5–3.0% to improve the texture and raise the protein content of bread, particularly 'slimming' bread, crispbread and speciality breads such as Vienna bread and hamburger rolls;
— to fortify weak flours, and to permit the use by millers of a wheat grist of lowered strong/ weak wheat ratio (particularly in the EC countries) by raising the protein content of the milled flour;
— in starch-reduced high protein breads (cf. p. 209), in which the gluten acts both as a source of protein and as a texturizing agent;
— in high-fibre breads (cf. p. 209) now being made in the U.S.A., to maintain texture and volume.

Vital gluten is also used as a binder and to raise the protein level in meat products, e.g. sausages, breakfast foods, pet foods, dietary foods and textured vegetable products (t.v.p.).

Vital gluten in bread

In the U.K., domestic bread consumption in 1988 averaged 30.28 oz/person/week. With a population of about 57 million, the total domestic bread consumption would have been about 2.5 million tonnes per year, or about 2.75 million tonnes total consumption, allowing for about 10% consumed non-domestically. With about 45,000 t of vital gluten used in bread in the U.K. in 1988, the average level of use would have been about 1.6%. The usual rates of addition are 1.5% for white bread and 4.5% for wholemeal bread.

In the U.S.A., about 70% of all vital wheat

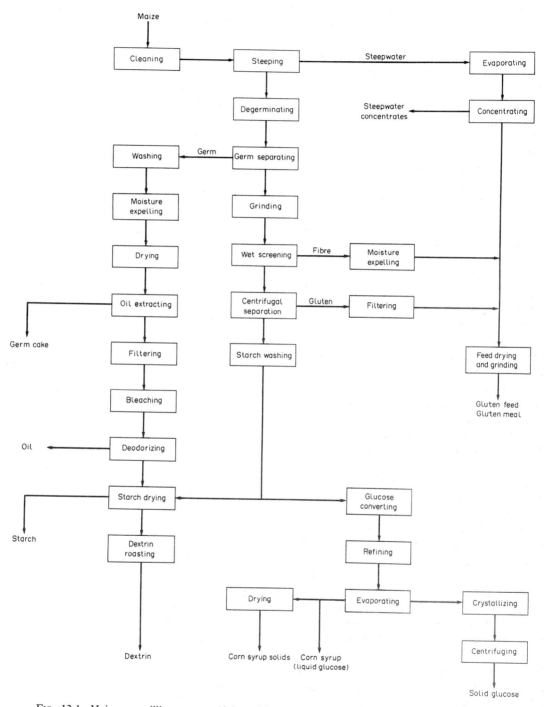

FIG. 12.1. Maize wet-milling process. (Adapted from Anon. (1958) *Food,* **27**: 291; *Corn in Industry,* 5th edn, Corn Industries Resarch Foundation Inc., New York, and S. A. Matz (Ed.) (1970) *Cereal Technology,* Avi Publ. Co. Inc., Westport, Conn., U.S.A.)

gluten is used in the manufacture of bread, rolls, buns and other yeast-raised products. The remainder finds uses in breakfast cereals (e.g. Kellogg's Special K), breadings, batter mixes and pasta products (Magnuson, 1985).

Gluten flour

Gluten flour is a blend of vital wheat gluten with wheat flour, standardized to 40% protein content in the U.S.A.

Maize

Maize is wet-milled to obtain starch, oil, cattle feed (gluten feed, gluten meal, germ cake) and the hydrolysis products of starch, viz. liquid and solid glucose and syrup.

Operations

The sequence of operations in wet milling of maize is shown in Fig. 12.1.

For safe storage, maize must be dried because the moisture content at harvest is generally higher than the desirable m.c. for storage. Drying temperature should not exceed 54°C; at higher temperatures changes occur in the protein whereby it swells less during steeping, and tends to hold the starch more tenaciously, than in grain not dried, or dried at lower temperatures. In addition, if dried at temperatures above 54°C the germ becomes rubbery and tends to sink in the ground maize slurry (whereas the process of germ separation depends on the floating of the germ), and the starch tends to retain a high oil content.

Steeping

The cleaned maize is steeped at a temperature of about 50°C for 28–48 h in water containing 0.1–0.2% of sulphur dioxide. Steeping is carried out in a series of tanks through which the steep water is pumped counter-current. The moisture content of the grain increases rapidly to 35–40%, and more slowly to 43–45%. The steeping softens the kernel and assists separation of the hull, germ and fibre from each other. The sulphur dioxide in the steep may disrupt the –SS– bonds in the matrix protein (glutelin), facilitating starch/protein separation.

After steeping, the steep water is drained off. It contains about 6% of solids, of which 35–45% is protein. The protein in the steep water is recovered by vacuum evaporation, allowed to settle out of the water in tanks, and dried as 'gluten feed' for animal feeding. The water recovered is re-used as steep water or, after concentration, as a medium for the culture of organisms from which antibiotics are obtained.

De-germing

The maize, after steeping, is coarsely ground in de-germing mills with the objective of freeing the germ from the remainder of the grain without breaking or crushing the germ. The machine generally used for this purpose is a Fuss mill, a bronze-lined chamber housing two upright metal plates studded with metal teeth. One plate rotates at 900 rev/min, the other is stationary. Water and maize are fed into the machine, which cracks open the grain and releases the germ. By addition of a starch–water suspension, the density of the ground material is adjusted to 8–10.5° Bé* (1.06–1.08 sp.gr.): at this sp.gr. the germs float while the grits and hulls settle.

Germ separation

The ground material flows down separating troughs in which the hulls and grits settle, while the germ overflows. More modern plants use hydrocyclones which require less space and are less costly to maintain than flotation equipment. Moreover, the germ separated on hydrocyclones is cleaner than that separated by flotation.

The germ is washed and freed of starch on reels, de-watered in squeeze presses and dried on rotary steam driers. The dry germ is cooked by

---

* Baumé Scale: a hydrometer scale on which 0° represents the sp. gr. of water at 12.5°C and 10° the sp. gr. of a 10% solution of NaCl at 12.5°C. It is also known as the Lunge Scale.

steam, and the oil extracted by hydraulic presses or by solvent extraction. The germ oil is screened, filtered and stored. The extracted germ cake is used for cattle feed.

Milling

The de-germed underflow from the germ separator is strained off from the liquor and finely ground on impact mills, such as an entoleter, or attrition mills, such as the Bauer mill. After this process, the starch and protein of the endosperm are in a very finely divided state and remain in suspension. The hulls and fibre, which are not reduced so much in particle size, can then be separated from the protein and starch on reels fitted with 18–20 mesh screens. Fine fibres, which interfere with the subsequent separation of starch from protein, are removed on gyrating shakers fitted with fine nylon cloth.

Separation of starch from protein

In the raw grain the starch granules are embedded in a protein network which swells during the steeping stage and tends to form tiny globules of hydrated protein (Radley, 1951–1952). Dispersion of the protein, which frees the starch, is accelerated by the sulphur dioxide in the steep water.

The effect of the sulphur dioxide, according to Cox et al. (1944), is due to its reducing, not to its acidic, property. The sulphur dioxide also has a sterlizing effect, preventing growth of microorganisms in the steep.

The suspension of starch and protein from the wet screening is adjusted to a density of 6°Bé (1.04 sp.gr) by de-watering over Grinco or string filters, and the starch separated from the protein in continuous high-speed centrifuges such as the Merco centrifugal separator.

The starch is re-centrifuged in hydrocyclones to remove residual protein and is then filtered and dried to 10–12% m.c. in kilns or ovens, or in tunnel or flash driers. The moisture content is further reduced by vacuum drying to 5–7% m.c. in the U.S.A., or to 1–2% m.c. in Britain.

The separated protein is filtered and dried in rotary or flash driers. Further fractionation to obtain the alcohol-soluble zein, which comprises about 50% of the maize gluten, by solvent extraction and precipitation may be carried out. Zein finds a use as a water-protective coating material for nuts and confectionery and as a binder for pharmaceuticals.

A combined dry-wet milling process for refining maize has been described (U.S. Pat. No. 4,181,748; 1980) in which the maize is dry-milled to provide endosperm, germ, hull and cleaning fractions. The endosperm fraction is wet-milled in two steps which respectively precede and follow an impact milling step. The principal products are prime maize starch, corn oil and an animal feed product.

Maize wet-milling in the U.S.A

The processing of maize into wet-processed products has increased greatly in the U.S.A. in recent years. In 1960/61, 3.94 Mt were processed into wet-processed products, but by 1984/85 the figure had increased to 20.7 Mt, representing 10.6% of the entire maize harvest in 1984 of 195 Mt. Of this, 7.9 Mt were used to produce high-fructose corn syrup (HFCS), 4.8 Mt for glucose and dextrose, 3.8 Mt for starch and 4.3 Mt for alcohol (Livesay, 1985).

The growth of the industry in the 1970s followed the development of a process to convert starch into high-fructose corn syrup (HFCS). The sweetness of HFCS allows it to be used as a substitute for sucrose in soft drinks and other processed foods.

Products of wet-milling

The wet-milling of maize yields about 66% of starch, 4% of oil, and 30% of animal feed, comprising about 24% of gluten feed of 21% protein content (made up of about 13% of fibre, 7% of steep water solubles and 4% of germ residue), plus about 5.7% of gluten meal of 60% protein content. The composition of products from the wet-milling of maize is shown in Table 12.1 (Wright, in Watson and Ramstad, 1987).

Most of the starch is further processed to make

TABLE 12.1
Composition of Maize Wet-Milled Products*

| | Moisture (%) | Protein† (%) | Fat (%) | Fibre | | Ash (%) | NFE§ (%) | Starch (%) |
				Crude (%)	NDF‡ (%)			
Maize	15.5	8.0	3.6	2.5	8.0	1.2	69.2	60.6
Corn gluten feed	9.0	22.6	2.3	7.9	25.4	7.8	50.1	low
Corn meal	10.0	62.0	2.5	1.2	4.1	1.8	22.5	low
Germ meal	10.0	22.6	1.9	9.5	41.6	3.8	52.2	low
Steep liquor	50.0	23.0	0	0	0	7.3	19.2	low

* Sources: Anon. 1982; Wright, K. N., 1987.
† $N \times 6.25$.
‡ Neutral detergent fibre.
§ Nitrogen-free extract.

modified starch, sweeteners and alcohol (Long, 1982).

Other uses for corn gluten include cork-binding agent, additive for printing dyes, and in pharmaceuticals. It is perhaps misleading that the protein product obtained from maize should be called 'gluten', because maize gluten in no way resembles the vital gluten that is obtained from wheat (cf. p. 260)

Ethanol can be made by yeast fermentation of maize starch, and has a particular advantage because the yeast can be re-cycled. About 85% of the ethanol produced from maize starch is blended with gasoline, in which it acts as an octane enhancer for unleaded fuel (May, 1987).

An edible film has been made from maize starch amylose obtained from high-amylose maize (cf. p. 99). Suggested uses for the film include the packing of gravies, sauces and coffee.

The starch obtained from the milling of waxy maize (cf. p. 99), called 'amioca', consists largely of amylopectin. Amioca paste is non-gelling and has clear, fluid adhesive properties.

Heated and dried maize starch/water slurries yield pregelatinized starch, known as 'instant starch', as it thickens upon addition of cold water.

Uses for wet-milled maize products

Uses for maize starch include paper manufacture, textiles, adhesives and packaged foods, and as the starting material for further processing by chemical treatment, to make various kinds of

modified starch for particular purposes, or by enzymic hydrolysis, to yield maltose, which can be further treated to make dextrose (D-glucose), regular corn syrup, high-fructose corn syrup (HFCS), and malto-dextrins.

In modifying starch, the objectives are to alter the physical and chemical characteristics in order to improve functional characteristics, by oxidation, esterification, etherification, hydrolysis or dextrinization. The methods used are acid thinning, bleaching or oxidation, cross-linking, substitution or derivatization, instantizing.

In acid thinning, or conversion, the glucosidic linkages joining the anhydroglucose units are broken, with the addition of water. The resulting thinning reduces the viscosity of the starch paste, and allows such starches to be cooked at higher concentrations than the native starch. In a wet process of conversion, the starch is treated with 1–3% of hydrochloric or sulphuric acids at about 50°C, then neutralized and the starch filtered off. Acid-thinned starches are used in confectionery products, particularly starch jelly candies (Moore *et al.*, 1984). Non-food uses include paper-sizing, calendering, coating applications (Sanford and Baird, 1983; Bramel, 1986). In a dry process conversion, in which dry starch powder is roasted with limited moisture and a trace of hydrochloric acid, the main product is dextrins, used for adhesives and other non-food purposes.

In bleaching, or oxidation, aqueous slurries of starch are treated with hydrogen peroxide, peracetic acid, ammonium persulphate, sodium hypochlorite, sodium chlorite, or potassium permanganate and then neutralized with sodium bisulphite. The xanthophyll and other pigments are bleached, thereby whitening the starch which then becomes suitable for use as a fluidizing agent in, e.g. confectioners' sugar. Starch, in aqueous slurry, can be oxidized with about 5.5% (on d.wt.) of chlorine as sodium hypochlorite. Hydroxyl (–OH) groups are oxidized, forming carboxyl (–C=O) or carbonyl (OH–C=O) groups, with cleavage of glucosidic linkages. The bulkiness of the –C=O groups reduces the tendency of the starch to retrograde. Bleached starch finds uses in batter and breading mixes in fried foods (Moore *et al.*, 1984), to improve adhesion. Oxidized starch finds non-food uses in paper sizing, due to its excellent film-forming and binding properties (Bramel, 1986).

Cross-linking improves the strength of swollen granules, preventing rupture. Granular starch is cross-linked by treatment with adipic acid and acetic anhydride, forming distarch adipate, or with phosphorus oxychloride, forming distarch phosphate. The slurry is neutralized, filtered, washed and dried. The viscosity of cross-linked starch is higher than that of native starch.

Derivatization or substitution consists of the introduction of substitution groups on starch by reacting the hydroxyl groups (–OH) with monofunctional reagents such as acetate, succinate, octenyl succinate, phosphate or hydroxypropyl groups. Derivatization retards the association of gelatinized amylose chains, improves clarity, reduces gelling and improves water-holding (Orthoefer, 1987). Substitution may be combined with cross-linking to yield thickeners with particular processing characteristics (Moore *et al.*, 1984). Starch esters — acetates, phosphates, octenyl succinates — find uses as thickeners in foods, e.g. fruit pies, gravies, salad dressings, filled cakes, because they withstand refrigeration and freeze/thaw cycles well. Non-food uses include warp sizing of textiles, surface sizing of paper, and gummed tape adhesives (Sanford and Baird, 1983; Bramel, 1986).

Phosphate mono-esters of starch, made by roasting starch with orthophosphates at pH 5–6.5 for 0.5–6 h at 120°–160°C, have good clarity, high viscosity, cohesiveness, and stability to retrogradation. They are used as emulsifiers in foods and for many non-food uses (Orthoefer, 1987).

Dextrose is made by enzymic hydrolysis of starch and crystallization; *HFCS* by partial enzymic isomerization of dextrose hydrolysates; *regular corn syrup* and *malto-dextrins* by partial hydrolysis of corn starch with acid, acid + enzyme, or enzyme only (Hebeda, 1987).

In making dextrose, thermostable bacterial alpha-amylase from *B. subtilis* or *B. licheniformis* is used for liquefying starch to 10–15 D.E. (dextrose equivalent) followed by saccharification with glucoamylase from *Aspergillus niger* to 95–96% dextrose (dry basis). The glucoamylase releases dextrose step-wise from the non-reducing end, cleaving both alpha-1→4 and alpha-1→6 bonds. The hydrolysate is clarified, refined, and processed to crystalline dextrose, liquid dextrose, high-dextrose corn syrup or HFCS feed.

Glucose and dextrose are used in beer, cider, soft drinks, pharmaceuticals, confectionery, baking and jams.

The refined dextrose hydrolysate can be treated with immobilized glucose isomerase of bacterial origin. This enzyme catalyzes the isomerization of dextrose to D-fructose. A product containing 90% of HFCS may be obtained by chromatographic separation.

Dextrose and HFCS are important sweeteners: dextrose has 65–76% of the sweetness of sucrose, while HFCS is 1.8 times as sweet as sucrose and 2.4 times as sweet as dextrose.

The use of maize in the U.S.A. to produce HFCS increased from 0.25 Mt in 1971/72 to 8.1 Mt in 1985/86. The explosive growth of HFCS production between 1972 and 1985 was due to the technical breakthrough in the process for making HFCS, the high cost of sugar in the U.S.A., and the availability of abundant stocks of relatively low-cost maize in the U.S.A. (May, 1987). By 1991, HFCS had become the largest product category of the U.S. corn wet-milling industry, accounting for more than one-third of the industrial grind. When reporting world production, however, the U.S. Dept. Agric. now refers to HFSS (high-frucose starch syrup) rather than HFCS, acknowledging that alternative starch sources, such as wheat and potatoes, can be converted to fructose-rich syrups using enzymic conversions, refining and separation technologies which are currently applied in the U.S.A. to starch separated from maize. On this basis, world production of HFSS increased from 0.65 Mt in 1975 to 7.1 Mt in 1988. Production of HFSS in the U.S.A. similarly increased from 0.48 Mt in 1975 to 5.3 Mt in 1988. Over this period, HFSS production as a percentage of total world sugar plus HFSS consumption increased from 0.9% in 1975 to 6.3% in 1988 (Meyer, 1991).

Sorghum

The methods used for the wet-milling of sorghum closely resemble those described for maize (cf. p. 262), but the process is more difficult with sorghum than with maize. The problems are associated with the small size and spherical shape of the sorghum kernel, the large proportion of horny endosperm, and the dense high-protein peripheral endosperm layer. Varieties with dark-coloured outer layers are not satisfactory for wet-milling because some of the colour leaches out and stains the starch.

Steeping

The cleaned sorghum is first steeped in water (1.61–1.96 1/kg) for 40–50 h in a counter-current process. Part of the water is charged with 0.1–0.16% of sulphur dioxide, which is absorbed by the grain and weakens the protein matrix in which the starch granules are embedded.

De-germing

The steeped grain, in slurry form, is ground in an attrition mill. The mill has knobbed plates, one static, one rotating at about 1700 rev./min. The milling detaches the germ and liberates about half of the starch from the endosperm. The germ, which contains 40–45% of oil, floats to the surface of the slurry, and is removed from the heavier endosperm and pericarp in a continuous liquid cyclone.

The de-germed endosperm and pericarp are wet-screened on a nylon cloth with 70–75 μm apertures, through which the free starch and protein pass. The residue, mostly horny endosperm, is re-ground in an entoleter, impact mill, or other suitable mill, to release more starch, and is again screened.

De-watering

The washed tailings of the screen are de-watered to about 60% m.c. by continuous screw presses. The product is known as 'fibre'. The fibre, blended with concentrated steep water (containing solubles leached out during the steeping) and spent germ cake, is dried in a continuous flash drier to yield 'milo gluten feed'.

Starch/protein separation

The de-fibred starch granules and protein particles are separated in a Merco continuous centrifuge by a process of differential sedimentation; the starch granules, with a density of 1.5, settle out of an aqueous slurry at a faster rate than the protein particles, density 1.1. The starch slurry is dried on a Proctor and Schwartz moving-belt tunnel drier, or by flash drying.

Sorghum 'gluten'

The 'gluten' (protein) is concentrated, filtered and flash dried. The protein content of milo gluten is 65–70% on dry basis. A blend of milo

TABLE 12.2
Composition of Products From Wet Milling of Sorghum* (dry basis)

Product	Yield† (%)	Protein (%)	Fat (%)	Starch (%)
Germ	6.2	11.8	38.8	18.6
Fibre	7.4	17.6	2.4	30.6
Tailings	0.8	39.2	—	25.3
Gluten	10.6	46.7	5.1	42.8
Squeegee	1.2	14.0	0.6	81.6
Starch	63.2	0.4	—	67.3
Solubles	6.6	43.7	—	—

* Source: Freeman and Bocan (1973).
† Percentage of dry substance in whole grain.

gluten with milo gluten feed, reducing the protein content to 45% d.b., is known as 'milo gluten meal'.

Products of sorghum wet milling

The yield and composition of products from the wet milling of sorghum are shown in Table 12.2.

Millet

The possibility of using pearl millet as raw material for a wet-milling process has been investigated (Freeman and Bocan, 1973). The small millet grains were much more difficult to de-germ than were sorghum or maize, although the potential yield of oil from millet exceeded those from the other cereals. Separation of protein from starch was also more difficult with millet than from sorghum or maize, and the products of separation were somewhat less pure than those obtained from the other cereals. The starch from millet resembled that from sorghum and maize in most respects, but the granule size was slightly smaller (cf. p. 57), and the starch had a slightly lower tendency to retrograde.

The composition of products from the wet-milling of pearl millet is shown in Table 12.3.

Rice

Solvent extraction milling (SEM)

This is a process applied to rice in order to obtain de-branned rice grain, and also rice bran and rice oil as separate products. The SEM process cannot strictly be described either as a dry-milling or as a wet-milling process, although it involves stages which could be included in each of these categories.

The customary method for milling rice uses abrasion of the rice grain in a dry condition to remove bran from the endosperm (cf. p. 160). In the SEM process (also called X-M) the bran layers are first softened and then 'wet-milled' in the presence of a rice oil solution. The separated bran has a higher protein content than that of the

TABLE 12.3
Composition of Products From Wet Milling of Pearl Millet* (dry basis)

Product	Yield† (%)	Protein (%)	Fat (%)	Starch (%)
Germ	7.5	10.4	45.6	10.4
Fibre	7.3	11.8	6.0	13.5
Tailings	1.5	34.1	1.9	34.2
Gluten	12.1	37.8	9.0	44.0
Squeegee	4.1	17.7	0.8	75.5
Starch	49.3	0.7	0.1	57.5
Solubles	16.8	46.1	—	—

* Source: Freeman and Bocan (1973).
† Percentage of dry substance in whole grain.

residual grain, is virtually fat-free, and is thus much more stable than that separated in the conventional dry milling process.

In the SEM process (U.S. Pat. No. 3,261,690), rice oil is applied to brown rice (de-hulled rough rice) in controlled amounts, and softening of the bran is accomplished. The bran is removed by milling machines of modified conventional design in the presence of an oil solvent — rice oil/hexane miscella. The miscella acts as a washing or rinsing medium to aid in flushing bran away from the endosperm, and as a conveying medium for continuously transporting detached bran from rice. The miscella lubricates the grains, prevents rise of temperature, and reduces breakage.

The de-branned rice is screened, rinsed and drained, and the solvent removed in two stages. Super-heated hexane vapour is used to flush-evaporate the bulk of the hexane remaining in the rice, and the rice is subjected to a flow of inert gas which removes the last traces of the solvent.

The bran/oil miscella slurry is pumped to vessels in which the bran settles, and is then separated centrifugally, while being rinsed with hexane to remove the oil. The last traces of solvent are removed from the bran by flash de-solventizing, and the bran is cooled.

The oil/hexane miscella from the bran-settling vessels is pumped to conventional solvent recovery plant, where the hexane is stripped from the oil.

SEM (X-M) plants are operated at Abbeyville, Louisiana, U.S.A. and in many Asian countries.

Advantages of the SEM process over the conventional milling process are:

1. An increase of up to 10% on head rice yield.
2. A decrease in the fat content of the rice, which improves its storage life.
3. An increase in stability of the bran product.
4. A yield of 2kg of rice oil from each 100kg of unmilled rice (2% yield on rice wt).

The bran product has potential application in breakfast cereals, baby foods, baked goods. The oil has edible and industrial applications, and is a rich source of a wax with properties similar to those of myricyl cerotate, or carnauba wax. Other applications are in margarine, cosmetics, paints (Edwards, 1967).

References

ANON. (1982) *Corn Wet-Milled Feed Products*, 2nd edn. Corn Refiners Assoc., Washington, D.C. U.S.A.

BRAMEL, G. F. (1986) Modified starches for surface coatings or paper. *Tappi*, **69**: 54–56.

COX, M. J., MacMASTERS, M. M. and HILBERT, G. E. (1944) Effect of the sulphurous steep in corn wet milling. *Cereal Chem.*, **21**: 447.

EDWARDS, J. A. (1967) Solvent extraction milling. *Milling*, 21 July: 48.

FREEMAN, J. E. and BOCAN, B. J. (1973) Pearl millet: a potential crop for wet milling. *Cereal Sci. Today*, **18**: 69.

GODON, B. (1988) Les débouches du gluten travaux de l'IRTAC et de l'INRA. *Inds. aliment. agric.*, **105**: 819–824.

HEBEDA, R. E. (1987) Corn sweeteners. In: *Corn: Chemistry and Technology*, pp 501–534 WATSON, S. A. and RAMSTAD, P. E. (Eds.). Amer. Assoc. Cereal Chem., St Paul, MN, U.S.A.

LIVESAY, J. (1985) Estimates of corn usage for major food and industrial products. In: *Situation Rep.*, pp 8–10 U.S. Dept. Agric., Econ. Res. Serv., Washington D.C., FdS–296, March.

LONG, J. E. (1982) Food sweeteners from the maize wet-milling industry. In: *Processing, Utilization and Marketing of Maize*. pp. 282–299 SWAMINATHAN, M. R., SPRAGUE, E. W. and SINGH, J. (Eds). Indian Council of Agric. Technol., New Delhi.

MAGNUSON, K. M. (1985) Uses and functionality of vital wheat gluten. *Cereal Fds Wld.*, **30** (2): 179–181.

MAY, J. B. (1987) Wet milling: process and products. In: *Corn: chemistry and technology*. pp. 377–397 WATSON, S. A. and RAMSTAD, P. E. (Eds). Amer. Assoc. Cereal Chem., St Paul, MN, U.S.A.

McDERMOTT, E. E. (1985) The properties of commercial glutens. *Cereals Fds Wld*, **30** (2): 169–171.

MEYER, P. A. (1991) High fructose starch syrup production and technology growth to continue. *Milling and Baking News*, March 26, **70** (4): 34.

MOORE, C. O., TUSCHHOFF, J. V., HASTINGS, C. W. and SCHANFELT, R. V. (1984) Applications of starches in foods. In: *Starch: chemistry and technology*, pp. 579–592 2nd edn. WHISTLER, R. L., BeMILLER, J. N. and PASCHALL, E. F. (Eds). Academic Press, Orlando, FL., U.S.A.

ORTHOEFER, F. T. (1987) Corn starch modification and uses. In: *Corn: chemistry and technology*. pp. 479–499 WATSON, S. A. and RAMSTAD, P. E. (Eds). Amer. Assoc. Cereal Chem., St Paul, MN, U.S.A.

RADLEY, J. A. (1951–2) The manufacture of maize starch. *Food Manuf.* **26**: 429, 488; **27**: 20.

SANFORD, P. A. and BAIRD, J. (1983) Industrial utilization of polysaccharides. Pp. 411–490 in: *The Polysaccharides*, Vol. 2. ASPINALL, G. O. (Ed.). Academic Press, New York, U.S.A.

WRIGHT, K. N. (1987) Nutritional properties and feeding value of corn and its by-products. In: *Corn: Chemistry and Technology*. pp. 447–478 WATSON, S. A. and RAMSTAD, P. E. (Eds). Amer. Assoc. Cereal Chem., St Paul, MN, U.S.A.

Further Reading

ANDRES, C. (1980) Corn – a most versatile grain. *Fd. Processing*, May: 78.

AUTRAN, J-C. (1989) Soft wheat: view from France. *Cereal Fds Wld*, **34** (Sept.): 667–668, 671–672, 674, 676.

Brit. Pat. Spec. No. 1,596,742 (Koninklijke Scholten-Honig NV).

FINNEY, P. L. (1989) Soft wheat: view from Eastern United States. *Cereal Fds Wld*, **34** (Sept.): 682, 684, 686–687.

LEATH, M. N. and HILL, L. D. (1987) Economics, production, marketing and utilization of corn. In: *Corn: chemistry and technology*. WATSON, S. A. and RAMSTAD, P. E. (Eds). Amer. Assoc. Cereal Chem., St Paul, MN, U.S.A.

LINKO, P. (1987) Immobilized biocatalyst systems in the context of cereals. In: *Cereals in a European Context*. pp. 107–118 MORTON, I. D. (Ed.) Ellis Horwood, London.

MITTLEIDER, J. F., ANDERSON, D. E., McDONALD, C. E. and FISHER, N. (1978) An analysis of the economic feasibility of establishing wheat gluten processing plants in North Dakota. Bull. 508, North Dakota Agricultural Experiment Station, Fargo, N.D. and U.S. Dept. Commerce.

RAO, G. V. (1979) Wet wheat milling. *Cereal Fds Wld*, **24** (8): 334–335.

U.S. Pat. Spec Nos. 3,261,690 (SEM process); 3,958,016 (Pillsbury Co.); 3,979,375 (Far-Mar-Co process); 4,181,748 (dry-wet milling); 4,201,708 (Far-Mar-Co process); 4,217,414 (Walon).

WATSON, S. A. and RAMSTAD, P. E. (Eds) (1987) *Corn: Chemistry and Technology*. Amer. Assoc. Cereal Chem., St Paul, MN, U.S.A.

13

Domestic and Small Scale Products

Introduction

Cereals are prepared for consumption by domestic processing on a small scale in many parts of the world, but particularly in the less industrialized countries. The types of cereal grains so used are principally wheat, maize, sorghum and the millets, each of which finds greatest use in those countries in which it grows indigenously. Thus, wheat and sorghum are widely used for domestic processing in the Indian subcontinent; maize is similarly used in Mexico and many African countries; sorghum and the millets are also used in many African countries.

The types of product made domestically are many and varied, and go by many names in the various countries. The categories of product include pancake-like flat breads, which may be unfermented (e.g. chapatti, roti, tortilla) or fermented (e.g. kisra, dosa, injera); porridges, which may be stiff porridges (e.g. ugali, tuwo, asida) or thin porridges (e.g. ogi, ugi, nasha, madida); steam-cooked dumpling-like foods (e.g. couscous, burabusko, kenkey); boiled products (e.g. acha, kali); snack foods, which may be popped, parched, puffed or fried (e.g. tortilla chips, corn chips, taco shells) and beverages, either alcoholic (e.g. burukutu, bantu beer, kaffir beer) or non-alcoholic (e.g. mahewu).

As countries become more industrialized, so commercial processes are introduced to make products that resemble or imitate those made by traditional domestic methods. Such products, now made commercially, include dry masa flour (from maize), kisra, wheat flour tortillas, while a process for making tortillas by extrusion cooking has been described. Thus, while traditional foods continue to be available, the local people are relieved of the daily tedium of domestic preparation.

Products made from wheat

Chapattis

In West Pakistan wheat comprises about 60% of the total cereal crop. None is exported, and about 90% of it is ground to make wholemeal called 'whole atta', a meal of near 100% extraction rate, from which chapattis — essentially wholemeal pancakes or flat bread — are made. Chapattis are also commonly eaten in India, Tibet, China and the Near East.

To provide for the increasing demand for white flour, and with rollermills replacing stone mills, the milling of wheat in India/Pakistan has been modified so as to produce white flour ('maida') and semolina (60–65%), bran (10–15%) and a residue called 'resultant atta' (25–35%) from which chapattis are made.

For making chapatti flour the wheat should have a high 1000 kernel weight, plump grains, light-coloured bran, and a protein content of 10.5–11.0%. A strong gluten is not required, but water absorption of the flour should be high. The alpha-amylase activity need not be very low: a Falling Number (cf. p. 184) of 65 is satisfactory.

Flour of fine granularity yields chapattis of superior quality.

Chapatti flour milled in the U.K. for the use of Asian immigrants is a granular fine wheatmeal of about 85% extraction rate, made by blending white flour with fine offal or bran so that the background colour is white and the large brown specks of bran are conspicuous.

Chapattis are made by mixing whole atta or resultant atta with water to form a dough, which is rested for 1 h. The dough is then divided into portions of 50–200 g, which are flattened by hand. The dough discs are baked on an iron plate over an open fire.

Types of chapatti include Tanoori Roti (baked inside a mud oven), Khameri Roti (containing yoghurt or buttermilk, sugar and salt, and the dough allowed to ferment), and Nan (made from white flour of 75% extraction rate by a yeasted sponge-and-dough process (cf. p. 202), with the addition of sugar, salt, skimmed milk, ghee and gram flour or eggs (Chaudhri and Muller, 1970)).

The hardening or firming of chapattis may be delayed by the inclusion of shortening (3%) or 0.5% of either glyceryl monostearate or sodium stearoyl-2-lactylate, thereby increasing the shelf-life to 72 h. The best results were obtained by a combination of shortening and glyceryl mono-stearate (Sidhu et al., 1989).

Germination of wheat leads to an increase in reducing sugar content, diastatic activity and production of damaged starch, while decreasing the Falling Number, gluten content and chapatti water absorption. The chapattis made from sprouted wheat had a better (sweetish) flavour but slightly harder texture. However, after storage for four days, the chapattis made from sprouted wheat had improved texture and overall quality (Leelavathi et al., 1988).

The inclusion of 10% of full fat soya flour made from steamed soya beans with 90% of wheat flour for making chapattis almost completely eliminated the activity of trypsin inhibitor (Verma et al., 1987).

The use of triticale flour in partial replacement of wheat flour for making chapattis has been suggested, the quality of the chapattis was not impaired (Khan and Rashid, 1987).

Tortillas

A flour tortilla is a flat, circular, light-coloured bread, about one sixteenth of an inch thick and 6–13 in. in diameter, made from wheat flour. Wheat flour tortillas are widely consumed in Mexico and the United States.

Traditionally, tortillas were made domestically by mixing wheat flour with water, lard and salt to make dough. The dough was divided and rolled or hand-shaped to make tortilla discs which were baked on a hot griddle (Serna-Saldivar et al., 1988a). More recently, tortillas have been produced commercially, using hot-press, die-cut or hand-stretch procedures. Hot-press tortillas are baked for a relatively longer time at lower temperature and puff while baking; they resist tearing and have a smooth surface. Die-cut tortillas use stronger doughs with greater water absorption, resulting in a product of lower moisture content and less resistance to cracking and breaking. Hand-stretch tortillas are irregular in shape and intermediate in quality (Serna-Saldivar et al., 1988a).

The loss of flexibility of tortillas during storage may be due to retrogradation of starch, and may be prevented by the use of plasticizers, which increase flexibility and extensibility. Water is the most important plasticizer for starch and polysaccharides and should contribute 34–45% of the dough by weight. The plasticizer components should also include glycerol or sorbitol, 5–7% by wt, also oil and fat, 7–9% by wt. Addition of yeast, 1–3% on dough wt, helps the development of flavour (U.S. Pat. No. 4,735,811).

Pretzels

Pretzels, crisp knot-shaped biscuits, flavoured with salt, are made from wheat flour plus shortening (1.25% on flour wt), malt (1.25%), yeast (0.25%), ammonium bicarbonate (0.04%) and water (about 42%). A dough made from these ingredients is rolled into a rope, then twisted and allowed to relax for 10 min. The rope is passed through rollers, to set the knots, and allowed to ferment for 30 min. The starch on the surface is then gelatinized by passing the rope through a bath of caustic soda (1%) for 25 sec at about 93°C. The dough pieces are then salted with 2% sodium chloride and baked in three stages: at 315°C for 10 min, then at 218°C to reduce the moisture content to 15%, and finally at 121°C for 90 min (Hoseney, 1986).

Products made from maize

Tortillas

In Central America, tortillas are made from masa, which is obtained by the stone-grinding of nixtamal, or lime-cooked maize. Traditionally, nixtamal, or hominy, was made by cooking maize in the leachate from wood ashes, the principal objective being to loosen and then remove the pericarp. In the modern process, whole maize is cooked in excess water containing 0.5–2.0% of hydrated lime (on maize basis) at 83°–100°C for 50–60 min, then cooled to about 68°C and allowed to steep for 8–24 h. This process is called nixtamalization, and the resulting product is called nixtamal. During this process the endosperm and germ are hydrated and softened, with partial gelatinization of the starch, and the alkali solubilizes cell walls leading to weakening of the pericarp, and facilitating its removal. The nixtamal made by the traditional process is washed to remove loose pericarp and excess lime and is then stone-ground to produce masa.

A modern commercial process for making hominy (nixtamal) uses lye (caustic soda) solution, in which the maize grains remain for 25–40 min until the pericarp is free. It is then boiled and washed to remove the pericarp and the lye. The hominy is then salted and canned.

An attrition milling process for grinding the nixtamal is essential, using synthetic, lava, or alumina stones that cut, knead and mash the nixtamal to form masa. The wet-ground product is obtained as a dough containing about 55% m.c.: it consists of pieces of endosperm, aleurone, germ, pericarp, free starch granules, free lipids and dissolved solids which form a 'glue-like' material that holds together the masa structure. There is no gluten in the masa dough: cohesion of the mass is due to the surface tension of the

water, and is most successful when the particle size of the material is fine, and the amount of water contained is just enough to fill the spaces between the particles. The masa is sheeted, cut into triangular shapes, and baked on a griddle for 39 sec at 280°C to make tortillas.

The temperature used for baking the tortillas is as high as possible, short of causing puffing. (Gomez et al., 1987, 1989; Hoseney, 1986; Rooney and Serna-Saldivar, 1987; Serna-Saldivar et al., 1988b).

Nixtamalization and pellagra

The disease pellagra has been associated with a deficiency of the vitamin niacin (nicotinic acid) or niacinamide, and is prevalent among peoples who rely upon maize for a large proportion of their daily food. Some 50–80% of the niacin in maize occurs in a bound form as niacytin or niacinogen, which is biologically unavailable, and renders the maize deficient in niacin (Mason et al., 1971). Pellagra is not suffered by Mexicans who consume maize meal in the form of tortillas. The alkaline conditions obtaining during the lime cooking (nixtamalization) release the bound niacin and make it biologically available.

However, there is some evidence that the onset of black tongue in dogs (a disease similar to pellagra in humans) may be due to excessive intake of the amino acid leucine, in which maize is rich. In trials with dogs on maize diets, black tongue developed only in dogs on high leucine diets. Alkaline treatment of maize to prepare masa or hominy has been shown to result in some loss of the amino acids arginine and cystine; commercially-made masa and tortillas contained lysinoalanine and lanthionine, which are breakdown products of cystine and arginine (Sanderson et al., 1978).

Snack products

Tortilla chips (tostados) were traditionally produced by frying stale, left-over tortillas, and are still prepared in this way in Mexican restaurants. Alternatively, freshly-made masa, after sheeting and cutting, can be fried for 1 min at 190°C to produce tortilla chips.

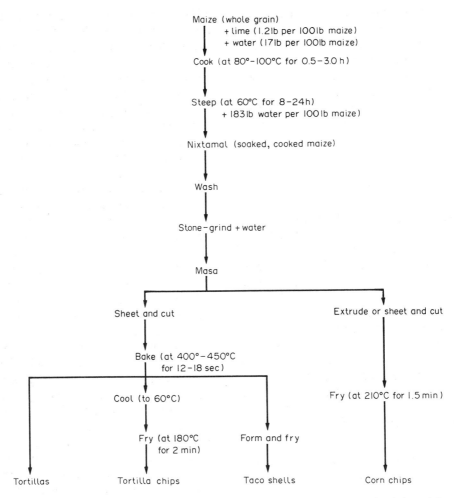

FIG. 13.1 Stages in the preparation of tortillas and corn snacks from whole maize (adapted from Gomez *et al.*, 1987; Rooney and Serna-Saldivar, 1987).

Corn chips can be made from unbaked masa by extrusion through a forming extruder into hot fat, while taco shells are made by frying tempered tortillas in deep fat. The tempering allows equilibration of the moisture to occur (Hoseney, 1986).

The stages in the preparation, from maize, of nixtamal, masa, tortillas and the snack products mentioned are shown in Fig. 13.1.

Dry masa flour

Commercial production of dry masa flour, from which tortillas and corn snacks can be made, has

greatly expanded in recent years because the dry masa flour is very convenient to use. In the preparation of dry masa flour, maize is cooked and steeped in lime water by the traditional batch method, or may be cooked in a continuous process by spraying the maize with a lime solution before insertion in a steam cooker conveyor. The cooked grain is washed to remove the pericarp and then stone-ground or hammer-milled. The masa is dried, e.g. by falling against a rising stream of hot air, then hammer-milled and size-graded by sieving, the oversize particles being reground. The various particle size fractions are then blended in appropriate proportions for particular uses, and packed (Gomez *et al.*, 1987).

Tortillas are eaten alone, like bread, or fried (taco shells), or with fillings or toppings (nachos, tostadas, enchiladas, taquitos, burritos, tamales). In Colombia, maize grain is pounded with potash and a little water in the domestic preparation of maize meal. The alkaline effect of the potash is to loosen the bran and also to unbind the bound niacin.

Kenkey

In Ghana, traditional foods prepared from maize are known as kenkey. Cleaned maize grain is steeped in water for 3 days, the grain ground to flour, dough prepared from flour and water, spontaneous fermentation of the dough allowed for 3 days, part of the fermented dough partly cooked and then mixed with the uncooked part, the dough mass moulded into balls, the balls wrapped in maize cob sheaths and boiled to produce kenkey. Levels of niacin and lysine increase during the period of dough fermentation, but fall to the original levels during the partial cooking. There is further loss of niacin during preparation of the kenkey (Ofosu, 1971).

Products made from sorghum and the millets

Most of the products made from maize, mentioned above, can also be made from sorghum, or from a blend of sorghum and maize, and some also from pearl millet. However, the traditional methods of preparation vary from country to country, as the following brief survey indicates.

The traditional method of preparation in African countries and India where sorghum and the millets form the staple food is simple pounding in a mortar to loosen the husk and to reduce the grain to wholemeal or semolina, followed by winnowing. The grain is stored as such, and from it the day's requirements of meal or flour are prepared. Preparation of larger quantities of meal is not practised because of the rapidity, in hot climates, with which the unde-germed meal becomes rancid. The yield of edible products is

85%, comprising 39% of coarse semolina, 16% of fine semolina, and 30% of flour.

Porridges made from cereal flours, such as sorghum flour, are the most important dishes consumed by the people living in Africa south of the Sahara. Both thick porridges and thin porridges are made, differing in the flour/water ratio required. Thick porridges use about 1 flour: 2 water, whereas thin porridges use 1 flour: 3–4 water. To make the porridge, the flour of sorghum, millet or maize, or a blend of these, is just boiled with water. Nasha is a thin porridge made from a fermented batter. Sorghum flour is mixed with a starter and water, and left to ferment for 12–16 h. The fermented batter is then diluted with water, cooked, and flavoured with spices or fruit juices.

Wholemeal flour from sorghum may be cooked with water to make semi-solid dumplings.

From the flour or meal, unleavened bread or chapattis are made, or the ground product may be used to make a beverage.

Millet may be consumed in the form of porridge (called kasha in the former Soviet Union) made from dry parched grain, or it may be cooked with sugar, peanuts or other foods to make desserts. Massa is made from millet flour by cooking a portion of the flour to make a thin porridge, while making a batter from the remainder of the flour, mixing it in, and leaving the mixture to ferment overnight. The mixture is flavoured with salt, pepper and onions, and fried. The product is spongy in texture, and eaten with sorghum, maize or millet porridge (Economic Commission for Africa, 1985).

In *Nigeria*, sorghum and pearl millet are used in four ways:

1. The dry grain is ground to make either (a) a meal or flour, from which porridge (tuwo) is prepared, or (b) grits, from which burabusko, a food resembling couscous, is prepared. The grits are agglomerated by blending with water, and then steamed for three successive periods of 15 min each (Galiba *et al.*, 1987). Pancakes may be made by frying a pasta.
2. The dry grain is roasted and then ground to make roasted meal or flour, from which snacks (guguru, adun) are prepared.

3. The grain is steeped in water and a lactic fermentation allowed to proceed for 1–4 days. The moist grain is then pounded and used to prepare a fermented porridge product (ogi, akamu).

4. The grain is soaked and allowed to sprout. The sprouted grains are dried and ground to make a malt from which beverages (pito, burukutu) are prepared.

In the *Sudan* and *Ethiopia* the flour or meal is used for making flat cakes (kisra, injera) or it may be mixed with cassava flour. Kisra, a staple food in the Sudan, is made by mixing sorghum flour, of 80–85% extraction rate, with water and a starter, and leaving it to ferment overnight, and then baking at 160–180°C for 30 sec. Kisra is now also made commercially, and has a shelf-life of 48 h at room temperature (Economic Commission for Africa, 1985).

The grain may be parched, popped or boiled whole.

In *Ethiopia*, injera are made from the flour of teff (*Eragrostis tef*), an indigenous type of millet. In the traditional domestic process, teff flour is mixed with water and allowed to ferment overnight by action of endogenous microflora to produce a sour dough, and then baked in the metad, or injera oven, to make injera, a pancake-like unleavened bread. The fermentation may be promoted by using a starter culture, called irsho, a thin paste saved from a previous fermentation. Fermented teff flour is also used for making porridge, beer (tella) and spirits (katikalla) (Umeta and Faulks, 1989).

In *Uganda*, sorghum grain is malted and sprouted, the radicle removed, and the remainder of the grain dried. Some of the pigment and the bitter tannins are thereby removed. The sugars produced by the malting make a sweet-tasting porridge. The grain is also used for brewing.

In *India* and *other Asian countries* wholemeal flour from sorghum or millet may be used to make dry unleavened pancakes (roti, chapatti, tortilla). It has been estimated that 70% of the sorghum grown in India is used for making roti, a proportion that increases to 95% in Maharashtra state (Murty and Subramanian, 1982).

If chapattis are made with cold water, the dough lacks cohesiveness because the protein in sorghum and millet is not gluten-like. The use of boiling water to make the dough results in partial gelatinization of the starch and imparts sufficient adhesiveness to permit the rolling out of thin chapattis. The water absorption of sorghum flour is higher than that of wheat flour; thus, the baking time for sorghum chapattis is longer than that for wheat chapattis. A blend of about 30% sorghum flour with 70% of wheat flour produces chapattis of improved eating quality.

In India and Africa, whole sorghum grain, or dehulled and polished sorghum grain (pearl dura) may be boiled to make balila, which is used in a similar way to rice. Sorghum may also be eaten as a stiff porridge.

In India, grain sorghum and pearl millet may be popped, but whereas maize is popped in hot oil, the sorghum and millet grains are popped in hot sand (Hoseney, 1986). A more detailed account of traditional foods made from sorghum in various countries, including methods of preparation, is given by Rooney *et al.* (1986).

Sorghum and maize react in the same way when subjected to the alkaline cooking process of nixtamalization. This process causes the hull or pericarp to peel away from the kernels, facilitating its subsequent removal. The starch granules throughout the kernel swell, but some of the granules in the peripheral endosperm are destroyed. Tortillas made from a blend of 80% pearled or unpearled sorghum plus 20% yellow maize by the alkaline cooking process had an acceptable flavour and a soft texture. The reduced cooking and steeping times required by sorghum as compared with maize are advantageous, and the cooking time is further reduced by using pearled, rather than unpearled, sorghum (Bressani *et al.*, 1977; Bedolla *et al.*, 1983; Gomez *et al.*, 1989).

Tortilla chips could be made from white sorghum by lime-cooking at boiling temperature for 20 min, using 0.5% lime, quenching to 68°C and then steeping the grains for 4–6 h to produce nixtamal, which was then stone-ground to masa (a coarse dough). The masa was sheeted, cut into pieces, baked at 280°C for 39 sec and then fried in oil at 190°C for 1 min. The tortilla chips thus made from sorghum had a bland flavour; a more acceptable product, with the traditional flavour, could be made similarly from a blend of equal parts of maize and sorghum (Serna-Saldivar *et al.*, 1988b).

References

BEDOLLA, S., DE PALACIOS, M. G., ROONEY, L. W., DIEHL, K. C. and KHAN, M. N. (1983) Cooking characteristics of sorghum and corn for tortilla preparation by several cooking methods. *Cereal Chem.* **60** (4): 263–268.

BRESSANI, R. I., ELIAS, L. G., ALLWOOD PAREDAS, A. E. and HUEZO, M. T. (1977) Processing of sorghum by lime-cooking for the preparation of tortillas. *Proceedings of Symposium on Sorghum and Millets for Human Food*, DENDY, D. A. V. (Ed.). Tropical Products Institute, London.

CHAUDRI, A. B. and MULLER, H. G. (1970) Chapattis and chapatti flour. *Milling* **152** (11): 22.

ECONOMIC COMMISSION FOR AFRICA (1985) *Technical compendium on composite flours*, KENT, N. L. (Tech. Ed.). United Nations Economic Commission for Africa, Addis Ababa.

GALIBA, M., ROONEY, L. W., WANISKA, R. D. and MILLER, F. R. (1987) The preparation of sorghum and millet couscous in West Africa. *Cereal Fds Wld* **32** (12): 878–884.

GOMEZ, M. H., ROONEY, L. W., WANISKA, R. D. and PFLUGFELDER, R. L. (1987) Dry corn masa flours for tortilla and snack food production. *Cereal Fds Wld* **32** (5): 372–377.

GOMEZ, M. H., MCDONOUGH, C. M., ROONEY, L. W. and WANISKA, R. D. (1989) Changes in corn and sorghum during nixtamalization and tortilla baking. *J. Fd Sci.*, **54** (2): 330–336.

HOSENEY, R. C. (1986) *Principles of Cereal Science and Technology*. Amer. Assoc. Cereal Chem., St Paul, MN, U.S.A.

KHAN, M. N. and RASHID, J. (1987) Nutritional quality and technological value of triticale. *Asian Fd J.*, **3** (March): 17–20.

LEELAVATHI, K. and HARIDAS RAO, P. (1988) Chapati from germinated wheat. *J. Fd. Sci. Technol.*, **25** (May/June): 162–164.

MASON, J. B., GIBSON, N. and KODICEK, E. (1971) The chemical nature of bound nicotinic acid. *Biochem. J.*, **125**: 117P.

MURTY, D. S. and SUBRAMANIAN, V. (1982) Sorghum roti: I. Traditional methods of consumption and standard procedures for evaluation. In: *Proceedings of International Symp. on Sorghum Grain Quality*, pp. 73–78, ROONEY, L. W. and MURTY, D. S. (Eds.). Int. Crop Res. Inst. Semi-Arid Tropics, Patancheru, A.P., India.

OFOSU, A. (1971) Changes in the levels of niacin and lysine during the traditional preparation of kenkey from maize grain. *Ghana J. agric. Sci.*, **4**: 153.

ROONEY, L. W., KIRLEIS, A. W. and MURTY, D. S. (1986) Traditional foods from sorghum: their production, evaluation and nutritional value. In: *Advances in Cereal Science and Technology*, Vol. VIII, Ch. 7, POMERANZ, Y. (Ed.) Amer. Assoc. Cereal Chem., St Paul, MN, U.S.A.

ROONEY, L. W. and SERNA-SALDIVAR, S. O. (1987) In: *Corn: Chemistry and Technology*. WATSON, S. A. and RAMSTAD, P. E. (Eds.) Amer. Assoc. Cereal Chem., St Paul, MN, U.S.A.

SANDERSON, J., WALL, J. S., DONALDSON, G. L. and CAVIUS, J. F. (1978) Effect of alkaline processing of corn on its amino acids. *Cereal Chem.* **55**: 204–213.

SERNA-SALDIVAR,S. O., ROONEY, L. W. and WANISKA, R. D. (1988a) Wheat flour tortilla production. *Cereal Fds Wld*, **33** (10): 855–864.

SERNA-SALDIVA, S. O., TELLEZ-GIRON, A. and ROONEY, L. W. (1988b) Production of tortilla chips from sorghum and maize. *J. Cereal Sci.*, **8**: 275–284.

SIDHU, J. S., SEIBEL, W. and BRUEMMER, J. M. (1989) Effect of shortening and surfactants on chapati quality. *Cereal Fds Wld*, **34** (March): 286–290.

UMETA, M. and FAULKS, R. M. (1989) Lactic acid and volatile (C_2–C_6) fatty acid production in the fermentation and baking of tef (*Eragrostis tef*). *J. Cereal Sci.*, **9**: 91–95.

U.S. PAT. SPEC. NO. 4,735,811 (Plasticizers for tortillas).

VERMA, N. S., MISHRA, H. N. and CHAUHAN, G. S. (1987) Preparation of full fat soy flour and its use in fortification of wheat flour. *J. Fd Sci. Technol.*, **24** (Sept./Oct.): 259–260.

Further Reading

ADRIAN, J., FRANGUE, R., DAVIN, A., GALLANT, D. and GAST, M. (1967) The problem of the milling of millet. Nutritional interest in the traditional African process and attempts at mechanisation. *Agronomie tropicale*, **22** (8): 687.

ASP, N.–G. and BJÖRCK, I. (1989) Nutritional properties of extruded foods. In: *Extrusion Cooking*, Ch. 14, MERCIER, C., LINKO, P. and HARPER, J. (Eds.). Amer. Assoc. Cereal Chem., St Paul, MN, U.S.A.

AYKROYD, W. R., GOPOLAN, C. and BALASUBRAMANIAN, S. C. (1963) *The nutritive value of Indian foods and the planning of satisfactory diets*. Indian Council of Medical Recearch, New Delhi, Spec. Rpt Ser. No. 42.

DESIKACHAR, H. S. R. (1975) Processing of maize, sorghum and millets for food uses. *J. sci. indust. Res.* **34** (4): 231.

DESIKACHAR, H. S. R. (1977) Processing of sorghum and millets for versatile food uses in India. *Proceedings of a Symposium on Sorghum and Millets for Human Food* DENDY, D. A. V. (Ed.) Tropical Products Institute, London.

HULSE, J. H., LAING, E. M. and PEARSON, O. E. (1980) *Sorghum and the millets*. Academic Press, London.

KODICEK, E. and WILSON, P. W. (1960) The isolation of niacytin, the bound form of nicotinic acid. *Biochem. J.*, **76**: 27P.

PERTEN, H. (1983) Practical experience in processing and use of millet and sorghum in Senegal and Sudan. *Cereal Fds Wld*, **28** (11): 680–683.

POMERANZ, Y. (Ed.) (1988) *Wheat: Chemistry and Technology*, 3rd edn, Vol. II, Amer. Assoc. Cereal Chem., St Paul, MN, U.S.A.

RAGHAVENDRA RAO, S. N., MALLESHI, N. G., SREEDHARA-MURTHY, S., VIRAKTAMATH, C. S. and DESIKACHAR, H. S. R. (1979) Characteristics of *roti*, *dosa* and vermicelli from maize, sorghum and bajra. *J. Fd Sci. Technol.*, **16**: 21.

WATSON, S. A. and RAMSTAD, P. E. (Eds.) (1987) *Corn: Chemistry and Technology*. Amer. Assoc. Cereal Chem., St Paul, MN, U.S.A.

14

Nutrition

Introduction

Nutrition in most adults is concerned with the supply and metabolism of those components of the diet needed to maintain normal functioning of the body (water and oxygen are also necessary but these are not generally regarded as nutrients). In the young and in pregnant and lactating mothers additional nutritional requirements are imposed by the need to support growth or milk production.

Nutrients — the substances that provide energy and raw materials for the synthesis and maintenance of living matter, in the diet of humans and other animals, comprise *protein, carbohydrate, fat*, all of which can provide energy, *minerals and vitamins*. Those nutrients that cannot be made in sufficient quantities by conversion of other nutrients in the body, are called *essential*. They include some vitamins, minerals, essential amino acids and essential fatty acids. An insufficiency of an essential nutrient causes a specific deficiency disease. Deficiency diseases are now rare in the West where food is plentiful, but they remain a problem in the Third world, where natural disasters and conflict frequently lead to malnutrition and even starvation. Aid provided for the relief of such disasters always includes a high proportion of cereals, demonstrating their high nutritional value.

Although cereals make an important contribution to the diet they cannot alone support life because they are lacking in *vitamins A* (except for yellow maize), B_{12} and *C*. Whole cereals also contain *phytic acid*, which may interfere with the absorption of iron, calcium and some trace elements. Also cereal proteins are deficient in certain essential amino acids, notably lysine. Cereals however are rarely consumed alone, and nutrients in foods consumed together, may mutually compensate for each other's deficiencies.

While it is indisputable that individuals and populations should consume the right amounts of nutrients to avoid symptoms of deficiency and excess, defining those 'right amounts' is not easy, not least because the requirements vary from one individual to another. The British Government, for more than 30 years issued standards in the form of *Recommended Intakes for Nutrients* (DHSS, 1969) and *Recommended Daily Amounts (RDA)* of food energy and nutrients (DHSS, 1979). In revising the recommendations for the dietary requirements of the *nation*, the Committee on Medical Aspects of Food Policy (COMA), noted that the standards were frequently used in a way that was never intended, that is they were used to assess the adequacy of the diets of *individuals*. In order to ensure that deficiencies were avoided the RDAs represented at least the minimum requirements of those individuals with the greatest need. In terms of the population as a whole, therefore, they were overestimates, and individuals ingesting less than the RDA for any nutrient may be far from deficient. Instead of revized RDAs, *Dietary Reference Values (DRV) for Food Energy and Nutrient for the United Kingdom* (DH, 1991) were issued. They applied to energy, proteins, fats, sugars, starches, non-starch polysaccharides, 13 vitamins and 15 minerals, and they comprised:

Estimated Average Requirement (EAR) — an

estimate of the *average* requirement or need for food energy or a nutrient.

Reference Nutrient Intake (RNI) — enough of a nutrient for almost *every* individual, even someone who has high needs for the nutrient.

Lower Reference Nutrient Intake (LRNI) — the amount of a nutrient that is enough for only a small number of people with *low* needs.

Safe Intake — a term normally used to indicate the intake of a nutrient for which there is not enough information to estimate requirements. A safe intake is one which is judged to be adequate for almost everyone's needs but not so large as to cause undesirable effects.

In recognition of the fact that people of different sexes, ages (e.g. infants, children, adults) and conditions (e.g. pregnant and lactating mothers) have different requirements, DRVs appropriate to the different groups were defined.

Deficiency diseases, such as rickets, stunting, deformities and anaemia, are now rare in Western countries and, in considering the relationship between food and disease, emphasis has shifted to other diseases that are thought to be diet related: these include cancer of the colon (associated with animal protein intake, particularly meat), breast cancer (associated with fat intake), and stroke and heart disease, associated with consumption of salt and animal (saturated) fat (Bingham, 1987).

Recommendations

Recommendations appropriate to the U.K. situation are that *cereals*, particularly whole grain, together with *potatoes*, should be eaten in generous amounts at each main meal to satisfy *appetite*, three or more portions of *fresh vegetables or fruit*, preferably green or yellow, should be eaten per day and two or more portions of *low fat foods* containing the *most protein*. *Low fat dairy foods* should be chosen in preference to high fat ones and all sugar, refined starches, and foods made from them, such as *biscuits*, *cakes*, *sweets*, etc. should be used *sparingly*.

More than 80g per day for men and more than 50g per day for women, of alcohol is considered excessive and should be avoided. So also should table salt and foods cooked or preserved in excessive salt.

In the U.S.A. the USDA/USDHHS published guidelines (1985) for a healthy diet including the following recommendations:

Eat a variety of foods.
Maintain desirable weight.
Avoid too much fat, saturated fat and cholesterol.
Eat foods with adequate starch and fibre.
Avoid too much sugar.
Avoid too much sodium.
If you drink alcoholic beverages, do so in moderation.

Both U.K. and U.S. recommendations acknowledge the importance of cereals, particularly as a source of energy and non-starch polysaccharides. In addition, however, cereals provide many other valuable nutrients, including proteins, vitamins and minerals as several of the tables and figures in this chapter demonstrate.

Cereals in the diet

For the majority of the world's human population, cereal-based foods constitute the most important source of *energy* and other *nutrients*. In the poorest parts of the world starchy foods, including cereals, may supply 70% of total energy. In the wealthiest nations the proportion obtained from cereals has declined fairly rapidly: in the U.S.A. during the present century the proportion of total energy provided by cereals has dropped from 40% to between 20 and 25%. The proportions of some important nutrients derived from cereals and products in Britain are shown in Table 14.1.

Positive attributes of cereals as foods

Starch

Cereals are a particularly rich source of starch as it constitutes by far the most abundant storage product in the endosperm. Starch is an important source of *energy*, it is found only in plants (although the related compound *glycogen* occurs in

TABLE 14.1
Contributions (%) Made by Cereal Products to the Nutritional Value of Household Food in Britain. 1990*

| | Cereals | Bread | | Cakes, pastries, biscuits | Breakfast cereals |
		White	Brown and wholemeal		
Energy	31.5	7.2	3.4	8.2	3.4
Fat	12.8	1.0	0.8	7.6	0.5
Fatty acids:					
Saturated	12.1	0.6	0.3	9.0	0.3
Polyunsaturated	13.7	2.2	1.4	5.0	1.4
Sugars	18.7	1.8	0.7	10.0	2.8
Starch	73.0	20.4	9.0	9.9	8.5
Fibre	45.5	7.4	11.6	5.0	9.9
Calcium	24.6	7.5	2.8	4.0	1.0
Iron	49.0	8.7	7.7	6.7	15.4
Sodium	38.3	12.6	6.5	5.0	5.4
Vitamin C	1.5	—	—	—	1.0
Vitamin A	1.1	—	—	0.5	—
Vitamin D	12.3	—	—	1.3	9.6

* Values calculated from appendix B, Table 14, *Household Food and Expenditure 1990* MAFF, HMSO, London, 1991.
— = nil. Reproduced with the permission of the Controller of Her Majesty's Stationery Office.

animal tissues). In the past, starch has been undervalued by nutritionists, who have emphasized its association with obesity, and recommended reduction in, for example, bread consumption by those wishing to control their weight. However, starch is preferable, as an energy source, to fat, and a further advantage of starch consumed as part of a cereal food is that it is accompanied by vitamins, minerals and protein. In the best balanced diet starch would probably contribute rather more than the 20% that it provides in the average U.K. diet today (nearly 40% comes from fat, and 13% from sugar). The value of 31.5% for energy contributed by cereals (Table 14.1) refers only to foods consumed in the home.

Most starch is consumed in cooked products, in the majority of which the starch granules are gelatinized, making them readily digestible by amylase enzymes pesent in the gut. For this to occur however, abundant water is required as starch can absorb more than 20 times its own mass during gelatinization. In some baked products, such as shortbread, much fat and little water are present; consequently few of the granules are gelatinized. Other factors, such as osmotic conditions, affect gelatinization, these are much affected by the amount of sugar in the recipe: in high sugar conditions water activity is

low and gelatinization takes place at an elevated temperature.

Energy is released from starch by digestion of starch polymers to produce glucose, which is absorbed into the bloodstream. Glucose yields *16 kilojoules* or *4 kilocalories* per gram (joules are now the preferred unit in which to express energy, 4.184 J = 1 cal).

Starches from 'amylo' mutant types of cereals (mainly maize), which have a higher than usual amylose content, are less readily digested. After cooking at high temperatures, the indigestibility may be enhanced, giving rise to a small proportion of *resistant starch*. Even in other cooked cereal products some resistant starch can arise; it behaves like fibre, passing unchanged through the gut. The method of cooking is important in determining the amount of resistant starch formed. In corn flakes produced by extrusion cooking the proportion of resistant starch is less than in conventionally produced flakes (Ch. 11).

Energy is a vital requirement of every healthy individual but energy that is not expended in physical or physiological activity is stored either as adipose tissue or glycogen. These provide a necessary store from which energy may be released when required. The superiority of starch as a dietary energy source does not derive from a particularly high calorific value; in fact that of *fat* is higher, at *37 kJ* (9 kcal) per gram, as is alcohol at *29 kJ* (7 kcal) per gram than that of starch.

Protein content and quality

Cereals including bread, contribute approximately 25% of the protein in the average adult diet in the U.K. Three thin slices of bread contribute as much protein as an egg.

In nutritional terms there are two factors of prime importance in relation to protein: the total protein content, and the contribution that essential amino acids make to the total.

There are eight essential amino acids (out of a total of 20 or so) — *methionine, tryptophan, threonine, valine, isoleucine, leucine, phenylalanine* and *lysine*. Two other amino acids are sometimes classified as essential but they can be made in the body — *tyrosine* from phenylalanine and *cysteine/cystine* from methionine. Their presence in foods reduces the requirements of the relevant 'parent' essential amino acids. In foods derived from plants in general, the *sulphur-containing amino acids* methionine and cysteine are most likely to be limiting, but this is not true of cereal grains. In cereals, *lysine* is the first to be limiting: rice, oat and rye are relatively rich among wholegrain cereals but they are deficient in relation to the FAO/WHO (1973) reference amino acid pattern, in which the lysine content is 5.5 g/16 g of N. *Maize protein* is also limiting in *tryptophan*, based on the reference value of 1.0 g/16 g of N, which the other cereals just reach. A comparison between wheat protein and protein from other food sources is shown in Fig. 14.1.

In the past, much was made of the superiority of animal-derived proteins, containing, as they do, the correct proportions of essential amino acids. However, protein types are rarely eaten alone and they tend to complement each other; for example, bread may be eaten with cheese, a good source of lysine. Even in vegetarian diets, many legumes and nuts supply essential amino acids. A good combination is rice and peas.

Both content (Fig. 14.2) and composition of protein are affected by the contributions of different

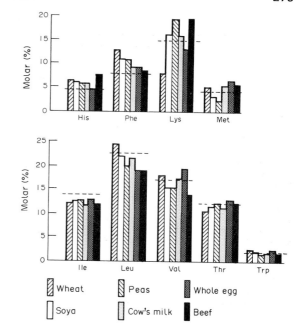

FIG 14.1 The essential amino acids in food proteins. The relative proportions of each essential amino acid are shown expressed as the percentage of the total essential amino acids. From Coultate, 1989, by courtesy of The Royal Society of Chemistry.

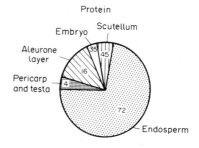

FIG. 14.2 Distribution of total protein in the wheat grain. The figures show the percentage of the total protein found in the various anatomical parts. (Based on micro-dissections by J. J. C. Hinton. From *The Research Association of British Flour Millers 1923–60*.)

anatomical parts of the grain; and in the endosperm, by the contributions of the different protein fractions (albumins, globulins, prolamins, glutelins).

The insoluble fractions are particularly deficient in lysine, as illustrated in the comparison of the solubility fractions of wheat endosperm (Table 14.2).

TABLE 14.2
Amino Acid Composition of Wheat Proteins: Glutenin, Gliadin, Albumin, Globulin (g amino acid/16 g nitrogen)

Amino acid	Glutenin*	Gliadin*	Albumin†	Globulin‡
Isoleucine	3.9	4.5	4.1	1.4
Leucine	6.9	7.2	10.7	9.2
Lysine	2.3	0.7	11.0	12.2
Methionine	1.7	1.5	0	0.4
Phenylalanine	4.8	5.6	5.0	3.2
Threonine	3.3	2.3	2.9	4.5
Tryptophan	2.1	0.7	n.d.	n.d.
Valine	4.5	4.4	8.1	2.2
Cystine	2.5	3.1	6.7	12.6
Tyrosine	3.6	2.6	3.4	2.3
Alanine	3.1	2.3	5.6	4.3
Arginine	4.2	2.7	7.5	14.5
Aspartic acid	3.9	3.0	7.9	6.3
Glutamic acid	34.1	40.0	17.7	5.9
Glycine	4.5	1.8	3.1	5.6
Histidine	2.4	2.3	4.3	2.2
Proline	11.0	14.7	8.4	3.3
Serine	5.9	5.1	4.7	9.1

* From Ewart (1967), recalculated. Original data are given as moles of anhydroamino acids per 10^5 of recovered anhydro amino acids.
† From Waldshmidt-Leitz and Hochstrasse (1961).
‡ From Fisher et al. (1968).
n.d. = not determined.

Among samples of the same cereal, there are considerable variations in protein content. Because there is a consistent relationship between protein content and the proportions of the fractions present, protein composition in whole grains also varies. Nevertheless, the differences among samples of the same cereal are generally less than differences among cereal species, and proteins characteristic of individual species can be described. It must however be remembered that values cited in comparisons are only at best averages, representing points within a range typical of the species.

Values for the proportions of classified amino acids in whole grains of cereals are given in Table 14.3.

Fibre

The laxative properties of fibre or 'roughage', as it was previously picturesquely described, have been well known for many hundreds of years (Hippocrates advocated it around 400 B.C.!) but within the last 20 or so, attention has been focussed on them by the assertion that the high fibre of African diets prevents many chronic non-infective diseases common in the West, where refined carbohydrates are more commonly consumed. Some of the more extreme claims for the beneficial effects of fibre, emanating from the surge of activity consequent upon these assertions have now been seriously challenged, as populations of Third-world states eating lower fibre diets but not showing high incidence of the relevant diseases have been discovered. Nevertheless, dietary fibre has been shown to have palliative effects on diseases, particularly those of the gut and *diabetes mellitus*.

Cholesterol

This is much publicized as an indicator of potential health problems, particularly heart disease, strokes and blocked arteries, but it is not all bad. Some cholesterol is necessary in the body as a precursor of hormones and bile acids. Cholesterol is transported in the blood in three principal forms: free cholesterol or bound to lipoprotein as either *High density lipoprotein* (HDL), or *Low density lipoprotein* (LDL). *HDL* even confers some *protection* against heart disease, so reduction below a threshold actually increases risk. About 80% of blood cholesterol is associated with *LDL*, however, and it is this form which is believed to deposit in the arteries. It is also this form that increases as a result of consumption of saturated fats. Considerable reduction in *blood cholesterol* levels have been reported in response to increased cereal fibre in the diet. Several mechanisms have been proposed to account for the hypocholesterolaemic effect of soluble fibres: viscous soluble fibre may exert an effect by physically entrapping cholesterol or bile acids in the digestive tract, thereby preventing their absorption and resulting in their increased excretion. Alternatively, β-glucans may be fermented by colonic bacteria to short chain fatty acids. Several of these compounds have been suggested as inhibitors of cholesterol synthesis. In relation to the first possibility, it has been found that different bile acids are bound more effectively by

TABLE 14.3
Amino Acid Content of Cereal Grains* (g amino acid/16 g nitrogen)

Amino acid	Wheat	Triticale	Rye	Barley	Oats	Rice
Essential						
Isoleucine	3.8	4.1	3.6	3.8	4.2	3.9
Leucine	6.7	6.7	6.0	6.9	7.2	8.0
Lysine	2.3	3.0	2.9	3.5	3.7	3.7
Methionine	1.7	1.9	1.2	1.6	1.8	2.4
Phenylalanine	4.8	4.8	4.5	5.1	4.9	5.2
Threonine	2.8	3.1	3.3	3.5	3.3	4.1
Tryptophan	1.5	1.6	1.2	1.4	1.6	1.4
Valine	4.4	5.0	4.9	5.4	5.6	5.7
Non-essential						
Cysteine/cystine	2.6	2.8	2.3	2.5	3.3	1.1
Tyrosine	2.7	2.3	1.9	2.5	3.0	3.3
Alanine	3.3	3.6	3.7	4.1	4.6	6.0
Arginine	4.0	4.9	4.2	4.4	6.6	7.7
Aspartic acid	4.7	5.9	6.5	6.1	7.8	10.4
Glutamic acid	33.1	30.9	27.5	24.5	21.0	20.4
Glycine	3.7	3.9	3.6	4.2	4.8	5.0
Histidine	2.2	2.5	2.1	2.1	2.2	2.3
Proline	11.1	10.7	10.4	10.9	4.7	4.8
Serine	5.0	4.6	4.3	4.2	4.8	5.2
Protein†	16.3	12.1	17.8	14.5	17.9	11.1

Amino acid	Maize	Sorghum	Millets		
			pearl	foxtail	proso
Essential					
Isoleucine	4.0	3.8	4.3	6.1	4.1
Leucine	12.5	13.6	13.1	10.5	12.2
Lysine	3.0	2.0	1.7	0.7	1.5
Methionine	1.8	1.5	2.4	2.4	2.2
Phenylalanine	5.1	4.9	5.6	4.2	5.5
Threonine	3.6	3.1	3.1	2.7	3.0
Tryptophan	0.8	1.0	1.4	2.0	0.8
Valine	5.2	5.0	5.4	4.5	5.4
Non-essential					
Cysteine/cystine	2.5	1.1	1.8	1.4	1.0
Tyrosine	4.4	1.5	3.7	1.6	4.0
Alanine	7.7	9.5	11.3		
Arginine	4.7	2.6	3.3	2.3	3.2
Aspartic acid	6.4	6.3	6.4		
Glutamic acid	18.8	21.7	22.2		
Glycine	3.9	3.1	2.3		
Histidine	2.8	2.1	2.3	1.2	2.1
Proline	8.8	7.9	6.9		
Serine	4.9	4.3	6.9		
Protein†	10.6	10.5	13.5	12.4	12.5

* Data for wheat, barley, oats, rye, triticale and pearl millet from Tkachuk and Irvine (1969); data for rice (except tryptophan) from Juliano *et al.* (1964); data for maize (except tryptophan) from Busson *et al.* (1966); data for sorghum from Deyoe and Shellenberger (1965); data for foxtail and proso millets from Casey and Lorenz (1977); tryptophan data for rice and maize calculated from Hughes (1967). All data are for whole grains, except oats and rice — hulled grains — and rye — dark rye flour, ash 1.1% d.b.

† $N \times 5.7$, d.b. Original data for maize and sorghum given as $N \times 6.25$, viz. 11.6% and 11.5% respectively.

different types of fibre; pentosans of different types of rice even vary in their effectiveness and their 'preferred' bile acids (Normand *et al.*, 1981).

It has been found that eating bran is an effective cure for constipation and diverticular disease. It is more doubtful whether dietary fibre is as effective in preventing problems other than constipation. An important factor, though not the only one, contributing to the beneficial effects of bran is its ability to hold water, thus increasing stool weight and colonic motility. The relative water-holding capacities of cereal brans and other sources of fibre, given by Ory (1991) are shown in Table 14.4.

TABLE 14.4
Water-holding Capacity of Cereal Brans and Other Fibre-Containing Foods

Fibre source	Water-holding capacity. g/100g d.m.
Sugar beet pulp	1449
Apple pomace	235–509
Apple, whole fruit	17–46
All bran	436
Wheat bran	109–290
Rice bran	131
Oat bran	66
Maize bran	34
Cauliflower	28
Lettuce	36
Carrot	33–67
Orange, whole fruit	20–56
Orange pulp	176
Onion, whole vegetable	14
Banana, whole fruit	56
Potato, minus skin	22

Sources: Gormley, 1977; Hetler and Hackler, 1977 (their values are reported as 'corrected' for fresh weight basis); and Chen *et al.*, 1984. Reprinted with permission from Ory, R. L., 1991. Copyright, 1991, American Chemical Society.

Fats

Apart from essential fatty acids, the liver is able to make all the fat that the body requires from carbohydrates and protein, provided these are eaten in sufficient quantities. About 10 g of essential fatty acids are needed every day by the human body, but in the U.K. the average consumption is 10-fold this.

Nearly all the fat in the diet is composed of triacylglycerols (triglycerides). Saturated fatty acids, found mainly in animal fats and hardened fats, include lauric, myristic (14:0) and palmitic (16:0) acids — the three which have been implicated in raising the levels of cholesterol in blood. Palmitic acid (16:0) is the most commonly occurring fatty acid, comprising 35% of animal fats and palm oil, and 17% of other plant oils and fish oils. The most commonly occurring mono-unsaturated fatty acid is oleic acid (18:1), it contributes 30–65% of most fats and oils. All cereal grain lipids are rich in unsaturated fatty acids (see Table 14.5).

Palmitic (16:0) is a major saturated, and linoleic (18:2) a major unsaturated fatty acid in most cereals, exceptions being brown rice and oats which are rich in oleic acid (18:1). Millets are richer in stearic acid (18:0) than are other cereals. *No plant oils contain cholesterol.*

Two advantages of rice bran oil are the low content of *linolenic acid* and its high content of *tocopherols*, both important from the point of view of oxidative stability. Its high content of *linoleic acid* makes it a good source of essential fatty acid.

Oat groats contain 7% oil, pearl millet 5.4%, maize 4.4%, sorghum 3.4%, brown rice 2.3%, barley 2.1% and wheat 1.9%.

The hard, high-melting fraction of rice bran wax has lustre-producing qualities similar to those of carnauba wax. It has been approved by the U.S. Food and Drug Administration as a constituent of food articles up to 50 mg/kg and for use as a plasticizer for chewing gum at 2%. (Juliano, 1985b)

Maize germ oil is rich in essential fatty acids (about half of its fatty acid content is linoleic). It is used as a salad oil and for cooking.

Minerals

At least 15 minerals are required by humans. Of these, deficiencies are unlikely to occur in *phosphorus, sodium, chlorine or potassium*, even though daily requirements are relatively high. *Anaemia*, due to *iron* deficiency, is one of the most common nutritional disorders, particularly in pre-menopause women. Iron from exhausted red blood cells is re-used in new cells, so that almost the only requirement is to replace blood that has been lost. Wholegrain cereals contain sufficient

TABLE 14.5
The Fatty Acid Composition of Cereal Lipids *

Material	Saturated			Unsaturated		
	Myristic $C_{14.0}$ (%)	Palmitic $C_{16.0}$ (%)	Stearic $C_{18.0}$ (%)	Oleic $C_{18.1}$ (%)	Linoleic $C_{18.2}$ (%)	Linolenic $C_{18.3}$ (%)
Barley						
6-row	3.3	7.7	12.6	19.9	33.1	23.1
2-row	1.0	11.5	3.1	28.0	52.3	4.1
Maize	—	14.0	2.0	33.4	49.8	1.5
Millet						
pearl	—	17.8	4.7	23.9	50.1	3.0
foxtail	0.6	11.0	14.7	21.8	38.2	6.4
proso	—	11.5	—	25.8	50.6	7.8
Oats	0.5	15.5	2.0	43.5	35.5	2.0
Rice	—		17.6	47.6	34.0	0.8
Rye	—	21.0	—	18.0	61.0	—
Sorghum	0.4	13.2	2.0	30.5	49.7	2.0
Triticale	0.7	18.7	0.9	11.5	61.2	6.2
Wheat						
grain	0.1	24.5	1.0	11.5	56.3	3.7
germ	—	18.5	0.4	17.3	57.0	5.2
endosperm	—	18.0	1.2	19.4	56.2	3.1

* Source: Kent (1983).

iron to supply a large proportion of an adult's daily requirement, but there is some doubt as to whether it can be absorbed from cereal and legume sources because of the presence of phytic acid. About 900 mg of calcium are present in the average U.K. diet, and of this, 25% is supplied by cereals. Growing children and pregnant and lactating mothers have a higher requirement of about 1200 mg per day. The aged have an enhanced requirement for calcium as it may be depleted by insufficient *vitamin D*. Adequate calcium consumed during growth affords some protection against *osteoporosis* in later years. A further protective function served by adequate calcium in the diet concerns the radioactive isotope strontium 90 (Sr^{90}), produced as part of the fall-out of nuclear explosions, which can arise from weapons or accidents in nuclear power stations. Sr^{90} can replace calcium in bones, causing irritation and disease, but in the presence of high calcium levels this is less likely. Flours, other than wholemeals, malt flours and self-raising flours (which are deemed to contain sufficient calcium), are required to be supplemented with chalk (calcium carbonate) in the U.K. but it is doubtful if this is necessary. If the exception of

wholemeal might seem illogical (since wholemeal, of all types of flour contains the largest amount of phytic acid, and would seem to require the largest addition of chalk), it must be remembered that consumers of this particular product are concerned to an exceptional extent with the concept of absence of all additions.

Bran and wheat germ are good sources of *magnesium* but, as with other minerals, absorption can be impaired by the phytate also present.

In addition to the above, the following elements are required by the body, but in much smaller 'trace' amounts: *iodine, copper, zinc, manganese, molybdenum, selenium* and *chromium* and even smaller quantities of *silicon, tin, nickel, arsenic* and *fluorine* may be needed.

Wholegrain cereals can contribute to the supply of *zinc*, although its absorption might be impaired by phytic acid. The *selenium* content of grain depends upon the selenium status of the soil on which the crop was grown. In North America many selenium-rich soils support cereals, and wheat imported from that continent has relatively large amounts present, enabling half the daily requirement to be met from cereals. Soils in the U.K. have less selenium and hence the grains produced on them are poorer in the element. Symptoms of selenium deficiency have been reported in countries with notably deficient soils, including New Zealand and some areas of China.

The mineral composition of cereal grains is shown in Table 3.6.

Vitamins

Vitamins comprise a diverse group of organic compounds. They are necessary for growth and metabolism in the human body, which is incapable of making them in sufficient quantities to meet its needs, hence the diet must supply them to maintain good health. Most vitamins are known today by their chemical descriptions, rather than the earlier identification as vitamin A, B, C etc. A table of equivalence relates the two methods of nomenclature (Table 14.6). Those in bold type occur in cereals in significant quantities (in relation to daily requirements).

TABLE 14.6
Vitamins and their Occurrence in Cereals

Vitamin	Chemical Name	Concentration in cereals
A	Retinol and Carotene	
B₁	**Thiamin**	Embryo (scutellum)
B₂	**Riboflavin**	Most parts
B₆	**Pyridoxin**	Aleurone
B₁₂		
	Nicotinic acid (niacin)	Aleurone (not maize)
	Folic acid	
	Biotin	
	Pantothenic acid	Aleurone, endosperm
	Choline	
	Carnitine	
C	Ascorbic acid	
D	Cholicalciferol and ergocalciferol	
E	**Tocopherol and tocotrienol**	Embryo
K	Phylloquinone	

Vitamins are sometimes classified according to solubility; thus A, D, E and K are fat soluble, and B and C are water soluble. Fat soluble vitamins are the more stable to cooking and processing.

It is clear from Table 14.6 that it is the B vitamins (more specifically thiamin, riboflavin, pyridoxine nicotinic acid and pantothenic acid)

and vitamin E, that are most important in cereal grains. The average contents of B-vitamins are shown in Table 14.7.

The table also includes values for inositol and p-amino-benzoic acid. Although essential for some micro-organisms, these substances are no longer considered essential for humans. Their status as vitamins is thus dubious. Choline and inositol are by far the most abundant but cereals are not an important source as many foods contain them and deficiencies are rare (Bingham, 1987).

Distribution of vitamins in cereals

Variation in content from one cereal to another is remarkably small except for niacin (nicotinic acid), the concentration of which is relatively much higher in barley, wheat, sorghum and rice, than in oats, rye, maize and the millets.

Details of the distribution in grains were worked out by Hinton and his associates, who assayed the dissected morphological parts of wheat, maize and rice. Their results for wheat are shown in Table 14.8.

The distribution of these vitamins in the wheat grain is also shown diagrammatically in Fig. 14.3.

The proportions of total thiamin and niacin are shown for rice and maize in Table 14.9.

The distributions of thiamin in rice and wheat are quite similar: it is concentrated in the scutellum, though not to the same degree as in rye and maize. The embryonic axis of rice, which has a relatively high concentration of thiamin, contains over one tenth of the total in the grain, a larger proportion than that found in the other cereals (see Table 14.10).

Cereals, except maize, contain *tryptophan*, which can be converted to *niacin* in the liver in the presence of sufficient thiamin, riboflavin and pyridoxin. The distributions of niacin in wheat, rice and maize are similar, it is concentrated in the aleurone layer. About 80% of the niacin in the bran of cereals occurs as niacytin, a complex of polysaccharide and polypeptide moieties

TABLE 14.7
B Vitamin Content of the Cereal Grains (µg/g)

Cereal	Thiamin	Riboflavin	Niacin	Pantothenic acid	Biotin
Wheat					
hard	4.3	1.3	54	} 10	
soft	3.4	1.1	45		0.1
Barley	4.4	1.5	72	5.7	0.13
Oats (whole)	5.8	1.3	11	10	0.17
Rye	4.4	2.0	12	7.2	0.05
Triticale	9.2	3.1	16	7.5	0.06
Rice (brown)	3.3	0.7	46	9	0.1
Maize	4.0	1.1	19	5.3	0.1
Sorghum	3.5	1.4	41	11	0.19
Millet					
pearl	3.6	1.7	26	11.4	—
foxtail	5.9	0.8	7	—	—
proso	2.0	1.8	23	—	—
finger	3.6	0.8	13	—	—

Cereal	Pyridoxin	Folic acid	Choline	Inostiol	*p*-Amino benzoic acid
Wheat					
hard	} 4.5	0.5	1100	2800	2.4
soft					
Barley	4.4	0.4	1000	2500	0.5
Oats (whole)	2.1	0.5	940	—	—
Rye	3.2	0.6	450	—	—
Triticale	4.7	0.7	—	—	—
Rice (brown)	4.0	0.5	900	—	—
Maize	5.3	0.4	445	—	—
Sorghum	4.8	0.2	600	—	—
Millet					
pearl	—	—	—	—	—
foxtail	—	—	—	—	—
proso	—	—	—	—	—
finger	—	—	—	—	—

* Sources of data are as quoted in Kent (1983). A dash indicates that reliable data have not been found.
The above data are similar to those given by Holland *et al.* (1991) who provide comprehensive tables of vitamin contents of cereal products also.

TABLE 14.8
Distribution of B Vitamins in the Wheat Grain. Concentration in µg/g and % of total in parts

Part of grain	Thiamin† Concn. µg/g	%	Riboflavin‡ Concn. µg/g	%	Niacin‡ Concn. µg/g	%	Pyridoxin‡ Concn. µg/g	%	Pantothenic acid‡ Concn. µg/g	%
Pericarp, testa, hyaline	0.6	1	1.0	5	25.7	4	6.0	12	7.8	9
Aleurone layer	16.5	32	10	37	741	82	36	61	45.1	41
Endosperm	0.13	3	0.7	32	8.5	12	0.3	6	3.9	43
Embryonic axis	8.4	2	13.8	12	38.5	1	21.1	9	17.1	3
Scutellum	156	62	12.7	14	38.2	1	23.2	12	14.1	4
Whole grain	3.75		1.8		59.3		4.3		7.8	

* Sources of data: Clegg (1958); Clegg and Hinton (1958); Heathcote *et al.* (1952); Hinton (1947); Hinton *et al.* (1953)
† Wheat variety Vilmorin 27.
‡ Wheat variety Thatcher.

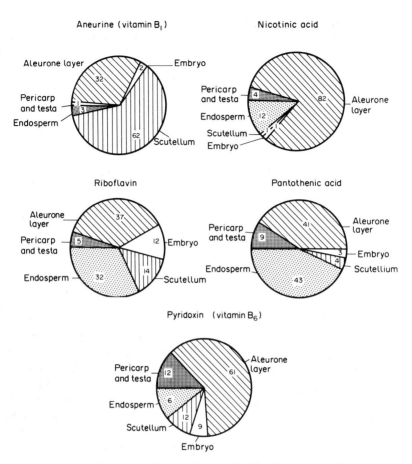

FIG. 14.3 Distribution of B vitamins in the wheat grain. The figures show the percentages of the total vitamins in the grain found in the various anatomical parts. (Based on micro-dissections by J. J. C. Hinton. From *The Research Association of British Flour Millers 1923–60.*)

TABLE 14.9
Distribution of Thiamin in Rice and of Niacin in Rice and Maize*

Part of the grain	Percentage of total thiamin in rice	Percentage of total niacin		
		in rice (var: Indian)	in maize Flint	Sweet
Pericarp, testa, hyaline layer	34	5	2	3
·Aleurone layer	—	80.5	63	59
Endosperm	8	12.3	20	26
Embryonic axis	11	0.6	2	2
Scutellum	47	1.6	13	10

* Sources of data: Heathcote *et al.* (1952); Hinton and Shaw (1953).

biologically unavailable to man unless treated with alkali (Carter and Carpenter, 1982).

Riboflavin and pantothenic acid are more uniformly distributed. Pyridoxin is concentrated in aleurone and other nonstarchy endosperm parts of the grain.

The uneven distribution of the B vitamins throughout the grain is responsible for considerable differences in vitamin content between the whole grains and the milled or processed products.

Vitamin E is essential for maintaining the orderly structure of *cell membranes*; it also behaves as an *antioxidant*, particularly of polyunsaturated fatty acids. These are abundant in *nervous tissue*, including the *brain*. Deficiency symptoms, which

TABLE 14.10
*Thiamin in Embryonic Axis and Scutellum of Cereal Grains**

Cereal	Wt of tissue (g/100g grain)		Thiamin concentration (μg/g)		Proportion of total thiamin in grain (%)	
	Embryonic axis	Scutellum	Embryonic axis	Scutellum	Embryonic axis	Scutellum
Wheat (average)	1.2	1.54	12	177	3	59
Barley (dehusked)	1.85	1.53	15	105	8	49
Oats (groats)	1.6	2.13	14.4	66	4.5	28
Rye	1.8	1.73	6.9	114	5	82
Rice (brown)	1.0	1.25	69	189	11	47
Maize	1.15	7.25	26.1	42	8	85

* Sources of data: Hinton (1944, 1948).

are rare as stores of the vitamin in the body are large, include failure of nervous functions. Other functions of vitamin E are claimed by some, but these are not well substantiated by experiment (Bingham, 1987).

In the U.K. cereals contribute about 30% of vitamin E requirement. Wheat contains α-, β-, γ- and δ-tocopherols, the total tocopherol content being 2.0–3.4 mg/100g. α-, β- and γ-tocotrienols are also present. The biological potency of β-, γ- and δ-tocopherols are 30%, 7.5% and 40% respectively, of that of the alpha-tocopherol. The total tocopherol contents of germ, bran, and 80% extraction flour of wheat are about 30, 6 and 1.6 mg/100g, respectively (Moran, 1959). α-Tocopherol predominates in germ, γ-tocopherol in bran and flour, giving α equivalents of 65%, 20% and 35% for the total tocopherols of germ, bran and 80% extraction flour, respectively.

Quoted figures for the total tocopherol content of other cereal grains are (in mg/100g): barley 0.75–0.9, oats 0.6–1.3, rye 1.8, rice 0.2–0.6, maize 4.4–5.1 (mostly as γ-tocopherol), millet 1.75 (mostly as γ-tocopherol) (Science Editor, 1970; Slover, 1971).

The oil of cereal grains is rich in tocopherols: quoted values (in mg/g) are: wheat germ oil 2.6, barley oil 2.4, oat oil 0.6, rye oil 2.5, maize oil 0.8–0.9 (Green *et al.*, 1955; Slover, 1971). Wheat tocopherols have particularly high vitamin E activity (Morrison, 1978).

Effects of processing

Some animals, including poultry, may be fed grains of various types in a totally unmodified form. Other stock may consume grains that have received only minimal modification such as crushing, and some elements of the human diet, such as muesli may also contain these. Nevertheless most cereals are consumed only after processing and this affects the nutritive value of the products. Changes in nutritional properties of cereals result from several types of processing, such as: *refinement, cooking* and *supplementation*.

Refinement

This incudes processes such as milling, that separate anatomical parts of the grain to produce a palatable foodstuff. The most palatable (lowest fibre), and most stable (lowest fat) parts of the grains are not necessarily the most nutritious, and if only these are consumed, much of the potential benefit can be lost. For example, 80% of the vitamin B_1 is removed when rice is milled and polished. This results from the fact that many of the nutrients reside in the embryo and outer parts of the grains (mainly in the aleurone tissue). The degree of change depends upon the degree of separation that occurs. The effects of varying *extraction rate* in wheat milling are illustrated diagramatically in Fig. 14.4, and the effects of varying extraction rate on wheat and rye milling products are shown in Table 14.11.

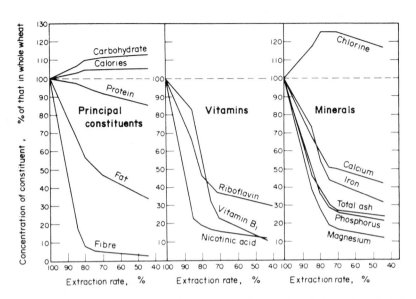

FIG. 14.4 Nutrient composition of flours of various extraction rates in relation to that of whole wheat. From Kent (1983).

TABLE 14.11
Composition of Flour and Milling by-products at Various Extraction Rates

Material	Yield (%)	Protein (%)	Oil (%)	Ash (%)	Crude fibre (%)	Thiamin (µg/g)[*]	Niacin (µg/g)[*]
Wheat†							
Flour	85	12.5	1.5	0.92	0.33	3.42	
85% extraction	80.5	12	1.4	0.72	0.20	2.67	19
80% extraction	70	11.4	1.2	0.44	0.10	0.7	10
70% extraction							
Fine wheatfeed							
85% extraction	10	12.6	4.7	5.1	10.6	6.0	
80% extraction	12.5	14.3	4.7	4.7	8.4	10.4	191
75% extraction	20	15.4	4.7	3.5	5.2	14.0	113
Bran							
85% extraction	5	11.1	3.7	6.1	13.5	4.6	
80% extraction	7	12.4	3.9	5.9	11.1	5.0	302
70% extraction	10	13	3.5	5.1	8.9	6.0	232
Rye‡							
Flour							
60% extraction	60	5.7	1.0	0.5	0.2		
75% extraction	75	6.9	1.3	0.7	0.5		
85% extraction	85	7.5	1.6	1.0	0.8		
100% extraction	100	8.2	2.0	1.7	1.6		
Brans§							
fine		14.0	3.2	4.2	5.0		
coarse		16.6	5.2	3.8	9.4		
Germ§		35.5	10.3	4.8	3.4		

[*] Naturally occuring.
Sources: † Jones (1958); ‡ Neumann *et al.* (1913); § McCance *et al.* (1945).

TABLE 14.12
Composition of Milling Products of Various Cereals (Percent, dry basis)

Cereal and fraction	Protein	Fat	Ash	Crude fibre	Carbohydrate
Barley					
Pearl barley[1]	9.5	1.1	1.3	0.9	85.9
Barley flour[1]	11.3	1.9	1.3	0.8	85.4
Barley husk[2]	1.6	0.3	6.2	37.9	53.9
Barley bran[3]	16.6	4.4	5.6	9.6	64.3
Barley dust[3]	13.6	2.5	3.7	5.3	74.9
Oats[2]					
Oatmeal	12.9	7.5	2.1	1.2	75.0
Rolled oats	13.3	7.6	2.0	0.9	74.7
Oat flour	14.2	7.9	2.0	1.1	73.4
Oat husk	1.4	0.4	4.5	37.8	0.9
Oat dust	10.6	5.0	6.3	22.8	10.6
Meal seeds	8.6	3.8	3.1	18.7	28.8
Oat feed meal	3.5	1.5	3.8	30.6	5.8
Rice					
Brown rice[4]	8.5	2.21	1.4	1.0	86.9
White rice[4]	7.6	0.4	0.6	0.3	91.0
Parboiled rice[4]	8.2	0.3	0.8	0.2	90.1
Hulls[5]	1.1	0.4	27.3	34.1	34.1
Bran[6]	14.4	21.1	8.9	10.0	45.6
Bran, extracted[6]	17.8	0.6	11.1	12.2	57.8
Polishings[7]	12.1	9.9	5.5	2.2	70.3
Maize					
Maize grain[8]	11.2	4.8	1.7	1.9	80.4
Dry milling products					
Grits[9]	10.5	0.9	—	—	—
Meal[8]	10.1	5.7	1.2	1.4	83.7
Flour[8]	8.1	1.5	0.7	1.0	88.7
Germ meal[10]	14.6	14.0	4.0	4.6	62.8
Hominy feed[8]	10.6	0.8	0.3	0.4	87.9
Wet milling products					
Corn flour[11]	0.8	0.07	0.1	—	99.0
High protein corn gluten meal[12]	68.8	6.2	2.0	1.3	21.7
Corn gluten feed[12]	24.8	4.2	8.0	8.9	54.1
Corn germ meal[12]	25.1	5.1	4.1	10.6	55.1
Sorghum					
Dry milling products[13]					
Whole sorghum	9.6	3.4	1.5	2.2	
Pearled sorghum	9.5	3.0	1.2	1.3	
Flour, crude	9.5	2.5	1.0	1.2	
Flour, refined	9.5	1.0	0.8	1.0	
Brewers' grits	9.5	0.7	0.4	0.8	
Bran	8.9	5.5	2.4	8.6	
Germ	15.1	20.0	8.2	2.6	
Hominy feed	11.2	6.5	2.7	3.8	
Wet milling products[14]					
Germ	11.8	38.8			18.6
Fibre	17.6	2.4			30.6
Tailings	39.2	—			25.3
Gluten	46.7	5.1			42.8
Squeegee	14.0	0.6			81.6
Starch	0.4	—			67.3
Solubles	43.7	—			—

TABLE 14.12
Continued

Cereal and fraction	Protein	Fat	Ash	Crude fibre	Carbohydrate
Millet[14]					
Wet milling products					Starch.
Germ	10.4	45.6			10.4
Fibre	11.8	6.0			13.5
Tailings	34.1	1.9			34.2
Gluten	37.8	9.0			44.0
Squeegee	17.7	0.8			75.5
Starch	0.7	0.1			57.5
Solubles	46.1	—			—

Sources: [1]Chatfield and Adams (1940); [2]Original 3rd edn; [3]Watson (1953); [4]U.S.D.A. (1963); [5]Houston (1972); [6]Australian Technical Millers (1980); [7]Fraps (1916); [8]Woods (1907); [9]Stiver Jr (1955); [10]Woodman (1957); [11]Boundy (1957); [12]Reiners *et al.* (1973); [13]Hahn (1969); [14]Freeman and Bocan (1973).

TABLE 14.13
Average Nutrients in 100g ($\frac{1}{6}$ pint) of Beer

Nutrient	Draft bitter	Draft mild	Keg bitter	Bottled lager	Bottled pale ale	Bottled stout
Energy, kJ	132	104	129	120	133	156
Protein, g.	0.3	0.2	0.3	0.2	0.3	0.3
Alcohol, g	3.1	2.6	3.0	3.2	3.3	2.9
Sugars, g	2.3	1.6	2.3	1.5	2.0	4.2
Sodium, mg	12	11	8	4	10	23
Potassium, mg	38	33	35	34	49	45
Calcium, mg	11	10	8	4	9	8
Magnesium, mg	9	8	7	6	10	8
Iron, mg	0.01	0.02	0.01	0	0.02	0.05
Copper, mg	0.08	0.05	0.01	0	0.04	0.08
Zinc, mg	—	—	0.02	—	—	—
Riboflavin, mg	0.04	0.03	0.03	0.02	0.02	0.04
Niacin, mg	0.60	0.40	0.45	0.54	0.52	0.43
Pyridoxin, mg	0.02	0.02	0.02	0.02	0.01	0.01
Vitamin B_{12} µg	0.17	0.15	0.15	0.14	0.14	0.11
Folic acid, µg	8.8	4.5	4.6	4.3	4.1	4.4
Pantothenic acid, mg	0.1	0.1	0.1	0.1	0.1	0.1
Biotin µg	0.5	0.5	0.5	0.5	0.5	0.5

Zero or trace amounts in all beers, of fat, starch, dietary fibre, vitamins A,C,D and thiamin.
Source: Bingham (1987).

Compositions of fractions obtained by milling other cereals are shown in Table 14.12.

In the filtrations and distillations involved in production of beers and spirits, respectively, many of the nutrients in the original grains are removed from the main product. Some nutrients are derived from other ingredients, notably vitamin B_{12} from yeast.

The nutrients in several types of beer are shown in Table 14.13.

Cooking

Changes resulting from cooking are complex as in only a few cases, such as boiled rice and pasta, are cereal products cooked substantially on their own. More frequently they are included in a recipe, so that differences between the raw cereal ingredient and the final product reflect not only changes due to cooking but also to dilution and interaction with other ingredients.

MAILLARD

In the traditional method of tortilla production, cooking and refinement are combined as the heating of whole maize grains in alkaline water loosens the pericarp and embryo, allowing the endosperm to be concentrated. Many nutrients, including vitamins and fibre, are lost from the principal product as a result. In spite of this, niacin availability in tortillas is higher than in uncooked maize. Protein availability is generally reduced due to a number of cross-linking reactions (Rooney and Serna-Saldivar, 1987). Fortification may more than compensate for the deficiencies.

In the case of parboiling of rice the nutritional properties of the refined product are improved as nutrients such as vitamins and minerals, in the parts of the grain that are subsequently removed by milling, migrate with water into the endosperm (see Table 10.3). The loss of vitamins on washing of milled rice is reduced but so is protein availability (Bhattacharya, 1985).

Cooking losses vary according to the amount of water used, they are greater when excess water is present and least when the double boiler is used (Table 14.14).

TABLE 14.14
B Vitamin losses from Milled Rice Cooked by Different Methods*

Nutrient	% loss on cooking		
	Excess water	Absorbable water	Double boiler
Thiamin	47	19	4
Riboflavin	43	14	7
Niacin	45	22	3

* Source: Juliano, 1985a.

As recipes involving cereal products abound, a comprehensive illustration of changes during cooking is not possible here. A selection of examples of raw and cooked products is given in Table 14.15

A more comprehensive analysis of the more common breads in the U.K. is given in Table 14.16.

Specific interactions

Some of the more important interactions among ingredients during cooking are described below.

Maillard reactions are reactions between sugars and amino groups which give rise to browning — important in the production of commercial caramel and in providing a colour to the crust of bread and other baked products. The amino acids which engage in Maillard reactions most readily are those with a free amino group in their side-chain, particularly lysine, followed by arginine, tryptophan and histidine. While the effects on product palatability are generally valued, there is a nutritional price to pay as the cross-linked sugar prevents access by proteolytic enzymes, obstructing digestion of the essential amino acids involved. Fortunately, the proportion of amino acids involved in crust browning is relatively small (Coultate, 1989).

Probably the most significant enzymic changes that occur during processing are those involved in conversion of starch into sugars and the subsequent fermentation mediated by micro-organisms, converting sugars to alcohol or lactic acid. Another important example is the reduction in phytin content during fermentation and proving of bread doughs, and the accompanying increase in the more readily absorbed inorganic phosphate, due largely to catalysis by phytases produced by yeasts.

Supplementation

This differs from the other types of processing in that change in nutritional properties is its primary purpose, rather than a consequence of an improvement in palatability. In general it changes the eating qualities as little as possible.

Because of the staple nature of cereal products, their contributions to the diets of the populations of most of the world's nations are frequently perceived by Governments as an important means of ensuring adequate nutritive standards, not only through their natural composition, but also through the addition of nutrients from other sources. Various terms are applied to such additions, including *restoration*, *fortification* and *enrichment*. Restoration implies the replacement of nutrients lost during processing, such as milling, to the level found in the original grain, while fortification and enrichment suggest addition of nutrients not originally present, or enhancement of originally present nutrients.

TABLE 14.15
Composition of Cereal Grains and some Products

Product	Water (%)	Protein (%)	Fat (%)	Carbohydrate (%)	Energy value (kJ per 100g)
Wheat (and wholemeal)	14	12.7	2.2	63.9	1318
Flour					
white breadmaking	14	11.5	1.4	75.3	1451
plain	14	9.4	1.3	77.7	1450
brown	14	12.6	1.8	68.5	1377
Bran	8.3	14.1	5.5	26.8	872
Germ	11.7	26.7	9.2	(44.7)	1276
Bread					
white sliced	40.4	7.6	1.3	46.8	926
toasted	27.3	9.3	1.6	57.1	1129
brown	39.5	8.5	2.0	44.3	927
toasted	24.4	10.4	2.1	56.5	1158
French stick	29.2	9.6	2.7	55.4	1149
Hamburger buns					
crusty	26.4	10.9	2.3	57.6	1192
soft	32.7	9.2	4.2	51.6	1137
croissants	31.1	8.3	20.3	38.3	1505
Cream crackers	4.3	9.5	16.3	68.3	1857
Semi-sweet biscuits	2.5	6.7	16.6	74.8	1925
Cake (cake-mix)	31.5	5.3	3.3	52.4	1951
Spongecake	15.2	6.4	26.3	52.4	1920
Madeira cake	20.2	5.4	16.9	58.4	1652
Pastry (short)	20.0	5.7	27.9	46.8	1874
Durum Wheat					
Macaroni					
raw	9.7	12.0	1.8	75.8	1483
boiled	78.1	3.0	0.5	18.5	365
Spaghetti					
raw	9.8	12.0	1.8	74.1	1465
boiled	73.8	3.6	0.7	22.2	442
Spaghetti, wholemeal, raw	10.5	13.4	2.5	66.2	1379
boiled	69.1	4.7	0.9	23.2	485
Rye					
grain and wholemeal	15	8.2	2.0	75.9	1428
rye bread	37.4	8.3	1.7	45.8	923
crispbread	6.4	9.4	2.1	70.6	1367
Rice					
brown, raw	13.9	6.7	2.8	81.3	1518
boiled	66.0	2.6	1.1	32.1	597
white, easy cook					
raw	11.4	7.3	3.6	85.8	1630
boiled	68.0	2.6	1.3	30.9	587
Oat					
oatmeal	8.2	11.2	9.2	66.0	1567
porridge (made with water)	87.4	1.5	1.1	9.0	1381

Data from Holland *et al.* (1991). Reproduced with permission of the Royal Society of Chemistry.

Supplementation policies usually arise in response to disasters which bring about shortages of essential foods through conflict or poverty. Thus, World War II was responsible for the formulation of policy in the U.K., and the Great Depression of the 1930s precipitated the institution of the U.S. fortification programme. Supplementation policy may also apply specifically to cereal products exported as part of an Aid Programme to populations in which a particular deficiency prevails.

In the case of the U.S. the *Food and Drugs Administration (FDA)* has defined in the *U.S. Code of Federal Regulations* the purposes for which addition of nutrients is appropriate, as:

TABLE 14.16
Nutrient Composition of Bread in the UK (per 100g)

Nutrient	White average	White sliced	Brown	Germ bread	Wholemeal
Water, g	37.3	40.4	39.5	40.3	38.3
Protein, g	8.4	7.6	8.5	9.5	9.2
Fat, g	1.9	1.3	2.0	2.0	2.5
Fatty acids					
Saturated, g	0.4	0.3	0.4	0.3	0.5
Monounsaturated, g	0.4	0.2	0.3	0.3	0.5
Polyunsaturated, g	0.5	0.4	0.6	0.7	0.7
Sugars, g	2.6	3.0	3.0	1.8	1.8
Carbohydrate, g	49.3	46.8	44.3	41.5	41.6
Starch, g	46.7	43.8	41.3	39.7	39.8
Dietary fibre, g					
Southgate	3.8	3.7	5.9	5.1	7.4
Englyst	1.5	1.5	(3.5)	3.3	(5.8)
Energy, kJ	1002	926	927	899	914
Thiamin, mg	0.21	0.2	0.27	0.8	0.34
Riboflavin, mg	0.06	0.05	0.09	0.09	0.09
Nicotinic acid, mg	1.7	1.5	2.5	4.2	4.1
Potential Nicotinic acid, mg	1.7	1.6	1.7	1.9	1.8
Pyridoxin, mg	0.07	0.07	0.13	0.11	0.12
Folic acid, µg	29	17	40	39	39
Pantothenic acid, mg	0.3	(0.3)	0.3	(0.3)	0.6
Biotin, µg	1	(1)	3	(2)	6
Vitamin E, mg	Tr	Tr	Tr	N	0.2
Sodium, mg	520	530	540	600	550
Potassium, mg	110	99	170	200	230
Calcium, mg	110	100	100	120	54
Magnesium, mg	24	20	53	56	76
Phosphorus, mg	91	79	150	190	200
Iron, mg	1.6	1.4	2.2	3.7	2.7

Data from Holland *et al.* (1991).

Tr: trace. ( ): estimated. N: The nutrient is present in significant quantities but there is no reliable information available.

- correction of a recognized dietary deficiency;
- restoration of nutrients lost during processing;
- balancing the nutrient content in proportion to the caloric content;
- avoidance of nutritional inferiority of new products replacing traditional foods;
- compliance with other programmes and regulations.

For any supplementation initiative to succeed it is necessary that the added nutrient is physiologically available; it does not create an inbalance of essential nutrients; it does not adversely affect the acceptability of the product; and there is reasonable assurance that intake will not become excessive.

Levels of addition

In the U.K. the composition of flour is controlled by the Bread and Flour Regulations 1984 (SI 1984, No. 1304), as amended by the Potassium Bromate (Prohibition as a Flour Improver) Regulations 1990 (SI 1990, No. 339). Flour derived from wheat and no other cereal, whether or not mixed with other flour, must contain the following nutrients in the amounts specified:

Nutrient	Required quantity (in mg/100g)
Calcium carbonate	not less than 235 and not more than 390
Iron	not less than 1.65
Thiamin	not less than 0.24
Nicotinic acid* or Nicotinamide*	not less than 1.60

* also known as niacin.

The requirement concerning calcium carbonate does not apply to wholemeal, self-raising flour (as they have a calcium content of not less than 0.2%) or wheat malt flour; while iron, thiamin and

TABLE 14.17
Standards for Flour and Bread in Canada, and Enrichment Standards for Cereals in U.S.A., Compared with Whole Cereal Complements (mg/100g)

Product	Thiamin	Riboflavin	Niacin	Iron	Calcium (+)
U.S.A.					
Enriched flour	0.64	0.4	5.3	4.4	(211)
Enriched SR flour	0.64	0.4	5.3	4.4	211
Enriched bread	0.4	0.24	3.3	2.75	(132)
Enriched farina	0.4 –0.6	0.26–0.33	3.5 –4.4	2.9	(110)
Maize meal and grits	0.44–0.66	0.26–0.4	3.52–5.3	2.86–5.73	(110–165)
Self raising maize meal	0.44–0.66	0.26–0.4	3.52–5.3	2.86–5.73	(110–385)
Rice	0.44–0.88	0.26–0.53	3.52–7.05	2.86–5.73	(101–220)
Pasta (macaroni and noodles)	0.88–1.10	0.37–0.48	5.95–7.49	2.86–3.63	(110–137)
Canada					
Flour and enriched flour	0.44–0.77	0.27–0.48	3.5 –6.4	2.9 –4.3	(110–140)
Enriched bread and Enriched white bread	0.24	0.18	2.20	1.76	(66)
Whole grain					
Wheat	0.43	0.13	5.0	4.6	48
Maize	0.4	0.11	1.9	3.1	20
Brown rice	0.33	0.07	4.6	1.9	22
White rice	0.07	0.03	1.6	0.9	12

(+) Brackets indicate optional.
Sources: Kent (1983); Ranum (1991).

nicotinic acid or nicotinamide should be naturally present in wholemeal and need to be added to flour other than wholemeal only where they are not already present in the specified amounts.

In Canada only enriched flour can be sold; in the U.S.A. the FDA does not require enrichment but, as many States require it, virtually all affected products sold across state borders are enriched. The FDA specifies what products advertized or labelled as 'enriched' must contain.

The current requirements for Canada and the U.S.A. are given in Table 14.17

Addition of vital wheat gluten to bread flour may be regarded as nutrient supplementation, but this is not its primary purpose. The rationale behind its addition is purely to improve the breadmaking properties of the flour.

In the U.K. and Canada, addition of nutrients to flour, where addition is required, is made at the flour mill. In the U.S.A. enrichment is permitted either at the mill or at the bakery.

Enrichment of whole grains requires a more subtle approach than that of ground cereal products, as nutrients added to the raw grains can easily be removed during cooking. In Japan several strategies have been adopted for addition of vitamins, minerals and lysine. In the multinutrient method two stages of addition are involved: the first consists of soaking for up to 24 h in acetic or hydrochloric acid solution (0.2, 1 or 5% at 39°C) containing thiamin, niacin, riboflavin, pantothenic acid and pyridoxin; the grains are then steamed and dried, after which they are coated with vitamin E, calcium and iron in separate layers. Finally a protective coating is applied: it consists of an alcoholic (or propan-2-ol) solution of zein, palmitic or stearic acid, and abietic acid. In addition to retaining the majority of the added nutrients during washing and cooking, coated rice retains the natural nutrients to an improved degree (Misaki and Yasumatsu, 1985). Losses are approximately 10% for coated rice and up to 70% for milled rice. Coated rice also provides a nutritious and palatable alternative to the unpopular brown or parboiled types in Puerto Rico.

Negative attributes

In spite of the fact that cereals are among the safest and most important foodstuffs, there are some anti-nutritious aspects of their composition (some of these affect only a small proportion of consumers). Also there are hazards associated with their storage and handling. Awareness of these allows monitoring to be carried out, providing a means of ensuring their safe and proper use.

Phytic acid

An important constituent of cereals, legumes and oilseed crops is phytic acid. The systematic name of phytic acid in plant seeds is myoinositol-1,2,3,5/4,6–hexakis (dihydrogen phosphate) (IUPAC-IUB, 1968). Phytic acid in the free form is unstable, decomposing to yield orthophosphoric acid, but the dry salt form is stable. The terms phytic acid, phytate, and phytin refer respectively to the free acid, the salt, and the calcium/magnesium salt, but some confusion arises in the literature where the terms tend to be used interchangeably. The salt form, phytate, accounts for 85% of the total phosphorus stored in many cereals and legumes. In cereals its distribution varies, in that in maize the majority lies in the embryo, while in wheat, rye, triticale and rice most of the phytate is found in the aleurone tissue. In pearl millet the distribution is apparently more uniform (Table 14.18)

TABLE 14.18
Distribution of Phytate in the Morphological Components of Cereals

Cereal	Fraction	Phytate %	% of total in grain
Maize	Endosperm	0.04	3 (4)
	Embryo	6.39	87 (95)
	Hull	0.07	< 0.1 (1)
Wheat	Endosperm	0.004	2 (1)
	Embryo	3.91	12 (29)
	Aleurone	4.12	86 (70)
Rice	Endosperm	0.01	1 (3)
	Embyro	3.48	7.6 (26)
	"Pericarp"	3.37	80.0 (71)
Pearl millet	Endosperm	0.32	(48)
	Embryo	2.66	(31)
	Bran	0.99	(11)

Values from Reddy *et al.* (1989). The values in parenthesis are calculated from the table values, using the proportions given by Kent (1983).

The negative nutritional attributes of phytate derive from the fact that it forms insoluble complexes with minerals, possibly reducing their bioavailability and leading to failure of their absorption in the gut of animals and humans, and thus to mineral deficiencies.

An estimate of daily phytate intake in the U.K. is 600–800 mg. Of this 70% comes from cereals, 20% from fleshy fruits and the remainder from vegetables and nuts (Davies, 1981).

The reduced bioavailability of minerals due to phytate depends on several factors, including the nutritional status of the consumer, concentration of minerals and phytate in the foodstuff, ability of endogenous carriers in the intestinal mucosa to absorb essential minerals bound to phytate and other dietary substances, digestion or hydrolysis of phytate by phytase and/or phosphatase in the intestine, processing operations, and digestibility of the foodstuff. The 'other substances' referred to include dietary fibre (non-starch polysaccharides) and polyphenolic compounds.

Many metal ions form complexes with phytate, including nickel, cobalt, manganese, iron and calcium; the most stable complexes are with zinc and copper. The phosphorus of the phytate molecule itself is also only partly available to non-ruminant animals and humans; estimates vary from 50 to 80% and depend upon several factors including the calcium-phytic acid ratio (Morris, 1986).

The presence of phytate may also influence the functional and nutritional properties of proteins, the nature of the phytate–protein complexing being dependent on pH. The mechanisms involved are complex and ill-understood but it is greatest at low pH because, under these conditions, phytic acid has a strong negative charge and many plant proteins are positively charged. Very little complexing between wheat proteins and phytate has been found and complexing between rice bran phytate and rice bran proteins occurred only below pH 2.0 (Reddy *et al.*, 1989).

Inhibition of the activity of enzymes such as trypsin, pepsin and *alpha*-amylase by phytic acid has been reported. In the case of the amylase enzyme it is not clear whether this results from the phytic acid complexing with the enzyme itself, or chelation of the Ca^{2+} ions required by the enzyme.

Tannins

Tannins are phenolic compounds of the flavonoid group, that is they are derivatives of flavone. They are considered in the 'negative' attributes

TABLE 14.19
Percentage of Tannin in Cereal grains at 14% Moisture Content

Brown Rice	Wheat	Maize	Rye	Millet	Barley	Oat	Sorghum
0.1	0.4	0.4	0.6	0.6	0.7	1.1	1.6–(5)★

Data from Juliano (1985a). ★ Collins (1986).

section of this chapter because it is alleged that they reduce protein digestibility through phenol–protein complexing. Some flavonoids have also been implicated as carcinogens. On the other hand, they have also been credited with the ability to stimulate liver enzymes which offset the effects of other carcinogens. Clearly we know little about the nutritional effects of tannins (Bingham, 1987). There is no doubt that they and their derivatives can adversely affect flavour and colour, thus reducing palatability.

Although present in all cereals, tannins are particularly associated with sorghums of the 'birdproof' type. All sorghums contain phenolic acids and other flavonoids, but only brown types contain procyanidin derivatives, otherwise known as condensed tannins. Table 14.19 shows their relative abundance in cereal species.

Harmful effects of alcoholic drinks

Most beers contain 2–4% w/w. of alcohol, barley wine contains 6%. Lagers contain less alcohol than ales and stouts but it is absorbed more rapidly from lagers because of their greater effervescence. Spirits whether produced from grains, other fruits or starch, contain about 33%.

Low to moderate consumption of alcohol increases life expectancy, reduces blood cholesterol concentration and (in France) is associated with low incidence of coronary vascular disease. However, alcohol is a drug which can become addictive, it is rapidly absorbed into the blood, one fifth of the amount being ingested, through the stomach wall and the remainder from the small intestine. Following ingestion of alcohol, blood vessels become dilated and body heat loss is accelerated, overriding the body's thermoregulatory mechanisms — sometimes dangerously. From the blood, alcohol diffuses widely; it is a *diuretic*, promoting the production of urine, giving rise to dehydration — one of the causes of the

hangover. It also has an inhibiting effect on the nervous system, including the brain, inducing euphoria and depressing judgement. The amount of alcohol required to induce intoxication depends on the individual, influential factors being body weight, sex and drinking habits, regular heavy consumers developing a higher threshold than infrequent drinkers. It is generally considered that symptoms of intoxication are apparent when blood contains 100 mg/100ml alcohol (a 70 kg man drinking 1.6 l (3 pints) of beer, containing 50 g ethanol, achieves this level). Reactions are impaired below this level however, and in the U.K. the limit for someone in charge of a motor vehicle is 80 mg/100ml, in some countries it is zero. The intoxicant effect on women is usually greater than on men as their bodies metabolize the alcohol more slowly and their body weight is generally lower than that of men. Moderate drinking is regarded as less than 50g of alcohol per day for men, and 30g for women.

Long term effects of alcohol consumption are damage to the heart, liver (*cirrhosis* is a condition in which some liver cells die) and brain. Cancers of the mouth, the throat and the upper digestive system particularly, are statistically associated with regular heavy consumption of alcohol.

Allergies

An *allergy* is an unusual immune response to a natural or man-made substance that is harmless to most individuals. *Hypersensitivity* is a non-immune response to a substance which generates responses whose symptoms can be similar to allergies. Allergies involve abnormally high reactions of the body's natural immune mechanisms against invasion by foreign substances. Substances that stimulate an immune response are known as *antigens*, and when the response is abnormal the antigens involved are called *allergens*. The subject of cereals as allergens was reviewed by Baldo and Wrigley (1984).

Allergies arise to substances that enter the body as a result of inhalation or ingestion, alternatively they can be a response to skin contact. Cereal pollens and fragments between 1 and 5 μm present in the dust raised when grains or other plant parts are being handled, constitute the inhaled antigens. While everyone is likely to inhale some pollen grains, of cereals and many other species, those people involved in handling cereals are more likely to inhale grain dust.

One of the best documented and longest established allergies associated with cereals is *bakers' asthma*. As the name implies, bakers' asthma and *rhinitis* are most prevalent in, but not exclusive to, those who habitually handle flour. It is now established as a IgE mediated reaction (i.e. it is a true allergy — IgE is an immunoglobulin associated with an allergic response) arising from inhalation of airborne flour and grain dust. As an occupational disease it is declining as a result of improved flour handling methods creating progressively less dust. Nevertheless significant proportions of bakery and mill workforces exhibit some sign, though not necessarily asthmatic symptoms, of the allergy.

Many studies have been made on the harmful effects of inhaling grain dusts, for example, comparisons of grain handlers with city dwellers in an American study showed that inhalation of grain dust gave rise to serious problems in the lungs. *Chronic bronchitis* was significantly higher in grain handlers (48%) than in controls (17%), as was wheezing at work ('occupational asthma'), and airway obstruction. Compared with smoking, grain dust exposure had a greater effect on the prevalence of symptoms but had the same or less effect on lung function (Rankin *et al.*, 1979). In other studies skin-prick tests, which are a means of detecting allergic responses, confirmed their occurrence following exposure to grain dust (Baldo and Wrigley, 1984).

While it is clear that inhalation of dusts is harmful, and that immune responses are demonstrable, the degree of responsibility for the symptoms attributable specifically to allergy cannot easily be established as similar symptoms have been detected in the absence of allergies.

Coeliac disease is a pathologial condition, leading to loss of villi and degeneration of the intestine wall, and induced by *gliadin* and *gliadin-like* proteins. Other names by which it is known are *gluten sensitive enteropathy*, *sprue*, *nontropical sprue* and *idiopathic steatorrhea*. Symptoms of the disease are *malabsorption* and the resulting deficiencies of vitamins and minerals, and a loss of weight.

Its cause remains unknown, although there is some evidence that it results from a genetic defect characterized by the absence of an enzyme necessary for gliadin digestion. Alternatively it may involve an allergic response in the digestive system. Most coeliac patients are childen, the symptoms showing when cereals are first introduced in their diet. In a normal condition cells lining the small bowel absorb nutrients which are duly passed into the blood, but, in coeliac patients, the cells are irritated and become damaged. Their resulting failure to absorb leads to vomiting, passing of abnormal stools containing the unabsorbed nutrients, and to symptoms such as pot belly, anorexia, anaemia, rickets and abnormal growth. Milder forms may pass unnoticed, but in adult life, may result in general ill health, weight loss, tiredness and *osteomalacia*. Newly diagnosed adults probably were born with the disease, but their symptoms presented only in later life.

Treatment consists of strict adherance to a gluten-free diet. All untreated wheat flour must be avoided, and rye, triticale, oat and barley products are also excluded. Maize and rice (and probably millet and sorghum) are accepted as being totally gluten-free. Because cereal flours are so versatile, gluten is unexpectedly present in many foods and labelling requirements are inadequate to indicate its presence in all cases. Manufacturers and consumers of gluten-free products can confirm the absence of gluten using antibody test kits. Some successful kits use antibodies raised to wheat ω-gliadins. Although these are not considered to be toxic (it is the α- and β-gliadins that are) they retain their antigenicity after heating and are thus suitable for use on cooked products as well as raw ingredients (Skerrit *et al.*, 1990).

Schizophrenia

Evidence of the implication of cereal prolamins as a contributory factor to the incidence of schizophrenia in genetically susceptible patients has been reviewed by Lorenz (1990). Much of the evidence in favour of the association is epidemiological, but some clinical trials support it. Coeliac disease is reported to occur considerably more frequently in schizophrenic patients than would be expected by chance alone, and administration of gluten-free diets has in some cases led to remission of both disorders. Not all studies support a relationship between gluten ingestion and schizophrenia, and the difficulties involved in designing definitive experiments in this area make a rapid conclusion to the question unlikely.

Dental cartes

This is an infectious disease that leads to tooth decay through the production of organic acids by bacteria present in the oral cavity. The bacteria are supported by fermentable sugars and starch, and foods containing high proportions of these nutrients have been investigated in view of their perceived potential to encourage development of the disease.

Lorenz (1984) reviewed work on the cariogenic and protective effects of diet patterns, with particular reference to cereals. There is a general concensus that sugary foods, particularly sticky ones, which remain in the mouth for a long time, do lead to increases in the occurrence and severity of caries in susceptible subjects. There are reports that the disease is prevalent among bakers and bakery workers, but bread is not considered to be cariogenic and no difference between white and wholemeal breads has been established in this context. Cereal products with a high sucrose content, such as cakes and sweet biscuits, can increase caries, but the relationship is not simple and many factors are involved. Corn- based breakfast cereals have been shown to be more cariogenic than those made from wheat or oats but conflicting results have emerged from different trials in which sucrose-coated breakfast cereals were compared with uncoated equivalents. Lorenz (1984) concluded

that few unquestionable guidelines can at present be offered as to the avoidance or cariogenic foods.

β-Glucans

A number of negative nutritional effects have been associated with beta-glucas. They may impair absorption of mineral and fat soluble vitamins, but additional research is required in this area. Perhaps the most extensively studied negative nutritional factor involves reduced growth rates of chickens fed barley-based diets. Barley is of low digestibility and low energy value when fed to chickens, resulting in poor growth and production. Low production values are attributed to β-glucans, as supplementation with β-glucanase improves the performance of barley-based diets (Pettersson et al., 1990). Presumably β-glucans in the endosperm cell wall physically limit the accessibility to starch and protein, and increase digesta viscosity.

Non-cereal hazards

Adverse effects on health can arise from the consumption of foods contaminated by toxic substances. In the handling of all food raw materials, some risk of contamination exists and cereals are no more susceptible than other natural products to these. Similarly, all natural products are associated with particular risks from, for example, infection with specific diseases. Hazards of this sort are considered here.

Ergotism

The disease gangrenous ergotism, caused by consumption of products containing 'ergot' — the sclerotia of the fungus *Claviceps purpurea* — is discussed in Ch. 1.

Mycotoxins

These are toxic secondary metabolites produced by *fungi*. Technically, the toxins in poisonous mushrooms and in ergot are mycotoxins, but poisoning by mushrooms or by ergot requires consumption of at least a moderate amount of the fungus tissues containing the toxins, whereas the mycotoxins in, for example nuts or cereal grains, may be present in the absence of any obvious mould. Strictly, substances like *penicillin* and *streptomycin* are mycotoxins, but the term is usually reserved for substances toxic to higher animals (including humans). The term '*antibiotics*' is generally applied to compounds produced by fungi that are toxic to bacteria.

Nearly 100 species of fungus were associated with mycotoxins giving rise to symptoms in domestic animals, by Brooks and White in 1966. Of these over half were in the genera *Aspergillus*, *Penicillium*, and *Fusarium*.

Mycotoxins cause relatively little concern in cereals, compared with *peanuts*, *Brazil nuts*, *pistachio nuts* and *cottonseed*, the risks depending to some extent on their place of origin. Nevertheless, maize is an excellent substrate for the growth of *Aspergillus flavus* and *A. parasiticus*, two fungal species that produce *aflatoxins*. The importance of mycotoxins in the cereal context was reviewed by Mirocha *et al.* (1980).

Aflatoxins

There are six important aflatoxin isomers: B_1, B_2, G_1, G_2, M_1 and M_2; the initial letter of the first four being that of the colour of the fluorescence which they exhibit under ultraviolet light, i.e. *B*lue, and *G*reen. The *M* indicates *M*ilk, in which these isomers were first found (this is systematic?). All isomers are toxic and carcinogenic but aflatoxin B_1 is the most abundant and presents the greatest danger to consumers. Several derivatives, produced by the microorganisms or in the consuming animal are also toxic.

In the United States it has been found that infection of maize with *Aspergillus* spp. occurs in the field, usually after damage to the grains by insects. Degree of infection varies geographically, growth of the organism being favoured by high temperatures. It is thus most common in the *southeast U.S.A.* and tropical maize-growing countries such as *Thailand* and *Indonesia* and relatively rare in the cooler Corn Belt of North America. Other genera are more associated with infection during storage. Aflatoxins have been detected in small amounts in wheat, barley, oats, rye and sorghum.

Other mycotoxins include *Zearalenones*, pro-duced mainly by species of *Fusarium*, the major producers being members of the *F. roseum* complex. Zearalenone is an oestrogen and its effects primarily involve the genital system. Most affected are pigs, the effects on humans are not well documented. Although reported in wheat, barley, oats and sorghum, maize is the cereal most affected.

Trichothecenes are a group of about 40 compounds associated with *Fusarium* spp. but also produced by other fungi. These mycotoxins are implicated as causes of *Alimentary toxic aleukia* in man. Occurrence of the disease in Russia between 1931 and 1943 was associated with consumption of proso millet that had been left in the field over winter and had become seriously infected with a number of fungal species. Trichothecenes have been isolated from other cereal species, but mainly maize stored on the cob.

Ochratoxin is associated with *spontaneous nephropathy* in pigs, this mycotoxin from *Penicillium virididicatum*, has been found in stored (often overheated) maize and wheat, in North America and Europe.

Citreoviridin, *Citrinin* and other mycotoxins are described as 'yellow-rice toxins' as they cause yellowing of rice in storage. They are associated with diseases of the heart, liver and kidneys and they are produced mainly by *Penicillium citreoviriden*, *P. citrinum* and *P. islandicum*.

Limits on aflatoxin levels between 5 and 50 μg/kg have been set by developed countries for cereals and other commodities intended for feed or food use. The actual exposure that will produce cancer in humans has not been established and limits for all mycotoxins are based, not on scientific principles but notional concepts of safety. While this state of affairs persists, international agreement on standards, though desirable, is unlikely (Stoloff *et al.*, 1991).

References

AUSTRALIAN TECHNICAL MILLERS ASSOCIATION (1980) Current use and development of rice by-products. *Australas. Baker Millers J.* April: 27

BALDO, B. A. and WRIGLEY, C. W. (1984) Allergies to cereals. *Adv. Cereal Sci. Technol.* 6: 289–356.

BENGTSSON, S., AMAN, P., GRAHAM, H. NEWMAN, C. W. and NEWMAN, R. K. (1990) Chemical studies on mixed linkage β-glucan in hull-less barley cultivars giving different hypocholesterolemic responses in chickens. *J. Sci. Food Agric.* **52**: 435–445.

BHATTACHARYA, K. R. (1985) Parboiling of rice, In: *Rice: Chemistry and Technology*, 2nd edn, Ch. 8, pp. 289–348, JULIANO, B. O. (Ed.) Amer. Assoc. of Cereal Chemists, St Paul, MN. U.S.A.

BINGHAM, S. (1987) *The Everyday Companion to Food and Nutrition.* J. M. Dent and Sons Ltd, London.

BOUNDY, J. A. (1957) Quoted in KENT-JONES, D. W. and AMOS, A. J. (1967).

BROOKS, P. J. and WHITE, E. P. (1966) Fungus toxins affecting mammals. *Ann. Rev. Phytopath.* **4**: 171–194.

BUSSON, F., FAUCONNAU, G., PION, N. and MONTREUIL, J. (1966) Acides aminés. *Ann. Nutr. Aliment.* **20**: 199.

CARTER, E. G. A. and CARPENTER, K. J. (1982) The bioavailability for humans of bound niacin from wheat bran. *Am. J. clin. Nutr.* **36**: 855–861.

CASEY, P. and LORENZ, K. (1977) Millet — functional and nutritional properties. *Bakers' Digest.* Feb: 45.

CHATFIELD, C. and ADAMS, G. (1940) *Food Composition*, U. S. Dept. Agric. Circ. 549.

CHUNG, O. K. (1991) Cereal lipids. In: *Handbook of Cereal Science and Technology*, pp. 497–553, LORENZ, K. J. and KULP, K. (Eds.) Marcel Decker, Inc. N.Y., U.S.A.

CLEGG, K. M. (1958) The microbiological determination of pantothenic acid in wheaten flour. *J. Sci Food Agric.* **9**: 366.

CLEGG, K. M. and HINTON, J. J. C. (1958) The microbiological determination of vitamin B₆ in wheat flour, and in fractions of the wheat graiin. *J. Sci. Food Agric.* **9**: 717.

COLLINS, F. W. (1986) Oat phenolics: structure, occurrence and function. In: *Oats: Chemistry and Technology*, Ch. 9, pp. 227–295, WEBSTER, F. H. (Ed.) Amer. Assoc. of Cereal Chemists Inc., St. Paul, MN. U.S.A.

COULTATE, T. P. (1989) *Food The Chemistry of its Components*, 2nd edn, Roy. Soc. of Chemistry. London.

DAVIDSON, M. H., DUGAN, L. D. BURNS, J. H. BOVA, J., SORTY, K. and DRENNAN, K. B. (1991) The hypocholesterolemic effects of β-glucan in oatmeal and oat bran. A dose controlled study. *JAMA* **265**: 1833–1839.

DAVIES, K. (1981) Proximate composition, phytic acid, and total phosphorus of selected breakfast cereals. *Cereal Chem.* **58**: 347.

DE RUITER, D. (1978) Composite flours. *Adv. in Cereal Sci. Technol.* **2**: 349–379.

DEYOE, C. W. and SHELLENBERGER, J. A. (1965) Amino acids and proteins in sorghum grain. *J. agric. Fd. Chem.* **13**: 446.

DHSS (Department of Health and Social Security) (1969) *Recommended Intakes of Nutrients for the United Kingdom.* London: HMSO (Reports on public health and medical subjects; 120).

DHSS (Department of Health and Social Security) (1979) *Recommended Daily Amounts of Food Energy and Nutrients for Groups of People in the United Kingdom.* London: HMSO (Reports on health and social subjects; 15).

DH (Department of Health) (1991) *Dietary Reference Values for Food Energy and Nutrients for the United Kingdom.* (Report on health and social subjects 41).

EWART, J. A. D. (1967) Amino acid analysis of glutenins and gliadins. *J. Sci. Fd Agric.* **18**: 111.

FISHER, N., REDMAN, D. G. and ELTON, G. A. H. (1968) Fractionation and characterization of purothionin. *Cereal Chem.* **45**: 48.

FRAPS, G. S. (1916) The composition of rice and its by-products. *Texas Agr. Exp. Sta. Bull.* **191**: 5.

FREEMAN, J. E. and BOCAN, B. J. (1973) Pearl millet: a potential crop for wet milling. *Cereal Sci. Today* **18**: 69.

GREEN, J., MARCINKIEWICZ, S. and WATT, P. R. (1955) The determination of tocopherols by paper chromatography. *J. Sci. Food Agric.* **6**: 274.

HAHN, R. R. (1969) Dry milling of sorghum grain. *Cereal Sci. Today.* **14**: 234.

HEATHCOTE, J. G., HINTON, J. J. C. and SHAW, B. (1952) The distribution of nicotinic acid in wheat and maize. *Proc. roy Soc.* **B. 139**: 276.

HINTON, J. J. C. (1944) The chemistry of wheat with particular reference to the scutellum. *Biochem J.* **38**: 214.

HINTON, J. J. C. (1947) The distrbution of vitamin B₁ and nitrogen in the wheat grain. *Proc. roy Soc.* **B134**: 418.

HINTON, J. J. C. (1948) The distribution of vitamin B₁ in the rice grain. *Br. J. Nutr.* **2**: 237.

HINTON, J. J. C. (1953) The distribution of protein in the maize kernel in comparison with that in wheat. *Cereal Chem.* **36**: 19.

HINTON, J. J. C. and SHAW, B. (1953) The distribution of nicotinic acid in the rice grain. *Br. J. Nutr.* **8**: 65.

HOLLAND, B., WELCH, A. A., UNWIN, I. D., BUSS, D. H., PAUL, A. A. and SOUTHGATE, D. A. T. (1991) *McCance and Widdowson's The Composition of Foods*, 5th edn, Roy. Soc. Chem. and MAFF. Cambridge.

HOME-GROWN CEREALS AUTHORITY (1991) Grain storage practices in England and Wales. *H-GCA Weekly Bull.* **25** (Suppl.): No. 38.

HOUSTON, D. F. (1972) *Rice: Chemistry and Technology*, Amer. Assn of Cereal Chemists. St. Paul, MN. U.S.A.

HUGHES, B. P. (1967) Amino acids. In: *The Composition of Foods.* McCANCE, R. A. and WIDDOWSON. E. M. (Eds.) Med. Res. Count., Spec. Rep. Ser. No. 297, 2nd Impression, H.M.S.O. London.

IUPAC-IUB, (1968) The nomenclature of cyclitols. *Eur. J. Biochem.* **5**: 1.

JONES, C. R. (1958) The essentials of the flour-milling process. *Proc. Nutr. Soc.* **17**: 7.

JULIANO, B. O. (Ed.) (1985a) Production and utilization of rice. In: *Rice: Chemistry and Technology*, Ch. 1, pp. 1–16, JULIANO, B. O. Amer. Assoc of Cereal Chemists Inc. St. Paul MN. U.S.A.

JULIANO, B. O. (Ed.) (1985b) Rice bran. In: *Rice: Chemistry and Technology*, Ch. 18, pp. 647–687, JULIANO, B. O. Amer. Assoc of Cereal Chemists Inc. St. Paul MN. U.S.A.

JULIANO, B. O., BAPTISTA, G. M., LUGAY, J. C. and REYES, A. C. (1964) Rice quality studies on physicochemical properties of rice. *J. agric. Food Chem.* **12**: 131.

KENT, N. L. (1983) *Technology of Cereals.* 3rd edn, Pergamon Press Ltd, Oxford.

KENT-JONES, D. W. and AMOS, A. J. (1967) *Modern Cereal Chemistry*, 6th edn, Food Trade Press, London.

KLOPFENSTEIN, C. F. (1988) The role of cereal beta-glucans in nutrition and health. *Cereal Foods World* **33**: 865–869.

LORENZ, K. (1984) Cereal and dental caries. *Adv. Cereal Sci. Technol.* **6**: 83–137.

LORENZ, K. (1990) Cereals and schizophrenia. *Adv. Cereal Sci. Technol.* **10**: 435–469.

McCANCE, R. A., WIDDOWSON, E. M., MORAN, T., PRINGLE, W. J. S. and MACRAE, T. F. (1945) The chemical composition of wheat and rye and of flours derived therefrom. *Biochem. J.* **39**: 213.

MINISTRY OF AGRICULTURE, FISHERIES AND FOOD (1991) *Household Food Consumption and Expenditure 1990.* Annual

Report of the National Food Survey Committee, HMSO, London.

MINISTRY OF AGRICULTURE, FISHERIES AND FOOD. (1991) *Annual Report of the Working Party on Pesticide residues: 1989–1990.* H.M.S.O. London.

MIROCHA, C. J., PATHRE, S. V. and CHRISTENSEN, C. M. (1980) Mycotoxins. *Adv. Cereal Sci. Technol.* 3: 159–225.

MISAKI, M. and YASUMATSU, K. (1985) Rice enrichment and fortification. In: *Rice: Chemistry and Technology,* Ch. 10, pp. 389–401, JULIANO, B. O. (Ed.) Amer. Assoc. of Cereal Chemists Inc. St Paul MN. U.S.A.

MORAN, T. (1959) Nutritional significance of recent work on wheat, flour and bread. *Nutr. Abs. Revs.* 29; 1.

MORRIS, E. R. (1986) Phytate and dietary mineral bioavailability. In: *Phytic acid: Chemistry and Applications,* pp. 57–76, GRAF, E. (Ed.) Pilatus Press, Minneapolis, MN. U.S.A.

MORRISON, W. R. (1978) Cereal lipids. *Adv. Cereal Sci. Technol.* 2: 221–348.

NEUMANN, M. P., KALNING, H. SCHLEIMER, A. and WEINMANN, W. (1913) Die chemische Zusammensetzung des Roggens und seiner Mahlprodukte. *Z. ges. Getreidew.* 5: 41.

NEWMAN, R. K. NEWMAN, C. W. and GRAHAM, H. (1989) The hypocholesterolemic function of barley β-glucan. *Cereal Foods World* 34: 883–886.

NORMAND, F. L., ORY, R. L. and MOD, R. R. (1981) Interactions of several bile acids with hemicelluloses from several varieties of rice. *J. Food Sci.* 46: 1159.

ORY, R. L. (1991) *Grandma called it roughage — Fibre, Facts and Fallacies.* Amer. Chem. Soc., Washington D.C.

PETTERSSON, D. GRAHAM, A. H. and AMAN, P. (1990) Enzyme supplementation of broiler chicken diets based on cereals with endosperm cell walls rich in arabinoxylans and mixed linkage β-glucans. *Animal Prod.,* 51: 201–207.

RANKIN, J., DOPICO, G. A., REDDAN, W. G. and TSIATIS, A. (1979) Respiratory disease in grain handlers. In: *Proceedings of the International Symposium on Grain Dust,* p. 91, MILLER, B. S. and POMERANZ, Y. (Eds.) Kansas State Univ. Manhattan. U.S.A.

RANUM, P. (1991) Cereal enrichment. In: *Handbook of Cereal Science and Technology,* Ch. 22, pp. 833–861, LORENZ, K. J. and KULP, K. (Eds.) Marcel Dekker, Inc. NY.

REDDY, N. R., PIERSON, M. D., SATHE, S. K. and SALUNKHE, D. K. (1989) *Phytates in Cereals and Legumes.* CRC Press Inc. Boca Raton. FL. U.S.A.

REINERS, R. A., HUMMEL, J. B., PRESSICK, J. C. and MORGAN, R. E. (1973) Composition of feed products from the wet-milling of corn. *Cereal Sci. Today* 18: 372.

ROONEY, L. R. and SERNA-SALDIVAR, S. O. (1987) Food uses of whole corn and dry-milled fractions. In: *Corn: Chemistry and Technology,* Ch. 13, pp. 399–429, WATSON, S. A. and RAMSTED, P. E. (Eds) Amer. Assoc. of Cereal Chemists, St Paul, MN. U.S.A.

SCIENCE EDITOR (1970) Tocopherols and cereals. *Milling* 152 (4): 24.

SKERRITT, J. H., DEVERY, J. M. and HILL, A. S. (1990) Gluten intolerance: chemistry, celiac-toxicity, and detection of prolamins in foods. *Cereal Foods World* 35: 638–639, 641–644.

SLOVER, H. T. (1971) Tocopherols in foods and fats. *Lipids* 6: 291.

STIVER, T. E., Jr (1955) American corn-milling systems for degermed products. *Bull. Ass. oper. Millers* 2168.

STOLOFF, L., VAN EGMOND, H. P. and PARK, D. L. (1991) Rationales for the establishment of limits and regulations for mycotoxins. *Food Additives and Contaminants* 8: 213–222.

TKACHUK, R. and IRVINE, G. N. (1969) Amino acid composition of cereals and oilseed meals. *Cereal Chem.* 46: 206.

U.S. CODE OF FEDERAL REGULATIONS Title 21, Part 104B, Fortification Policy.

USDA (1963) *Agricultural Handbook 8.* USDA Agricultural Research Service.

USDA/USDHHS (1985) *Nutrition and Your Health: Dietary Guidelines for Americans.* U.S. Dept. of Agriculture and U.S. Dept. of Health and Human Services, Washington D.C.

WALDSCHMIDT-LEITZ, E. and HOCHSTRASSE, K. (1961) Grain proteins. VII. Albumin from barley and wheat. *Z. physiol. Chem.* 324: 423.

WATSON, S. J. (1953) The quality of cereals and their industrial uses. The uses of barley other than malting. *Chemy Ind.* 95.

WOODMAN, H. E. (1957) *Rations for Livestock.* Ministry of Agriculture, Fisheries and Food, Bull. 48, H.M.S.O., London.

WOODS, C. D. (1907) Food value of corn and corn products. *U.S. Dept. Agric. Farmers Bull.* 298.

Further Reading

ANON. *Diet, Nutrition and the Prevention of Chronic Disease.* World Health Organization, Geneva.

BENDER, A. E. (1990) *Dictionary of Nutrition and Food Technology.* 6th edn. Butterworth and Co. (Publ.) London.

DREHER, M. L. (1987) *Handbook of Dietary Fibre.* Marcel Dekker Inc. N.Y.

FRØLICH, W. (1984) *Bioavailability of Minerals from Unrefined Cereal Products.* In vitro *and* in vivo *studies.* University of Lund Sweden.

JAMES, W. P. T., FERRO-LUZZI, A., ISAKSSON B. and SZOSTAK, W. B. (1987) *Healthy Nutrition* World Health Organization, Copenhagen.

KUMAR, P. J. and WALKER-SMITH, J. A. (1978) *Coeliac Disease: 100 years.* St. Bartholomews Hospital, London.

MCLAUGHLIN, T. (1978) *A Diet of Tripe.* David and Charles. Newton Abbot, Devon.

O'BRIEN, L. O. and O'DEA, K. (1988) *The Role of Cereals in the Human Diet.* Cereal Chem Division, Royal Australian Chem. Inst. Parkville, Victoria.

RAYNER, L. *Dictionary of Foods and Food Processes.* Food Sci. Publ. Ltd. Kenley, Surrey.

SANDERS, T. and BAZALGETTE, T. (1991) *The Food Revolution.* Transworld Publishers Ltd. London.

SLATTERY, J. (1987) *The Healthier Food Guide.* Chalcombe Publications, Marlow, Bucks.

SOUTHGATE, D. A. T., WALDRON, K., JOHNSON, I. T. and FENWICK, G. R. (Eds.) (1990) *Dietary Fibre.* Roy Soc. Chemistry, Cambridge.

TRUSWELL, A. S. (1992) *ABC of Nutrition* 2nd edn. British Medical Journal, London.

15

Feed and Industrial Uses for Cereals

Introduction

The worldwide usage of all the cereals, gathered together, as revealed in the FAO Food Balance Sheets, is about 4% for seed, and the remainder almost equally shared between human food use (49%) and animal feed plus 'processing and other' (principally industrial) use (47%). The last category divides into about 37% for animal feed plus 10% for industrial use. There is, however, considerable variation among the eight principal cereals, as shown in Table 15.1.

TABLE 15.1
*World Usage of the Principal Cereals, 1984–86 average**

Cereal	Food (%)	Feed (%)	Processing and other (%)	Seed (%)
Wheat	66.5	20.2	6.7	6.6
Barley	5.4	73.1	14.9	6.6
Oats	5.3	78.3	7.3	9.1
Rye	33.4	43.9	14.3	8.4
Rice	88.0	1.8	7.0	3.2
Maize	20.7	63.9	13.9	1.5
Sorghum	35.2	56.6	6.9	1.3
Millet	74.7	10.8	11.8	2.7
Total	49.0	36.9	9.8	4.3

* *Source*: FAO Food Balance Sheets 1984–86. Food and Agriculture Organization, Rome, 1990.

Thus, very little rice or millet is used for feed, while over 70% of the entire crops of barley and oats are so used, with the use of maize and sorghum for animal feed not far behind.

For industrial use, the cereals seem to fall into two categories: wheat, oats, rice and sorghum, of which about 7% of the entire crop in each case is used for industrial purposes, and barley, rye, maize and millet, of which nearly twice as much (12–15%) is used industrially.

Raw materials used for feed and industrially

The raw materials used for these purposes fall into three categories:

1. The whole grain as harvested, or perhaps with a minimum of processing.
2. Certain components of the grain which provide the starting material for chemical processing. These would include starch, for production of ethanol, and pentosans, for manufacture of furfural.
3. By-products of the milling process, which are not usually suitable for human food, but which can be used for animal feed and for a wide range of industrial uses, including fillers, adhesives, abrasives, etc., besides the manufacture of furfural.

Animal feed

Apart from usage for human food, animal feed is by far the largest use for cereals — both whole grains and milling by-products.

Thus, in 1986/87, out of a total world production of all cereals of 1,830 million tonnes, 892 million tonnes (48.7%) were used for animal feed (USDA, 1987b).

Maize is easily the most widely used cereal, with about 282 million tonnes being used for animal feed (annually) worldwide in 1984–1986, followed by barley (about 127 million tonnes) and wheat (about 103 million tonnes) (FAO, 1990).

A large proportion of the cereal grains fed to animals passes through the hands of 'animal feed processors'. By way of example, of the 5.2 million tonnes of wheat used for animal feed in the UK in 1988/89, 2.87 million tonnes (55%) were used by animal feed processors (a figure based on returns from compounders only, in Great Britain) (H-GCA, 1990). The remainder, still a considerable quantity, would presumably have been fed directly from the farm to the animals, not via processors.

In the same year, 1988/89, 5.1 million tonnes of wheat were used by flour millers in the UK which, besides yielding 3.954 million tonnes of flour, also produced 1.132 million tonnes of milling by-products — bran and middlings (fine offal), most of which would have been used for animal feed. Thus, the total quantity of wheat plus wheat milling by-products used for feeding animals in the UK in 1988/89 must have been about 6.3 million tonnes (NABIM, 1991).

A similar state of affairs probably exists for wheat in other countries, and also, to varying extents, for other cereal grains worldwide.

Processing cereals for animal feed

The treatments applied to cereals by animal feed processors are both expensive and time-consuming, and obviously would not be undertaken unless such treatments offered considerable advantages over the feeding of untreated whole grain, and were cost-effective. Both cold and hot, dry and wet, mechanical and chemical methods of treatment may be used, with the objectives of improving palatability, avoiding wastage, and encouraging consumption, thus leading to a greater efficiency of food usage and perhaps faster growth. Other objectives would be to improve digestibility and/or nutritive value, to prevent spoilage, and to detoxify poisons and to inactivate anti-nutritional factors.

The actual treatment used will depend on the kind of cereal involved and the proportion of that cereal in the feed; also on the species of animal for which the feed is intended, particularly whether for ruminants or for monogastric animals and probably also on the stage in the animal's life cycle, e.g. thinking of poultry, whether for young chicks, for broilers, or for laying hens.

Some of the treatments applied by animal feed processors to cereal grain, and the resulting benefits are described below.

Grinding

This is the commonest treatment, and relatively inexpensive. Roller mills or hammer mills may be used, but hammer mills are favoured because, by choice of a screen of suitable size, the hammer mill can yield ground material of any particular size from cracked grain to a fine powder. The objective of grinding is to improve the digestibility. Coarsely-ground grain is preferred for ruminants; more finely-ground grain for swine and poultry.

Soaking

Grain may be soaked in water for 12–24 h, followed by rolling, for livestock feeding. The soaking softens the grain and causes it to swell, thereby improving palatability.

Reconstitution

A process in which grain is moistened to 25–30% m.c. and then stored in an oxygen-limiting silo for 14–21 days. This process is successful with maize and sorghum, and improves the feed/growth ratio for beef cattle.

Steam-rolling and steam-flaking

Grain is treated with steam for 3–5 min (for steam-rolled) or for 15–30 min (for steam-flaked) and then rolled between a pair of smooth rollers. These processes improve the physical texture and soften the grain. Steam-flaking makes thinner flakes than steam-rolling. The heat treatment may improve protein utilization by ruminants. In the steam-flaking process there will be some rupturing of the starch granules, and partial gelatinization of the starch, resulting in more efficient use of the feed by ruminants.

Pelleting

Steamed grain is ground and the mass is forced through a die to make pellets — a form which all domestic animals seem to prefer to meal. However, pelleted cereals are not recommended for ruminants because a decrease in food intake may ensue. On the other hand, pelleting leads to increased consumption by swine and poultry, possibly because pelleting masks the flavour of unpalatable ingredients in the diet. Pelleting also improves the utilization of amino acids by swine. The heat used in pelleting may be effective in inactivating heat-labile toxins.

Popping and micronizing

In grain that has been popped and then rolled, rupture of the endosperm improves utilization of the starch in the digestive tract. Micronizing is a similar process to popping, but uses infrared radiation for heating the grain.

Treatment of high-moisture grain

Grain harvested at a relatively high moisture content, e.g. 20–35% m.c., can be chemically treated to prevent the development of moulds during storage, to produce excellent feed.

Recommended treatment of high-moisture grain is with acids, used at a rate of 1.0–1.5%. Such acids could be propionic alone, or with acetic or formic acids. Maize and barley, thus treated, can be fed to swine, maize and sorghum to beef cattle (Church, 1991).

Feeding maize to animals

Maize (corn) provided 85% of the cereals fed to broilers in the U.S.A. in 1984, with sorghum providing 11% and wheat 4% (USDA, 1987a). In 1986, the cereals contributing to the feed for livestock in the U.S.A. were: maize 75%, sorghum 9%, oats 6%, barley 4%, wheat plus rye 6%. For beef cattle, maize is generally fed with another cereal: the greatest benefit comes from combining slowly digested grains (maize, sorghum) with rapidly digested grains (wheat, barley, high-moisture maize). A combination of 67% of wheat plus 33% of dry-rolled maize gave a 6% complementary effect as compared with feeding 100% dry-rolled maize (Kreikmeier, 1987). Cattle fed 75% of high-moisture maize plus 25% of dry-rolled grain sorghum or dry maize gained more rapidly and used the feed more efficiently than those fed either grain alone (Sindt et al., 1987). For cows, a typical diet would contain 41% of high-moisture maize (along with alfalfa and soya-bean meal, etc.) (Schingoethe, 1991). Grinding, cracking or rolling the grains, or steam-flaking, may improve the digestibility. A calf-grower feed might contain 65% of maize, sorghum or barley, plus 10% of rolled oats and 20% of soyabean meal (Morrill, 1991).

Typical feeds for early-weaned lambs would include 67% of ground shelled maize plus 10% of cottonseed hulls, or 74% of ground ear maize plus soyabean meal and supplements (Ely, 1991).

For pigs of all ages, maize might provide 85% of the grain in the rations. Use of high-lysine maize would allow a reduction in the amount of soyabean meal needed. The feed grains must be ground, e.g. through a hammer mill with $3/16$–$3/8$ in. screen. Finely-ground maize is used more efficiently than coarsely-ground, but very fine grinding, making a dusty meal, is to be avoided. The proportion of ground maize in the feed for swine at various stages of growth could be 80% for pregnant sows and gilts, 76% for lactation diets for sows and gilts, 63–71% for young pigs, 78% for growing pigs and 84% for finishing pigs (Cromwell, 1991).

For feeding poultry, maize, sorghum, wheat and barley are the most important cereals and should be ground and perhaps pelleted. Pelleting prevents the sorting out of constituents of the diet, and is recommended for chicks and broilers. Pelleting minimizes wastage and improves palatability. Maize could provide 57% of the feed for broiler starters, 62% for broiler finishers, 45% for chicks, 57% for growers (7–12 weeks) and developers (13–18 weeks), 48% for laying hens (Nakaue and Arscott, 1991).

For horses, 25–39% of the feed could be cracked maize, along with 45–30% of rolled and 7–10% of wheat bran (Ott, 1991).

By-products of the milling of maize are also used for animal feeding. A product known as hominy feed comprises the entire by-product streams from the dry milling of maize. It is a relatively inexpensive high-fibre, high-calorie material which is high in carotenoids (yellow pigments desirable for chicken feed) and vitamins A and D. Hominy feed is an excellent source of energy for both ruminants and monogastric animals, in this respect being equal or superior to whole maize. Hominy feed competes with other maize by-products — corn gluten feed and spent brewers' grains — as an animal feed. Hominy feed may partially replace grain in diets for horses, provided the feed is pelleted (Ott, 1991). Gluten feed is recovered from the steeping water in which maize is steeped as a stage in wet-milling (q.v.). After the separation of the germ, in the wet-milling of maize, and extraction of the oil, the residue — germ cake — is used for cattle feed.

Maize cobs

The maize cob (corn cob in the U.S.A.) is the central rachis of the female inflorescence of the plant to which the grains are attached, and which remains as agricultural waste after threshing. As about 180 kg of cobs (d.b.) are obtained from each tonne of maize shelled, the annual production of cobs in the U.S.A. alone is of the order of 30 million tonnes.

Cobs consist principally of cellulose 35%, pentosans 40% and lignin 15%. Agricultural uses for maize cobs, listed by Clark and Lathrop (1953), include litter for poultry and other animals; mulch and soil conditioner; animal and poultry feeds. The feeding value of corncobs is about 62% of that of grains. Up to 67% of ground corncobs, with 14% of ground shelled maize and some soyabean meal and molasses-urea provided a suitable feed for cattle. For poultry, a feed containing corncob meal plus ground maize is

preferred to one in which ground maize is the sole cereal because it results in better plumage, less feather-picking, and less cannibalism. On the other hand, the corncob plus maize feed gives a reduced egg production and less body-weight gain (Clark and Lathrop, 1953).

Barley for animal feed

Apart from its use in malting, brewing and distilling (c.f. Ch. 9), the next most important use for barley is as food for animals, particularly pigs, in the form of barley meal.

As whole barley contains about 34% of crude fibre, and is relatively indigestible, the preferred type of barley for animal feeding is one with a low husk content. Low protein barleys are favoured for malting and brewing, but barley of high protein content is more desirable for animal feed.

The total digestible nutrients in barley are given as 79%. Digestible coefficients for constituents of ground barley are 76% for protein, 80% for fat, 92% for carbohydrate and 56% for fibre (Morrison, 1947).

The feeding value of barley is said to be equal to that of maize for ruminants (Hockett, 1991) and 85–90% of that of maize for swine (Cromwell, 1991). For swine, barley can replace all the maize in the feed; indeed, barley is preferred to maize for certain animals, e.g. pigs. The feeding value of barley for pigs is improved by grinding, pelleting, cubing, rolling or micronizing (Hockett, 1991). It is also used extensively in compound feeds.

For poultry, a feed containing barley and maize improved egg production and feed efficiency as compared with either cereal fed alone (Lorenz and Kulp, 1991).

Swine fed barley grew faster and had a more efficient feed/gain ratio if the barley was pelleted than if fed as meal. Feed for pregnant sows and gilts can contain up to 85% of ground barley, up to 65% for lactating sows, 80% for growing pigs and 86% for finishing pigs (Cromwell, 1991).

The barley is normally fed either crushed or as a coarse meal, thereby avoiding wastage that could result from the passage of undigested grains through the alimentary tract. The widespread use of barley by pig feeders is related to its effect on the body fat, which becomes firm and white if the ration contains a large amount of barley meal (Watson, 1953).

'Hiproly' (i.e. hi-pro-ly) barley is a mutant two-row barley from Ethiopia containing the '*lys*' gene, which confers high lysine content (cf. p. 71). Hiproly barley contains 20–30% more lysine than is found in normal barley. High-lysine barley has been shown to improve the growth rate of pigs (Hockett, 1991). A recent high-lysine barley mutant originating in Denmark is Risø 1508, with 50% more lysine than in Hiproly barley. Risø 1508 is intended to provide a feedstuff with an improved amino acid balance for the pig and dairy industries, one objective being to avoid the necessity of feeding fishmeal, which gives a taint to the product.

By-products from the dry milling of barley to make pearl barley are used for animal feed, particularly for ruminants and horses, as constituents of compound feeds. Brewers' grains and distillers' dried grains are by-products from the brewing and distilling industries that can be incorporated in feeds for ruminants; they are too fibrous for pigs and poultry.

Wheat for animal feeding

The animal-feed use of wheat 1980–1984, world-wide, averaged 19.8 million tonnes per annum. In the U.S.A. alone, in 1988/89, 270 million bu (about 7.35 million tonnes) were used for livestock feed. The use of wheat for animal feed is influenced by price, location and nutrient value (Mattern, 1991).

The importance of wheat as an animal feedstuff is further illustrated by the establishment, by the Home-Grown Cereals Authority, in the U.K., of quality specifications for 'standard feed wheat', in association with the National Farmers' Union and the U.K. Agricultural Supply Trade Association, in 1978, and subsequently updated. These quality specifications, which apply to grain destined for the feed compounder, mention moisture content (max. 16%) and content of impurities — ergot, max. 0.05%; other cereals 5%; non-grain impurities 3% (H-GCA, 1990). Wheat fed directly

to animals, viz. not via a feed processor, could include parcels that did not meet these standards and also wheat that was unfit for milling.

When wheat was fed to cattle, the efficiency of feed usage was greater for dry-rolled wheat than for wholewheat. Dry milling increased grain digestibility from 63 to 88%. Further processing, e.g. steam-flaking or extruding, gave no further improvement (Church, 1991). For beef cattle, wheat is best used in combination with other feed grains, e.g. maize or grain sorghum. A blend of 67% wheat plus 33% dry-milled maize improved feed efficiency as compared with either wheat or maize alone (Ward and Klopfenstein, 1991).

When fed to finishing lambs for market, wheat had 105% of the feeding value of shelled maize when the wheat comprised up to 50% of the total grain (Ely, 1991). For pigs, wheat is an excellent food, but is often too expensive. Wheat is similar to maize on an energy basis, but has a higher content of protein, lysine, and available phosphorus, and wheat can replace all or part of the maize in the diet for pigs. Non-millable wheat, damaged moderately by insects, disease, or containing garlic, can be fed to swine (Cromwell, 1991).

For feeding poultry, wheat should be ground, and preferably pelleted, to avoid sorting out of feed constituents by the birds. For poultry, the feed efficiency of wheat is 93–95% of that of maize (Nakaue and Arscott, 1991).

Wheat milling by-products — bran and middlings — provide palatable food for animals. Wheat middlings can replace grain in the feed, provided the diets are pelleted — otherwise they are too dusty. The energy of wheat middlings is utilized better by ruminants than by monogastric animals. Cows fed rations containing 60% of concentrate did well if 40% of the concentrate was wheat middlings; swine did well when wheat middlings replaced up to 30% of the maize in the rations. Middlings are also fed to poultry. Wheat bran is the favoured feedstuff for horses and for all ruminants (Church, 1991).

Oats for feeding animals

The usage of oats for livestock feeding in the U.S.A. in 1986 was 9.6 million tonnes, exceeding the feed usage of barley (7.0 million tonnes) in that year (USDA, 1987b), although worldwide the usage of oats for animal feed, at 38.3 million tonnes per annum, was less than one-third of the amount of barley so used worldwide, viz. 126.8 million tonnes, in 1984–1986 (FAO, 1990).

Oats have a unique nutritional value, particularly for animals which require feed having a relatively high level of good quality protein, but with lower energy content. The level of protein in oat groats is higher than that in other cereals; moreover, the quality of oat protein, particularly the amino acid balance, surpasses that of the protein of other cereals, as shown by feeding tests (Webster, 1986; McMullen, 1991).

The good value of high-protein oats has been shown in diets for swine and poultry, although the nutritive value for these non-ruminants can be further improved by supplementation of the oats with lysine and methionine (Webster, 1986).

For feeding to animals, oats are first ground or rolled. Rolled oats can provide 10% of the feed for calves (along with 65% of maize, sorghum or barley) (Klopfenstein et al., 1991), and is a good starter feed for pigs, although too expensive for other pigs. Ground oats can provide 25% of the feed for pregnant sows, 20% for lactating sows, 10% for young pigs, 15% for growing and finishing pigs (with maize, wheat or barley supplying most of the remainder of the feed) (Cromwell, 1991). For feeding to pigs, the oats should be ground through a hammer mill, using a ½ in. screen. Pelleting of the ground oats gives faster growth than unpelleted meal for swine (Cromwell, 1991).

The feeding value for swine, relative to maize, is 100% for oat groats, and 80% for whole oats. For poultry, oats have 93% of the value of maize for broilers, 89% for layers (Cromwell, 1991; Nakaue and Arscott, 1991).

For feeding to finishing lambs for the market, oats have 80% of the feeding value of maize (Ely, 1991).

Historically, oats are regarded as the ideal feed for horses, and in North America this view still obtains. For young or poor-toothed horses, the whole oats are best rolled or crushed. As compared with whole oats, crushed oats gave a 5% feeding advantage for working horses, and a 21%

advantage for weanlings and yearlings (Ott, 1973). Oats that are musty should not be used (Ott, 1991).

The by-products of the dry milling of oats — oat dust, meal seeds, oat feed meal — are of reasonably good feeding value. Oat feed meal (= oat mill feed in the U.S.A.) is a feed of low nutritive value suitable for ruminants, used to dilute the energy content of maize and other grains.

Feed oats — the lights, doubles and thin oats removed during the cleaning of oats — are almost equally nutritious to normal oats, and are used for livestock feeding (Webster, 1986).

Sorghum for animal feed

Sorghum is a major ingredient in the feed for swine, poultry and cattle, particularly in the Western hemisphere. From the worldwide production of sorghum of 66 million tonnes in 1984–1986, 56.6% (37.3 million tonnes) were used for animal feed. In the same period, the U.S.A. alone produced 14.9 million tonnes, of which 14.8 million tonnes went for animal feed (FAO, 1990).

For feeding to animals, the sorghum is hammer-mill ground and then generally steam-flaked, using high moisture steam for 5–15 min to raise the moisture content to 18–20%, followed by rolling to make thin flakes (Rooney and Serna-Saldivar, 1991). Steam-flaking improves the feed efficiency of sorghum.

For swine, low-tannin types of sorghum have a nutritive value equal to that of maize, but brown, high-tannin types, grown for their resistance to attack by birds, and their decreased liability to weathering and fungal infestation, have a reduced nutritive value (Cromwell, 1991).

Sorghum provided 11% of all the cereal grain fed to broilers in the U.S.A. in 1984, and 9% of all the cereal grain fed to livestock in the U.S.A. in 1986 (USDA, 1987a,c).

As compared with dry-rolled sorghum, reconstituted sorghum (moistened to 25–30% m.c. and then stored for 14–21 days in a silo in a low oxygen atmosphere before feeding) produced a better daily weight gain in feed-lot cattle, and also produced a considerable improvement in feed/

gain ratio (Stock *et al.*, 1985). When fed to swine, reconstituted sorghum gave a slight improvement only in the case of high-tannin sorghum.

For beef cattle, grain sorghum has 85–95% of the feeding value of maize. The sorghum is digested slowly in the rumen and has a relatively lower total tract digestibility (Klopfenstein *et al.*, 1991).

Ground sorghum can provide up to 80% of the feed for pregnant sows, 76% for lactating sows, 71% for young pigs, 78% for growing pigs, and 84% for finishing pigs (Cromwell, 1991). For poultry, suggested rations include 18% of sorghum for chick starters, 13% for growers (7–12 weeks), 14% for developers (13–18 weeks), and 20% for layer-breeders (fed as all-mash in a warm climate) (Nakaue and Arscott, 1991).

Rye in animal feed

Of the annual total world usage of rye for animal feed, of 14.5 million tonnes, in the period 1984–1986, nearly 94% was used in Europe and the USSR, the USSR using 39.3%, Poland 30.3% and Germany (FRD plus GDR) 15.2%. Usage in the whole of north and central America was only 3.4% of the total.

Rye is used in areas where it is cheaper than barley, but, although rye is high in energy, growth of animals on rye is slower than on other cereals, possibly because its unpalatability restricts intake. Rye contains a high level of pectin (a carbohydrate), which reduces its feeding value. Thus, rye is compounded with other cereals for animal feed. Rye also contains a resorcinol — 5–alkyl-resorcinol — which was once thought to be toxic to animals. Attempts are being made to breed lines of rye with lower levels of resorcinol. Horses feed on rye grain show no ill effects from possible toxic constituents (Antoni, 1960), and rye can be successfully fed to swine and cattle when it contributes up to 50% in a mixed feed.

The presence of ergot in rye is a risk if the rye is fed to swine, as the ergot can cause abortion in sows, and reduce the performance of growing pigs (Drews and Seibel, 1976; Cromwell, 1991; Lorenz, 1991).

Rice in feed for animals

A total of 6.5 million tonnes of rice was used annually, worldwide, in the period 1984–86 for animal feeding, nearly all (6.0 million tonnes) being used in Asia (China 3.2 Mt; Thailand 0.7 Mt) (FAO, 1990).

For feeding swine, rice if pelleted can replace 50% of the maize in the feed, or 35% if fed as meal. For young pigs, the feed could contain 20% of rice bran, if pelleted (Sharp, 1991). The feeding value of pelleted, broken rice for swine is 96% of that of maize (Cromwell, 1991).

Considerable use for animal feeding is made of the by-products of rice milling. Rice pollards — a mixture of rice bran and rice polishings — is a high energy, high protein foodstuff comparing well with wheat. It contributes a useful amount of biotin, pantothenic acid, niacin, vitamin E and linoleic acid to mixed feeds, thereby reducing the requirement for supplementation with vitamin/minerals premix. The contribution of linoleic acid in rice pollards is of particular value in rations for laying hens, where it has a beneficial effect on egg size (Australian Technical Millers, 1980). For growing pigs, up to 30% of rice pollard can be fed in balanced rations without adverse effects on growth rate or carcase quality (Roese, 1978).

Extracted rice bran (the residue left after oil extraction from rice bran) has an increased content of protein and a good amino acid profile for monogastric animals, also good protein and phosphorus contents for ruminants. However, it is not a good source of fatty acids.

Rice mill feed is a mixture of rice pollards and ground rice hulls used for animal feed. In 1986, 0.72 million tonnes of rice mill feed were used for animal feed in the U.S.A. (USDA, 1987c).

Ground rice hulls are a highly fibrous, low energy foodstuff, suitable for diluting the energy level in rations for cattle, sheep, goats, pigs and poultry (Australian Technical Millers, 1980). The total digestible nutrients (at 14% m.c.) in rice hull are 15% for cattle and 25% for sheep (Juliano, 1985).

Rice hulls contain 9–20% of lignin, thereby limiting their use for animal feed. Various delignification processes have been suggested, e.g.

treatment with alkali or with acid (see Juliano 1985, Ch. 19 for details). Treatment with 12% caustic soda also reduced the high silica content of rice hulls, while treatment of the hulls with anhydrous ammonia plus monocalcium phosphate at elevated temperature and pressure increased the crude protein equivalent, broke down the harsh silica surface, and softened the hulls, thus providing an acceptable feedstuff for cattle and sheep (Juliano, 1985).

An even more successful treatment of rice hulls was incubation with *Bacillus* spp. for several days: this treatment reduced the lignin and crude fibre contents to a greater degree than soaking in caustic soda (Juliano, 1985).

The millets for animal feeding

Of the 3.1 million tonnes of millet used worldwide annually for animal feed in 1984–1986, 1.3 million tonnes were used in the former Soviet Union, 1.1 Mt in Asia (including 0.7 Mt in China), and 0.5 Mt in Africa (including 0.3 Mt in Egypt) (FAO, 1990). There was no recorded use of millet for animal feed in the U.S.A. during this period, although it is reported that proso millet (*Panicum miliaceum*) is grown in the U.S.A. for birdseed (Serna-Saldivar et al., 1991).

Feeding trials have shown that millets have nutritive values comparable to, or better than, those of the other major cereals: proso millet has 89% of the value of maize for feeding swine, and can replace 100% of the maize in rations for swine (Cromwell, 1991). Animals fed millet perform better than those fed sorghum; they produce better growth because the millet has a higher calorific content and better quality protein (Serna-Saldivar et al., 1991).

Poultry produced better gains when fed millet than when fed sorghum or wheat; the efficiency of feed conversion was better for chicks fed pearl millet (*Pennisetum americanum*) than on wheat, maize or sorghum; proso millet was equivalent to sorghum or maize in respect of egg production and weight, and efficiency of feed use.

For swine, finger millet (*Eleusine coracana*) was as good as maize for pig finishing diets; proso millet had a slightly lower feed efficiency than maize for swine, but became equal to maize when supplemented with lysine.

Pearl millet (rolled) provided excellent protein for beef cattle, and steers gained as well on rolled pearl millet as on sorghum (Serna-Saldivar et al, 1991).

Production of ethanol from cereals

Ethanol (ethyl alcohol) is produced by the enzymic action of yeast on sugars, which are themselves produced by the hydrolysis of starch. In as much as all cereals contain a large proportion of starch, it should be possible to obtain ethanol from any cereal. This happens when cereal grains are malted and then brewed to make beer, which is an aqueous solution of alcohol. The production of ethanol can be regarded as a modification of the brewing process, in which starch separated from the grains is the starting material, and pure ethanol, rather than an aqueous solution, is the final product.

The principal reasons for making ethanol from cereals are that ethanol can be used as a partial replacement of gasoline as a fuel for internal combustion engines and that the process is a useful way of dealing with surplus grain whenever it arises. Interest in the process increases when fuel shortages occur and/or when prices of feed grain are depressed.

Thus, a wheat surplus in Sweden in 1984 was dealt with by establishing a plant that separated the starch from a residue to be used for animal feed. About half of the starch (the best quality) was to be used by the paper industry, the remainder for production of ethanol. The addition of 4% of ethanol to gasoline does not lead to starting problems or to an increase in fuel consumption and, in fact, increases the octane number (Wadmark, 1988).

The possible yield of ethanol varies according to the type of cereal used: 430 l/t from rice, 340–360 l/t from wheat, 240–250 l/t from barley and oats (Dale, 1991).

The process of manufacture of ethanol from cereals starts by grinding the grain and then cooking it with water and acid or alkali. Amylase enzyme is added to the cooled mash to promote the hydrolysis of starch to glucose, and the mash is then fermented with yeast, releasing carbon dioxide gas, and producing alcohol. The wort is treated with steam in a beer still, and the alcohol is finally separated in a rectification column, yielding 95% ethanol and leaving a protein-enriched residue, suitable for animal feeding (Dale, 1991).

As a motor fuel, ethanol has various advantages over gasoline: it has a very high octane number; it increases engine power; it burns more cleanly, producing less carbon monoxide and oxides of nitrogen. On the other hand, there may be difficulties in starting the engine on ethanol alone, and accordingly a blend of ethanol with gasoline is generally used.

Other uses for ethanol made from cereals, besides motor fuel, include use as a solvent in antifreeze and as the raw material for the manufacture of various chemicals, e.g. acetaldehyde, ethyl acetate, acetic acid, glycols (Dale, 1991).

The carbon dioxide evolved during the fermentation stage finds uses in oil fields, for recovery of additional oil, in the manufacture of methanol, as a refrigerant and in carbonated beverages (Dale, 1991).

A process for the continuous production of ethanol from cereals, involving screening, filtering, saccharification, fermentation and distillation stages, has recently been patented (Technipetrol SPA, 1989), while a dual-purpose flour mill has been described in which the flour is air-classified to produce a high protein fraction (particle size: 2–5 µm) and a residual protein-depleted fraction for use as the starting material for production of ethanol (Bonnet and Willm, 1989).

Corncobs can be used for the production of ethanol, and also of furfural (*vide infra*). By treating the cobs with dilute sulphuric acid, 80% of the pentosans in the cobs are converted to pentoses, from which furfural is obtained, while the residual cellulose can be hydrolyzed to glucose in 65% yield (Clark and Lathrop, 1953).

Extrusion cooking has been suggested as a method for pretreating grain to be used for the production of ethanol. The thermomechanical effects of extrusion cooking produce gelatinization and liquefaction of the starch so that no liquefying enzyme is needed for the subsequent saccharification with glucoamylase. Ethanol yields from wet-extruded and from steam-cooked grain were almost equal, but the extrusion method uses less energy. Roller-milled whole barley, wheat or oats can be used in this process, with or without the addition of thermostable alpha-amylase, which appears to have little effect during extrusion cooking. The fermentation stage is carried out using either yeast (*Saccharomyces cerevisiae*) or the bacterium *Zymomonas mobilis*, the latter producing an increased initial rate of fermentation (Linko, 1989a,b).

Furfural production from cereals

Corncobs, the hulls of oats and rice, and the fibrous parts of other cereals are rich in pentosans, condensation products of pentose sugars, which are associated with cellulose as constituents of cell walls, particularly of woody tissues.

Thus, the pentosan content of oat hulls is given as 29%, along with 29% of cellulose and 16% of lignin (McMullen, 1991). Pentosans are the starting material for the manufacture of furfural, a chemical with many uses. Indeed, commercial utilization of oat hulls and other pentosan-rich cereal materials lies in the manufacture of furfural (MacArthur-Grant, 1986).

Furfural was first produced commercially in 1922. By 1975, oat hulls were providing about 22% of the annual demand for furfural in the U.S.A., but thereafter the demand for furfural and other furan chemicals far outstripped the supply of oat hulls, and increasing use was then made of other sources of pentosans, viz. rice hulls, corncobs, bagasse (Shukla, 1975).

Plants for the commercial production of furfural from agricultural residues have been established in the U.S.A. The plant at Cedar Rapids, Iowa, uses oat hulls and corncobs; the one at Memphis, TN uses rice hulls, corncobs and cottonseed hulls; while the plant at Omaha, NB uses corncobs only (Clark and Lathrop, 1953).

The commercial process for manufacturing furfural involves the boiling of the pentosan-containing material with strong acid (sulphuric or hydrochloric) and steam for 7–9 h at 70 psi

$$(C_5H_8O_4)_n + nH_2O \rightleftharpoons nC_5H_{10}O_5$$

Pentosan   Pentose

$$C_5H_{10}O_5 \rightleftharpoons C_4H_3O \cdot CHO + 3H_2O$$

Pentose   Furfural

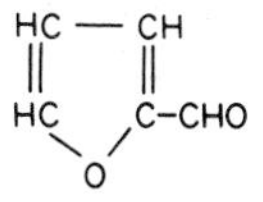

Furfural

FIG. 15.1 Chemical reactions in the production of furfural from pentosans.

pressure. Previous grinding of the hulls is not required. A sequence of reactions takes place. The pentosans are dissociated from the cellulose; then the pentosans are hydrolyzed to pentose sugars, and finally the pentose sugars undergo cyclo-hydration to form furfural, a heterocyclic adehyde, which is removed continuously by steam distillation (see Fig. 15.1) (Dunlop, 1973; Johnson, 1991).

The theoretical yield of furfural from pentose is 64% (plus 36% of water), so the theoretical yield of furfural from oat hulls containing 29% of pentosans would be 22%, although a yield of only about 13% is achieved in practice. The yield from corncobs is similar, while that from rice hulls is somewhat lower, at 12% theoretical, 5% in practice (Juliano, 1985; Pomeranz, 1987).

A large proportion of the cost of the process is accounted for by the need to raise high pressure steam: for every 1 lb of furfural produced, 15–26 lb of high pressure steam at 188°C are required.

Furfural finds uses as a selective solvent for refining lubricating oils and petroleum spirit, and for refining animal and vegetable oils in the manufacture of margarine. It is also used for the purification of butadiene, which is needed for the manufacture of synthetic rubber.

One of the most important uses for furfural, however, is in the manufacture of nylon. Nylon, a synthetic fibre, defined chemically as a polyamide, was first produced in 1927 by the firm E. I. du Pont de Nemours and Co., and was introduced to the industry in 1939. 'Polyamides' are formed by the condensation of a diamide and dibasic acid, and those most often used in the manufacture of nylon are hexamethylene diamine and adipic acid. The value of furfural arises from the fact that it is an important source of hexamethylene diamine.

Other uses for furfural include: production of formaldehyde furfural resins for making pipes and tanks; production of tetrahydrofurfural alcohol, a solvent for dyes, paints, etc.; production of polytetramethylene ether glycol for making thermoplastics; manufacture of D-xylose, phenolic resin glues and adhesives; production of anti-skid tread composition; filter aid in breweries (MacArthur-Grant, 1986).

Other industrial uses for cereals

There are many other industrial uses, besides the production of ethanol and furfural, for cereal grains, their milled products, and the by-products of milling. Use of cereals in malting, brewing and distilling have been discussed in Ch. 9. Other industrial uses are dependent on either the chemical characteristics or the physical properties of the raw material. They make use of cereals or cereal products as absorbents, abrasives, adhesives, binders and fillers, carriers, and for such purposes as filter aids, litter for animals, fertilizers, floor-sweepings, fuel, soil conditioners, oil well drilling aid. They are also used in the paper and mineral processing industries.

Wheat

Industrial uses for the milling products of wheat have been listed by Pomeranz (1987). Both wheat flour and wheat starch are used in paper sizing and coating, and as adhesives in the manufacture of paper, boards, plywood, etc. Starch is also used for finishing textiles.

Gluten separated from wheat flour finds uses in paper manufacture, as an adhesive, and as the starting material for the preparation of sodium glutamate and glutamic acid.

Wheat germ is used in the production of antibiotics, pharmaceuticals and skin conditioners, while wheat bran may be used as a carrier of enzymes, antibiotics and vitamins (Pomeranz, 1987).

Starch is also used to make rigid urethane foam for insulation and paints, in plastics and to process crude latex in the manufacture of rubber.

Rye

The gums, both soluble and insoluble, make rye a good substitute for other gums in wet-end additives. The starch in rye flour has a high water-binding capacity and finds use as adhesives, for example in the glue, match and plastics industries, and as pellet binders and foundry core binders (Drews and Seibel, 1976; Lorenz, 1991).

Maize starch

Starch is modified in various ways for use in the paper-making industry. Such modifications include acid modification and hydroxyethylation (with ethylene oxide) for paper coating, oxidation (with sodium hypochlorite) and phosphated (with sodium phosphate) for paper sizing and improvement of paper strength, cationic derivatives for paper strengthening, stiffening and improved pigment retention by the paper.

Dextrinized starch, pregelatinized starch, and succinate derivatives are used as adhesives, while carboxymethylated starch is used in paints, oil drilling muds, wall-paper adhesives and detergents (Johnson, 1991).

Maize and sorghum

Maize grits are used in the manufacture of wall paper paste, and of glucose by 'direct hydrolysis'.

Coarse or granulated maize meal is used for dance floor wax and for handsoap; fine meal or corn (maize) cones finds use as a dusting agent and as an abrasive in hand soap.

Corn (maize) gluten is used as a cork-binding agent, as an additive for printing dyes, and in pharmaceuticals.

Acid-modified flours of maize and sorghum are used as binders for wall board and gypsum board, providing a strong bond between gypsum and the liner. In the building industry, maize flour is used to provide insulation of fibre board and plywood, and wafer board. In the pharmaceutical industry, maize flour is used for the production of citric acid and other chemicals by fermentation processes.

Extrusion-cooked maize flour and sorghum flour are used as core-binders or foundry binders in sand–cereal–linseed oil systems, while a thermosetting resin has been made by combining an acid-modified, extruded maize flour with glyoxal or a related polyaldehyde, the mixture binding the sand particles. As an adhesive, maize flour and sorghum grits are used in the production of charcoal briquets, corrugated paper, and animal feed pellets.

Both maize flour and sorghum flour find uses in ore-refining, e.g. in the refining of bauxite (aluminium ore), and as binders in pelletizing iron ore. Another use for precooked maize flour or starch and for sorghum grits is in oil well drilling, where the flour or starch reduces loss of water in the drilling mud which cools and lubricates the drilling bit. For this purpose, the flours may be precooked on hot rolls or by extrusion cooking.

Maize flour has been used as an extender in polyvinyl alcohol and polyvinyl chloride films for use as agricultural mulches, and as extenders in rigid polyurethane resins for making furniture (Alexander, 1987).

Corn (maize) cobs consist chiefly of cellulose, hemicellulose (pentosan), lignin and ash. Their industrial uses have been listed by Clark and Lathrop (1953). They include:

Agricultural uses: as litter for poultry, preferably reduced to particles of 0.25–0.75 in. in size. Their use reportedly reduces the mortality of chickens from coccidiosis.

As mulch around plants, for retaining moisture and controlling weeds, and as a soil conditioner for improving soil texture.

As carriers and diluents for insecticides and pesticides, preferably ground to pass a U.S. standard No. 60 sieve. Partly rolled crushed cobs have been used as material on which to grow mushrooms.

Industrial uses: those based on physical properties include use for corncob pipes; in the manufacture of vinegar; as an abrasive, when finely ground, for cleaning fur and rugs; for cleaning moulds in the rubber and glass industries; for soft-grit air-blasting, burnishing and polishing the parts of airplane engines and electric motors. Soft-grit blasting can remove or absorb rust, scale from hard water, oil, grease, wax and dirt from a variety of metals.

Finely ground cobs can replace sawdust (when mixed with sand and paraffin oil) as a floor sweeping compound, and can be used as an abrasive in soaps.

Ground corncobs find uses in the manufacture of building materials, including asphalt shingles and roofing, brick and ceramics; and as fillers for explosives, e.g. in the manufacture of dynamite, for concrete, for plastics (to replace wood flour), for plywood glues and adhesives — in which the corncob flour improves the spreading and binding properties, and for rubber compounds and tyres — in which the corncob material adds non-skid properties.

The lignin separated from corncobs by alkaline treatment could be used as a filler for plastic moulding compounds; as a soil stabilizer in road building; in the manufacture of leather and as an adhesive, but it is more expensive than lignin obtained from wood. Thus, economic uses for corncobs depend on those properties in which cobs show a superiority over wood waste or other materials (Clark and Lathrop, 1953).

Industrial uses based on chemical properties (besides the manufacture of ethanol and furfural, see above) include the manufacture of fermentable sugars, solvents and liquid fuel; production of charcoal, gas and other chemicals by destructive distillation; use as a solid fuel (oven-dry cobs have a calorific value of about 18.6 MJ/kg; 8000 Btu/lb); and in the manufacture of pulp, paper and board (Clark and Lathrop, 1953; Klabunde, 1970).

Oats

Suggested uses for oat starch are as coating agents in the pharmaceutical industry; in photocopy papers; and as adhesives; but the starch would be competing, in these uses, with rice starch and wheat starch.

Oatmeal is used in the cosmetics industry as a component of facial masks and soap. The cleaning effect of the oatmeal may be due to the beta-glucan component, or to the oat oil (Webster, 1986).

Hulls of rice and oats

As the composition of the hulls (husks) of oats and of rice somewhat resembles that of corncobs, many of the industrial uses for the hulls duplicate some of those mentioned above for corncobs.

Rice hulls differ, however, in yielding a large amount of ash (22%) upon incineration, of which 95% is silica, most of the rest being lime and potash. Thus, rice hills can be used as a source of high-grade silica in various manufacturing industries.

Oat husk is used as a filter aid in breweries, where it is mixed with the ground malt and water in the mash tun in order to keep the mass porous. It is also used, when finely ground, as a diluent or filler in linoleum, and as an abrasive in air-blasting for removing oil and products of corrosion from machined metal components, as an antiskid tread component, and as a plywood glue extender (Hutchinson, 1953; McMullen 1991).

The cariostatic properties of oat hulls suggest possible uses as components of chewing gum and other products (McMullen, 1991).

Uses for rice hulls include chicken litter, soil amendment for potting plants, ammoniated for fertilizers, filter aid, burnt for floor sweepings, binder for pelleted feeds, insulating material, filler for building materials (e.g. wallboard: see Brit. Pat. No. 1,403,154), binder and absorbent for pesticides, explosives (Juliano, 1985; Sharp, 1991).

Rice hulls may be used as an abrasive in soft-grit blasting of metal parts, but as the hulls contain 18–20% of silica they are too abrasive when used alone, and are preferably mixed with ground corncobs in the ratio of 40% rice hulls to 60% of cob particles (Clark and Lathrop, 1953). Rice hulls have also been used successfully for mopping up oil spillages on the surface of water. After skimming off the hulls, the water was left clean enough to drink.

Fine sieve fractions of ground rice hulls can be used as excipients (carriers) for nutrients, pharmaceuticals, biological additives in animal feed premixes, and for poison baits.

The high silica content of rice hull ash is the reason for the use of the ash as a constituent of cement, together with slaked lime: rice hull ash cement is more acid-resistant than portland cement (Juliano, 1985). Rice hull ash can be used as a silica source in glass and ceramics industries, e.g. for making Silex bricks (in Italy) and Porasil bricks (in Canada). Slaked lime reacts with the silica during firing to provide a vitreous calcium silicate bond.

Rice hull ash can be used as a source of sodium silicate for the manufacture of water glass, reinforcing rubber compounds, as an absorbent for oil, as an insulator for steel ingots, as an abrasive for tooth paste, as an absorbent and as a water purifier (Australian Technical Millers Association, 1980).

Rice hulls fired at temperatures below 700°C yield amorphous silica which has been used for making solar-grade silicon, which can be used for solar cells. The silica can be chlorinated to silicon tetrachloride, which is then reacted with metallurgical grade silicon to produce trichlorsilane for making solar-grade silicon.

References

ALEXANDER, R. J. (1987) Corn dry milling: processes, products, applications. In: *Corn: Chemistry and Technology*, Ch. 11, WATSON, S. A. and RAMSTAD, P. E. (Eds), Amer. Assoc. Cereal Chem., St Paul, MN, U.S.A.

ANTONI, J. (1960) Rye as feedmeal. *Landbauforschung* **10**: 69–72.

AUSTRALIAN TECHNICAL MILLERS ASSOCIATION (1980) Current use and development of rice by-products. *Australas. Baker Millers J.* April: 27.

BONNET, A. and WILLM, C. (1989) Wheat mill for production of proteins and ethanol. *Ind. Cer.* Nov/Dec.: 37–46.

CHURCH, D. C. (1991) Feed preparation and processing. In: *Livestock Feeds and Feeding*, 3rd edn, Ch. 11, CHURCH, D. C. (Ed.) Prentice-Hall International Inc., Englewood Cliffs, N.J., U.S.A.

CLARK, T. F. and LATHROP, E. C. (1953) *Corncobs — Their Composition, Availability, Agricultural and Industrial Uses*. U.S. Dept. Agric., Agric. Res. Admin., Bur. Agric. Ind. Chem., AIC–177, revised 1953.

COR TECH RESEARCH LTD. (1972) Resin-coated rice hulls and the production of composite articles therefrom. *Brit. Pat. Spec.* No. 1,403,154.

CROMWELL, G. L. (1991) Feeding swine. In: *Livestock Feeds and Feeding*, 3rd edn, Ch. 21, CHURCH, D. C. (Ed.), Prentice-Hall International, Inc., Englewood Cliffs, N.J., U.S.A.

DALE, B. E. (1991) Ethanol production from cereal grains. In: *Handbook of Cereal Science and Technology*, Ch. 24, LORENZ, K. J. AND KULP, K. (Eds) Marcel Dekker, Inc., N.Y., U.S.A.

DREWS, E. and SEIBEL, W. (1976) Bread-baking and other uses around the world. In: *Rye: Production, Chemistry, Technology*, BUSHUK, W. (Ed.) Amer. Assoc. Cereal Chem., St Paul, MN, U.S.A.

DUNLOP, A. P. (1973) The furfural industry. In: *Industrial Uses of Cereals*, pp. 229–236, POMERANZ, Y. (Ed.) Amer. Assoc. Cereal Chem., St Paul, MN, U.S.A.

ELY, D. G. (1991) Feeding lambs for market. In: *Livestock Feeds and Feeding*, 3rd edn, C. 19, CHURCH, D. C. (Ed.)

Prenctice-Hall International Inc., Englewood Cliffs, N.Y., U.S.A.

FOOD AND AGRICULTURE ORGANISATION OF THE UNITED NATIONS (1990) *Food balance sheets, 1984–86*. F.A.O., Rome.

HOCKETT, E. A. (1991) Barley. In: *Handbook of Cereal Science and Technology*, Ch. 3. LORENZ, K. J. and KULP, K. (Eds.), Marcel Dekkor, Inc., N.Y., U.S.A.

HOME-GROWN CEREALS AUTHORITY (1990). *Cereal Statistics*. H-GCA, London.

HUTCHINSON, J. B. (1953) The quality of cereals and their industrial uses. *Chemy Ind.* 578.

JOHNSON, L. A. (1991). Corn: production, processing, utilization. In: *Handbook of Cereal Science and Technology*, Ch. 2, LORENZ, K. J. and KULP, K. (Eds.), Marcel Dekker, Inc., N.Y., U.S.A.

JULIANO, B.O. (1985) *Rice: Chemistry and Technology*, 2nd edn. Amer. Assoc. Cereal Chem., St Paul, Minn., U.S.A.

KLABUNDE, H. (1970) Various methods for the industrial processing of corn. *Northw. Miller* May: 83.

KLOPFENSTEIN, T. J., STOCK, R. and WARD, J. K. (1991) Feeding growing-finishing beef cattle. In: *Livestock Feeds and Feeding*, Ch 14. CHURCH, D. C. (Ed.) Prentice-Hall International, Inc., Englewood Cliffs, N.J., U.S.A.

KREIKMEIER, K. (1987) *Nebr. Beef Cattle Rpt.* MP 52:9

LINKO, P. (1989a) The twin-screw extrusion cooker as a versatile tool for wheat processing. In: *Wheat is Unique*, Ch. 22. POMERANZ, Y. (Ed.) Amer. Assoc. Cereal Chem., St Paul, MN, U.S.A.

LINKO, P. (1989b) Extrusion cooking in bioconversions. In: *Extrusion Cooking*, Ch. 8, MERCIER, M., LINKO, P. and HARPER, J. (Eds.), Amer. Assoc. Cereal Chem., St Paul, MN, U.S.A.

LORENZ, K. J. (1991) Rye. In: *Handbook of Cereal Science and Technology*, LORENZ, K. J. and KULP, K. (Eds) Marcel Dekker Inc., N.Y., U.S.A.

LORENZ, K. J.. and KULP, K. (1991) *Handbook of Cereal Science and Technology*. Marcel Dekker, Inc., N.Y., U.S.A.

MACARTHUR-GRANT, L. A. (1986) Sugars and nonstarchy polysaccharides in oats. In: *Oats: Chemistry and Technology*, Ch. 4. WEBSTER, F. H. (Ed.), Amer. Assoc. Cereal Chem., St Paul, MN, U.S.A.

MARTIN, J. H. and MACMASTERS, M. M. (1951) Industrial uses for grain sorghum. *U.S. Dept. Agric., Yearbook Agric., 1951*, p. 349.

MATTERN, P. J. (1991) Wheat. In: *Handbook of Cereal Science and Technology*, Ch. 1, LORENZ, K. J. and KULP, K. (Eds) Marcel Dekker, Inc., N.Y., U.S.A.

McMULLEN, M. S. (1991) Oats. In: *Handbook of Cereal Science and Technology*, Ch. 4, LORENZ, K. J. and KULP, K. (Eds.) Marcel Dekker, Inc., N.Y., U.S.A.

MORRILL, J. L. (1991) Feeding dairy calves and heifers. In: *Livestock feeds and feeding*, 3rd edn, Ch. 16, CHURCH, D. C. (Ed.) Prentice-Hall International Inc., Englewood cliffs, N.J., U.S.A.

MORRISON, F. B. (1947). *Feeds and Feeding*. 20th edn. Morrison Publ. Co., Ithaca, N.Y., U.S.A.

NAKAUE, H. S. and ARSCOTT, G. H. (1991) Feeding poultry. In: *Livestock Feeds and Feeding*, 3rd edn, Ch 22. CHURCH, D. C. (Ed.) Prentice-Hall International, Inc., Englewood Cliffs, N.J., U.S.A.

NATIONAL ASSOCIATION OF BRITISH AND IRISH MILLERS (1991) *Facts and Figures 1991*. N.A.B.I.M., London.

OTT, E. A. (1973) *Symposium on effect of processing on the nutritional value of feeds*. Washington, D.C., Nat. Acad. Sci., p. 373.

OTT, E. A. (1991) Feeding horses. In: *Livestock Feeds and Feeding*, 3rd edn, Ch. 23, CHURCH, D. C. (Ed.), Prentice-Hall International Inc., Englewood Cliffs, N.J., U.S.A.

POMERANZ, Y. (1987) *Modern Cereal Science and Technology*. VCH Publishers, Inc., N.Y., U.S.A.

ROESE, G. (1978) Rice pollard good value. *Rice Mill News*, June.

ROONEY, L. W. and SERNA-SALDIVAR, S. O. (1991) Sorghum. In: *Handbook of Cereal Science and Technology*, Ch. 5. LORENZ, K. J. and KULP, K. (Eds.), Marcel Dekker, Inc., N.Y., U.S.A.

SCHINGOETHE, D.J. (1991) Feeding dairy cows. In: *Livestock Feeds and Feeding*, 3rd edn., Ch. 15. CHURCH, D. C. (Ed.) Prentice-Hall International Inc., Englewood Cliffs, N.J., U.S.A.

SERNA-SALDIVAR, S. O. MCDONOUGH, C. M. and ROONEY, L. W. (1991) The millets. In: *Handbook of Cereal Science and Technology*, Ch. 6, LORENZ, K. J. and KULP, K. (Eds.), Marcel Dekker Inc., N.Y., U.S.A.

SHARP, R. N. (1991) Rice: production, processing, utilization. In: *Handbook of Cereal Science and Technology*, Ch. 7. LORENZ, K. J. and KULP, K. (Eds), Marcel Dekker, Inc., N.Y., U.S.A.

SHUKLA, T. P. (1975) Chemistry of oats: protein foods and other industrial products. *Crit. Rev. Food. Sci. Nutr.*, October: 383–431.

SINDT, M. *et al.* (1987). *Nebr. Beef Cattle Rpt.* MP 52: 9.

STOCK, R. *et al.* (1985). *Nebr. Beef Cattle Rpt.* MP 48: 32.

TECHNIPETROL SPA (1989) Process and apparatus for the continuous production of ethanol from cereals, and method of operating said apparatus. *World Intellectual Property Organization Patent* 89/01522.

UNITED STATES DEPARTMENT OF AGRICULTURE (1987a) *Agricultural Statistics 1986*. U.S.D.A., Washington, D.C.

UNITED STATES DEPARTMENT OF AGRICULTURE (1987b) *World grain situation and outlook. Foreign Agr. Ser. Circ.* Series FG-2-87 (November), Washington, D.C.

UNITED STATES DEPARTMENT OF AGRICULTURE (1987c) Feed situation and outlook. *Econ. Res. Ser.* FDS-304 (November), Washington, D.C.

WADMARK, G. (1988) Fuel ethanol from cereals: a Swedish pilot project. *World Grain* 6 (June): 12–13.

WARD, J. K. and KLOPFENSTEIN, T. J. (1991) Nutritional management of the beef cow herd. In: *Livestock Feeds and Feeding*, 3rd edn, Ch. 13. CHURCH D. C. (Ed.), Prentice-Hall International Inc., Englewood Cliffs, N.J., U.S.A.

WATSON, S. J. (1953) The quality of cereals and their industrial uses. *Chemy Ind*, Jan. 31: 95–97.

WEBSTER, F. H. (1986) *Oats: chemistry and technology*. Amer. Assoc. Cereal Chem., St Paul, MN, U.S.A.

Further Reading

BRIT. PAT. SPEC. No. 585,772 (1944). Improvements in the manufacture of furfural.

BROWN, I., SYMONS, E. F. and WILSON, B. W. (1947) Furfural: a pilot plant investigation of its production from Australian raw materials. *J. Coun. Sci. Industr. Res.* 20: 225.

CHURCH, D. C. (1991) *Livestock Feeds and Feeding*, 3rd edn. Prentice-Hall International Inc., Englewood Cliffs, N.J., U.S.A.

HITCHCOCK, L. B. and DUFFEY, H. R. (1948) Commercial production of furfural in its 25th year. *Chem. Engng. Prog.* 44: 669.

LATHROP, A. W. and BOHSTEDT, G. (1938) Oat mill feed: its usefulness and value in livestock rations. *Wisconsin Agr. Exp. Sta. Res. Bull.* 135.

MEUSER, F. and WIEDMANN, W. (1989) In: *Extrusion Cooking*, MERCIER, M., LINKO, P. and HARPER, J. (Eds.) Amer. Assoc. Cereal Chem., St Paul, MN, U.S.A.

PETERS, F. N. (1937) Furfural as an outlet for cellulosic waste material. *Chemy Engng News* 15: 269.

RACHIE, K. O. (1975) *The Millets — Importance, Utilization and Outlook*. International Crop Research Institute for Semi-Arid Tropics, Begumpet, Hyderabad, India.

SMITH, A. (1989) Extrusion cooking: a review. *Food Sci. Technol. Today* 3 (3): 156–161.

WOLF, M. J., MACMASTERS, M. M., CANNON, J. A., ROSEWALL, E. C. and RIST, C. E. (1953) Preparation and properties of hemicelluloses from corn hulls. *Cereal Chem.* 30: 451.

INDEX